Scintillation Dosimetry

IMAGING IN MEDICAL DIAGNOSIS AND THERAPY

Series Editors: Andrew Karellas and Bruce R. Thomadsen

Published titles

Quality and Safety in Radiotherapy
Todd Pawlicki, Peter B. Dunscombe,
Arno J. Mundt, and Pierre Scalliet, Editors
ISBN: 978-1-4398-0436-0

Adaptive Radiation Therapy
X. Allen Li, Editor
ISBN: 978-1-4398-1634-9

Quantitative MRI in Cancer
Thomas E. Yankeelov, David R. Pickens,
and Ronald R. Price, Editors
ISBN: 978-1-4398-2057-5

Informatics in Medical Imaging
George C. Kagadis and Steve G. Langer, Editors
ISBN: 978-1-4398-3124-3

Adaptive Motion Compensation in Radiotherapy
Martin J. Murphy, Editor
ISBN: 978-1-4398-2193-0

Image-Guided Radiation Therapy
Daniel J. Bourland, Editor
ISBN: 978-1-4398-0273-1

Targeted Molecular Imaging
Michael J. Welch and William C. Eckelman,
Editors
ISBN: 978-1-4398-4195-0

Proton and Carbon Ion Therapy
C.-M. Charlie Ma and Tony Lomax, Editors
ISBN: 978-1-4398-1607-3

Comprehensive Brachytherapy: Physical and Clinical Aspects
Jack Venselaar, Dimos Baltas, Peter J. Hoskin,
and Ali Soleimani-Meigooni, Editors
ISBN: 978-1-4398-4498-4

Physics of Mammographic Imaging
Mia K. Markey, Editor
ISBN: 978-1-4398-7544-5

Physics of Thermal Therapy: Fundamentals and Clinical Applications
Eduardo Moros, Editor
ISBN: 978-1-4398-4890-6

Emerging Imaging Technologies in Medicine
Mark A. Anastasio and Patrick La Riviere, Editors
ISBN: 978-1-4398-8041-8

Cancer Nanotechnology: Principles and Applications in Radiation Oncology
Sang Hyun Cho and Sunil Krishnan, Editors
ISBN: 978-1-4398-7875-0

Monte Carlo Techniques in Radiation Therapy
Joao Seco and Frank Verhaegen, Editors
ISBN: 978-1-4665-0792-0

Image Processing in Radiation Therapy
Kristy Kay Brock, Editor
ISBN: 978-1-4398-3017-8

Informatics in Radiation Oncology
George Starkschall and R. Alfredo C. Siochi,
Editors
ISBN: 978-1-4398-2582-2

Cone Beam Computed Tomography
Chris C. Shaw, Editor
ISBN: 978-1-4398-4626-1

Tomosynthesis Imaging
Ingrid Reiser and Stephen Glick, Editors
ISBN: 978-1-4398-7870-5

Stereotactic Radiosurgery and Stereotactic Body Radiation Therapy
Stanley H. Benedict, David J. Schlesinger, Steven
J. Goetsch, and Brian D. Kavanagh, Editors
ISBN: 978-1-4398-4197-6

Computer-Aided Detection and Diagnosis in Medical Imaging
Qiang Li and Robert M. Nishikawa, Editors
ISBN: 978-1-4398-7176-8

IMAGING IN MEDICAL DIAGNOSIS AND THERAPY
Series Editors: Andrew Karellas and Bruce R. Thomadsen

Scintillation Dosimetry

Edited by

Sam Beddar
Luc Beaulieu

CRC Press
Taylor & Francis Group
Boca Raton London New York

CRC Press is an imprint of the
Taylor & Francis Group, an **informa** business

A TAYLOR & FRANCIS BOOK

CRC Press
Taylor & Francis Group
6000 Broken Sound Parkway NW, Suite 300
Boca Raton, FL 33487-2742

First issued in paperback 2021

ISBN 13: 978-0-367-78305-1 (pbk)
ISBN 13: 978-1-4822-0899-3 (hbk)

Library of Congress Cataloging-in-Publication Data

Names: Beddar, Sam, editor. | Beaulieu, Luc, 1969- editor.
Title: Scintillation dosimetry / edited by Sam Beddar and Luc Beaulieu.
Other titles: Imaging in medical diagnosis and therapy.
Description: Boca Raton, FL : CRC Press, Taylor & Francis Group, [2016] |
"2016 | Series: Imaging in medical diagnosis and therapy | Includes
bibliographical references and index.
Identifiers: LCCN 2015039050| ISBN 9781482208993 (alk. paper) | ISBN
1482208997 (alk. paper)
Subjects: LCSH: Ionizing radiation--Measurement. | Scintillators. | Radiation
dosimetry.
Classification: LCC QC795.42 .S45 2016 | DDC 539.7/75--dc23
LC record available at http://lccn.loc.gov/2015039050

Visit the Taylor & Francis Web site at
http://www.taylorandfrancis.com

and the CRC Press Web site at
http://www.crcpress.com

To Lilia, Lena Sophie, and Alexander Samy
And to Chantal, Alexandre, Catherine, and Gabriel
—The scintillating elements in our lives.

Contents

Series Preface

Advances in the science and technology of medical imaging and radiation therapy have been more profound and rapid than ever before since their inception over a century ago. Further, the disciplines are increasingly cross-linked as imaging methods become more widely used to plan, guide, monitor, and assess treatments in radiation therapy. Today the technologies of medical imaging and radiation therapy are so complex and so computer driven that it is difficult for the persons (physicians and technologists) responsible for their clinical use to know exactly what is happening at the point of care, when a patient is being examined or treated. The persons best equipped to understand the technologies and their applications are medical physicists, and these individuals are assuming greater responsibilities in the clinical arena to ensure that what is intended for the patient is actually delivered in a safe and effective manner.

The growing responsibilities of medical physicists in the clinical arenas of medical imaging and radiation therapy are not without their challenges, however. Most medical physicists are knowledgeable in either radiation therapy or medical imaging, and expert in one or a small number of areas within their discipline. They sustain their expertise in these areas by reading scientific articles and attending scientific talks at meetings. In contrast, their responsibilities increasingly extend beyond their specific areas of expertise. To meet these responsibilities, medical physicists periodically must refresh their knowledge of advances in medical imaging or radiation therapy, and they must be prepared to function at the intersection of these two fields. How to accomplish these objectives is a challenge.

At the 2007 annual meeting of the American Association of Physicists in Medicine in Minneapolis, this challenge was the topic of conversation during a lunch hosted by Taylor & Francis Group and involving a group of senior medical physicists (Arthur L. Boyer, Joseph O. Deasy, C.-M. Charlie Ma, Todd A. Pawlicki, Ervin B. Podgorsak, Elke Reitzel, Anthony B. Wolbarst, and Ellen D. Yorke). The conclusion of this discussion was that a book series should be launched under the Taylor & Francis banner, with each volume in the series addressing a rapidly advancing area of medical imaging or radiation therapy of importance to medical physicists. The aim would be for each volume to provide medical physicists with the information needed to understand technologies driving a rapid advance and their applications to safe and effective delivery of patient care.

Each volume in the series is edited by one or more individuals with recognized expertise in the technological area encompassed by the book. The editors are responsible for selecting the authors of individual chapters and ensuring that the chapters are comprehensive and intelligible to someone without such expertise. The enthusiasm of volume editors and chapter authors has been gratifying and reinforces the conclusion of the Minneapolis luncheon that this series of books addresses a major need of medical physicists.

The series *Imaging in Medical Diagnosis and Therapy* would not have been possible without the encouragement and support of the series manager, Luna Han of Taylor & Francis Group. The editors and authors, and most of all I, are indebted to her steady guidance of the entire project.

AUTHOR

William Hendee Rochester, Minneapolis

Preface

This book marks an important milestone toward the establishment of scintillation dosimetry and its greater acceptance in the clinical and research medical physics communities. It was created largely for medical physics graduate students and clinical medical physicists, but it should be a welcome addition for any scientist interested in increasing their knowledge of this field. This book also should serve as a reference for those who are or will be interested in engaging in scintillation dosimetry research. The intent is to provide the reader with a strong scientific basis in plastic scintillation dosimetry in particular as well as a wide scope of technical implementations (from point source dosimetry scaling up to 3D volumetric dosimetry and 4D scintillation dosimetry) and clinical applications (from machine quality assurance [QA] to small field and *in vivo* dosimetry). The book also covers related optical techniques such as optically stimulated luminescence (OSL) or Cerenkov luminescence, as well as other dosimetry topics. Overall, this book represents an up-to-date reference on the status of this technology in the field of radiation dosimetry.

Before the publication of two seminal papers by Beddar, Mackie, and Attix in 1992,* citing the use of scintillation detectors for accurate dose measurements was uncommon. Figure P.1 illustrates the results of a PubMed search (http://www.ncbi.nlm.nih.gov/pubmed) for published articles based on the keywords *scintillation* or *scintillator*, and *dosimetry* over five 5-year intervals from 1990 through 2014.

Growing interest in the field is clear from the nearly 10-fold increase in peer-reviewed publications during the most recent 5-year interval compared to the early 1990s. Elements that spurred this interest, in particular during 2000–2004, were related to the ongoing activities of two important collaborations, one between Sam Beddar and Natalka Suchowerska from Australia and the other between Sam Beddar and Luc Beaulieu from Canada. Although the first one ended by mid-2000, the latter activity remains active to this day. The collaboration between the Beddar and Beaulieu laboratories has *pushed the envelope* of plastic scintillation dosimetry far beyond single-point measurements of photon and electron beams to methods

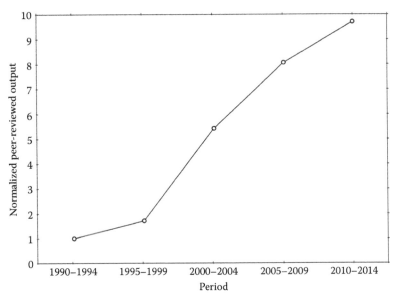

Figure P.1 Published peer-reviewed manuscripts on scintillation dosimetry (normalized to the 1990–1994 period) for five different 5-year intervals.

---

* Water-equivalent plastic scintillation detectors for high-energy beam dosimetry: I. Physical characteristics and theoretical consideration. *Phys. Med. Biol.* 1992; 37(10):1883–1900; Water-equivalent plastic scintillation detectors for high-energy beam dosimetry: II. Properties and measurements. *Phys. Med. Biol.* 1992; 37(10):1901–1913.

and detector systems for real-time one-, two-, and three-dimensional dose measurements. This work led to development and demonstration of new applications for accurate quality control in small field, *in vivo*, and proton dosimetry. For the first time, plastic scintillators were used to experimentally extract ion chamber and diode effective point of measurements for electron beams and diode correction factors for CyberKnife small field dosimetry. This group further demonstrated a hyperspectral approach to light processing that allowed multipoint scintillation fiber dosimetry, a first in the field. Finally, this collaboration is at the heart of the first commercially available plastic scintillation dosimeter, the Exradin W1.

We also gratefully acknowledge other major contributors to publications in the field of scintillation dosimetry, particularly those of Claus Andersen (Technical University of Denmark at Risø) on OSL dosimetry and a framework for error detection when the technology is applied for *in vivo* dosimetry; Bongsoo Lee (Konkuk University, Korea) for radiation therapy applications; Brian Pogue and David Gladstone (Darthmouth University) for their work on Cerenkov dosimetry; and Daniel Hyer (University of Iowa) for radiology applications. All of these researchers have contributed to this book, either directly or through the involvement of their former students.

Note that throughout this book, we allowed the use of all commonly used spelling forms for Cerenkov (Cherenkov, Čerenkov, Cerenkov) as it has been used in the literature. This particular type of electromagnetic radiation was named after scientist Pavel Alekseyevich Cherenkov, who was the first to detect it experimentally and shared the Nobel Prize in Physics in 1958 with Ilya Frank and Igor Tamm for the discovery of Cherenkov radiation made in 1934.

We sincerely thank all of the authors that contributed to the chapters of this book, and we appreciate their patience in making this book a reality. Their efforts have brought all of the chapters to a very high standard, and one that allows them to *stand alone* as much as possible.

AUTHORS

Sam Beddar The University of Texas MD Anderson Cancer Center, Houston, Texas
Luc Beaulieu Université Laval, Québec, Canada

Editors

Sam Beddar, PhD, is a tenured professor at the University of Texas MD Anderson Cancer Center, Houston, Texas, in the Division of Radiation Oncology and professeur adjoint in the Département de Physique, Génie Physique et Optique, Université Laval, Québec, Canada and adjunct professor in the Department of Medical Physics, University of Wisconsin School of Medicine, Madison, Wisconsin. He is also the Chief of Clinical Research and Service Chief of Gastrointestinal Service in the Department of Radiation Physics at the UT MD Anderson Cancer Center. His research interests include scintillation dosimetry, intraoperative radiation therapy, four-dimensional computerized tomography, four-dimensional MR imaging techniques for radiation therapy and prompt gamma imaging for proton therapy. He has served as a mentor for many graduate students, postdoctoral research fellows, and clinical residents. Dr. Beddar has published over 140 scientific papers and book chapters, and served as a reviewer for National Institutes of Health study section review panels. Dr. Beddar has been PI on NIH R01s, SBIR phase I and phase II grants, Co-investigator on a P01, R21s and project leader on a NIH T-32 grant, and numerous industrial grants. He has served as an associate editor for *Medical Physics*, section editor for the *Journal of Applied Clinical Medical Physics*, and a reviewer for numerous scientific journals, including the *International Journal of Radiation Oncology, Biology, Physics* and *Physics in Medicine and Biology*.

Luc Beaulieu, PhD, is a professor in the department of physics, a director of the graduate medical physics program at Université Laval (part of Canada's top 10 research universities), and a medical physicist and head of the Medical Physics Research Group in the department of radiation oncology, Centre hospitalier universitaire de Québec, Québec, Canada. He is the recipient of numerous awards and has organized international conferences. Dr. Beaulieu has been the author or coauthor of over 185 articles published in refereed journals. He has served as president of the Canadian Organization of Medical Physicists and is an active member of the American Association of Physicists in Medicine, the American Society for Therapeutic Radiology and Oncology, and the American Brachytherapy Society. Dr. Beaulieu research interests encompass image-guided brachytherapy, Monte Carlo dose calculation methods and detector development.

Contributors

Claudine Nì. Allen
Centre d'optique, photonique et laser
Département de physique, de génie physique et
    d'optique
Université Laval
Québec, Canada

Claus E. Andersen
Center for Nuclear Technology
Technical University of Denmark
Roskilde, Denmark

Louis Archambault
Centre de recherche sur le cancer
Département de physique, de génie physique et
    d'optique
and
Centre de recherche du CHU de Québec
Département de radio-oncologie
Université Laval
Québec, Canada

Luc Beaulieu
Centre de recherche sur le cancer
Département de physique, de génie physique et
    d'optique
and
Centre de recherche du CHU de Québec
Département de radio-opncologie et axe oncologie
Université Laval
Québec, Canada

Sam Beddar
Department of Radiation Physics
The University of Texas MD Anderson Cancer
    Center
Houston, Texas

Josephine Chen
Department of Radiation Oncology
University of California, San Francisco
San Francisco, California

Chinmay Darne
Department of Radiation Physics
The University of Texas MD Anderson Cancer
    Center
Houston, Texas

Marie-Ève Delage
Centre de recherche sur le cancer
Département de physique, de génie physique et
    d'optique
and
Centre de recherche du CHU de Québec
Département de radio-oncologie et axe oncologie
Université Laval
Québec, Canada

Marion Eichmann
Technische Universität Dortmund
Dortmund, Germany

Ryan F. Fisher
Department of Radiology
Cleveland Clinic
Cleveland, Ohio

Dirk Flühs
Klinische Strahlenphysik
Universitätsklinikum Essen
Essen, Germany

Anne-Marie Frelin
Grand Accélérateur National d'Ions Lourds
Caen, France

David J. Gladstone
Department of Radiation Oncology
Geisel School of Medicine
Dartmouth-Hitchcock Medical Center
Lebanon, New Hampshire

Adam K. Glaser
Thayer School of Engineering
Dartmouth College
Hanover, New Hampshire

Maxime Guillemette
Department of Physics
Université Laval
Québec, Canada

Mathieu Guillot
Department of Radiation Oncology
Centre hospitalier universitaire de Sherbrooke
Québec, Canada

Daniel E. Hyer
Department of Radiation Oncology
University of Iowa
Iowa City, Iowa

David A. Jaffray
Department of Radiation Physics
Princess Margaret Cancer Centre
and
Department of Radiation Oncology
University of Toronto
Toronto, Ontario, Canada

Kyoung Won Jang
School of Biomedical Engineering
College of Biomedical and Health Science
Konkuk University
Seoul, Republic of Korea

Assen Kirov
Memorial Sloan-Kettering Cancer Center
New York City, New York

David Klein
Department of Radiation Physics
The University of Texas MD Anderson Cancer
    Center
Houston, Texas

Sébastien Lamarre
Centre d'optique, photonique et laser
Département de physique, de génie physique et
    d'optique
and
Centre de recherche sur les matériaux
    avancés
Département de chimie
Université Laval
Québec, Canada

Jamil Lambert
Westdeutsches Protonentherapiezentrum
    Essen
Essen, Germany

Dominic Larivière
Laboratoire de radioécologie
Département de chimie
Université Laval
Québec, Canada

Marie-Ève Lecavalier
Centre d'optique, photonique et laser
Département de physique, de génie physique et
    d'optique
and
Laboratoire de radioécologie
Département de chimie
Université Laval
Québec, Canada

Bongsoo Lee
School of Biomedical Engineering
College of Biomedical and Health Science
Konkuk University
Seoul, Republic of Korea

Geordi G. Pang
Sunnybrook Health Sciences Centre
Toronto, Ontario, Canada

Brian W. Pogue
Thayer School of Engineering
and
Department of Physics and Astronomy
Dartmouth College
Hanover, New Hampshire

Madison Rilling
Centre de recherche sur le cancer
Département de physique, de génie physique et
    d'Optique
and
Centre de recherche du CHU de Québec
Département de radio-oncologie et axe oncologie
Université Laval
Québec, Canada

Alexandra Rink
Department of Radiation Physics
Princess Margaret Cancer Centre
and
Department of Radiation Oncology
University of Toronto
Toronto, Ontario, Canada

Daniel Robertson
Department of Radiation Physics
Division of Radiation Oncology
The University of Texas MD Anderson Cancer
    Center
Houston, Texas

Dany Theriault
Centre hospitalier universitaire de Québec
Lévis, Québec, Canada

François Therriault-Proulx
Department of Radiation Physics
The University of Texas MD Anderson Cancer
    Center
Houston, Texas

Landon Wootton
Department of Radiation Physics
The University of Texas MD Anderson Cancer
    Center
Houston, Texas

Kamil M. Yenice
Department of Radiation and Cellular Oncology
The University of Chicago
Chicago, Illinois

Wook Jae Yoo
School of Biomedical Engineering
College of Biomedical and Health Science
Konkuk University
Seoul, Republic of Korea

Rongxiao Zhang
Department of Physics and Astronomy
Dartmouth College
Hanover, New Hampshire

Part I

Basic principles and theory

Scintillation of organic materials

Marie-Ève Delage and Luc Beaulieu

Contents

1.1 INTRODUCTION: LUMINESCENCE, FLUORESCENCE, AND PHOSPHORESCENCE

Different scintillating materials (organic, inorganic, and organometallic materials) have the property of emitting visible light after being excited by one of the many possible sources. This property is related to a generic term, *luminescence*, which is defined as a prompt emission of light following the excitation of the substance. The book *The Glossary of Terms Used in Photochemistry* gives a more precise definition in refining the physical characteristics of the substance itself and in relation with its environment: Luminescence is a "spontaneous emission of radiation from an electronically excited species or from a vibrationally excited species not in thermal equilibrium with its environment" (Braslavsky 2007, p.367). Several sources of excitation can be used to generate luminescence. For example, photoluminescence is associated with an excitation process using visible photons. In addition to the mode of excitation, physical characteristics of the electronic states involved in the luminescence process can split the latter in different categories. Taking again photoluminescence as an example, two subdivisions appear, characterized by the spin multiplicities implied in the electronic states of the luminescent transition: fluorescence and phosphorescence (Knoll 2010). These two phenomena are of particular interest when depicting the scintillation process in organic materials. The former arises from transitions in the energy levels of a molecular entity with retention of spin multiplicity. The latter is however a luminescent process characterized by a change in spin multiplicity, typically from triplet to single states. Moreover, phosphorescence gives a light emission of longer wavelength than fluorescence due to the smaller energy difference of the transition implied in the process. Both definitions of fluorescence and phosphorescence apply to organic molecules, which is not the case for nanocrystalline semiconductors and metallic nanoparticles, because of the irrelevance of the concept of spin multiplicity to these materials. Thus, more general definitions are needed for these emitters. Fluorescence can be defined as the emission from an excited state reached by direct photoexcitation, whereas phosphorescence is an emission from another excited state, the transition associated having the particularity of being forbidden. In this section, the scintillation process of organic compounds will be discussed by overseeing the main concepts involved. At first, the scintillation physics will be addressed with a description of the excitation and de-excitation processes. Physical concepts related to Stokes shift and wavelength shifters will also be presented. Section 1.2 will end with an overview of other de-excitation modes possible in organic scintillators. Section 1.3 will get into the notions surrounding the scintillation light emission such as the rise time and signal falloff, and the light–energy relationship. Then, a brief comparison of scintillators and scintillating fibers will be made, presenting the main characteristics of each type of organic scintillators. A presentation of the properties and spectrum of selected organic scintillators will further be presented.

1.2 SCINTILLATION PHYSICS

Many phenomena can occur when an organic material is excited. This section intends to explain those that are happening on a molecular level, focusing on the transitions producing visible light. An important concept to keep in mind when depicting excitation and de-excitation processes is quenching, which regroups all non-radiative de-excitation processes and competes with the production of light. Chapter 2 will cover this topic at length. Another particularity to be aware of is the presence of a light component contributing to the fluorescence signal on a larger time scale than the main signal: the slow component. This particularity will be explained in Section 1.3.1.

1.2.1 EXCITATION, IONIZATION, AND DE-EXCITATION PROCESSES

As mentioned previously, various sources can be used to generate the excitation responsible for the light emission of a species. Luminescence can thus be categorized according to the mode of excitation of the substance: *photoluminescence* is a well-known term referring to an excitation by visible photons, while tribo- and sonoluminescence may sound less familiar, referring to an excitation by frictional and electrostatic forces and by ultrasound, respectively. As organic scintillators may be used for dosimetric applications, it is relevant to mention here that luminescence originating from excitation by ionizing radiation is called *radioluminescence* or *scintillation*. Other denominations for luminescence are reported in Table 1.1 (Valeur and Berberan-Santos 2013).

Two types of molecules have a specific electronic structure giving the inherent property to the substance of emitting light: conjugated and aromatic organic molecules, also named *organic fluors*. Typical examples with some of their properties can be found in Table 1.2 (The PubChem Project 2015).

On a molecular level, it is the π-electrons of these molecules that are involved in the excitation process preceding the luminescence. These nonlocalized electrons are weakly bound to their parent (carbon) atom than the localized, strongly bounded σ-electrons. Thus, less energy is required for transitions to occur in the π-electronic absorption band. Figure 1.1 shows a scheme of an anthracene molecule with the π-electrons localization identified.

Table 1.1 **Specific names of luminescence depending on the mode of excitation of the scintillating material**

EXCITATION	APPELLATION
In vivo biochemical reactions	Bioluminescence
Electron beams	Cathodoluminescence
Chemical reaction	Chemiluminescence
Electric field	Electroluminescence
Heating (with a previous energy storage)	Thermoluminescence

TABLE 1.2 **Properties of typical organic fluors**

MOLECULE NAME	CHEMICAL FORMULA	TYPE OF SCINTILLATOR ASSOCIATED	PROPERTIES
Anthracene	$C_{14}H_{10}$	Pure crystal	Colorless when crystallized from benzene; blue fluorescence (Considine 1974)
Stilbene	$C_{14}H_{12}$		Colorless or slightly yellow crystals; main emission peak at 350 nm (Tzeli et al. 2012)
Naphtalene	$C_{10}H_8$		Colorless to brown solid
p-Therphenyl	$C_{18}H_{14}$	Fluor in liquid scintillator	Colorless or light-yellow solid
PBD	$C_{20}H_{14}H_2O$		–
PPO	$C_{15}H_{11}N_2O$		–
3-HF	$C_{15}H_{10}O_3$	Plastic	Yellow; violet fluorescence in concentrated sulfuric acid

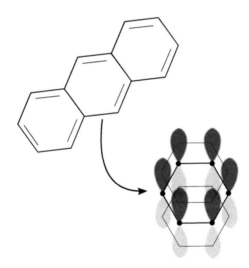

Figure 1.1 Anthracene chemical structure with representation of electronic p-orbitals for one typical aromatic cycle. π-electron delocalized orbitals are represented by the dotted lines.

In the case where the energy is sufficient, molecules are excited into the π-electronic *singlet* states. Each state has many vibrational levels to which they are attached, having energy values differing from each other by approximately 0.15 eV. A schematic view of the energy levels of a π-electronic structure is presented in Figure 1.2. A molecule can be excited in one of these possible levels. However, as a net effect of the excitation, a major population of excited molecules is found in the S_{10} state (1 = first excited state, 0 = ground level of vibrational energy for this state) because states with an excess of vibrational energy, in the order of 10^{-12} s, quickly lose this additional energy in order to reach thermal equilibrium (Knoll 2010).

Other excitation processes are possible but essentially do not contribute to the fluorescence emission. Excitation can also cause π-electrons ionization for which a subsequent ion recombination is observed mainly (75%) in the triplet states (Buck 1960), contributing to a slow component of scintillation light. Some molecules experience ion recombination from the singlet π-states, making a small contribution (12%) to the scintillation. The excitation of other electron-excited states (e.g., σ-electron of carbon 1s electron)

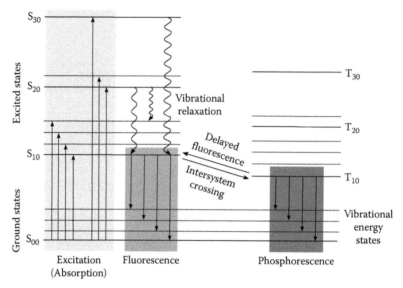

Figure 1.2 Jablonski diagram presenting the π-electronic atomic levels of organic molecules.

is also possible but the energy absorbed by the molecule is dissipated thermally, hence not producing fluorescence. Ionization of electrons other than π-electrons can happen as well, resulting in temporary or permanent damages. This will be further discussed in Section 1.3.6.

Once excited, molecules have multiple ways of spending the absorbed energy. The main de-excitation processes for unitary systems, that is, systems with only one type of molecules, are described in this section. These processes are represented in Figure 1.3 showing energy levels for two identical molecules and are identified by numbers associated with the process described in the text. It should be noted that they also occur for binary and ternary systems (described in Section 1.2.3).

Fluorescence (tag 1) is the main process of interest and results from a singlet–singlet transition from S_{10} to S_{0x}, where x stands for the vibrational level number. It can be followed by radiative migration (tag 5)—for example, the fluorescence photon is absorbed by another molecule—a phenomenon important for large scintillator because more molecules are available for this process or the escape of the emission, that is, production of visible light (tag 6). The second possible phenomenon is internal conversion (tag 2) described as a nonradiative transition between two energy levels of the same multiplicity. Conversion of vibrational energy is rapid (10^{-12} s) compared to transition from S_1 to S_0 (10^{-9} s), so the transition finishes with a vibrational relaxation to the first level of the singlet state (e.g., $S_{13} \rightarrow S_{11} \rightarrow S_{10}$). Third, the molecule can undergo non-radiative migration (tag 3), where the energy of the first singlet state migrates, through the exciton diffusion, a nonradiative process, from molecules to molecules prior to emission. The fourth possible de-excitation process is called *internal quenching* (tag 4) and happens when the energy of the first excited singlet state is dissipated nonradiatively via a quenching transition toward S_{0x} or an intersystem crossing transition toward T_{1x}. Finally, a process called *intersystem crossing* (ISC) can occur giving place to phosphorescence or delayed fluorescence (see Figure 1.2). Spin-orbit coupling enables ISC, a radiationless

Basic principles and theory

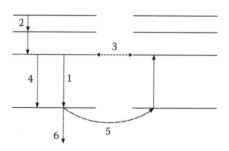

Figure 1.3 De-excitation processes for unitary systems.

transition between levels of the same energy but related to states with different spin multiplicities. The excited electron has to change its spin going from being the same or the inverse of its paired electron in the ground state. The more the spin–orbit interaction is strong, the more it facilitates spin inversion of the electron, thus promoting ISC. In the case of phosphorescence, a transition occurs in transferring the electron from the absorption singlet band to the triplet state first level (T_1). After this transition, the molecule de-excites into one of the vibrational levels of the singlet ground state emitting a visible photon. The electron can also go back from T_1 to S_1, especially if the energy difference between the two levels is small and the T_1 level has a lifetime long enough to allow the transition. Then, a $S_1 \rightarrow T_1$ transition can produce what is called *delayed fluorescence*, having the same spectrum as fluorescence, but happening over a longer period of time (10^{-6} s or longer).

1.2.2 STOKES SHIFT

Each process, absorption and emission, has a spectrum representative of the vibrational spacing in an associated singlet state. S_1-level spacing is related to the absorption spectrum, while S_0 spacing is related to the emission spectrum. Most of the time, the spectra are the mirror images of one another because of the similarity of the vibrational levels distribution in either the ground or excited states, as seen in Figure 1.4.

The maximum of the first absorption band and the maximum of the fluorescence spectrum for the same electronic transition are separated by a distance named the *Stokes shift*, quantified in terms of the wave numbers, wavelength, frequency, or energy between the peaks. An interesting feature of organic scintillators is that almost all fluorescent transitions have energy lower (higher wavelength) than the value necessary for excitation, making the scintillator mostly transparent to its own emission spectrum. Nevertheless, absorption and emission spectra can overlap, a result from the fact that molecules have a thermal motion providing the additional energy required for the absorption spectrum to overlap the emission spectrum. Observing fluorescence at cryonic temperatures can minimize the apparition of this overlap. It is preferred to have two well-distinguished spectra, that is, large Stokes shift, in order to minimize self-absorption—associated with the radiative migration described—by the scintillator. Many studies have focused on developing a scintillator with large Stokes shift in order to minimize the loss of scintillation signal. Among the possibilities, a molecule presented in Table 1.2, the 3-HF (3-hydroflavone), was put forward, having a Stokes shift of 191 nm when dissolved in xylene (Renschler and Harrah 1985). Other derivatives of methyl and methoxy substituted 1,3-diphenyl-2-pyrazolines showed a good displacement between the absorption and the emission peaks, making them candidates of interest for large-size scintillator where self-absorption is an issue (Güsten et al. 1978). In Figure 1.5 are represented spectra for the original 1,3-diphenyl-2-pyrazolines (a) and for its derivative (b), both in cyclohexane.

1.2.3 WAVELENGTH SHIFTERS AND *THREE COMPONENTS* SYSTEMS

Organic scintillators can be classified in three categories regarding their composition. Unitary systems are made of pure crystals (e.g., anthracene, *trans*-stilbene), binary systems are composed of a scintillating material incorporated in a solvent (e.g., liquid solution of *p*-terphenyl in toluene), and ternary systems are made of two scintillating materials (primary scintillator and a wavelength shifter) incorporated in a solvent

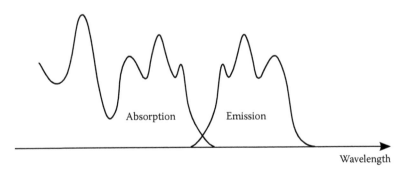

Figure 1.4 Mirror image of absorption and emission spectra due to the similarity of the vibrational levels for S_1 and S_0 states.

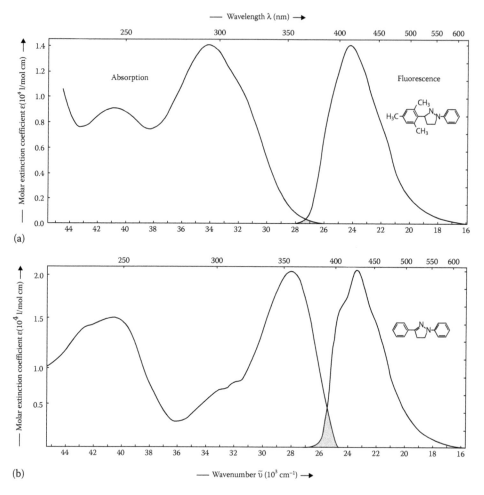

Figure 1.5 Electronic absorption and absolute fluorescence spectrum of 1,3-diphenyl-2-pyrazoline (a) and 1-phenyl-3-mesityl-2-pyrazoline (b) in cyclohexane. (Reprinted with permission from Güsten, H. et al., Organic scintillators with unusually large Stokes' shifts, *J. Phys. Chem.*, 82, 459, 1978. Copyright 2015 American Chemical Society.)

(e.g., plastic solution of *p*-terphenyl and POPOP in polystyrene). Sometimes, the use of an organic additive is desired in order to better match the sensitivity range of a photodetector. The wavelength shifter is of interest in that case, as it absorbs the light emitted by the primary substance of the scintillator and reemits it at a longer wavelength. In addition to a better spectral match with the photodetector, wavelength shifters give isotropic reemitted light, assuring a better light collection (Knoll 2010).

In the *three components* systems, the three species of molecules present in these types of systems offer other types of de-excitation than those described with unitary systems. Figure 1.6 illustrates well the processes possible when a ternary system is de-excited. The processes are numbered following the description in the text.

Nonradiative transfer is possible from a molecule of the main constituent to either another molecule of this species (tag 1) or a molecule of the primary solute (tag 2). Then this kind of transfer can occur from the primary solute to the secondary solute (wavelength shifter) (tag 3). Furthermore, radiative transfer is also possible between the three components (tags 4–6): the visible photon emitted by the main constituent is absorbed instead of escaping the system and contributing to the scintillation signal. This phenomenon happens obviously when the absorption spectrum of one constituent overlaps (at least partially) the emission spectrum of the other constituent concerned. As in unitary systems, internal quenching can happen for binary and ternary systems, but in this case, it can happen in any of the constituents, reducing the scintillation efficiency (tags 7–9). Another consequence of the presence of three types of molecules is that

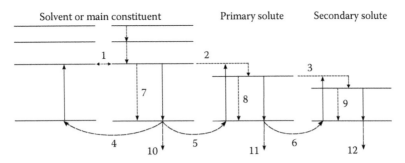

Figure 1.6 Processes of de-excitation for ternary systems.

the emission of visible light can come from one or other components of the system (tags 10–12). It then produces more or less extended spectra on the wavelength axis, with one or multiple peaks. Ternary systems must therefore be chosen carefully, based on the intended application.

1.2.4 OTHER DE-EXCITATION MODES

Molecules have the possibility of undergoing many other de-excitation processes, some being related intimately to the environment conditions. A competition between these de-excitation pathways and fluorescence may occur if the molecule stays in the excited state during the lifetime comparable to the time during which the competitive processes take place. Detailed description of these phenomena is beyond the scope of this book. More information can be found in *Molecular Fluorescence* (Valeur and Berberan-Santos 2013). However, a brief description of these modes will be given for completeness.

Once excited, an organic molecule can spend its newly acquired energy by undergoing intramolecular charge transfer, which is defined as a "process that changes the overall charge distribution in a molecule" according to the *Compendium of Chemical Terminology* (Nič et al. 2009, p.753). The transfer of charge is done between two regions of a polarized molecular entity and a dependence on the environment is observed, causing shifts in emission and absorption spectra. There is also a possibility that a molecule will change its conformation using the absorbed energy. Interactions between an excited molecule and other molecules may further compete with fluorescence such as electron transfer, proton transfer, energy transfer, and excimer of exciplex formation. An excimer is an excited dimer formed by two identical molecules, one excited and the other unexcited, as a result of a collision between the two molecules. An exciplex is similar to an excimer with the difference that the unexcited molecule is not of the same nature that of the excited molecule. At last, photochemical transformation can occur. It is a process in which organic photochemical reactions lead to bonds breaking and new bonds formation. This transformation results in a loss of recovery of the ground state, enhancing luminescence quenching. It should be taken into consideration that conformational change, electron transfer, proton transfer, energy transfer, and excimer or exciplex formation may lead to a fluorescent emission superimposing that of the initially excited molecule. However, differentiation of this emission with the primary one should be physically made; for example, excimer and exciplex have their own fluorescence spectrum and decay time. Excimers may dissociate thermally, regenerating the excited (S1) monomer fluorescence, which can then produce fluorescence.

1.3 SCINTILLATION LIGHT EMISSION

Light emission of organic scintillators depends on the factors related to the nature of the scintillator. As a consequence, different spectral bands and response times can be encountered when looking through the properties of organic scintillating materials. Each of these materials will have its proper fluorescence intensity exponential decay over time. The response of organic molecules to excitation also depends on the nature of the scintillator, the type of ionizing particles (directly or indirectly ionizing), and the energy of the incident particles, allowing the possibility of using these dependencies to our advantage in applications such as pulse shape discrimination. The light response is also dependent on the degree of damage inflicted to the scintillator. This damage can be caused by a prolonged exposure to ionizing radiation; an exposure to light and oxygen, which for plastics leads to polymer degradation; or surface crazing that compromises the internal total reflection, a phenomenon on which relies the transmission of the scintillation light. Key concepts related to these dependences of the efficiency of the organic scintillators light emission will be presented in addition to a description of the selected organic materials (Section 1.3.6).

1.3.1 RISE-TIME AND SIGNAL FALLOFF

Pure crystals are considered as instant light emitters. Hence, no time has to be taken into account for the scintillation light to reach a maximum. For their part, organic solutions scintillators need a finite time (τ_R) for their scintillation response to be at its maximum value, which is of the order of 10^{-9} s. Both types of scintillator have a signal falloff that is characterized by the decay time (τ), typically of the order of 10^{-8}–10^{-9} s (Birks 1964), representing the time for a relative loss of $1/e$. The light intensity for the fluorescence is thus represented by an exponential decay.

A typical time distribution of the fluorescence signal is represented in Figure 1.7 for different scintillators sizes (Moszyński and Bengtson 1979). For the pure crystals, one only needs these parameters to compute the intensity of the scintillation emission:

$$n(t) = \frac{N}{\tau} e^{-t/\tau}; \quad n(t) \text{ is in photons/s} \tag{1.1}$$

For other organic scintillators, rise time has to be taken into consideration, especially for plastic scintillators, which present a very fast decay. Therefore, the intensity is expressed as

$$n(t) = \frac{N}{(\tau - \tau_R)} \left[e^{-t/\tau} - e^{-t/\tau_R} \right] \tag{1.2}$$

However, this expression can be simplified to the first case expression considering that the rise time for most practical scintillators (other than plastics) is much more rapid then the decay time, so the former

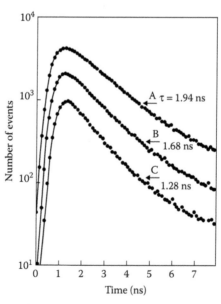

Figure 1.7 Time distribution spectra of light pulses from the binary liquid scintillator (BIBUQ in xylene) excited by beta-rays. Curves A, B, and C correspond to different sample heights (5, 1, and 0.2 cm, respectively). (Reprinted from *Nucl. Instrum. Methods Phys. Res. A*, 350, Moszyński, M. et al., Study of n-γ discrimination with NE213 and BC501A liquid scintillators of different size, 226–234, 1994. Copyright 2015, with permission from Elsevier.)

can be taken as zero. Returning to the case of plastic scintillators, it appears that the best description for fluorescence light intensity decay would be a convolution of a Gaussian with an exponential:

$$n(t) = N \cdot f(\sigma, t) \cdot e^{-t/\tau} \tag{1.3}$$

where f is the Gaussian function with a standard deviation σ.

Two types of fluorescence were presented previously, having for only difference the intermediate transitions implied in the light emission, hence their decay time. *Straightforward* fluorescence has a decay time of some nanoseconds, while delayed fluorescence decay time is more of some hundred of nanoseconds. For most scintillators, we can then observe two fluorescent components: the fast and the slow components. For a more detailed model of the light intensity decay, the slower component has to be taken into consideration. Scintillation intensity can then be expressed as a combination of exponential decays with relative magnitudes (A and B) dictated by the nature of the material

$$A \cdot e^{-t/\tau_f} - B \cdot e^{-t/\tau_s} \tag{1.4}$$

with decay time for fast (f) and slow (s) components.

Usually, the fast component is used as the main scintillation signal as it dominates over the slow one.

The fraction of light produced via both the fast and the slow components depends on the nature of the exciting particle. Hence, it is possible to recognize the kind of particle that deposited its energy in a detector by the shape of the emitted light pulse, as seen in Figure 1.8 for stilbene (Attix 1986). This recognition method is called *pulse shape discrimination*. It is based on the fact that the fast and the slow components arise from a de-excitation involving different electronic states. The population of these states differs in proportion depending on the *dE/dx* (specific energy loss) of the incident particle. For example, a heavier particle, having a large *dE/dx*, results in a larger density of triplet states along the track of the particle because the higher density of molecules excited give rise to more intermolecular interactions. This phenomenon hinders fluorescence and increases the contribution to delayed fluorescence.

Pulse shape discrimination is commonly used to count neutrons in the presence of a gamma radiation background with a liquid scintillator. The neutrons are detected by the scattering of protons. For their part, gamma rays interact via the usual processes (photoelectric, Compton, and pair production). This way, the background signal generated by the gamma radiation can be suppressed and only the neutron signal

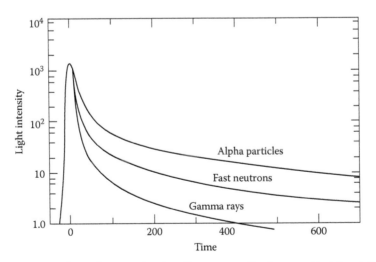

Figure 1.8 Time dependence of scintillation pulses in stilbene, normalized to equal heights at time zero, when excited by radiations of different LET. (Reprinted with permission from Bollinger, L.M. and Thomas, G.E., Measurement of the time dependence of scintillation intensity by a delayed-coincidence method, *Rev. Sci. Instrum.*, 32, 1044–1050, 1961. Copyright 2015, American Institute of Physics.)

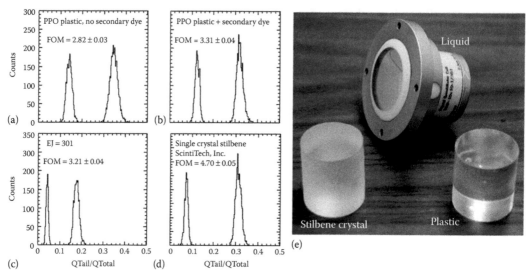

Figure 1.9 Neutron/gamma PSD FOMs for (480 ± 75) keVee energy range obtained with (a) binary and (b) ternary plastics in comparison with commercial (c) liquid and (d) single crystal scintillators and (e) measurements made with the same size samples using 252Cf source. (Reprinted from *Nucl. Instrum. Methods Phys. Res. A*, 668, Zaitseva, N. et al., Plastic scintillators with efficient neutron/gamma pulse shape discrimination, 88–93, 2012. Copyright 2015, with permission from Elsevier.)

can be withdrawn (Winyard et al. 1971). Nowadays, plastic scintillators can do efficient pulse shape discrimination. Figure 1.9 (Zaitseva et al. 2012) compares the efficiency of the plastic scintillators to those of liquid and single crystal scintillators.

1.3.2 SCINTILLATION EFFICIENCY

Only a small fraction of the energy absorbed by the scintillator is converted into light, the rest of it being dissipated nonradiatively. This degree of conversion can be characterized by the absolute scintillation efficiency, which is defined as the fraction of the energy originally deposited in a material that was used for the scintillation light (Craun and Smith 1970). Anthracene has a scintillation efficiency of 5% in the blue region of the visible spectra and serves as a calibration landmark for other organic scintillators because it has the highest practical scintillation efficiency. Typical plastic scintillators have light yield values around 40%–65% of anthracene, while liquid scintillators light efficiency is more than 60% going up to almost 80% of anthracene (Saint-Gobain Crystals & Detectors, Paris, France). Scintillation efficiency can also be characterized in terms of the number of photons produced per unit (MeV) of the energy absorbed. As for scintillating fibers, the datasheet for the BCF-12 scintillating fiber states that 8000 photons in the range of 400–575 nm (average of 432 nm) are emitted per MeV. This leads to a *W/e* equivalent value of 125 eV per photon (Saint-Gobain Scintillation Brochure). Scintillation efficiency is also strongly dependent on the particle type and energy, the response for electrons being higher than that for heavier particles such as proton and alpha particles for equivalent energies. More details on this dependence will be provided in Section 1.3.3.

1.3.3 LIGHT–ENERGY RELATIONSHIP AND BIRKS' FORMULA

Considerations on quenching have been made in saying that this phenomenon reduces the scintillation efficiency by dissipating the energy non-radiatively and forms a high ionization density along the track of the particle incident on the scintillator. Passing over the fact that quenching is happening in the

Basic principles and theory

scintillator (issue tackled in Chapter 2), one can relate easily the fluorescent energy emitted per unit of path length (*dL/dx*) to the specific energy loss for the charged particle (*dE/dx*) that depends on the particle type and energy, with the scintillation efficiency (*S*)

$$\frac{dL}{dx} = S\left(\frac{dE}{dx}\right)$$

(1.5)

The response of organic scintillators is then quite linear for particle energy above 100 keV (Smith et al. 1968), as seen in Figure 1.10. In reality, it is a more complex function, supporting in a better way experimental data at lower energies. Birks explained this deviation from the ideal linearity of the organic scintillators light output by quenching effects (Birks 1964).

Considering that the density of damaged molecules is proportional to the ionization density, he proposed a ratio of these two quantities represented by the following expression: *B(dE/dx)*, where *B* is a proportionality constant. In presence of quenching, Equation 1.1 is modified, so that the fluorescence emission expression becomes

$$\frac{dL}{dx} = \frac{S(dE/dx)}{1 + kB(dE/dx)}$$

(1.6)

Equation 1.6 introduces the parameter *k* accounting for the fraction of the molecules that will undergo quenching. This equation is referred to as the Birks formula.

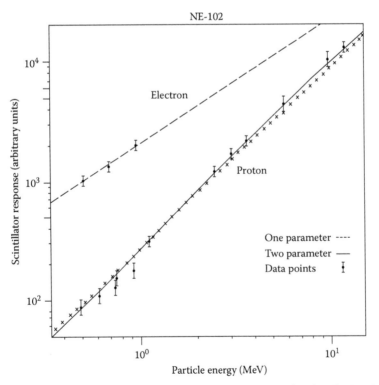

Figure 1.10 Light output vs. particle energy for electrons and protons stopped in the plastic scintillator NE-102. The light output is proportional to electron energy, but not to proton energy. (Reprinted from *Nucl. Instrum. Methods*, 80, Craun, R.L. and Smith, D.L., Analysis of response data for several organic scintillators, 239–244, 1970. Copyright 2015, with permission from Elsevier.)

Two approximations of this equation can be made, depending on the type of ionizing particle. For electrons (directly incident or from X-rays), if dE/dx is assumed to be small for sufficiently large values of E, the Birks differential equation can be simplified to $L = SE$. For alpha (or heavier particles) particles, dE/dx is very large, so the saturation occurs along the particle track. In that regime, the Birks formula reduces to the simple form: $dL/dx = S/kB$, where the light production does not depend on the variation of dE/dx anymore.

An extended version of the Birks formula has been proposed in order to better fit the data, approaching the simple formula for small dE/dx, with an empirically fitted parameter C (Leo 1994):

$$\frac{dL}{dx} = \frac{S(dE/dx)}{1 + kB(dE/dx) + C(dE/dx)^2} \tag{1.7}$$

1.3.4 SCINTILLATORS VERSUS SCINTILLATING FIBERS

Plastic scintillators as well as scintillating plastic fibers are composed of an organic scintillator dissolved in a solvent that is subsequently polymerized: the solvent, at first composed of monomers, undergoes solidification when heated and forms a solid thermoplastic polymer. Their matrix (or bulk medium) containing the scintillating substance is however different; scintillators are made of polyvinyltoluene (PVT) and scintillating fibers of polystyrene. Plastic fibers are composed of a core, where the light is produced, and a cladding, having multiple functions, while simple scintillators only have what would correspond to a core. The cladding (habitually made of poly[methyl methacrylate]) allows total internal reflection since the refractive index of the core is chosen to be larger than the index of the cladding. The light can thus be transmitted over a substantial distance. The cladding is also useful to protect the core from abrasion or accumulation of foreign material that would compromise the efficiency of the light pipe effect (see Chapter 3). Different shapes of scintillator can be made: rods, cylinders, flat sheets, and fibers are available in different diameter sizes (few tenths of millimeter to several millimeter) and shapes. Regarding the transmission of the produced light, self-absorption is non-negligible if the size of the scintillator is large for both types of scintillators. Properties of some selected plastic scintillators and scintillating fibers will be presented in the following section.

1.3.5 PROPERTIES AND SPECTRUM OF KEY ORGANIC SCINTILLATORS

Let us first begin by a brief description of the reference scintillator mentioned before: anthracene. Its characteristic absorption and emission spectra are presented in Figure 1.11 (Kunnil et al. 2005). This pure crystal has the advantage of being very durable and has a high light output.

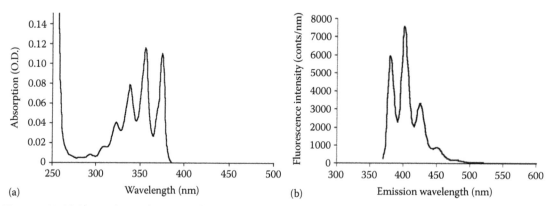

Figure 1.11 (a) Absorption and emission (b) spectra of a 14 µM anthracene solution in ethanol with excitation at 360 nm. (Reprinted from Kunnil, J. et al., Fluorescence quantum efficiency of dry *Bacillus globigii* spores, *Opt. Express*, 13, 8969, 2005. Copyright 2015, with permission of Optical Society of America.)

Basic principles and theory

However, its anisotropic response causes problems when collecting the light output, degrading the energy resolution obtainable with this crystal; the scintillation efficiency is dependent on the orientation of the incoming ionizing particle relative to the crystal axis, causing various directions of particles track within the crystal.

Figure 1.12 describes this dependence showing the energy loss for two orientations of an electron beam over anthracene thin foils (Kunstreich and Otto 1969). Moreover, this scintillator has a relatively long decay time (30 ns) compared to plastic and liquid scintillators (few nanoseconds). Anthracene is not easily machined or available in large sizes. All these limitations make this scintillator's utilization unfavorable nowadays over other organic scintillators such as plastic scintillators.

A relevant study (Archambault et al. 2005) compared different kinds of plastic scintillators and scintillating fibers in order to guide the reader toward a choice of an appropriate scintillator, depending on the application underlying. For the fibers, four models were tested: Saint-Gobain Crystals and Detectors (Paris, France) BCF-12 and BCF-60 and Kuraray (Tokyo, Japan) SCSF-78 and SCSF-3HF. They are composed of a core of polystyrene and a cladding of acrylic. Two plastic scintillators (BC-400 and BC-408, Saint-Gobain) made of PVT were also characterized in order to carry on a comparison with the scintillating fibers. Both types of scintillators were of cylindrical shape. Each of these scintillators and scintillating fibers has a characteristic emission spectrum that was measured (Figure 1.13). The emission peak wavelengths were compared to the nominal values given by the manufacturer, showing a good agreement. One of the observations made by the authors was that the scintillating fibers' spectra are wider than those of the plastic scintillator, and that the total light output of the BC-400 or BC-408 is less than that of the blue scintillating fiber.

Scintillation light measurements were also performed, and it was noted that the signal was rising linearly with both the radius and the length of the scintillator. However, there is always a trade-off between light production and attenuation: a bigger scintillator offers a higher probability of light production, but increasing the size of a scintillator also increases the attenuation the light will undergo. The light produced far from the coupling interface has also a smaller chance of reaching the light guide than the light produced near the interface.

Manufacturers offer multiple possibilities regarding liquid scintillators. Some examples of liquid scintillators presented here (Saint-Gobain Crystals and Detectors, Paris, France) have different decay times all around a few nanoseconds, and a theoretical maximum emission wavelength of 425 nm. It is further

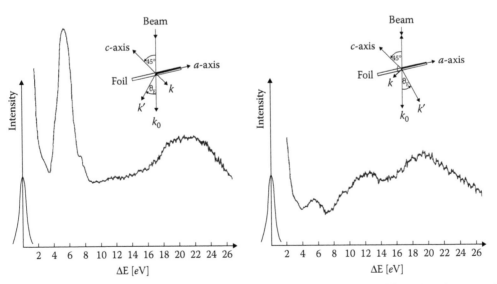

Figure 1.12 Anisotropy of energy loss spectra in anthracene single crystals. (Reprinted from *Opt. Commun.*, 1, Kunstreich, S. and Otto, A., Anisotropy of electron energy loss spectra in anthracene single crystals, 45–46, 1969. Copyright 2015, with permission from Elsevier.)

Basic principles and theory

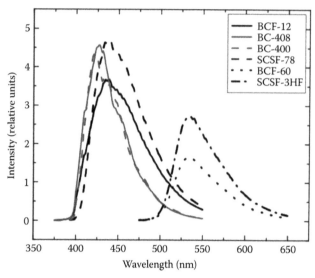

Figure 1.13 Experimental measurement of the emission spectra of four scintillating fibers (BCF-12, BCF-60, SCSF-78, SCSF-3HF) and two plastic scintillators (BC-400, BC-408). (Reprinted from Archambault, L. et al., *Med. Phys.*, 32, 2271, 2005. Copyright 2015, with permission from American Association of Physicists in Medicine.)

possible to add a loading element (such as boron or gadolinium) to provide a better detection of neutron and making possible neutron spectrometry. Liquid scintillators are useful when large detection volumes are needed, but it was reported that the discrimination quality is intimately related to the size of the scintillator, the larger ones providing a poorer quality (Moszyński et al. 1994).

Scintillators BC-501A, BC-505, and BC-509 are intended principally for gamma and fast neutron detection. Their emission spectra are given in Figure 1.14a. The first two are known for

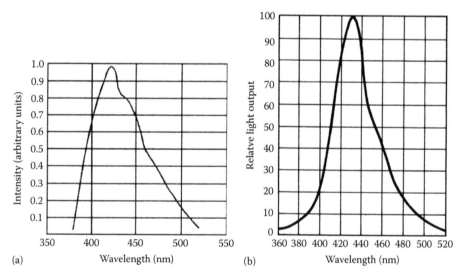

Figure 1.14 Fluorescene emission spectrum of (a) plastic scintillators BC-501A, BC-505, and BC-509 and (b) BC-531.

their capability to distinguish gamma and neutron scintillation pulses. It was demonstrated that BC-501A could achieve discrimination down to 40 keVee (1 keVee standing for the amount of energy converted to light by a 1 keV electron) (Swiderski et al. 2008). The last scintillator, BC-509, has a lower hydrogen content, avoiding the thermalizing properties of habitual hydrogenous detector if desired for the intended application. Also from Saint-Gobain, the BC-531 scintillator (Figure 1.14b) allows neutron detection but offers a high content in hydrogen and a high light output. It was used for proton beam as well as for photon beam dosimetric measurements (Beddar 2010).

Loaded scintillators are also available, such as the BC-523A, which has a loading element boron. In this type of scintillator, the neutrons are at first slowed down by scattering with hydrogen nuclei producing a prompt signal. They are then captured by ^{10}B that generates the ^{10}B $(n, \alpha)^7Li$ reaction emitting a definite amount of light. This way, it is possible to define a second pulse accounting for the capture of thermal neutrons. Loaded scintillators are also used to do pulse shape discrimination of fast neutrons in a background of gamma radiation.

1.3.6 EFFECTS OF RADIATION ON ORGANIC SCINTILLATORS

Different studies have reported radiation damage on organic scintillators. It is not of our intention to present a detailed literature review here but to put forward major findings. A general conclusion that can be drawn from the publications related to this subject is that organic scintillators light output is decreasing with irradiation dose but recovery in light output is possible. Anthracene scintillation efficiency was found to deteriorate when exposed to photon, electron, and alpha irradiation, with no recovery even when stored in the dark or heated for hours. Only Attix (1959) contradicts this observation with his report of a slight recovery when heating the crystal for an hour up to 100°C. The hypothesis of oxidation as the cause of radiation damage was put aside with the observation that gamma-ray irradiation in air, vacuum, or helium gave similar degradation effects. For typical plastics, the loss of signal becomes significant irradiated with a cumulated dose in the range of 10^3–10^4 Gy, while some radiation-resistant plastics show little change in scintillation intensity for a dose of around 10^5 Gy (Senchishin et al. 1995). Changes observed vary from a decrease in the light output due to damage done to the fluorescent component to a decrease of the light transmission due to the creation of optical absorption centers by radical species. Most of these species will eventually be quenched, reestablishing some of the light output but a portion of the radicals can react and form structures causing permanent damage (Biagtan et al. 1996). Degradation of the scintillation output is dependent on the dose rate, the presence of oxygen, and the nature of the radiation. Less recovery in light output was found when the dose rate is decreased and no recovery at all occurred from a certain dose rate and below, regardless of the dose (Biagtan et al. 1996) (see Figure 1.15).

This is true when the sample is being irradiated in air. No such changes were observed under inert atmosphere (Ar or N_2). Oxygen is a determinant factor affecting the scintillator properties. During the irradiation of a sample, oxygen available is free to react with some radical species formed under irradiation producing radiation-induced damage. The more oxygen is available throughout the plastic, the more damaging will be the irradiation, increasing the permanent radiation-induced absorption centers (Zorn 1993). After an exposure of 10 kGy, a yellow or brown color can be observed on certain scintillators (Wick et al. 1991). This coloration may be the result of the production of free radicals, trapped electrons, or conjugated double bonds. In the presence of oxygen, radicals can react with O_2 molecules diffusing into the organic material, bleaching the surface colored by radiation. This bleaching process is known as *annealing*. The effect of radiation damage on dosimetry with plastic scintillation detectors is covered in detail in Chapter 4.

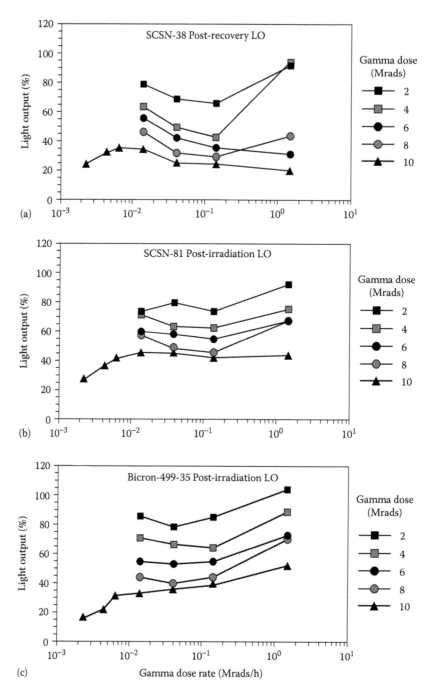

Figure 1.15 Post-recovery light outputs versus gamma dose rate at various doses: (a) SCSN-38, (b) SCSN-81, and (c) Bicron-499-35. (Reprinted from *Nucl. Instrum. Methods Phys. Res. B*, Biagtan, E. et al., Gamma dose and dose rate effects on scintillator light output, 125, Copyright 1996, with permission from Elsevier.)

1.4 CONCLUSION

The scintillation of organic materials is a well-understood process. The transitions implied in the scintillation phenomena are clearly described. This chapter reviewed the main concepts of the organic scintillation materials for anyone interested in having an overall preview of such phenomenon. Possibilities using organic scintillators are numerous in terms of scintillation applications. The flexibility in the choice of a scintillator offers a chance to the user to adapt the scintillating material to the needs of the application. This feature is particularly interesting in the field of scintillation dosimetry in medical physics where many requirements have to be met, such as water equivalence, short-time response, reduced dimensions, and nontoxicity of a material.

REFERENCES

Archambault, L., Arsenault, J., Gingras, L., Beddar, A.S., Roy, R., and Beaulieu, L., 2005. Plastic scintillation dosimetry: Optimal selection of scintillating fibers and scintillators. *Medical Physics*, 32 (7), 2271–2278.

Attix, F.H., 1959. High-level dosimetry by luminescence degradation. *Nucleonics (U.S.) Ceased Publication*, 17 (4), 142.

Attix, F.H., 1986. *Introduction to Radiological Physics and Radiation Dosimetry*. New York: Wiley.

Beddar, S., 2010. A liquid scintillator system for dosimetry of photon and proton beams. *Journal of Physics: Conference Series*, 250 (1), 012038.

Biagtan, E., Goldberg, E., Stephens, R., Valeroso, E., and Harmon, J., 1996. Gamma dose and dose rate effects on scintillator light output. *Nuclear Instruments and Methods in Physics Research Section B: Beam Interactions with Materials and Atoms*, 108 (1–2), 125–128.

Birks, J.B., 1964. *The Theory and Practice of Scintillation Counting*. New York: Pergamon Press.

Bollinger, L.M. and Thomas, G.E., 1961. Measurement of the time dependence of scintillation intensity by a delayed-coincidence method. *Review of Scientific Instruments*, 32 (9), 1044–1050.

Braslavsky, S.E., 2007. Glossary of terms used in photochemistry, 3rd edn. (IUPAC Recommendations 2006). *Pure and Applied Chemistry*, 79 (3), 367.

Buck, W.L., 1960. The origin of scintillations in organic materials. *IRE Transactions on Nuclear Science*, 7 (2–3), 11–16.

Considine, D.M., 1974. *Chemical and Process Technology Encyclopaedia*. New York: McGraw-Hill.

Craun, R.L. and Smith, D.L., 1970. Analysis of response data for several organic scintillators. *Nuclear Instruments & Methods*, 80 (2), 239–244.

Güsten, H., Schuster, P., and Seitz, W., 1978. Organic scintillators with unusually large Stokes' shifts. *The Journal of Physical Chemistry*, 82 (4), 459–463.

Knoll, G.F., 2010. *Radiation Detection and Measurement*, 4th edn. Hoboken, NJ: Wiley.

Kunnil, J., Sarasanandarajah, S., Chacko, E., and Reinisch, L., 2005. Fluorescence quantum efficiency of dry *Bacillus globigii* spores. *Optics Express*, 13 (22), 8969.

Kunstreich, S. and Otto, A., 1969. Anisotropy of electron energy loss spectra in anthracene single crystals. *Optics Communications*, 1 (2), 45–46.

Leo, W.R., 1994. *Techniques for Nuclear and Particle Physics Experiments: A How-To Approach*. Germany: Springer Science & Business Media.

Moszyński, M. and Bengtson, B., 1979. Status of timing with plastic scintillation detectors. *Nuclear Instruments & Methods*, 158, 1–31.

Moszyński, M., Costa, G.J., Guillaume, G., Heusch, B., Huck, A., and Mouatassim, S., 1994. Study of n-γ discrimination with NE213 and BC501A liquid scintillators of different size. *Nuclear Instruments and Methods in Physics Research Section A: Accelerators, Spectrometers, Detectors and Associated Equipment*, 350 (1–2), 226–234.

Nič, M., Jirát, J., Košata, B., Jenkins, A., and McNaught, A., eds., 2009. *IUPAC Compendium of Chemical Terminology: Gold Book*, 2.1.0 edn. Research Triangle Park, NC: IUPAC.

Renschler, C.L. and Harrah, L.A., 1985. Reduction of reabsorption effects in scintillators by employing solutes with large stokes shifts. *Nuclear Instruments and Methods in Physics Research Section A: Accelerators, Spectrometers, Detectors and Associated Equipment*, 235 (1), 41–45.

Senchishin, V.G., Markley, F., Lebedev, V.N., Kovtun, V.E., Koba, V.S., Kuznichenko, A.V., Tizkaja, V.D. et al., 1995. A new radiation stable plastic scintillator. *Nuclear Instruments and Methods in Physics Research Section A: Accelerators, Spectrometers, Detectors and Associated Equipment*, 364 (2), 253–257.

Smith, D.L., Polk, R.G., and Miller, T.G., 1968. Measurement of the response of several organic scintillators to electrons, protons and deuterons. *Nuclear Instruments & Methods*, 64 (2), 157–166.

Swiderski, L., Moszynski, M., Wolski, D., Batsch, T., Nassalski, A., Syntfeld-Kazuch, A., Szczesniak, T. et al., 2008. Boron-10 loaded BC523A liquid scintillator for neutron detection in the border monitoring. *IEEE Transactions on Nuclear Science*, 55 (6), 3710–3716.

The PubChem Project [online], 2015. Available from: http://pubchem.ncbi.nlm.nih.gov/ [Accessed March 16, 2015].

Tzeli, D., Theodorakopoulos, G., Petsalakis, I.D., Ajami, D., and Rebek, J., 2012. Conformations and fluorescence of encapsulated stilbene. *Journal of the American Chemical Society*, 134 (9), 4346–4354.

Valeur, B. and Berberan-Santos, M.N., 2013. *Molecular Fluorescence: Principles and Applications*, 2nd edn. Weinheim, Germany: Wiley-VCH.

Wick, K., Paul, D., Schröder, P., Stieber, V., and Bicken, B., 1991. Recovery and dose rate dependence of radiation damage in scintillators, wavelength shifters and light guides. *Nuclear Instruments and Methods in Physics Research Section B: Beam Interactions with Materials and Atoms*, 61 (4), 472–486.

Winyard, R.A., Lutkin, J.E., and McBeth, G.W., 1971. Pulse shape discrimination in inorganic and organic scintillators. I. *Nuclear Instruments & Methods*, 95 (1), 141–153.

Zaitseva, N., Rupert, B.L., PaweŁczak, I., Glenn, A., Martinez, H.P., Carman, L., Faust, M., Cherepy, N., and Payne, S., 2012. Plastic scintillators with efficient neutron/gamma pulse shape discrimination. *Nuclear Instruments and Methods in Physics Research Section A: Accelerators, Spectrometers, Detectors and Associated Equipment*, 668, 88–93.

Zorn, C., 1993. A pedestrian's guide to radiation damage in plastic scintillators. *Nuclear Physics B—Proceedings Supplements*, 32, 377–383.

Quenching of scintillation light

Daniel Robertson and Sam Beddar

Contents

2.1 INTRODUCTION

Accurate dosimetry requires an understanding of the relationship between the energy deposited in the medium by the radiation source and the signal produced by the detector. In scintillators, the conversion of energy from radiation into visible light depends on the radiation interaction mechanism and the physical and temporal distribution of the radiation on the microscopic level, also known as the *quality* of the radiation (ICRU 1970). The physical mechanism of scintillation in organic materials was described in detail in Chapter 1. This chapter focuses solely on the effect of the quality of the radiation on the production of scintillation light.

The response of organic scintillators varies with the ionization density of the incident radiation, which is usually quantified as the linear energy transfer (LET). In therapeutic photon beams, the ionization density is approximately uniform throughout the beam, and the rate of scintillation light emission is proportional to the radiation dose at any given location. However, in proton and heavy ion beams, the ionization density increases with depth. This causes a corresponding decrease in the scintillation response, commonly known as *quenching*. Many other detector types also suffer from LET-dependent quenching, including films, silicon diodes, thermoluminescent and optically stimulated luminescent detectors, alanine, diamond, and polymer gel detectors (Jirasek and Duzenli 2002, Karger et al. 2010).

Quenching is problematic for scintillation dosimetry of proton and heavy ion beams, limiting the usefulness of scintillator detectors for measurements of dose distributions. The LET variation is primarily in the depth direction, leading to a depressed signal in the Bragg peak of ion beams. Because the LET is relatively invariant perpendicular to the beam direction, scintillators can be used effectively for lateral profile measurements (Boon et al. 2000, Wilkens and Oelfke 2004).

Scintillation quenching in ion beams is unfortunate, because scintillator detectors have unique properties that would be beneficial for ion beam dosimetry, including high spatial and temporal resolution, near water equivalence, and the potential for rapid 2D and 3D measurements. An ideal solution to quenching would be to develop scintillator materials that are not LET dependent. However, despite many years of study, the physical processes behind quenching are poorly understood, and the development of quenching-free organic scintillators seems unlikely in the near future. One alternative solution is to add a depth-dependent correction factor to the scintillation signal. Early work in this area has suggested that this approach has the potential to provide accurate measurements of ion beam dose distributions, regardless of depth or spatial position (Kim et al. 2012, Wang et al. 2012, Robertson et al. 2013).

The aims of this chapter are to describe the phenomenon of scintillation quenching, explore various quenching models, and discuss methods that have been developed for quenching correction in therapeutic radiation beams. First, the radiation quality will be defined and evaluated for different radiation sources. This will be followed by an overview of experimental data showing variations in scintillator response with particle species and energy. Several models describing scintillation quenching will be discussed, followed by a description of quenching correction strategies that have been developed for scintillation dosimetry. The scope of this chapter is limited to quenching in organic scintillators, as these are the most commonly used scintillators for dosimetry of therapeutic radiation beams.

2.2 LINEAR ENERGY TRANSFER

A common metric for radiation quality is the LET, also known as the restricted linear collision stopping power, denoted by L_Δ. It is defined as the ratio of the average collisional energy loss dE of a charged particle traversing a distance dl, for energy transfers below a cutoff value Δ.

$$L_\Delta = \left(\frac{dE}{dl}\right)_\Delta \tag{2.1}$$

Because LET varies with the species and energy of the charged particle, practical radiation beams have a spectrum of LET values at any given spatial location. Therefore, for the purposes of scintillation dosimetry, it is necessary to use an average value of the LET at the location of interest. Two methods of LET averaging are commonly used: track-averaged and dose-averaged LET. The track-averaged LET in a small region is defined as

$$\bar{L}_{\Delta,t} = \int_0^\Delta T_\Delta(L) L \, dL \tag{2.2}$$

where:
$T_\Delta(L)$ is the distribution of track lengths as a function of LET below the cutoff value

The dose-averaged LET is similarly defined as

$$\bar{L}_{\Delta,d} = \int_0^\Delta D_\Delta(L) L \, dL \tag{2.3}$$

where:
$D_\Delta(L)$ is the dose distribution as a function of LET below the cutoff value (ICRU 1970)

The LET of the electrons set in motion by high-energy therapeutic photon beams is fairly constant, ranging between 0.2 and 2 keV/μm, regardless of the location in the photon beam. However, in proton and heavy ion beams, the LET varies with depth. As illustrated in Figure 2.1, the LET in high-energy proton beams

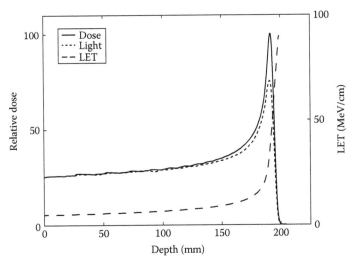

Figure 2.1 The central-axis depth-dose curve of a 161.6-MeV proton pencil beam, the corresponding light signal measured with an organic scintillator detector, and the LET. The quenching of the scintillation light in the Bragg peak is related to the sharp increase in the LET. The dose and LET were calculated using a Monte Carlo model of the beamline at the MD Anderson Proton Therapy Center, Houston, Texas.

is nearly constant at shallow depths, whereas at the distal edge of the Bragg peak, the LET increases dramatically. The same pattern is found in other heavy ion beams, but the magnitude of the LET increases with the mass of the particle. For example, alpha particles reach a peak LET of 260 keV/μm, whereas carbon ions may approach 1000 keV/μm at the distal end of the Bragg peak.

LET is a difficult property to measure directly, so most studies obtain LET values via Monte Carlo calculations (ICRU 1970, Berger 1993). Popular Monte Carlo packages for LET calculations include GEANT4 (Agostinelli et al. 2003), MCNPX (Waters et al. 2002), and FLUKA (Ferrari et al. 2005, Böhlen et al. 2014). An analytical LET calculation method has also been developed for broad proton beams in water (Wilkens and Oelfke 2003, 2004). This approach uses simplifying assumptions and approximations to enable the calculation of dose-averaged and track-averaged LET values in a fraction of the time required for Monte Carlo calculations. This analytical method has been shown to produce depth-LET distributions that agree with validated Monte Carlo calculations within ±5% on the central beam axis (Robertson et al. 2013).

2.3 LET DEPENDENCE OF SCINTILLATION LIGHT EMISSION

Early experiments using anthracene scintillator crystals for nuclear physics applications revealed that the magnitude of the scintillator response varies with the species of the irradiating particle. Further investigations showed that even for a given particle, the specific scintillation response (also called the luminescence yield) dL/dx varies with the specific energy loss dE/dx. This trend is illustrated in Figure 2.2, which contrasts a linear dose-light response with a fit of a quenching model (Birks 1951) to data from several anthracene response experiments with electrons, protons, and alpha particles (Brooks 1956).

Recent studies of organic scintillator response to high-energy proton beams show a similar trend. Figure 2.3 combines data from three recent studies using plastic and liquid scintillator solutions. Torrisi (2000) performed an accurate measurement of the luminescence yield of fluorine-doped polyvinyltoluene scintillators by measuring the light emitted from a thin (0.1 mm) sheet of scintillator exposed to proton beams of varying energies. The proton energy was varied by placing a polyethylene degrader of variable thickness in two different beams with nominal energies of 24 and 62 MeV. Kim et al. (2012) measured the scintillation response of plastic scintillating fibers intended for use in proton dosimetry. Their detector consisted of two orthogonal layers of BCF-60 scintillating fibers, with a square fiber cross section of 1 × 1 mm^2. The detector was placed in a 45-MeV proton beam, and the beam energy

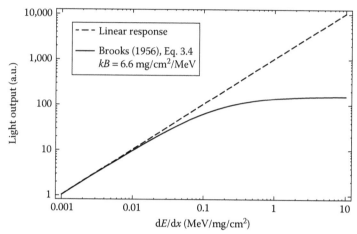

Figure 2.2 Light output as a function of specific energy loss dE/dx. A hypothetical scintillator with a linear response is compared with the response curve of anthracene. The anthracene curve is derived from a semiempirical quenching model (Equation 2.4), which was fit to measurements of the response of anthracene to electrons, protons, and alpha particles with various energies. (From Brooks, F.D., Organic scintillators, in: *Progress in Nuclear Physics*, vol. 5, Pergamon Press, London, 1956, pp. 252–313.)

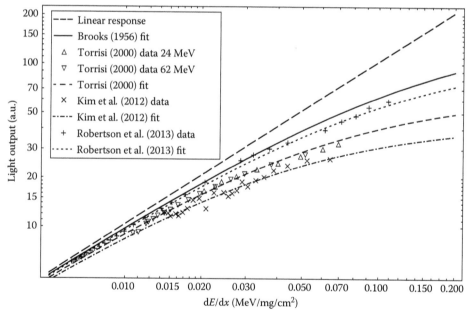

Figure 2.3 Measurements of scintillator light output as a function of specific energy loss dE/dx, and curves showing semiempirical fits to the measured data. The measurements were performed in therapeutic proton beams with organic plastic or liquid scintillators.

was varied by inserting a variable-thickness absorber made of polymethyl methacrylate and polyethylene terephthalate. Robertson et al. (2013) measured the scintillation response of the liquid scintillator BC-531 using miniature detectors consisting of a small volume of scintillator in a cap at the end of a 1-mm-diameter plastic optical fiber. The miniature detectors were placed in a therapeutic 161.6-MeV proton beam at different depths in a water-equivalent plastic phantom.

Despite the varying conditions and experimental setups, these measurements of scintillation response show a similar departure from linearity with increasing LET. This behavior appears to be common across all organic scintillators. In order to differentiate this phenomenon from other types of scintillation quenching caused by different physical mechanisms (e.g., oxygen quenching and concentration quenching), it is sometimes referred to as ionization quenching.

In the context of proton or heavy ion beam measurements, quenching is manifested by a decrease in the height of a Bragg peak measured with the scintillator. Figure 2.1 illustrates the quenching of a proton Bragg peak by comparing the central-axis depth-dose curve of a proton beam with the depth-light curve measured with a scintillator. The LET is also plotted to show the correlation between quenching and increased LET. In the following sections, we will review several models that have been used to describe the quenching process and account for its effects.

2.4 PHYSICAL MECHANISM OF QUENCHING

Despite decades of research in the area of scintillator response, there is still no comprehensive description of the physical mechanism underlying ionization quenching. However, most quenching models are based on assumptions about the physical processes involved. We will briefly introduce a few of these physical processes.

One of the oldest hypotheses about the mechanism of ionization quenching is that of saturation of scintillation centers. The basic principle here is that, if the energy deposition density in a given region of scintillator grows high enough, all scintillating centers in that region will be excited. Once this point is reached, additional energy deposition in that region will not yield more scintillation light, but will be dissipated by other mechanisms. This is a simple and intuitive conclusion, based on the information that quenching becomes more significant as LET (and therefore ionization density) increases. Unfortunately, it is incorrect. According to the saturation mechanism, an increase in the concentration of scintillating centers in a region should decrease the degree of quenching. However, this has not been borne out in experiments (Blue and Liu 1962, Gwin and Murray 1963).

Another proposed mechanism for ionization quenching, and one whose predictions are more in line with experimental results, is that quenching is caused by a reduction in the primary excitation efficiency of the scintillator. In this explanation, regions of the scintillator with a high ionization density are less efficient in their transfer of electronic excitation energy to the excited state responsible for scintillation light emission. It has been proposed that this reduction in efficiency is the result of *damage* to the scintillator molecules via ionization or excitation (Birks 1951).

Ionization quenching described by a reduction in the primary excitation efficiency can be categorized by the number of scintillator molecules that are involved in a given quenching interaction (Birks 1964). *Unimolecular quenching* assumes that the quenching process is isolated to a single molecule and proceeds independent of neighboring molecules. *Bimolecular quenching* involves an interaction of two neighboring excited molecules whose interaction results in quenching. This is typically envisioned as an interaction by which one excited molecule releases its energy in a nonradiative pathway, whereas the other is raised to an excited state from which it quickly fluoresces. Quenching interaction involving three or more excited molecules is termed j-*molecular quenching*, where j denotes an arbitrary number of excited molecules in the interaction. Both bimolecular and j-molecular quenching are dependent on the local concentration of excited molecules, whereas unimolecular quenching depends only on the local energy deposition characteristics.

Some physical mechanisms used to explain quenching explicitly consider the shape and size of the ionization column. One explanation relates quenching to a *thermal flash* in the ionization column, where quenching is caused by a brief temperature increase along the ion track (Galanin 1958). This explanation imputes ionization quenching solely to the known temperature dependence of scintillator response.

Additional insight into the physical mechanism of quenching can be gained by considering the temporal aspects of the light emission. Although the ionization density of different particles varies significantly, the light emission time remains fairly constant for a given scintillator regardless of the particle species. In addition, measurements indicate that the majority of quenching occurs within the first nanosecond of the passing of the ionizing particle, which is well before most of the scintillation light is emitted (Kallmann and Brucker 1957). After this initial quenching, the scintillation light emission curve for high-LET particles is similar to that of low-LET particles.

These experimental findings lend support to the theory that ionization quenching is due to a reduction in the primary excitation efficiency in the scintillator. This is consistent with the idea that quenching takes place at transient quenching centers that consist of temporarily ionized or excited molecules. The short time period of the quenching process is likely due to ion recombination, dissipation of thermal energy by conduction, and migration of excitation energy away from the ionization column (Birks 1964).

Quenching can be divided into a fast *static* component and a longer term *dynamic* component (Galanin and Chizhikova 1958, Berlman 1961). The static component relates to quenching of the primary scintillation efficiency and takes place on a timescale shorter than the transfer of energy between the solvent and the solute (fluor). It is considered to be the major source of quenching. Dynamic quenching has a smaller magnitude but occurs over a longer time period. Although the static component takes place primarily before the transfer of energy from the solvent to the fluor, the dynamic quenching mechanism occurs over a similar timescale as the energy transfer and is viewed as a competing process to energy transfer. Dynamic quenching appears to be related to the action of free radicals (Berlman et al. 1963).

2.5 QUENCHING MODELS

A few different models have been developed to describe the quenching phenomenon. Because a complete physical description of the quenching process is still lacking, these models are semiempirical, starting from basic assumptions about the underlying physics and then fitting parameters to match the measured data. For the most part, the models focus directly on the relationship between scintillation quenching and LET. In this section, we will introduce several scintillation quenching models that are particularly applicable to organic scintillators and therapeutic ion beams.

2.5.1 BIRKS MOLECULAR DAMAGE MODEL

2.5.1.1 Mathematical formalism

The oldest, simplest, and most commonly used quenching model was developed by Birks in the early 1950s (Birks 1951). It is a semiempirical model relating the scintillation light yield per unit path length dS/dx to the differential energy deposition of a charged particle dE/dx by way of the scintillation efficiency ε and the quenching parameter kB.

$$\frac{dS}{dx} = \frac{\varepsilon(dE/dx)}{1 + kB(dE/dx)} \tag{2.4}$$

The parameter $B(dE/dx)$ is the specific density of excited and ionized molecules, and the parameter k is the quenching parameter. Although k and B have different physical interpretations, they are difficult to measure separately, and in practice, they are commonly treated as a single material-specific quenching parameter kB.

The physical rationale of the model is that the primary excitation of the scintillator molecules is quenched due to the presence of a high density of ionized and excited molecules. Only unimolecular quenching is assumed in this model. A linear dependence between the ionization density and the degree of quenching is assumed, as can be seen by examining the denominator of Equation 2.4. The Birks model can also be written in terms of migration of excitation energy away from the ionization column (Kallmann and Brucker 1957).

When applied in a scintillation dosimetry context, the light emission dS/dx and the differential energy deposition dE/dx of a single particle track are of less interest than the total light emission and energy deposition in a given small volume v (i.e., the detector volume). In this case, the Birks equation can be rewritten as

$$S_v = \frac{\varepsilon E_v}{1 + kB \cdot \bar{L}_\Delta} \tag{2.5}$$

where:
S_v is the total scintillation light emission
E_v is the total energy deposition
$\bar{L}_\Delta$ is the average LET within volume v

The track-averaged or dose-averaged LET can be used in Equation 2.5, but the *kB* factor will differ depending on the type of LET averaging.

2.5.1.2 Measurements and fits to data

The Birks quenching model was first used to characterize the response of anthracene crystals. Figure 2.2 includes a fit of the Birks equation to the collected anthracene light yield measurements of several groups using electrons, protons, and alpha particles of varying energies (Brooks 1956). Figure 2.3 also includes fits of the Birks model to the measured data. The Birks model has been widely used to characterize ionization quenching in a variety of scintillators. Table 2.1 lists the measured *kB* factors for several organic scintillators. Interestingly, the values vary over a rather small range despite the differences in scintillator materials and experimental techniques.

2.5.2 EXPANSIONS ON THE BIRKS MODEL

The Birks model may be considered the *standard model* of ionization quenching in scintillators. Its simplicity, its long history of use, and its success at predicting the response of a variety of scintillators to various particles in diverse settings all stand as testaments to its power and utility. However, despite the successes of the Birks model, it is by no means perfect, and there have been numerous attempts to revise it in order to expand its scope of usefulness and address discrepancies between the model and measured data. We will now examine some of these variations.

2.5.2.1 Extended Birks model

The addition of a second-order term to the Birks model was first proposed by C. N. Chou (1952), for the purpose of taking into account "the variation of the quenching effect with the various modes and degrees of excitations which in turn depend upon the initial energy loss of the irradiating particle." This model can be written as

Table 2.1 **Quenching parameters (kB) from fits of Equation 2.4 to experimental data from several sources**

MATERIAL	dE/dx RANGE (MeV/g/cm²)	kB (g/cm²/MeV)	SOURCES
Anthracene	1.3–2790	6.60E–3	Brooks (1956)
Anthracene	30.3–591	1.46E–2	Craun and Smith (1970) and Smith et al. (1968)
BC-531	6.2–109	9.22E–3	Robertson et al. (2013)
BCF-12	5.0–90	9.40E–3	Wang et al. (2012)
BCF-60	14.3–64.5	2.38E–2	Kim et al. (2012)
NE-102	>97	9.10E–3	Prescott and Rupaal (1961)
NE-102	>34	1.00E–2	Evans and Bellamy (1959)
NE-102	30.3–494	1.31E–2	Craun and Smith (1970) and Smith et al. (1968)
NE-213	30.4–601	1.25E–2	Craun and Smith (1970) and Smith et al. (1968)
NE-230	62.7–760	1.10E–2	Craun and Smith (1970) and Smith et al. (1968)
Pilot B	30.3–494	1.59E–2	Craun and Smith (1970) and Smith et al. (1968)
Stilbene	30.2–489	9.55E–3	Craun and Smith (1970) and Smith et al. (1968)

$$\frac{dS}{dx} = \frac{\varepsilon(dE/dx)}{1 + kB(dE/dx) + C(dE/dx)^2}$$

(2.6)

where:

C is a second-order coefficient similar to kB

The parameter C can be attributed to various processes in the scintillator, including bimolecular quenching, but essentially it is an additional degree of freedom that can be used to obtain an improved fit to measured data.

The usefulness of the second-order term has been debated. Birks himself, in his book on scintillation counting, referred to Chou's work, but stated that the equation provided the best fit to the anthracene data for $C = 0$, or when Equation 2.6 reverts to Equation 2.4 (Birks 1964). We have found similar results in our work on quenching in liquid scintillators irradiated by proton beams.

However, Torrisi (2000) found that the second-order term provided an improved fit to their plastic scintillator measurements of proton beams. Their use of a very thin scintillator resulted in finer energy resolution than some previous measurements, which may account for the presence of a nonzero second-order quenching term. The contribution of the second-order term was an order of magnitude less than the primary term, and it is possible that some other experiments lacked the precision to identify a second-order term at this level.

Generally speaking, the importance of the second-order term may be evaluated by its lack of use in the decades since its development. Although the second-order Birks model is more general and provides additional latitude for data fitting, the first-order model is used much more frequently and seems to be sufficient to describe most quenching data. However, in some cases, the second-order term can be a useful way to increase the agreement between the model and measurements, and some scintillator formulations may feature more second-order quenching mechanisms than others. It is also possible that second-order quenching processes are common, but small enough in most cases that they are washed out by various sources of measurement uncertainty.

2.5.2.2 Nuclear energy loss terms

Some very interesting work on quenching in organic scintillators has been performed recently in the particle physics community. The development of detectors for neutrinos and other weakly interacting particles has led to a renewed effort to improve organic scintillator response models with the goal of reducing background events and improving sensitivity to very rare interactions. The goal of these detectors is to measure the recoils of hydrogen and carbon nuclei in the scintillator material caused by interactions with the particles of interest.

The calibration of the scintillator for these detectors is typically done by irradiating a sample of the scintillator with a mono-energetic neutron beam and measuring the intensity of scintillation light from nuclear recoils as well as the direction (and hence the residual energy) of the scattered neutrons. This allows the energy of the nuclear recoils to be inferred and related to the scintillator response. The calibration process is typically a search for *quenching factors*, which are the ratio of the total emitted light from a nuclear recoil to the expected signal with no quenching:

$$Q_i(E) = \frac{S_i(E)}{S_e(E)} = \frac{\int_0^E \left(\frac{dE}{(1 + kB(dE/dx)_i)} \right)}{\int_0^E \left(\frac{dE}{(1 + kB(dE/dx)_e)} \right)}$$

(2.7)

where:

The i subscripts refer to quantities for the given ion of the nuclear recoil

The e subscripts refer to quantities for electrons

One study entitled "The scintillation efficiency of carbon and hydrogen recoils in an organic liquid scintillator for dark matter searches" (Hong et al. 2002) found that the standard Birks equation did not effectively describe their measurements of scintillator response to low-energy carbon recoils. They hypothesized that a model accounting separately for the energy loss of the carbon recoils due to nuclear and electronic interactions could provide better agreement with their measurements. The motivation for this hypothesis was that the physics of energy deposition is quite different for nuclear and electronic interactions, and the scintillator might respond differently to these two energy loss mechanisms. Their model can be written as

$$\frac{dE}{dx} = \frac{\varepsilon_e(dE/dx)_e + \varepsilon_n(dE/dx)_n}{1 + kB_e(dE/dx)_e + kB_n(dE/dx)_n} \tag{2.8}$$

where:
    The e subscripts refer to electronic energy loss
    The n subscripts refer to nuclear energy loss

This model has four free parameters, but two constraints were added, bringing the number of parameters down to two. First, the ratio of quenching factors kB_n/kB_e was assumed to be proportional to the ratio of nuclear to electronic energy loss per collision, which is about 3250 for head-on collisions. Second, the value of kB_e was set to 10 mg/cm^2/MeV, in agreement with measurements by other groups (see Table 2.1). This revised quenching model agreed well with the carbon and proton recoil measurements over the energy range from 50 keV to 1 MeV.

Another study expanded on this approach to include a second-order term in a manner similar to Equation 2.6 (Yoshida et al. 2010). The objective was to estimate the background energy spectrum of the $^{13}C(\alpha,n)^{16}O$ reaction in the KamLAND neutrino detector. The authors investigated the responses of recoil protons between 424 keV and 10.5 MeV and carbon recoils between 171 keV and 2.2 MeV. They found that Equation 2.6 was required to obtain a good fit to the proton recoil data, and that, in addition to the second-order term, separate nuclear and electronic energy loss terms were also required to obtain a good fit to the carbon recoil data, resulting in the following equation:

$$\frac{dE}{dx} = \frac{\varepsilon_e(dE/dx)_e + \varepsilon_n(dE/dx)_n}{1 + kB_e(dE/dx)_e + C_e(dE/dx)_e^2 + kB_n(dE/dx)_n} \tag{2.9}$$

This work suggests a new approach that could lead to improvements in scintillation dosimetry for heavy ion therapy. Based on these two studies, it appears that Equation 2.8 or 2.9 may be necessary for an accurate description of quenching in therapeutic ion beams, particularly those using heavier particles such as carbon ions.

2.5.2.3 Kinetics of ionization quenching

Blanc et al. (1962) developed a formulation accounting for the temporal aspects of quenching, assuming radial diffusion of the deposited energy:

$$\frac{\partial n}{\partial t} = D\nabla^2 n - f(r,t)n - g(r,t)n^2 - \sum_{j=3}^{\infty} h_j(r,t)n^j + R(r,t) \tag{2.10}$$

where:
    $n(r,t)$ is the density of excitations in the scintillator at a given point, as a function of the distance r from the ion path and the time t
    D is the diffusion coefficient of the excitations
    f, gn, and $h_j n^{j-1}$ are the probabilities of single, double, and multiple-molecular quenching interactions
    R is the rate of fluorescence absorption at location r and time t

This is a general equation, with solutions that reduce to Equation 2.6 when the j-molecular term is neglected, and Equation 2.4 when the bimolecular and j-molecular terms are neglected (Birks 1964). In addition to providing a more general formulation of these equations, this model also yields predictions of the temporal aspects of the scintillation signal.

One unexpected result from early studies of scintillator response was the finding that a given scintillator can exhibit different kB factors when irradiated with different heavy ions (Newman et al. 1961, Birks 1964). This is not explained by the Birks model.

Papadopoulos (1999) developed a method to predict the scintillation response of a given scintillator to different ions, based on the rate of energy dissipation by excitation and ionization when a heavy ion passes through a material. The energy deposition rate is given by

$$\frac{dE}{dt} = -\frac{\sqrt{2}\lambda \ln(\mu E)}{\sqrt{ME}} \tag{2.11}$$

with λ and μ substituted for the physical constants pertaining to the ion and material:

$$\lambda = \frac{2\pi e^4 z^2 M}{m_0} n \tag{2.12}$$

$$\mu = \frac{4m_0}{M\overline{I}} \tag{2.13}$$

where:
  e is the electron charge
  m_0 is the electron mass
  z is the atomic number of the incident particle
  M is the atomic mass of the incident particle
  n is the number of electrons per unit volume of the material
  $\overline{I}$ is the mean excitation and ionization potential of the material

The Birks formula (Equation 2.4) can be used to determine the time dependence of the light emission (Papadopoulos 1997):

$$\frac{dL}{dt} = \sqrt{\frac{2E}{M}} \frac{\lambda S \ln(\mu E)}{E + kB\lambda \ln(\mu E)} \tag{2.14}$$

where:
  S is the absolute scintillation efficiency

Equations 2.11 and 2.14 can then be combined to obtain

$$dL = -S \frac{E}{E + kB\lambda \ln(\mu E)} dE \tag{2.15}$$

which can be evaluated using experimental scintillation versus energy data to obtain the value of kB for a given ion and material. This equation was used to determine the scintillation response of several ions in organic and inorganic scintillators, and the results were found to agree well with measured values (Papadopoulos 1999).

2.5.3 ENERGY DEPOSITION BY SECONDARY ELECTRONS MODEL

We will conclude our survey of quenching models with one of the more physics-based approaches, which relates ionization quenching to the energy deposition by secondary electrons (EDSE) in the scintillator. The impetus for this approach was the unexpected finding that the degree of quenching was different for two particles with identical dE/dx but different atomic number (Newman and Steigert 1960, Newman et al. 1961). As the scintillation emission was traditionally associated only with dE/dx, this led to a search for other energy deposition properties that differed between particles with different z but similar dE/dx. Meyer and Murray (1962) concluded that the difference was due to the energy distribution of scattered secondary electrons. Particles with higher z values produce a more energetic secondary electron spectrum. The higher energy electrons travel farther from the ion track, which should lead to a lower local energy deposition density and consequently lower quenching. Meyer and Murray proposed a two-part scintillation model based on this principle, with a primary column of ionization around the ion track and a z-dependent contribution from delta rays.

Kobetich and Katz (1968) further developed this model by formulating an expression for the energy density deposited by scattered electrons as a function of the radial distance from ion track. This was expressed as the energy flux through a cylindrical surface perpendicular to the ion path. A related model proposed that the quenching is complete in a small high-ionization density cylinder around the ion track, which produces no scintillation light (Luntz 1971, Luntz and Heymsfield 1972). The observed degree of quenching was then related to the radius of complete quenching and the shape of the transition from complete quenching to no quenching. These models found some success in matching experimental results, but left room for improvement.

More recently, another model relating to the energy deposition of secondary electrons has been developed by Michaelian and Menchaca-Rocha (1994). This model assumes that there is a maximum energy density ρ_q in the scintillator, above which quenching dominates, resulting in a maximum constant energy carrier density. The parameter ρ_q is considered to be an inherent constant of a given scintillator. The model also considers the diffusion of energy carriers out of the ionization column. An expression is derived for the electron energy deposition density ρ per unit path length of the ion in terms of the radial distance r from the ion track:

$$\rho(r) = N \frac{e^4}{nm_e} \frac{z^{*2}}{V^2} \frac{1}{r^2} \left[1 - \frac{r}{R_{max}} \right]^{d+1/n} \tag{2.16}$$

where:

N is the number of electrons per unit volume in the medium
e is the electron charge
n is the power for a power law approximating the electron range and is taken to be 5/3
m_e is the mass of the electron
z^* is the effective ion charge
V is the ion velocity
R_{max} is the maximum range of the ionized electrons
$d = 0.045Z_{eff}$, with Z_{eff} being the effective atomic number of the medium [for the derivation of this expression, see Michaelian and Menchaca-Rocha (1994)]

This description of the electron energy deposition density distribution is similar to those developed by Kobetich and Katz and Luntz, with the exception of the term in square brackets, which was included to account for electron backscattering.

Taking quenching into account, the energy carrier density per unit path length of the ion is then given by

$$\frac{dN_e}{dx} = K \left[\pi r_q^2 \rho_q + \int_{r_q}^{R_{max}} \rho(r) 2\pi r dr \right] \tag{2.17}$$

where:

K is a constant factor relating the number of energy carriers formed by the deposited energy
r_q is determined from Equation 2.16 by replacing $\rho(r)$ with ρ_q and solving for r

The light output in a given region is then given by

$$\frac{dL}{dx} = C\frac{dN_e}{dx} \tag{2.18}$$

where:

C is a proportionality constant incorporating K as well as the ratio of energy carrier density to light output and the experimental collection efficiency

The EDSE model identifies the ion velocity and effective charge as the primary variables related to the incident ion, which affect the scintillation response. The model has one material-specific free parameter, ρ_q (in addition to the overall normalization constant C), which is the critical energy deposition density that determines the radius of the quenching region. In a later publication, ρ_q was calculated for organic scintillators in terms of the unimolecular and bimolecular interactions of their singlet and triplet states (Michaelian et al. 1995). The model predictions yielded excellent agreement with experimental data from many incident ions and a wide range of energies.

At energies below about 10 MeV/A, the original EDSE model is less accurate due to the use of a classical impact parameter, based on an impulse approximation for transfer of momentum between the incident ion and the electron. The substitution of a quantum impact parameter leads to substantial improvement in the agreement of the model and data at low incident ion energies (Cruz-Galindo et al. 2002).

Despite the effectiveness of the EDSE model in predicting scintillation response for a wide variety of scintillators, ions, and ion energies, it has seen relatively little use, and the Birks equation is still the primary tool used to account for quenching in scintillator measurements. Concerned that this lack of use might be due to the EDSE model's mathematical complexity, Menchaca-Rocha (2009) developed a simplified version of the model that relies only on the local LET (analogous to the Birks formula), removing the need to perform a radial integration of the deposited energy density.

In order to obtain the best agreement of the EDSE model with experimental data, the scaling factor C must vary as a function of the atomic number of the incident ion (Michaelian and Menchaca-Rocha 1994, Michaelian et al. 1995). The mechanism behind this variation is not well understood, but it is smooth and changes slowly enough that an analytical function of C can provide accurate results across a range of Z. However, the addition of these (typically three) parameters to the EDSE model decreases its utility, such that it may not have an advantage over the Birks model for measurements involving fewer than five different ion species (Menchaca-Rocha 2009).

2.6 QUENCHING CORRECTION FOR SCINTILLATION DOSIMETRY OF THERAPEUTIC ION BEAMS

Most work in the area of scintillation quenching has been done in the context of nuclear and particle physics, where the goal is accurate detection and analysis of individual particle interactions. The objective in the realm of scintillation dosimetry for radiotherapy beams is somewhat different. Rather than investigating single interactions to determine the properties of particles and nuclei, we desire to measure the net result of billions of interactions in order to determine the macroscopic energy distribution of a therapeutic ion beam. This requires different tools and analysis and a different approach than most previous quenching-related studies.

A few recent studies have investigated the use of the Birks model to correct for quenching in scintillator measurements of therapeutic proton beams. In each case, the response of a scintillator detector was measured as a function of water-equivalent depth in a therapeutic proton beam, and the first-order Birks model quenching parameter (kB) was determined by comparing the scintillator response to the LET distribution in the proton beam, as calculated using Monte Carlo methods. In each case the second-order Birks equation was also considered, but the second-order term was near enough to zero as to be considered unimportant.

Kim et al. (2012) measured a 45-MeV proton beam with a detector consisting of two layers of polystyrene-based plastic scintillating fibers (BCF-60). They calculated the LET using the Monte Carlo package Geant4 and found a Birks parameter of $kB = (2.38 \pm 0.17) \times 10^{-2}$ g/cm^2/MeV for their system.

Wang et al. (2012) measured single-energy passive scattering proton beams with energies of 100, 180, and 250 MeV using a polystyrene-based plastic scintillating fiber detector (BCF-12). They calculated the LET using the Monte Carlo package MCNPX and found a Birks parameter of $kB = 9.40 \times 10^{-3}$ g/cm^2/MeV for their system. They applied the Birks model quenching correction factors to their data and found that it agreed with ionization chamber measurements to within $\pm 5\%$ (Figure 2.4).

Robertson et al. (2013) measured mono-energetic proton pencil beams ranging from 85.6 to 161.6 MeV with a linear alkyl benzene-based miniature liquid scintillator detector (BC-531). They calculated the LET using the Monte Carlo package MCNPX and found a quenching parameter of $kB = 9.22 \times 10^{-3}$ g/cm^2/MeV for their system. They measured depth-light distributions in a large volume scintillator detector using the same scintillator and applied the Birks model quenching correction factors to their data. They compared their corrected scintillation measurements with validated Monte Carlo depth-dose

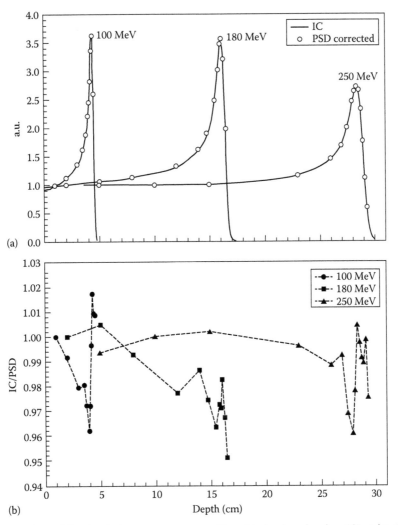

Figure 2.4 (a) Comparison of the depth-dose curves measured by a Markus ion chamber (IC) and a plastic scintillating fiber detector after making quenching corrections (PSD corrected). (b) The ratio of the ion chamber dose to the scintillator reading. (From Wang, L.L.W. et al., Determination of the quenching correction factors for plastic scintillation detectors in therapeutic high-energy proton beams, *Phys. Med. Biol.* 57, 7767, 2012. Copyright Institute of Physics and Engineering in Medicine. Reproduced by permission of IOP Publishing. All rights reserved.)

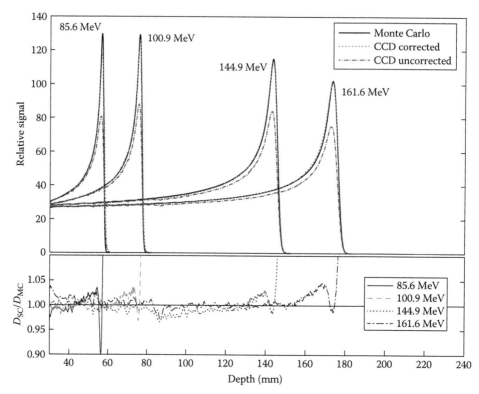

Figure 2.5 (Top) Central-axis depth-dose profiles for proton pencil beams. The dose calculated by a validated Monte Carlo model is compared with the corrected and uncorrected scintillation signal. (Bottom) The ratio of the corrected scintillation signal (D_{SC}) to the Monte Carlo dose (D_{MC}). (From Robertson, D. et al., Quenching correction for volumetric scintillation dosimetry of proton beams, *Phys. Med. Biol.*, 58, 261, 2013. Copyright Institute of Physics and Engineering in Medicine. Reproduced by permission of IOP Publishing. All rights reserved.)

calculations and found that the Bragg peak doses agreed to within ±3% for the higher energy beams and ±10% for the 85.6 MeV beam (Figure 2.5).

In each of the above-mentioned studies, the largest errors were encountered in the Bragg peak. This leads us to a common difficulty in quenching correction for ion beams. Quenching models take the LET (or the energy density in the EDSE model) as an input and use it to calculate the degree of quenching. In order to correct for quenching at a given location, the LET must be known at that location. However, it can be difficult to precisely align measured scintillation data with calculated LET data. Not only are the dose gradients steep in the Bragg peak, but the LET gradient in the Bragg peak is also very sharp (see Figure 2.1). Small inaccuracies in the alignment of the scintillation and LET data can lead to rather large errors in the quench-corrected dose values. This problem becomes more severe at lower proton beam energies, which typically feature sharper, narrower Bragg peaks. Although not insurmountable, this difficulty must be adequately addressed in order to provide accurate quenching correction in scintillation dosimetry of ion beams.

2.7 CONCLUSION

In summary, we have reviewed the concept of LET and shown its relation to the response of organic scintillators. We have reviewed several models developed to describe quenching and predict or parameterize its effects over a range of scintillating materials, ion species, and energies. The Birks model in its several forms has proven to be the simplest and most utilitarian of the quenching models, and it is by far the most commonly-used. Other models incorporating the dynamics of scintillator response and the distribution of energy deposition by secondary electrons give greater physical insight into the quenching process. Although these models are more mathematically challenging, they have the potential to provide insight into the finer details of ionization quenching, which may ultimately lead to more accurate quenching correction methods or the design of scintillator materials that are less sensitive to variations in LET. Finally, we have reviewed recent efforts to account for quenching in therapeutic ion beams and identified areas of improvement, which could lead to more accurate and clinically useful scintillation dosimetry tools.

REFERENCES

Agostinelli, S., J. Allison, K. Amako et al. 2003. GEANT4-a simulation toolkit. *Nuclear Instruments and Methods in Physics Research Section A: Accelerators, Spectrometers, Detectors and Associated Equipment* 506 (3): 250–303.

Berger, M. J. 1993. Penetration of proton beams through water I. Depth-dose distribution, spectra, and LET distribution. NIST Interagency/Internal Report. Gaithersburg, MD: National Institute of Standard and Technology.

Berlman, I. B. 1961. Luminescence in a scintillation solution excited by α and β particles and related studies in quenching. *The Journal of Chemical Physics* 34 (2): 598–603.

Berlman, I. B., R. Grismore, and B. G. Oltman. 1963. Study of [small alpha]/[small beta] ratios of organic scintillation solutions. *Transactions of the Faraday Society* 59 (0): 2010–2015.

Birks, J. B. 1951. Scintillations from organic crystals: Specific fluorescence and relative response to different radiations. *Proceedings of the Physical Society. Section A* 64 (10): 874.

Birks, J. B. 1964. *The Theory and Practice of Scintillation Counting.* In D.W. Fry and W. Higinbotham, eds., International Series of Monographs on Electronics and Instrumentation. Oxford: Pergamon Press.

Blanc, D., F. Cambou, and Y. G. De Lafond. 1962. Kinetics the fast component of scintillation in a pure organic medium. Application to anthracene. *Comptes rendus de l'Académie des Sciences Paris* 254: 3187.

Blue, J. W. and D. C. Liu. 1962. Scintillation response of aikali iodides to alpeha particles and protons. *IRE Transactions on Nuclear Science* 9 (3): 48–51.

Böhlen, T. T., F. Cerutti, M. P. W. Chin et al. 2014. The FLUKA code: Developments and challenges for high energy and medical applications. *Nuclear Data Sheets* 120: 211–214.

Boon, S. N., P. van Luijk, T. Bohringer et al. 2000. Performance of a fluorescent screen and CCD camera as a two-dimensional dosimetry system for dynamic treatment techniques. *Medical Physics* 27 (10): 2198–2208.

Brooks, F. D. 1956. Organic scintillators. In O.R. Frisch, ed., *Progress in Nuclear Physics*, vol. 5, pp. 252–313. London: Pergamon Press.

Chou, C. N. 1952. The nature of the saturation effect of fluorescent scintillators. *Physical Review* 87 (5): 904.

Craun, R. L. and D. L. Smith. 1970. Analysis of response data for several organic scintillators. *Nuclear Instruments & Methods* 80 (2): 239–244.

Cruz-Galindo, H. S., K. Michaelian, A. Martinez-Davalos, E. Belmont-Moreno, and S. Galindo. 2002. Luminescence model with quantum impact parameter for low energy ions. *Nuclear Instruments and Methods in Physics Research Section B: Beam Interactions with Materials and Atoms* 194 (3): 319–322.

Evans, H. C. and E. H. Bellamy. 1959. The response of plastic scintillators to protons. *Proceedings of the Physical Society* 74 (4): 483.

Ferrari, A., P. R. Sala, A. Fassò, and J. Ranft. 2005. FLUKA: A multi-particle transport code. Stanford, CA: Stanford Linear Accelerator Center.

Galanin, M. D. 1958. Reasons for the dependence of the yield of luminescence of organic substances on the energy of ionizing particles. *Optika i Spektroskopiya* 4: 758.

Galanin, M. D. and Z. A. Chizhikova. 1958. The quenching of luminescence of organic substances during excitation by alpha-particles. *Optika i Spektroskopiya* 4: 196.

Gwin, R. and R. Murray. 1963. Scintillation process in CsI(Tl). I. Comparison with activator saturation model. *Physical Review* 131 (2): 501–508.

Hong, J., W. W. Craig, P. Graham et al. 2002. The scintillation efficiency of carbon and hydrogen recoils in an organic liquid scintillator for dark matter searches. *Astroparticle Physics* 16 (3): 333–338.

ICRU. 1970. Linear energy transfer. Bethesda, MD: International Commission on Radiation Units and Measurements.

Jirasek, A. and C. Duzenli. 2002. Relative effectiveness of polyacrylamide gel dosimeters applied to proton beams: Fourier transform Raman observations and track structure calculations. *Medical Physics* 29 (4): 569–577.

Kallmann, H. and G. Brucker. 1957. Decay times of fluorescent substances excited by high-energy radiation. *Physical Review* 108 (5): 1122–1130.

Karger, C. P., O. Jakel, H. Palmans, and T. Kanai. 2010. Dosimetry for ion beam radiotherapy. *Physics in Medicine and Biology* 55 (21): R193.

Kim, C., B. Hong, G. Jhang et al. 2012. Measurement of scintillation responses of scintillation fibers for dose verification in proton therapy. *Journal of the Korean Physical Society* 60 (5): 725–730.

Kobetich, E. and R. Katz. 1968. Energy deposition by electron beams and delta rays. *Physical Review* 170 (2): 391–396.

Luntz, M. 1971. Track-effect theory of scintillation efficiency. *Physical Review B* 4 (9): 2857–2868.

Luntz, M. and G. Heymsfield. 1972. Track-effect account of scintillation efficiency for random and channeled heavy ions of intermediate velocities. *Physical Review B* 6 (7): 2530–2536.

Menchaca-Rocha, A. 2009. A simplified scintillator-response formula for multiple-ion energy calibrations. *Nuclear Instruments and Methods in Physics Research Section A: Accelerators, Spectrometers, Detectors and Associated Equipment* 602 (2): 421–424.

Meyer, A. and R. Murray. 1962. Effect of energetic secondary electrons on the scintillation process in alkali halide crystals. *Physical Review* 128 (1): 98–105.

Michaelian, K. and A. Menchaca-Rocha. 1994. Model of ion-induced luminescence based on energy deposition by secondary electrons. *Physical Review B* 49 (22): 15550–15562.

Michaelian, K., A. Menchaca-Rocha, and E. Belmontmoreno. 1995. Scintillation response of nuclear-particle detectors. *Nuclear Instruments and Methods in Physics Research Section A: Accelerators, Spectrometers, Detectors and Associated Equipment* 356 (2–3): 297–303.

Newman, E., A. Smith, and F. Steigert. 1961. Fluorescent response of scintillation crystals to heavy ions. *Physical Review* 122 (5): 1520–1524.

Newman, E. and F. Steigert. 1960. Response of NaI(Tl) to energetic heavy ions. *Physical Review* 118 (6): 1575–1578.

Papadopoulos, L. 1997. Rise time of scintillation emission in inorganic and organic scintillators. *Nuclear Instruments and Methods in Physics Research Section A: Accelerators, Spectrometers, Detectors and Associated Equipment* 401 (2–3): 322–328.

Papadopoulos, L. 1999. Scintillation response of organic and inorganic scintillators. *Nuclear Instruments and Methods in Physics Research Section A: Accelerators, Spectrometers, Detectors and Associated Equipment* 434 (2–3): 337–344.

Prescott, J. R. and A. S. Rupaal. 1961. The specific fluorescence of plastic scintillator NE102. *Canadian Journal of Physics* 39 (1): 221–227.

Robertson, D., D. Mirkovic, N. Sahoo, and S. Beddar. 2013. Quenching correction for volumetric scintillation dosimetry of proton beams. *Physics in Medicine and Biology* 58 (2): 261–273.

Smith, D. L., R. G. Polk, and T. G. Miller. 1968. Measurement of the response of several organic scintillators to electrons, protons, and deuterons. *Nuclear Instruments & Methods* 64 (2): 157–166.

Torrisi, L. 2000. Plastic scintillator investigations for relative dosimetry in proton-therapy. *Nuclear Instruments and Methods in Physics Research Section B: Beam Interactions with Materials and Atoms* 170 (3–4): 523–530.

Wang, L. L. W., L. A. Perles, L. Archambault et al. 2012. Determination of the quenching correction factors for plastic scintillation detectors in therapeutic high-energy proton beams. *Physics in Medicine and Biology* 57 (23): 7767–7782.

Waters, L. S., Hendricks, J. and McKinney, G. 2002. Monte Carlo N-Particle Transport Code system for Multiparticle and High Energy Applications 2.7d. Los Alamos National Laboratory, Los Alamos, NM.

Wilkens, J. J. and U. Oelfke. 2003. Inverse planning with RBE: A new concept for the optimization of proton therapy. *Medical Physics* 30 (6): 1447.

Wilkens, J. J. and U. Oelfke. 2004. Three-dimensional LET calculations for treatment planning of proton therapy. *Zeitschrift fur Medizinische Physik* 14 (1): 41–46.

Yoshida, S., T. Ebihara, T. Yano et al. 2010. Light output response of KamLAND liquid scintillator for protons and C-12 nuclei. *Nuclear Instruments and Methods in Physics Research Section A: Accelerators, Spectrometers, Detectors, and Associated Equipment* 622 (3): 574–582.

3 Optical fibers, light-guides, and light transmission

Bongsoo Lee and Wook Jae Yoo

Contents

3.1 INTRODUCTION

In recent years, there have been many studies on the development of various plastic scintillation dosimeters (PSDs) exploiting optical fibers for measuring real-time *in situ* dose information in both diagnostic radiology and radiotherapy. To avoid complicated dose calibration processes during scintillation dosimetry, all dosimetric materials of a PSD system should have a tissue or water-equivalent characteristic. In this respect, the dosimeter probe of the PSD system is composed of an organic scintillator and an optical fiber, and these components have a nearly water-equivalent characteristic. Here, an optical fiber, which is made of a transparent plastic or silica glass core surrounded by a cladding material, is used as a light guide.

In general, since an optical fiber is employed to transmit light signals generated or transduced in a sensing probe induced by certain physical parameters, fiber-optic sensors have many advantages over other conventional sensors, such as high spatial resolution due to the small size of the optical fiber, substantial flexibility, and real-time monitoring capabilities without significant diminution of the light signal. In addition, optical fiber can transmit light signals from a measuring site to a light-measuring device at a remote distance with immunity to medical and nuclear environmental influences, including high electromagnetic interference and radiofrequency interference.

In this chapter, basic principles of optical fiber, optical and physical properties of optical fiber to be considered for applying remote dosimetry, and reasons for selecting optical fiber for scintillation dosimetry are described.

3.2 BASIC PRINCIPLES OF OPTICAL FIBER

3.2.1 PRINCIPLE OF LIGHT TRANSMISSION IN OPTICAL FIBER

As a dielectric waveguide, an optical fiber can transmit light signals from a light source to a light-measuring device at a long distance using total internal reflection. Under the condition of total internal reflection, light can be reflected at a dielectric interface region between two transparent media that have different refractive indices, without any reflectors. Here, the ray of light must be incident on a boundary from the medium with a high refractive index to another with a low refractive index. If a ray of light transmits with a certain incidence angle (θ_1) that is less than the critical angle (θ_c) from medium 1 with a high refractive index (n_1) to medium 2 with a low refractive index (n_2), a portion of light will be reflected back to medium 1 with a reflection angle (θ_1), and another portion of light will be refracted into medium 2 with a refraction angle (θ_2), as shown in Figure 3.1a. The relationship between the refractive index and the angle of light is expressed by Snell's law [1,2]:

$$n_1 \sin \theta_1 = n_2 \sin \theta_2 \qquad (3.1)$$

Here, the refractive index (n) is the ratio of the velocity of light in a vacuum (c) to its velocity in a specific medium (v) and is given by the following relation:

$$n = \frac{c}{v} \qquad (3.2)$$

From Equation 3.2, the velocity of light slows down according to the increase of the refractive index of the medium that the light is traveling in, and Equation 3.1 can be changed using Equation 3.2 as follows:

$$\frac{\sin \theta_2}{\sin \theta_1} = \frac{n_1}{n_2} = \frac{V_2}{V_1} \qquad (3.3)$$

When the incidence angle (θ_1) is increased to the critical angle (θ_c), the refractive angle (θ_2) reaches 90°, as shown in Figure 3.1b and thereby, a ray of light travels along the interface region between two contacted media. Here, the critical angle (θ_c) is defined as

$$\theta_c = \sin^{-1} \left(\frac{n_2}{n_1} \right) \qquad (3.4)$$

If the incidence angle (θ_3) is increased to a value greater than the critical angle (θ_c), the light is totally reflected back into medium 1 with a reflection angle (θ_3) that is equal to the incidence angle (θ_3), as shown in Figure 3.1c. This phenomenon is called *total internal reflection*, and it is a basic principle of light transmission in an optical fiber [2–4].

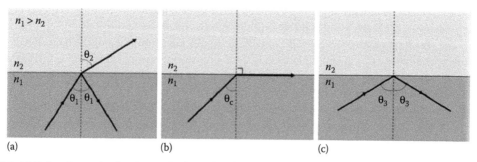

Figure 3.1 (a) Refraction and reflection at the boundary between two media with different refractive indices, (b) critical angle, and (c) total internal reflection.

3.2.2 REFRACTIVE INDEX PROFILES AND PROPAGATION MODES

In general, optical fibers have a cylindrical shape and a core/clad structure, as shown in Figure 3.2. To allow complete internal reflection, the cladding covers the core and light is propagated through the core. Here, the refractive index of the core material (n_1) is always higher than that of the cladding material (n_2) to transmit the light. Furthermore, the optical fiber is surrounded by a jacket to shield light noise and to protect the fiber from ambient contaminants.

According to the number of modes supported by a fiber, optical fibers can be simply classified as a multi-mode fiber (MMF) and a single-mode fiber (SMF). In the case of MMF, it has a large core diameter allowing for a large number of modes. MMF may be categorized as step-index (SI) and graded-index (GI) fiber. If the refractive index of the core is uniform with the core radius (i.e., the fiber has a step index refractive profile), the optical fiber is called a *SI fiber*; and if the refractive index of the core gradually decreases farther from the center of the core (i.e., the refractive index of the core is parabolic), it is a GI fiber.

Consider four different rays of light incident on the face of a SI fiber, as shown in Figure 3.3. At angle (θ_c'), a ray is refracted along the interface of the core and the cladding (i.e., the first mode); this incident angle (θ_c') is referred to as the *maximum acceptance angle* and θ_c is the critical angle for the internal reflection in an optical fiber. Next, at angle (θ_0') that is less than the maximum acceptance angle (θ_c') by the fiber axis, a ray will propagate inside the core by a series of total internal reflections (i.e., the second mode). At other incident angles (θ_1') higher than the angle (θ_c'), a ray will be refracted into the cladding and then lost as it escapes into the air (i.e., the third mode). Finally, if a ray incident on the face of the SI fiber perpendicularly, the light will travel parallel to the core (i.e., the fourth mode). From the light guiding in a multi-mode SI fiber, it can be seen that the second mode travels longer than the fourth mode. Because each velocity of the two modes inside the core of the SI fiber is identical and constant, the two modes will be arrived at the distal end of the SI fiber with separate times. This disparity between arrival times of each ray is known as *dispersion*.

To perform internal reflection that can transmit light signals via SI fiber, the maximum acceptance angle (θ_c') and the critical angle (θ_c) are determined by the refractive indices of the core and cladding. The relationship between the refractive index and the angle of light can be derived by Snell's law. At the point (P_1) of light incident on the face of an optical fiber, as shown in Figure 3.3,

$$n_0 \sin\theta_c' = n_1 \sin\left(90° - \theta_c\right) = n_1 \cos\theta_c \tag{3.5}$$

At the point (P_2) of light refracted along the interface of the core and the cladding,

$$n_1 \sin\theta_c = n_2 \sin 90° = n_2 \tag{3.6}$$

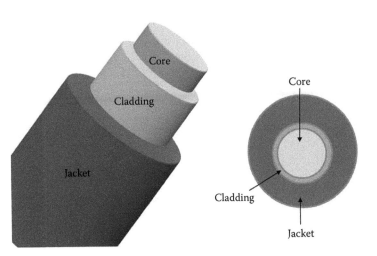

Figure 3.2 Basic structure of an optical fiber.

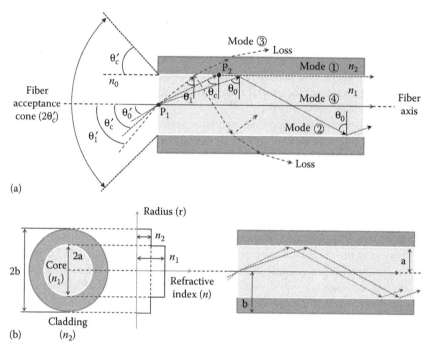

(a)

(b)

Figure 3.3 (a) Maximum acceptance angle and (b) light propagation in a multi-mode step-index fiber.

Combining Equations 3.5 and 3.6 and using Pythagorean theorem,

$$\sin^2 \theta_c + \cos^2 \theta_c = 1 = \left(\frac{n_2}{n_1}\right)^2 + \left(\frac{n_0 \sin \theta'_c}{n_1}\right)^2 \tag{3.7}$$

There is a fiber acceptance cone with a maximum acceptance angle (θ'_c) related to a parameter called the *numerical aperture* (NA) of the SI fiber. The acceptance angle limits the light accepted into the fiber for transmission. NA is defined as the light-gathering capability of an optical fiber and is given as

$$n_0 \sin \theta'_c = n_1 \cos \theta_c = \sqrt{n_1^2 - n_2^2} = NA \tag{3.8}$$

To compensate inherent dispersion of a SI fiber, a GI fiber has been developed. Because the refractive index decreases from the center of the core to the margin, a ray of light transmits with gradient changes in the GI fiber, while the light travels in a zigzag manner between the core/cladding boundaries on each side of the fiber axis in the SI fiber. Here, the gradient of the transmitting light ray bends gradually according to the variation of the refractive index in the core, as shown in Figure 3.4. Since the velocity of light in a specific

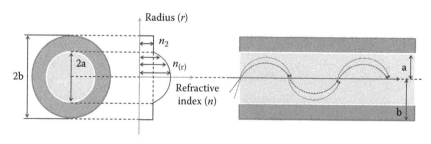

Figure 3.4 Light propagation in a multi-mode graded-index fiber.

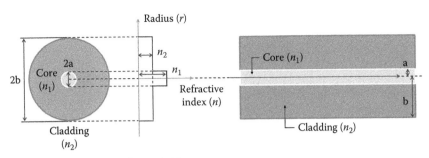

Figure 3.5 Light propagation in a single-mode fiber.

medium is inversely proportional to the refractive index of the medium and the central refractive index of the core is higher than the outer refractive index of the core, the higher refractive index in the center of the core slows the velocity of some light rays. This phenomenon allows most of the light rays to reach the receiving end at about the same time and thus reduces dispersion [3–6].

To characterize light propagation in optical fiber, the V-parameter is conveniently used and defined as

$$V = \frac{2\pi a}{\lambda} \sqrt{n_1^2 - n_2^2} = \frac{2\pi a}{\lambda}(NA) \qquad (3.9)$$

where:

a is the core radius
λ is the wavelength of the light injected into the fiber

From Equation 3.9, the V-parameter varies directly with the core diameter (2*a*) and the NA, and inversely with the wavelength of the incident light. For a particular wavelength (λ), if the V-parameter is lower than 2.405, either the core diameter or the NA has to be made smaller. With a decrease of the V-parameter, the number of allowed modes also decreases and thus, the cut-off condition of various modes is determined by this parameter. For a V-parameter less than 2.405, only one mode is permitted; accordingly, the V-parameter is very important since it defines the condition required for SMF [3,5].

Compared to MMF, SMF has a very thin core and only one mode (i.e., one light path), as shown in Figure 3.5. Accordingly, the light transmits straightly via the core, and thus SMF exhibits no dispersion caused by multiple modes. However, its small core diameter makes coupling light into the core and connecting to other fibers difficult. SMFs are normally used as a waveguide for communication because of their capacity to transmit optical information over a long distance.

3.3 OPTICAL AND PHYSICAL PROPERTIES OF OPTICAL FIBER TO BE CONSIDERED FOR APPLYING REMOTE DOSIMETRY

3.3.1 SCATTERING AND ABSORPTION

In a fiber-optic system, attenuation is an optical power loss of the light that is propagated along the length of the transmitting optical fiber, and thus it is generally defined as a logarithmic relationship between the optical output power and the optical input power [4,5,7].

$$P_{out} = P_{in} \exp(-\alpha L) \qquad (3.10)$$

where:

P_{out} is the optical power at distance *L* (km) from the fiber input
P_{in} is the optical power at the fiber input

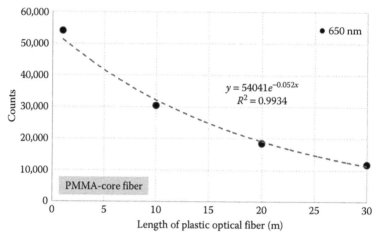

Figure 3.6 Attenuation of a plastic optical fiber according to the fiber length.

The attenuation coefficient (α) is expressed in terms of (dB/km) following the relation:

$$\alpha = \frac{10}{L} \log_{10}\left(\frac{P_{in}}{P_{out}} \right) \tag{3.11}$$

Figure 3.6 shows the variation of light intensity according to the length of a SI MMF (GH4001, Mitsubishi Rayon, New York) with a polymethyl methacrylate (PMMA)-based core when light with a central wavelength of 650 nm is incident to the fiber. As expected, the light intensity decreases with a negative exponential function with increasing fiber length. The attenuation coefficient varies with the wavelength and core material of the fiber. Therefore, for a particular wavelength of light transmitting via the optical fiber, the amount of attenuation in the fiber is mostly determined by the fiber length. A commonly used technique to measure total fiber attenuation is referred to as the *cutback method*. Power P_1 is measured at the distal end of the long length L_1 of the fiber. The fiber then is cut back to a smaller length L_2 and the power P_2 is measured again. At a particular wavelength, the attenuation coefficient is given as

$$\alpha = \frac{10}{L_1 - L_2} \log_{10}\left(\frac{P_2}{P_1} \right) \tag{3.12}$$

Attenuation is also dependent on the core diameter since it increases with decreasing core diameter. According to the decrease of the core diameter, light rays strike more frequently on the interface between the core and the cladding, and this gives rise to higher attenuation.

In an optical fiber, attenuation is caused by *intrinsic* and *extrinsic* factors. As a main *intrinsic* factor, Rayleigh scattering is caused by nonuniformities in the optical fiber and is the most common form of scattering. As wavelength (λ) decreases, the attenuation coefficient increases and Rayleigh scattering varies proportionally to λ^{-4}. Therefore, Rayleigh scattering is the primary attenuation mechanism over the visible light (VIS) range in the transmission wavelength range of optical fiber. On the other hand, absorption, which can be caused by the molecular structure of the core material and impurities in the fiber, becomes the dominant *intrinsic* factor at longer wavelengths. In scintillation dosimetry, however, *intrinsic* absorption has little effect on the transmission of scintillating light because most absorption by impurities causes optical loss of the fiber at the infrared (IR) region. Both *intrinsic* factors depend on the composition of the optical fiber and thus determine the ultimate fiber attenuation and cannot be removed. Figure 3.7 shows the typical spectral response of a SI MMF (BCF-98, Saint-Gobain Ceramic & Plastics, Malvern, PA) with a polystyrene (PS)-based core.

Jang et al. fabricated a PSD using a plastic scintillating fiber (BCF-60, Saint-Gobain Ceramic & Plastics, Malvern, PA), plastic optical fibers (POFs: SH4001, Mitsubishi Rayon, New York), and an optical spectrometer

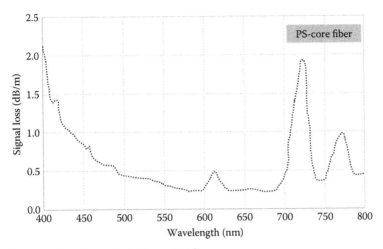

Figure 3.7 Spectral transmission loss of a plastic optical fiber.

(QE65000, Ocean Optics, Dunedin, FL) to measure both scintillating light and Cerenkov radiation generated by therapeutic electron beams [8]. In their experiment, the spectrum of Cerenkov radiation from a POF irradiated by a 6 MeV electron beam of a clinical linear accelerator (Linac: Clinac® 2100C/D, Varian Medical Systems) was measured and the wavelength at the peak was about 500 nm, in contrast to a previous work by Beddar et al., where the peak wavelength of Cerenkov radiation was found in a range of 400–480 nm [9]. Jang et al. noted that this discrepancy might be attributable to *intrinsic* transmission loss of the 20-m long POF used as a transmitting fiber in their experiment.

3.3.2 BENDING AND COUPLING LOSSES

Physical bends in the fiber are the main *extrinsic* factors that cause attenuation, and they can be classified into two categories: micro-bends and macro-bends. In a fiber-optic system, micro-bending loss inevitably occurs from microscopic imperfections in the geometry of the optical fiber, such as micro-defects at the fiber surface, changes of the core diameter, and results of the manufacturing process itself. However, the micro-bending loss has little effect on light transmission through the fiber (Figure 3.8).

Generally, a ray of light is propagated with total internal reflection in an optical fiber ($\theta_0 > \theta_c$). However, if the ray is traveling in the direction of propagation at an angle (θ_1) less than the critical angle (θ_c) in the bend region of the optical fiber, it will be partially transmitted out of the core, as shown in Figure 3.9, because the condition to perform total internal reflection is no longer satisfied [3]. This optical loss in the bent optical fiber is called *macro-bending loss* and is the most common form among the *extrinsic* causes of attenuation. Usually, the bending loss is unnoticeable if the finite radius of the fiber curvature is larger than 10 cm [4]. However, sometimes this leakage of optical power from an optical fiber can occur during real-time dose measurement using a PSD, and thus it has potential to affect the accuracy and reproducibility of the PSD system.

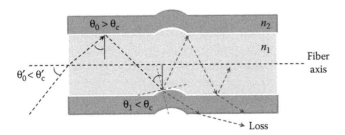

Figure 3.8 Leakage of optical power in the micro-bend region of an optical fiber.

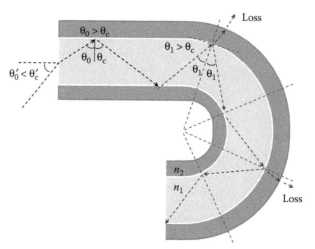

Figure 3.9 Leakage of optical power in the macro-bend region of an optical fiber.

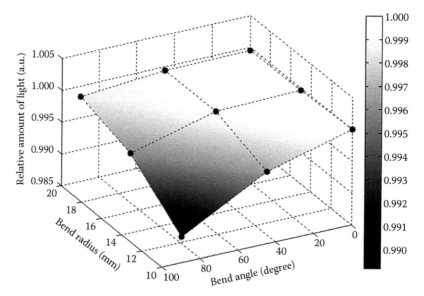

Figure 3.10 Optical loss in the macro-bend region of a plastic optical fiber.

Figure 3.10 shows the optical loss in the macro-bend region of a SI MMF (GH4001, Mitsubishi Rayon, New York) with a PMMA-based core. The amount of light reaching the light-measuring device decreases as the bend angle increases and the bend radius of the fiber curvature decreases. However, the amount of light does not decrease significantly until a small radius of 1 cm, where the bend angle is lower than 45°.

The macro-bending loss of a PSD for measuring absorbed dose in diagnostic radiology was described by Jones and Hintenlang, and Hyer et al. [10,11]. First, the performance of a PSD system (in their study, the dosimeter is referred to as a *fiber optic-coupled dosimeter*, FOCD) using a Cu^{1+}-doped quartz fiber, 15 m long MMF patch cable, and a charge-coupled device (charge-coupled device [CCD]: model S7031-1007, Hamamatsu Photonics, New Jersey) was evaluated by Jones and Hintenlang according to four bend radii (18.5, 11.5, 6.5, and 3.3 cm) [10]. When the transmitting fiber is bent through a large radius (~11.5 cm), the output signal of the PSD, which is irradiated by an X-ray beam emitted from computed tomography, is significantly decreased in a response of ~15%. Next, Hyer et al. described the response of a PSD based on a plastic scintillating fiber (BCF-12, Saint-Gobain Ceramic & Plastics, Malvern, PA), an optical fiber (400-UV, Ocean Optics, Dunedin, FL), and a photomultiplier tube (PMT: H7467, Hamamatsu Photonics,

New Jersey) according to five bend radii (20, 10, 7.5, 4, and 2.5 cm) [11]. In the experimental results, the output signal of the PSD remained relatively constant until a radius of 7.5 cm.

In a fiber-optic system, *extrinsic* optical losses can additionally occur due to restrictions of the NA, axial or angular misalignment with each other, flatness of the fiber end face, and Fresnel's reflection at the coupling interface between two media with different refractive indices. When light passes from one medium with a specific refractive index (n_1) to another with a different refractive index (n_2), a portion of the light is reflected back into the first medium. This phenomenon is referred to as *Fresnel's reflection*. Here, the fraction of optical power (R) that is reflected at the end face of optical fiber, is given as [5]

$$R = \frac{P_r}{P_i} = \left(\frac{n_1 - n_2}{n_1 + n_2}\right)^2 \tag{3.13}$$

where P_r and P_i are the reflected and incident light powers, respectively. The fraction of optical power (T) that is transmitted at the interface is given as

$$T = 1 - R = 1 - \left(\frac{n_1 - n_2}{n_1 + n_2}\right)^2 = \frac{4k}{(k+1)^2} \tag{3.14}$$

where $k = n_1/n_2$. At the coupling interface between two optical fibers, the coupling efficiency (η_F) is determined by the effect of the presence of Fresnel's reflection at the core-medium-core interface and can be obtained as follows:

$$\eta_F = \left[\frac{4k}{(k+1)^2}\right]\left[\frac{4k}{(k+1)^2}\right] = \frac{16k^2}{(k+1)^4} \tag{3.15}$$

Therefore, the optical loss (L_F), in decibels (dB), due to Fresnel's reflection is given by

$$L_F = -10\log_{10}(\eta_F) \tag{3.16}$$

In order to minimize optical loss and increase the transmission efficiency of the light reaching the light-measuring device, surface preparation and the coupling method of the scintillator with the optical fiber are important. As can be seen in Figure 3.11, the fiber end face preparation is achieved by polishing the fiber end with various types of lapping films in a repetitive figure-eight pattern [4].

Generally, the dosimeter probe of a PSD system is constructed by connecting an organic scintillator and an optical fiber for applications in medical dosimetry. Figure 3.12 shows the internal structure of a dosimeter probe. In fabricating the dosimeter probe, first, both ends of the scintillator and the optical fiber are cut using a fiber-cutter. Second, both ends of the scintillator and optical fiber are polished until they are perfectly flat with various types of lapping films in a regular sequence to maximize the transmission

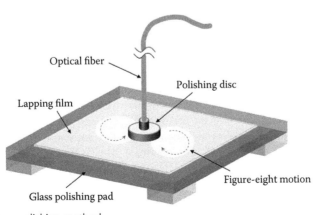

Optical fiber

Polishing disc

Lapping film

Figure-eight motion

Glass polishing pad

Figure 3.11 Fiber end face polishing method.

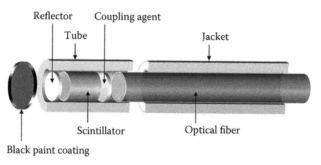

Figure 3.12 Internal structure of a dosimeter probe of a PSD system.

efficiency of the scintillating light. Third, to increase the collection efficiency of the scintillating light, the uncoupled end of the scintillator is normally coated with a white reflector material, such as polytetrafluoroethylene (PTFE, Teflon®)-based reflector tape and titanium dioxide (TiO_2)-based reflector paint. Increasing the amount of scintillating light reaching a light-measuring device improves the signal-to-noise ratio of the PSD system. Fourth, index-matching oils or coupling agents can be additionally used in the gap at the coupling interface to minimize optical loss due to the presence of Fresnel's reflection. Fifth, an organic content-based tube or jacket is used to connect the scintillator and the optical fiber considering alignment of both the outer diameters and central axis. All materials of the dosimeter probe should have a nearly water-equivalent characteristic to avoid complicated dose calibration processes necessitated by material differences between the dosimeter probe and the surrounding material, such as patient body and water phantom. Finally, interconnection devices, such as fiber-optic connectors, adapters, and mating sleeves, are used to easily, quickly, and accurately connect an optical fiber to another fiber or to a light-measuring device [11–13].

Ayotte et al. carried out an experimental study on surface preparation and coupling in plastic scintillator dosimetry [12]. To characterize the optical losses that resulted from coupling of scintillating fibers and clear optical fibers, they analyzed fiber polishing with aluminum oxide sheets, coating fibers with magnesium oxide, and the use of eight different coupling agents (air, three optical gels, an optical curing agent, ultraviolet, cyanoacrylate, and acetone). In their experiment, plastic scintillating fibers (BCF-12, Saint-Gobain Ceramic & Plastics, Malvern, PA) and plastic optical fibers (ESKA series, Mitsubishi Rayon, New York) were used as an organic scintillator and as a light guide, respectively. The scintillating light was measured by a CCD camera (Apogee Alta U2000C, Apogee Instruments, Logan, UT). From the experimental results, the absence of fiber polishing reduced the light-collection efficiency by approximately 40%. Also, they obtained the best results by using one of the optical gels as a coupling agent.

3.3.3 RADIATION-INDUCED ATTENUATION AND FADING EFFECT

When optical fiber is exposed to ionizing radiations, light attenuation in the fiber itself increases due to ionization and resultant point defects (e.g., defect center or color center). It is reported that the point defect plays a main role in the radiation-induced attenuation (RIA) for silica-based optical fibers and absorbs the propagating light signal [14–16]. The point defects cause absorption in ultraviolet (UV), VIS, or near IR spectral region and the resultant RIA is affected by the manufacturing conditions, the dopants present in the optical fiber, the type of radiation, the irradiation conditions, and the temperature applied to the optical fiber during or post irradiation. In recent years, there have been many experimental studies on the characterization of various defect centers at the origin of the RIA and the fading effect in optical fiber [17–21]. However, the mechanism of defect formation is not clear until now.

The characteristics of the defect centers E' in silica optical fiber (SOF) irradiated with gamma rays from a Co-60 source were investigated by Luo et al. [16]. In their experiment, the defect centers were created mainly from the oxygen vacancy in SOF and the defect concentration was increased linearly with the increase of radiation doses from 1 to 50 kGy at room temperature. However, Luo et al. noted that irradiation-induced defect centers can be efficiently decreased by thermal annealing of over 500°C.

Girard et al. also investigated the radiation-induced effects on the transmission of silica-based optical fibers [19]. In their study, two optical fibers were specially designed and fabricated to study the influence of fluorine (F) and germanium (Ge) doping elements on the radiation sensitivity of silica-based glasses. Using the deuterium-halogen source (DH2000, Ocean Optics, Dunedin, FL) and spectrometer (HR4000, Ocean Optics, Dunedin, FL), they characterized the conditions of two fibers before, during, and after 10 keV X-ray irradiation. For the Ge-doped fiber, three radiation-induced point defects, such as Ge(1), GeX, and Ge form of non-bridging oxygen hole centers (Ge-NBOHC), were responsible for the signal decrease in the 300–900 nm range. On the other hand, light losses in the 200–320 nm range were affected by a combination of Si-ODC, Si-NBOHC and SiE defects for the F-doped fiber.

The radiation response behavior of high phosphorous (P)-doped SI MMFs under low-dose gamma irradiation from a Co-60 projector was described by Paul et al. [21]. Various SOFs with core composition of SiO_2–P_2O_5 were characterized by a linear dose dependence of RIA. Two broad RIA peaks of 400–420 nm and 500–560 nm were found from the SOFs with different P_2O_5 contents due to the formation of phosphorous-oxygen hole center defects. The recovery nature of high P-doped fiber containing 16 mol% of P_2O_5 was evaluated after irradiation up to a cumulative dose of 1.0 Gy with a dose rate of 1 Gy/h at 540 and 560 nm transmission wavelengths. The recovery was taken for 2 h after irradiation and the fabricated P-doped fibers showed low fading behavior at room temperature.

Lee et al. characterized the gamma-ray induced Cerenkov radiation within SOFs (BFL37, BFH37, APCH, SPCH, BFL48, and BFH48, Thorlabs), a POF (SH4001, Mitsubishi Rayon, New York), and a plastic wavelength shifting fiber (PWSF: BCF-92, Saint-Gobain Ceramic & Plastics, Malvern, PA) and described a fading effect when these optical fibers are irradiated by gamma-rays from a Co-60 teletherapy unit (Theratron 780, AECL) with an activity of about 3000 Ci [22]. As a light-measuring device, they used a photomultiplier tube (PMT: H9305-03, Hamamatsu Photonics, New Jersey) with a measurable wavelength range from 185 to 900 nm. Generally, it is reported that POFs and SOFs with a high content of OH^- ions have high radiation-resistance characteristics. From their experimental results, the optical intensities of Cerenkov radiation generated in low hydroxyl (low-OH) SOFs (BFL37, APCH, BFL48) were much higher than those of high hydroxyl (high-OH) SOFs (BFH37, SPCH, BFH48). This result means that low-OH SOFs were more easily excited or ionized by the irradiation of high energy radiations and accordingly, they also produced more Cerenkov radiation compared to high-OH SOFs. However, while optical intensities of Cerenkov radiation from the low-OH SOFs decreased as the irradiation time increased, those of the high-OH SOFs were almost uniform, as shown in Figures 3.13 and 3.14. Also, the optical intensities of Cerenkov radiation generated from high-OH SOF (SPCH), POF (SH4001), and PWSF (BCF-92) were almost uniform during irradiation time, as shown in Figure 3.15. In conclusion, the species of optical fiber should be considered to reduce the fading effect for irradiation by high-energy

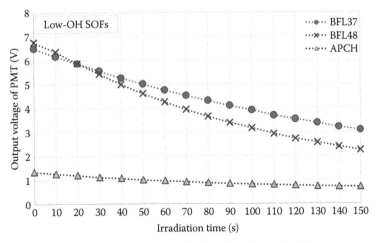

Figure 3.13 Intensities of Cerenkov radiation generated in low-OH silica optical fibers.

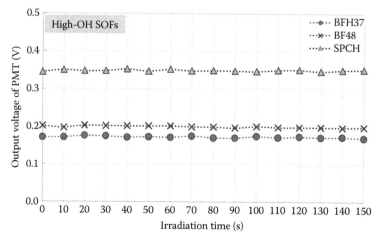

Figure 3.14 Intensities of Cerenkov radiation generated in high-OH silica optical fibers.

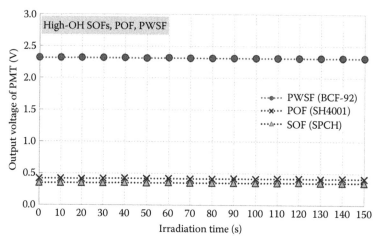

Figure 3.15 Intensities of Cerenkov radiation generated in high-OH silica, plastic, and plastic wavelength shifting optical fibers.

radiation beams. In scintillation dosimetry, the fading effect by the degradation of Cerenkov radiation according to the irradiation time can be reduced or removed by a subtraction method using another identical optical fiber.

3.4 SELECTION OF OPTICAL FIBER FOR SCINTILLATION DOSIMETRY

3.4.1 PLASTIC OPTICAL FIBER

Nearly all commercial optical fibers are classified into POF and SOF according to the core constituent material. Generally, POF is used instead of SOF at a short distance of less than 1 km because of its relatively high attenuation. As POFs are made of polymer, they have a maximum operation temperature range up to 85°C–105°C. Above this limit, POFs begin to lose their transparency. However, medical applications do not require long-distance transmission and high temperature. As a light guide of a PSD, POFs have many advantages over commercially available SOFs, including low cost, ease of use, light weight, large core diameter, and good flexibility, to transmit light signals having dose information from a sensing probe of a PSD in the examination or therapy room to a light-measuring device in the control room [23,24].

In general, PMMA and PS are commonly used for the core material of conventional POFs because of their transparency. The refractive indices of PMMA and PS are about 1.49 and 1.60, respectively. The cladding material of the PMMA-core fibers is a fluorinated polymer and the PS-core fibers normally use a PMMA as the cladding material. For the PMMA-core fiber, the smallest amount of attenuation can be observed around the green region while that of the PS-core fiber is located near the red region of the spectrum. Total optical loss of the PMMA-core fiber is lower than that of the PS-core fiber in the VIS range [25]. For radiotherapy dosimetry, all materials of the dosimeter probe should have a nearly water-equivalent characteristic to avoid complicated calibration processes. The mass densities, electron densities, atomic compositions, mean-energy absorption coefficient, and the mean-mass collision stopping powers of the PMMA and PS closely correspond with those of water [26–29]. In addition, no significant degradation was observed in light attenuation of POF up to 10–15 kGy of irradiation [30]. POFs are thus appropriate for use as the light guide of a PSD system for obtaining real-time dose information.

However, Cerenkov radiation can be generated in an optical fiber of a PSD system during irradiation of high-energy radiation. The Cerenkov threshold energy (CTE, E_{Th}) depends on the refractive index of the core material, because the threshold energy for electrons increases as the refractive index decreases. The CTEs of the electron to produce Cerenkov radiation in PMMA and PS are 178 and 146 keV, respectively. From the viewpoint of the CTE, the PMMA-core fiber therefore has an advantage over the PS-core fiber in terms of reducing unwanted noise that can be induced by low-energy charged particles, because the refractive index of PMMA is lower than that of PS [31–33].

In recent years, various square-type POFs and plastic scintillating fibers with PS-core and PMMA-cladding have been developed as a clear waveguide and an organic scintillator, respectively. These square-type fibers have a core/clad structure with a rectangular cross section and they have many favorable characteristics over general round-type fibers to obtain multidimensional dose distributions and scintillation images. By employing square-type fibers in the fabrication of an optical fiber array, dose measurement errors arising from air-gaps between the closely packed fibers can be avoided without requiring complicated fabrication processes. In addition, an optical fiber array or bundle can be easily fabricated using square-type fibers due to their rectangular cross section, as shown in Figure 3.16 [34–36].

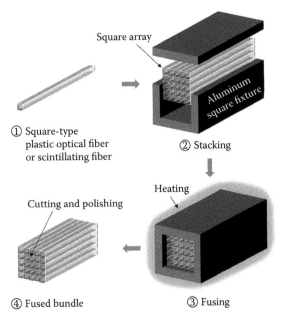

Figure 3.16 Schematic diagram of the sequential processes for fabrication of a fiber-optic bundle.

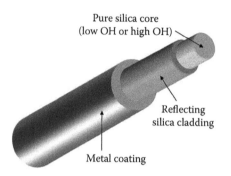

Pure silica core
(low OH or high OH)

Reflecting
silica cladding

Metal coating

Figure 3.17 Structure of a metal-coated optical fiber.

3.4.2 SILICA OPTICAL FIBER

The core materials of SOFs are divided into pure silica and silica glasses doped with dopants, such as germanium (Ge), phosphorus (P), and fluorine (F). There are principally two types of pure silica core materials: high OH and low OH. The concentration of OH⁻ ions in the fiber is very important because it affects the spectral transmission of the optical fiber in the UV-blue and near-IR spectral regions. First, low-OH fibers can transmit near-IR wavelength up to about 2400 nm, and they are known to have higher transmission rate in the VIS range due to their lower attenuation, compared to high-OH fibers. Next, the use of high-OH fibers for IR transmission is limited because of the absorption of OH⁻ ions. However, because the high concentration of OH⁻ ions favorably affects the structure of silica glass for UV transmission, high-OH fibers are more efficient than low-OH fiber to transmit light in the UV-blue region [7]. When SOFs are exposed to radiation, fiber attenuation increases due to ionization and the resultant color centers. Fortunately, high-OH fibers provide stable light signals due to their high radiation-resistance characteristics [22,37,38]. As mentioned in Section 3.3.3 regarding a fading effect for irradiation of therapeutic radiation beams, the species of optical fiber should be considered to reduce the fading effect of the PSD system during irradiation of high-energy radiation beams and both POF and high-OH SOF can serve as light guides to overcome this problem.

Recently, many kinds of metal-coated silica optical fibers with a silica core/silica cladding structure have been developed for use in harsh and extreme environments (Figure 3.17). These fibers have metal coating applied to the silica cladding during the drawing of the fiber and the coating materials are titanium (Ti), copper (Cu), and aluminum (Al). Therefore, they can remain stable in corrosive chemicals and withstand high temperature conditions with a humidity range of up to 100%. The maximum operating temperature of Ti-, Cu-, and Al-coated fibers is 230°C, 600°C, and 400°C, respectively, and temperature of up to 750°C can be achieved if a double layer of Cu and nickel (Ni) is applied. In addition, metal-coated silica fibers with a high-OH silica core have a radiation-resistance characteristic and thus, they can be used to transmit scintillation light signals in high temperature and radiation environmental conditions.

REFERENCES

1. Cutnell J D, Johnson K W. *Essentials of Physics*. Hoboken, NJ: John Wiley & Sons, 2006.
2. Hecht E. *Optics*. 2nd edn. Reading, MA: Addison-Wesley, 1987.
3. Davis C M, Carome E F, Weik M H, Ezekiel S, Einzig R E. *Fiberoptic Sensor Technology Handbook*. Herndon, VA: Optical Technologies, 1986.
4. Goff D R. *Fiber Optic Reference Guide: A Practical Guide to Communications Technology*, 3rd edn. Woburn, MA: Focal Press, 2002.
5. Khare R P. *Fiber Optics and Optoelectronics*. New Delhi, India: Oxford University Press, 2004.
6. Pedrotti F L, Pedrotti L S, Pedrotti L M. *Introduction to Optics*, 3rd edn. Upper Saddle River, NJ: Pearson Education, 2007.
7. Laurin T C et al. *The Photonics Design & Applications Handbook*, 38th edn. Pittsfield, MA: Laurin Publishing Company, 1992.

8. Jang K W, Yoo W J, Seo J K, Heo J Y, Moon J, Park J-Y, Kim S, Park B G, Lee B. Measurements and removal of Cerenkov light generated in scintillating fiber-optic sensor induced by high-energy electron beams using a spectrometer. *Opt. Rev.* **18** 176–179 (2011).
9. Beddar A S, Mackie T R, Attix F H. Cerenkov light generated in optical fibres and other light pipes irradiated by electron beams. *Phys. Med. Biol.* **37** 925–935 (1992).
10. Jones A K, Hintenlang D. Potential clinical utility of a fibre optic-coupled dosemeter for dose measurements in diagnostic radiology. *Radiat. Prot. Dosim.* **132** 80–87 (2008).
11. Hyer D E, Fisher R F, Hintenlang D E. Characterization of a water-equivalent fiber-optic coupled dosimeter for use in diagnostic radiology. *Med. Phys.* **36** 1711–1716 (2009).
12. Ayotte G, Archambault L, Gingras L, Lacroix F, Beddar A S, Beaulieu L. Surface preparation and coupling in plastic scintillator dosimetry. *Med. Phys.* **33** 3519–3525 (2006).
13. Yoo W J, Shin S H, Jeon D, Hong S, Kim S G, Sim H I, Jang K W, Cho S, Lee B. Simultaneous measurements of pure scintillation and Cerenkov signals in an integrated fiber-optic dosimeter for electron beam therapy dosimetry. *Opt. Express* **21** 27770–27779 (2013).
14. Griscom D. Trapped-electron centers in pure and doped glassy silica: A review and synthesis. *J. Non-Cryst. Solids* **357** 1945–1962 (2011).
15. Tomashuk A L, Salgansky M Y, Kashaykin P F, Khopin V F, Sultangulova A I, Nishchev K N, Borisovsky S E, Guryanov A N, Dianov E M. Enhanced radiation resistance of silica optical fibers fabricated in high O_2 excess conditions. *J. Lightwave Technol.* **32** 213–219 (2014).
16. Luo W, Xiao Z, Wen J, Yin J, Chen Z, Wang Z, Wang T. Defect center characteristics of silica optical fiber material by gamma ray radiation. *SPIE* **8307** 83072H (2011).
17. Girard S, Ouerdane Y, Boukenter A, Meunier J-P. Transient radiation responses of silica-based optical fibers: Influence of modified chemical-vapor deposition process parameters. *J. Appl. Phys.* **99** 023104 (2006).
18. Girard S, Keurinck J, Boukenter A, Meunier J-P, Ouerdane Y, Azais B, Charre P, Vie M. Gamma-rays and pulsed X-ray radiation responses of nitrogen-, germanium-doped and pure silica core optical fibers. *Nucl. Instrum. Methods Phys. Res. Sect. B* **215** 187–195 (2004).
19. Girard S et al. Radiation effects on silica-based preforms and optical fibers-I: Experimental study with canonical samples. *IEEE Trans. Nucl. Sci.* **55** 3473–3482 (2008).
20. Okamoto K, Toh K, Nagata S, Tsuchiya B, Suzuki T, Shamoto N, Shikama T. Temperature dependence of radiation induced optical transmission loss in fused silica core optical fiber. *J. Nucl. Mater.* **329–333** 1503–1506 (2004).
21. Paul M C, Bohra D, Dhar A, Sen R, Bhatnagar P K, Dasgupta K. Radiation response behavior of high phosphorous doped step-index multimode optical fibers under low dose gamma irradiation. *J. Non-Cryst. Solids* **355** 1945–1962 (2011).
22. Lee B, Jang K W, Yoo W J, Shin S H, Moon J, Han K-T, Jeon D. Measurements of Cerenkov lights using optical fibers. *IEEE Trans. Nucl. Sci.* **60** 932–936 (2013).
23. O'Keeffe S, Fitzpatrick C, Lewis E, Al-Shamma'a A I. A review of optical fibre radiation dosimeters. *Sens. Rev.* **28** 136–142 (2008).
24. Yoo W J, Shin S H, Jeon D, Han K-T, Hong S, Kim S G, Cho S, Lee B. Development of a fiber-optic dosimeter based on modified direct measurement for real-time dosimetry during radiation diagnosis. *Meas. Sci. Technol.* **24** 094022 (2013).
25. Zubia J, Arrue J. Plastic optical fibers: An introduction to their technological processes and applications. *Opt. Fiber Technol.* **7** 101–140 (2001).
26. Beddar A S. Water equivalent plastic scintillation detectors in radiation therapy. *Radiat. Prot. Dosim.* **120** 1–6 (2006).
27. Beddar A S. Plastic scintillation dosimetry and its application to radiotherapy. *Radiat. Meas.* **41** S124–S133 (2007).
28. Khan F M. *The Physics of Radiation Therapy*, 2nd edn. Baltimore, MD: Williams & Wilkins, 1994.
29. Task Group 21, Radiation Therapy Committee, AAPM. A protocol for the determination of absorbed dose from high-energy photon and electron beams. *Med. Phys.* **10** 741–771 (1983).
30. Protopopov Y M, Vasil'chenko V G. Radiation damage in plastic scintillators and optical fibers. *Nucl. Instrum. Methods Phys. Res. Sect. B* **95** 496–500 (1995).
31. Jang K W, Yagi T, Pyeon C H, Yoo W J, Shin S H, Jeong C, Min B J, Shin D, Misawa T, Lee B. Application of Cerenkov radiation generated in plastic optical fibers for therapeutic photon beam dosimetry. *J. Biomed. Opt.* **18** 027001 (2013).
32. Ross H H. Measurement of β-emitting nuclides using Cerenkov radiation. *Anal. Chem.* **41** 1260–1265 (1969).
33. Yoo W J, Han K-T, Shin S H, Seo J K, Jeon D, Lee B. Development of a Cerenkov radiation sensor to detect low-energy beta-particles. *Appl. Radiat. Isot.* **81** 196–200 (2013).
34. Moon J, Jang K W, Yoo W J, Han K-T, Park J-Y, Lee B. Water-equivalent one-dimensional scintillating fiber-optic dosimeter for measuring therapeutic photon beam. *Appl. Radiat. Isot.* **70** 2627–2630 (2012).

35. Shin S H, Yoo W J, Seo J K, Han K-T, Jeon D, Jang K W, Sim H I, Cho S, Lee B. Measurements of planar and depth dose distributions using a scintillating fiber-optic image sensor system for dosimetry in radiotherapeutic applications. *Opt. Rev.* **20** 178–181 (2013).

36. Yoo W J, Moon J, Jang K W, Han K-T, Shin S H, Jeon D, Park J-Y, Park B G, Lee B. Integral T-shaped phantom-dosimeter system to measure transverse and longitudinal dose distributions simultaneously for stereotactic radiosurgery dosimetry. *Sensors* **12** 6404–6414 (2012).

37. Jang K W, Yoo W J, Shin S H, Shin D, Lee B. Fiber-optic Cerenkov radiation sensor for proton therapy dosimetry. *Opt. Express* **20** 13907–13914 (2012).

38. Lu P, Bao X, Kulkarni N, Brown K. Gamma ray radiation induced visible light absorption in P-doped silica fibers at low dose levels. *Radiat. Meas.* **30** 725–733 (1999).

Plastic scintillation detectors: Basic properties

Chinmay Darne, Landon Wootton, and Sam Beddar

Contents

4.1 INTRODUCTION

In this chapter, the fundamental dosimetric characteristics of plastic scintillation detectors (PSDs) are presented in detail. The chapter begins with a description of the original PSD, with which many of these characteristics were initially established (Beddar et al. 1992a). This detector used plastic scintillator as its light-emitting element (as opposed to plastic scintillating fiber, a subtle but important difference). PSDs were the focus of scintillation dosimetry during the 1990s. In the early 2000s, the focus shifted to plastic scintillating fibers, which are distinguished from plastic scintillators by the addition of cladding to improve light propagation and the use of a different base material (typically polystyrene instead of polyvinyltoluene). This chapter draws largely on work done with the original PSD and its derivatives; the properties of other PSDs, including those built with scintillating fibers, have been studied by others (Archambault et al. 2005, Beierholm et al. 2014, Carrasco et al. 2015, Fontbonne et al. 2002, Frelin et al. 2005) and found to be largely identical to those of the original PSD. Thus, the properties presented in this chapter may be assumed to apply generally to PSDs using either plastic scintillators or scintillating fibers unless otherwise noted.

4.2 DESCRIPTION OF THE ORIGINAL PSD

The design of the original PSD, published in 1992 by Beddar et al. (1992a), is displayed in Figure 4.1. This detector consisted of (1) an active volume composed of BC-400 plastic scintillator, embedded in the center of a cylinder of polystyrene; (2) parallel-paired fiber light guides; and (3) two identical

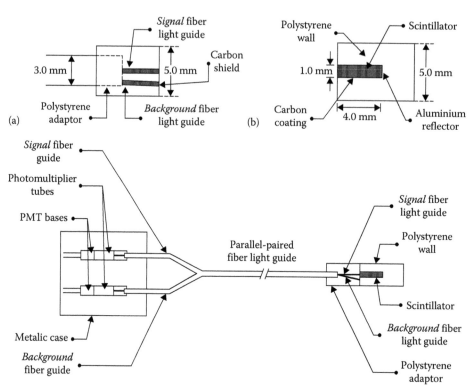

Figure 4.1 Diagram of the PSD system outlined by Beddar et al. and longitudinal cross sections of (a) the proximal end of the detector probe and (b) polystyrene enclosure that contains the scintillator. (Reproduced from Beddar, A.S. et al., Water-equivalent plastic scintillation detectors for high-energy beam dosimetry: I. Physical characteristics and theoretical considerations, *Phys. Med. Biol.*, 37, 1883–1900, 1992a. With permission from IOP Publishing.)

photomultiplier tubes (PMTs). The detector probe containing the scintillator incorporated two optical fibers, one of which was optically coupled to the scintillator with a clear epoxy, whereas the other was concealed from scintillation light with a layer of carbon shielding. The second fiber served to isolate background signal, primarily in the form of Cerenkov light, from the total light signal. As the two optical fibers were side by side, it was assumed that the same background was generated in both, and thus, by taking the difference in light output between the two, the scintillation signal could be recovered. Both fibers were enclosed in a black polyethylene jacket to shield them from external light. PMTs with absorption spectra matched to the scintillator emission were chosen to maximize sensitivity. This allowed the difference in DC output between the two PMTs, operated at the same gain, to serve as a precise proxy of the scintillation light emitted, and thus the dose absorbed by the detector probe. Various measures were used to maximize light collection, including reflective coating of the scintillator, polished fiber ends, and optical coupling epoxy for better transmission (Ayotte et al. 2006); conical polymethyl methacrylate (PMMA) light guides to diffuse light from the fibers to the PMTs, and silicon grease to couple the light guides to the faces of the PMTs.

The next three sections of this chapter cover PSD properties that were theoretically derived for the original PSD and subsequently experimentally verified.

4.2.1 WATER EQUIVALENCE

The water equivalence of organic scintillators is one of the primary motivations for using PSDs. An ideal radiation detector should be well matched to the medium in which the absorbed dose will be measured. Charged particles crossing such a detector will interact in a manner similar to that within the medium, leaving their overall path unaffected. This allows the detector to record the energy deposited by

Table 4.1 **Physical parameters for polystyrene, plastic scintillator, and water**

PARAMETER	POLYSTYRENE	SCINTILLATOR (BC-400)	WATER
Density (g/cm³)	1.060	1.032	1.000
Ratio of number of e⁻ in the compound to molecular weight (Z/A)	0.5377	0.5414	0.5551
Electron density (10^{23} e⁻ g⁻¹)	3.238	3.272	3.343
Composition (Z: fraction by weight [%])	1: 7.74 6: 92.26	1: 8.47 6: 91.53	1: 11.19 8: 88.81

unperturbed ionizing radiation, facilitating more accurate measurement. Water is often the dose-absorbing medium of choice in dosimetry because it is similar in composition to tissue. Thus, a water-equivalent detector is highly desirable.

Water equivalency requires that the constituent materials of a detector closely match the absorption and scattering properties of water over a clinically relevant range of energies. A typical PSD consists of a radiation-sensitive material (polyvinyltoluene or polystyrene with secondary/tertiary fluors) surrounded by a matching protective *wall* (polystyrene or PMMA). Table 4.1 shows close matching of various physical parameters for polystyrene, the plastic scintillator used in the original PSD, and water. For low-Z materials such as these, the dominant mode of interaction with photons of therapeutic energies is through Compton scattering. Thus, the parameters that govern Compton scattering, such as mass energy absorption coefficients, and those that govern the resulting secondary charged particles, such as mass collision stopping powers and mass angular scattering powers, are the primary considerations with respect to the water equivalence of a material.

The mass energy absorption coefficients for water, polystyrene, and the plastic scintillator for incident photons with energies between 10 keV and 20 MeV are displayed in Figure 4.2. For compounds and mixtures, such as plastic scintillators, Bragg's rule is used to calculate mass energy absorption coefficients. When the photon energy is above 100 keV, which is typical in radiotherapy, the mass energy absorption coefficients of the scintillator are closer to those of water than those of polystyrene. Good matching of the mass energy absorption coefficients of all three materials is evident in the figure, particularly above 100 keV.

Mass collision stopping powers and mass angular scattering powers for water, polystyrene, and the plastic scintillator for electrons with energies between 10 keV and 50 MeV are displayed in Figures 4.3 and 4.4. The mass collision stopping powers of the scintillator and polystyrene are well matched with those of water. Similarly, the mass angular scattering powers for all three materials are well matched for electron energies between 10 keV and 50 MeV. The similarities of these physical parameters imply excellent water equivalency for plastic scintillator and polystyrene.

In addition to the materials used, the density states of the detector constituents also play an important role in radiation interactions. This is referred to as the polarization or density effect. Consider the example of an ionization chamber where the detector volume is gas, although the surrounding wall is in a condensed state. This density mismatch triggers dipole distortion in the wall in close proximity to the photon travel path. This in turn weakens the Coulomb force field interaction experienced by distant wall atoms causing reduction in the energy loss cross section. This consequently decreases mass collision stopping power within the wall, and hence alters energy deposition within the detector. Quantitative studies have shown that this percentage reduction in mass collision stopping power for a condensed state medium versus gas of the same atomic composition is approximately 3% at 1 MeV, 12% at 10 MeV, and 20% at 50 MeV (Attix 1986, Berger and Seltzer 1983). For plastic scintillators, both the wall and the detector volume have the same density state in addition to possessing similar atomic structure, so polarization will be homogeneous and the radiation field unperturbed, preserving water equivalence.

Finally, Monte Carlo simulations have been used to establish the water equivalency of PSDs. Francescon et al. (2014) calculated correction factors for various detectors in small fields to convert dose readings to dose to water (a correction factor of unity would indicate water equivalence). Included in this study was a commercial PSD, the Exradin W1. Monte Carlo simulations of detector responses in small fields (5–25 mm

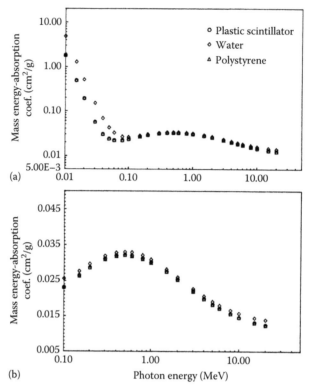

Figure 4.2 Mass energy absorption coefficients for plastic scintillator compared to water and polystyrene as a function of photon energy: (a) logarithm–logarithm plot; (b) linear–logarithm plot. (Reproduced from Beddar, A.S. et al., Water-equivalent plastic scintillation detectors for high-energy beam dosimetry: I. Physical characteristics and theoretical considerations, *Phys. Med. Biol.*, 37, 1883–1900, 1992a. With permission from IOP Publishing.)

circular fields) in comparison to the dose absorbed by an equivalent volume of water were used to generate the correction factors. This PSD deviated minimally from a correction factor of unity except for the smallest field where the correction factor was still less than 2%, indicating a high degree of water equivalency and spatial resolution.

4.2.2 APPLICATION OF BURLIN CAVITY THEORY TO PSDs

The Burlin cavity theory can be used to relate the dose absorbed by plastic scintillator to that absorbed by water (Burlin 1966). It is appropriate for cavities of intermediate size, where the cavity is not only too large to apply the Bragg–Gray theory but also too small to assume that the charged particles generated by photon interactions within the cavity deposit their energy completely therein. The dose deposited in a Burlin cavity composed of scintillator is expressed by the following equation (Attix 1986, Beddar et al. 1992a):

$$\frac{\overline{D_{sci}}}{D_{med}} = d_m (\overline{S})_{med}^{sci} + (1-d)\left(\overline{\frac{\mu_{en}}{\rho}}\right)_{med}^{sci} \tag{4.1}$$

where:

D_{sci} is the mean dose absorbed by the scintillator
D_{med} is the dose in charged particle equilibrium (CPE) at the same location in an unperturbed medium
$_m S_{med}^{sci}$ is the mean ratio of mass collision stopping powers for the scintillator and the medium
$(\mu_{en})_{med}^{sci}$ is the mean ratio of the mass energy absorption coefficients for the scintillator and the medium
d is related to the cavity size
ρ is the density in kg/m^3 or g/cm^3

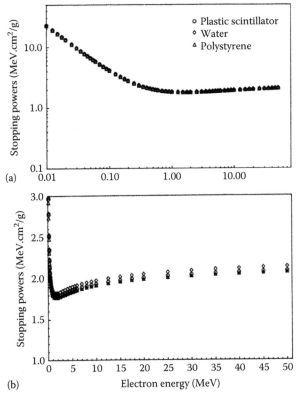

Figure 4.3 Mass collision stopping powers of plastic scintillator compared to water and polystyrene as a function of electron energy: (a) logarithm–logarithm plot; (b) linear–logarithm plot. (Reproduced from Beddar, A.S. et al., Water-equivalent plastic scintillation detectors for high-energy beam dosimetry: I. Physical characteristics and theoretical considerations, *Phys. Med. Biol.*, 37, 1883–1900, 1992a. With permission from IOP Publishing.)

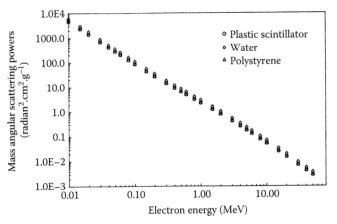

Figure 4.4 Mass angular scattering powers of plastic scintillator compared to water and polystyrene as a function of electron energy. (Reproduced from Beddar, A.S. et al., Water-equivalent plastic scintillation detectors for high-energy beam dosimetry: I. Physical characteristics and theoretical considerations, *Phys. Med. Biol.*, 37, 1883–1900, 1992a. With permission from IOP Publishing.)

Basic principles and theory

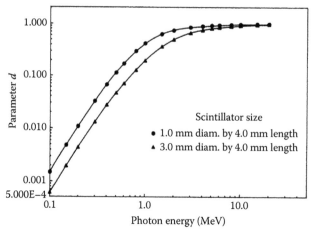

Figure 4.5 Burlin cavity theory parameter *d* as a function of photon energy for two plastic scintillator detector sizes. (Reproduced from Beddar, A.S. et al., Water-equivalent plastic scintillation detectors for high-energy beam dosimetry: I. Physical characteristics and theoretical considerations, *Phys. Med. Biol.*, 37, 1883–1900, 1992a. With permission from IOP Publishing.)

If the equilibrium charged particle fluence is approximately the same in the two materials,

$$d = \frac{(1 - e^{\beta L})}{\beta L} \qquad (4.2)$$

where:

β (cm⁻¹) is the effective absorption coefficient of the electrons in the cavity
L (cm) is the mean path length of the electrons across the cavity

For small cavities, *d* approaches unity, whereas for large cavities, it nears zero. Thus, *d* represents the fraction of the dose deposited in the cavity due to electrons generated outside the cavity. Note that Equation 4.2 is applicable only to diffuse (or isotropic) fields of electrons, such as secondary electrons generated by X-rays, but not to electron beams. For primary electron beams, the Burlin cavity theory is applied by omitting the last term of Equation 4.1.

Figure 4.5 shows the plot of *d* varying as a function of monoenergetic photon energy from 100 keV to 20 MeV for two PSDs with 1 and 3 mm diameters, both 4 mm long. For radiation in the diagnostic range (<250 keV), the value of *d* approaches zero, and therefore, a large cavity approximation is justified. For energies exceeding 20 MeV, *d* approaches unity, thus making a small cavity approximation applicable. For all other energies between the above two extremes, the value of *d* falls between 0 and 1, and therefore justifies the use of the Burlin cavity theory.

4.2.3 ENERGY DEPENDENCE IN THE EXTERNAL BEAM RADIOTHERAPY ENERGY RANGE

Insofar as PSDs are water-equivalent detectors, they should also be energy independent detectors. That is to say, a PSD should accurately measure dose for different quality photon beams and different energy electron beams without correction factors. The energy dependence of PSDs has been treated both theoretically and experimentally, and both approaches are presented in this section.

Beddar et al. (1992a) applied the Burlin cavity theory to theoretically evaluate the energy dependence of the original PSD for two limiting cases: (1) zero wall thickness (i.e., no polystyrene wall) and (2) a wall sufficiently thick to provide CPE. The ratio of dose absorbed by the detector to absorbed dose in water was calculated for a hypothetical pair of PSDs with 4 mm-long active volumes with 1 and 3 mm diameters for each limiting case. Figure 4.6 shows the ratio of the mean dose absorbed by the scintillator to that absorbed by water between 200 keV and 20 MeV. For the zero wall thickness scenario, the ratio of absorbed dose is almost constant with only 2% fluctuation over the analyzed energy range for both detectors. For the thick wall, the ratio is similar at low energies (200 keV to 3 MeV) but then gradually

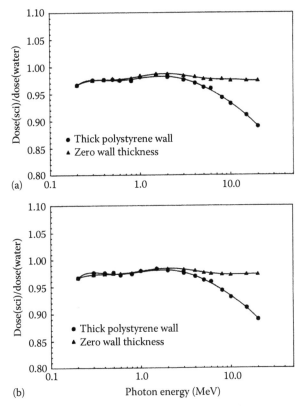

(a)

(b)

Figure 4.6 Ratio of absorbed dose in the plastic scintillator to water for scintillator sizes of (a) 1.0 mm in diameter × 4.0 mm in length and (b) 3.0 mm in diameter × 4.0 mm in length for thick polystyrene wall and a wall-less cavity. (Reproduced from Beddar, A.S. et al., Water-equivalent plastic scintillation detectors for high-energy beam dosimetry: I. Physical characteristics and theoretical considerations, *Phys. Med. Biol.*, 37, 1883–1900, 1992a. With permission from IOP Publishing.)

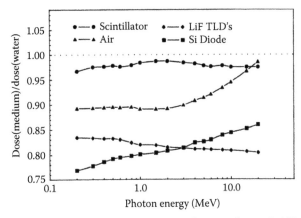

Figure 4.7 Ratio of absorbed dose in four different detector media (scintillator, air, LiF, and silicon) with respect to water. The detecting volume for the first three detectors was taken as 1.0 mm diameter × 4.0 mm in length. (Reproduced from Beddar, A.S. et al., Water-equivalent plastic scintillation detectors for high-energy beam dosimetry: I. Physical characteristics and theoretical considerations, *Phys. Med. Biol.*, 37, 1883–1900, 1992a. With permission from IOP Publishing.)

decreases by approximately 9% at 20 MeV. As these are limiting cases, the true response lies between the two. However, the wall thickness likely to be used in practice is on the order of 2 mm or less, much smaller than needed for establishing CPE. Therefore, the energy dependence for a typical PSD should adhere closely to that of the zero wall thickness scenario. To illustrate the advantage of PSDs in terms of energy response, Figure 4.7 displays the absorbed dose ratio for scintillator as well as for air,

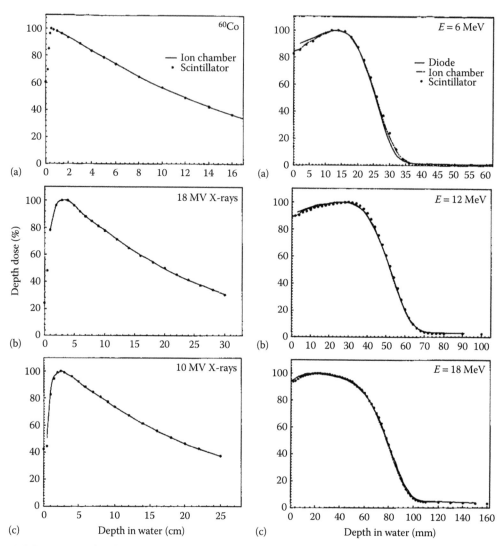

Figure 4.8 Percentage depth-dose curves measured in water for various photon beam qualities as seen in left column: (a) ^{60}Co gamma rays, (b) 18 MV X-rays, (c) 10 MV X-rays and for different electron energies as seen in right column: (a) 6 MeV electrons, (b) 12 MeV electrons, (c) 18 MeV electrons, in comparison with an ionization chamber and an electron diode. (From Beddar, A.S. et al., Water-equivalent plastic scintillation detectors for high-energy beam dosimetry: II. Properties and measurements, *Phys. Med. Biol.*, 37, 1901–1913, 1992b. With permission from IOP Publishing.)

thermoluminescent dosimeters, and diodes as determined with cavity theory. The PSD ratio is both closest to unity and varies the least of the range of therapeutic energies.

The energy independence of the original PSD was experimentally validated by measuring depth dose curves for a variety of photon and electron beams in a water phantom and comparing the curves to ones obtained with an ion chamber (Beddar et al. 1992b). Good agreement was seen at all energies for both modalities, indicating a high degree of energy independence. The results are exhibited in Figure 4.8.

4.3 GENERAL DOSIMETRIC CHARACTERISTICS OF PSDs

The characteristics outlined in the remaining sections were experimentally derived. Many of the results presented are drawn from the 1992 papers of Beddar et al. for the original PSD, but results generated by others are presented as well.

4.3.1 STABILITY

The short-term stability of a detector (i.e., the variation in the detector response to the same stimulus) is important for determining the number of measurements required to achieve a given overall accuracy. Ideally, the detector response will be constant for a given dose, but in practice some variation is unavoidable. For PSDs, the source of such variation could come from noise in the photodetector (LaCroix et al. 2010), stochastic variations in the number of photons produced by the scintillator for a given dose, and a variety of other sources.

Various groups have examined the short-term stability of PSDs with different designs and under a range of conditions. Beddar et al. (1992b) assessed the original PSD (active volume of 0.003 cm^3) by irradiating it in water with a ^{60}Co beam. PMTs were used to integrate the light produced by the scintillator over 60 s. Twenty consecutive readings resulted in a percentage standard deviation of 0.08%. Figure 4.9 shows the series of 20 readings with detector variations centered on the calculated mean value and enveloped completely within ±0.2% of it. The same experiment was performed using 6 and 10 MV photon beams from a linear accelerator, and resulted in standard deviations of 0.05% and 0.09%, respectively. Finally, the stability of the response to 6 and 18 MeV electron beams was determined by performing repeated 200 monitor unit irradiations of the scintillator. The electron beam stability study was divided into separate performance assessment for both fiber guides as electrons produce a greater amount of background in the form of Cerenkov radiation. Figure 4.10 shows that standard deviations in detection of a 6 MeV electron beam for the signal and background channel outputs were 0.08% and 0.06%, respectively; for an 18 MeV beam, the standard deviation of both channel outputs was 0.09%. Upon addition in quadrature, the net percentage standard deviations for 6 and 18 MeV beams were 0.1% and 0.13%, respectively.

Another study by Beddar (1994) examined a PSD constructed with BC-430 (active volume of 1.57 cm^3) and read out with a silicon photomultiplier. This detector also exhibited good stability with standard deviation of 0.09% for a ^{60}Co gamma source after 10 consecutive readings. Irradiating this detector with 6 and 18 MV photon beams, and 6–20 MeV electron beams resulted in standard deviations varying between 0.05% and 0.18%. An example of these results is depicted in Figure 4.11 for 18 MV X-rays and 20 MeV electron beams, and show 0.11% and 0.08% standard deviations, respectively.

More recently, Archambault et al. (2010) have investigated the use of PSDs for real-time dosimetry. As part of their study, they delivered 2–200 cGy to five PSDs (active volume of 0.0004 cm^3) and used a charge-coupled device camera in continuous acquisition mode to measure the dose to each detector in

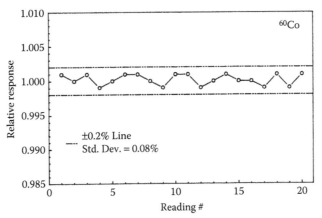

Figure 4.9 Stability of the scintillator detector system over 20 measurements under ^{60}Co irradiation. (Reproduced from Beddar, A.S. et al., Water-equivalent plastic scintillation detectors for high-energy beam dosimetry: II. Properties and measurements, *Phys. Med. Biol.*, 37, 1901–1913, 1992b. With permission from IOP Publishing.)

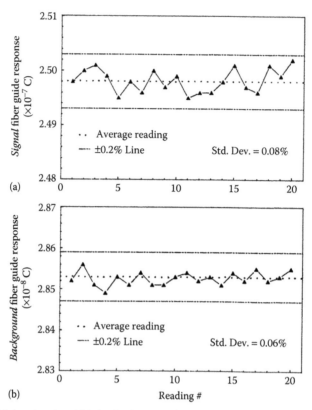

(a)

(b)

Reading #

Figure 4.10 Stability of (a) the *signal* and (b) the *background* fiber guide outputs of the scintillation detector system in a 6 MeV electron beam over 20 measurements. (Reproduced from Beddar, A.S. et al., Water-equivalent plastic scintillation detectors for high-energy beam dosimetry: II. Properties and measurements, *Phys. Med. Biol.*, 37, 1901–1913, 1992b. With permission from IOP Publishing.)

150 ms intervals. The standard deviations for measurements of 200, 20, and 2 cGy irradiations were 0.4%, 1.0%, and 2.3%, respectively. This illustrates that even PSDs with very small active volumes can possess excellent stability.

Beierholm et al. (2014) compared a commercial PSD (the Exradin W1 with an active volume of 0.023 cm³) with an in-house PSD (built with 0.016 cm³ of BCF-60) under 6, 10, and 15 MV photon irradiation (including flattening filter-free 6 and 10 MV beams), and found excellent stability. The standard deviation for the commercial PSD varied between 0.03% and 0.32%, and between 0.05% and 0.39% for the in-house PSD.

As exhibited by these studies, PSDs are capable of highly stable measurements. Variations in the response are often largely attributable to photodetector noise. The stability of a given PSD will therefore depend on the photodetector used and the dose being measured.

4.3.2 REPRODUCIBILITY

The consistency of a detector response over mid- to long-term time spans, here denoted as reproducibility, is critical when measurements taken at different times are to be compared to one another. An example of such a situation would be monitoring the output of a machine on a monthly basis. If variations in the output are observed, the reproducibility of the detector informs how much of the variation may be confidently attributed to actual variation in the machine as opposed to variation in the detector. Better reproducibility therefore allows the detection of smaller actual variations and is desirable for that reason.

Beddar (1994) investigated the reproducibility of a BC-430-based detector by irradiating it on a daily basis over 10 days, and on a weekly basis over 14 weeks. A ^{60}Co therapy unit was used for its stable output (after accounting for decay), allowing variation in the measured output to be attributed to the detector itself. Over the 10 days of daily irradiations, the standard deviation of the detector measurements was 0.3%,

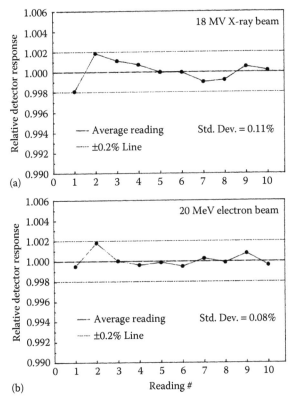

Figure 4.11 Stability of the scintillator detector system when exposed to high-energy beams: (a) 18 MV X-ray beam; (b) 20 MeV electron beam. (Reproduced from Beddar, A.S., A new scintillator detector system for the quality assurance of ^{60}Co and high-energy therapy machines, *Phys. Med. Biol.*, 39, 253–263, 1994. With permission from IOP Publishing.)

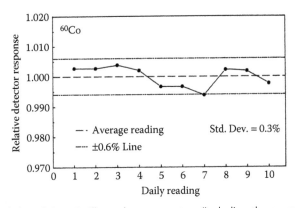

Figure 4.12 Daily reproducibility of the scintillator detector system (including the reproducibility of the experimental setup from one day to the next). (Reproduced from Beddar, A.S., A new scintillator detector system for the quality assurance of ^{60}Co and high-energy therapy machines, *Phys. Med. Biol.*, 39, 253–263, 1994. With permission from IOP Publishing.)

and all measurements fell within ±0.6% of the average value. Over the 14 weekly irradiations, the standard deviation of the measurements was 1.1%, with all measurements falling within 1.5%. These results are displayed in Figures 4.12 and 4.13, respectively.

Note that reproducibility deals with random fluctuations in response over mid- to long time spans. As demonstrated here, PSDs provide a very reproducible response. However, it is well known that PSDs experience a nonrandom unidirectional change in response over time when cumulatively exposed to very high quantities of radiation. This is due to radiation damage and is considered separately in Section 4.3.8.

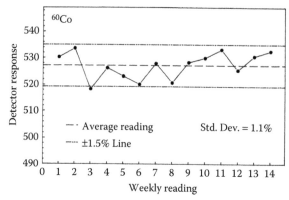

Figure 4.13 Weekly reproducibility over a period of 14 weeks demonstrating the long-term stability of the scintillator detector system. (Reproduced from Beddar, A.S., A new scintillator detector system for the quality assurance of ^{60}Co and high-energy therapy machines, *Phys. Med. Biol.*, 39, 253–263, 1994. With permission from IOP Publishing.)

4.3.3 DOSE LINEARITY

The use of calibration curves to accurately measure dose is unnecessary for detectors exhibiting linearity. Linearity therefore allows straightforward dose measurements with a constant sensitivity over the entire dynamic range.

The linearity of the original PSD was tested under photon and electron irradiation (Beddar et al. 1992b). For photon irradiation, the detector was submerged in water at the depth of maximum dose buildup (d_{max}) and irradiated with a 6 MV X-ray beam. The response of the detector to doses ranging from 40 to 400 cGy is plotted in Figure 4.14 and is linear. This indicates that no detector fatigue has set in over the given dose range. For electron irradiation, a 12 MeV electron beam from a linear accelerator delivered between 40 and 800 cGy to the detector, which was once again positioned at d_{max} in water. The results are plotted in Figure 4.15, and once again, the response is linear. The signals from both the background and scintillation fibers, as well as their sum, are displayed, and interestingly even these individual signals are linear, indicating that the background produced by Cerenkov radiation was far greater than the dark current of the PMT.

A separate study tested the linearity of a PSD built with BC-430 under ^{60}Co irradiation (Beddar 1994). In addition to confirming the linearity of PSDs, this detector proved sensitive enough to measure the end effect (sometimes called timer error) that results from the finite transition time of the radioactive source in a

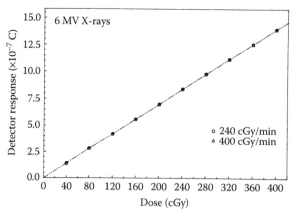

Figure 4.14 Signal linearity of the scintillation detector system for a 6 MV X-ray beam at two discrete dose-rates. (Reproduced from Beddar, A.S. et al., Water-equivalent plastic scintillation detectors for high-energy beam dosimetry: II. Properties and measurements, *Phys. Med. Biol.*, 37, 1901–1913, 1992b. With permission from IOP Publishing.)

Basic principles and theory

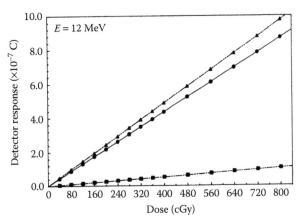

Figure 4.15 Signal linearity of the scintillation detector system for a 12 MeV electron beam. Plotted are the *signal* fiber guide response (▲), the *background* fiber guide response (■), and the net system response (●). (Reproduced from Beddar, A.S. et al., Water-equivalent plastic scintillation detectors for high-energy beam dosimetry: II. Properties and measurements, *Phys. Med. Biol.*, 37, 1901–1913, 1992b. With permission from IOP Publishing.)

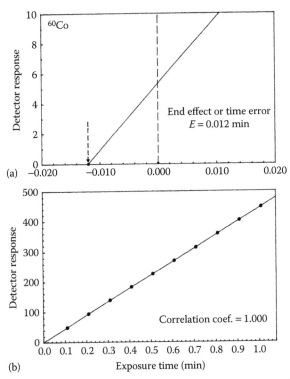

Figure 4.16 Sensitivity of the scintillator detector system showing (a) its ability in the determination of end effects (or timer errors) and (b) its linear response as a function of exposure time corrected for end effect. (Reproduced from Beddar, A.S., A new scintillator detector system for the quality assurance of ^{60}Co and high-energy therapy machines, *Phys. Med. Biol.*, 39, 253–263, 1994. With permission from IOP Publishing.)

cobalt unit (Figure 4.16). The measured end effect was compared with an ion chamber measured value and found to be in good agreement.

Carrasco et al. (2015) established the dose linearity for a commercial PSD (the Exradin W1) exposed to doses of 15–600 cGy. They observed no deviations from linearity greater than 0.5% except at the lowest doses where it was slightly larger, presumably due to a low signal-to-noise ratio. As a result of the measurement at low doses, the root mean square (RMS) deviation from linearity over the entire range of doses measured was 0.6% with an uncertainty of ±0.2%.

4.3.4 DOSE RATE PROPORTIONALITY

Ideally, a detector should be independent of the dose rate as well as the overall dose. The dose rate response of the original PSD was tested by subjecting it to irradiations by 6 and 10 MV beams using dose rates between 80 and 400 cGy/s (Beddar et al. 1992b). A Farmer-type ionization chamber was used to verify the dose rates. As displayed in Figure 4.17, the dose rate normalized detector response did not vary more than 0.5% for any dose rate, indicating that the detector was dose rate independent. A similar setup was used to establish the dose rate independence of the PSD under electron irradiation for the same range of dose rates (Figure 4.18).

Carrasco et al. (2015) established dose rate independence for a commercial PSD by irradiating it with 6 and 15 MV photon beams for a fixed setup at dose rates from 100 to 600 MU/min in 100 MU/min intervals. An RMS deviation of 0.53% in the detector response was measured when considering all dose rates. The linear accelerator used for this study produced different dose rates by varying the repetition rate of dose pulses rather than the dose per pulse. As a result, the dose rate on a very short timescale (i.e., during a pulse) does not change when the overall dose rate is changed. Therefore, the group further investigated dose rate independence by irradiating the PSD at a variety of source-to-surface distances for two different buildup materials, guaranteeing a varied instantaneous dose rate caused by an inverse square falloff and differential attenuation. Any deviation from inverse square fall off in the measured output was assumed to indicate a deviation from dose rate independence. An RMS deviation of 0.38% in the detector response from the inverse square law was measured, further establishing the dose rate independence of the detector.

4.3.5 SPATIAL RESOLUTION

High spatial resolution is required to accurately measure dose distributions with steep gradients such as those delivered in stereotactic radiosurgery. Scintillators can be shaped arbitrarily and possess high sensitivity such that very small volumes can produce usable signal. PSDs can therefore be constructed with excellent spatial resolution.

By way of example, Figure 4.19 displays the measured beam profiles of a half-blocked photon beam measured with two Farmer-type ion chambers (0.6 and 0.1 cm³ active volumes, respectively), a film, and a scintillator of 2.5 mm in diameter and 4.0 mm in length (an active volume of 0.02 cm³). The longitudinal axis of the scintillator was placed orthogonally to both the beam central axis and the scanning direction to minimize its profile. As can be seen in the figure, the scintillator data agrees well with the film due to minimal volume averaging. Figure 4.20 displays transverse profiles for three different field sizes of a 6 MV photon beam measured with a 0.1 cm³ active volume Farmer-type chamber, a diode, and a scintillator

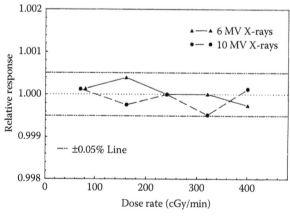

Figure 4.17 Dose rate independence of the scintillation detector system for X-rays, normalized to 240 cGy/min. (Reproduced from Beddar, A.S. et al., Water-equivalent plastic scintillation detectors for high-energy beam dosimetry: II. Properties and measurements, *Phys. Med. Biol.*, 37, 1901–1913, 1992b. With permission from IOP Publishing.)

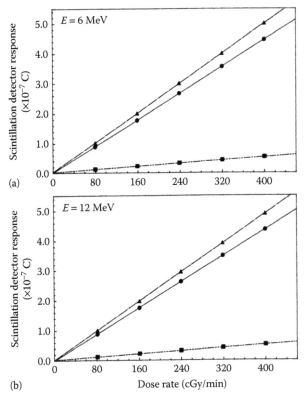

Figure 4.18 Dose rate proportionality of the scintillation detector system for (a) 6 MeV electrons and (b) 12 MeV electrons. Plotted are the *signal* fiber guide response (▲), the *background* fiber guide response (■), and the net system response (●). (Reproduced from Beddar, A.S. et al., Water-equivalent plastic scintillation detectors for high-energy beam dosimetry: II. Properties and measurements, *Phys. Med. Biol.*, 37, 1901–1913, 1992b. With permission from IOP Publishing.)

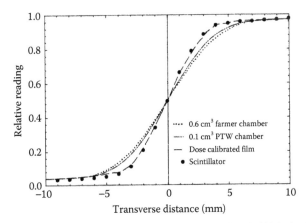

Figure 4.19 Spatial resolution determination for radiation detectors. A ^{60}Co half-blocked beam profile measured in water using scintillation detector and compared to two ion chambers and film. (Reproduced from Beddar, A.S. et al., Water-equivalent plastic scintillation detectors for high-energy beam dosimetry: II. Properties and measurements, *Phys. Med. Biol.*, 37, 1901–1913, 1992b. With permission from IOP Publishing.)

of 1 mm in diameter and 4.0 mm in length (an active volume of 0.03 cm^3). The scintillator and diode measurements agree well and show steeper falloff of the dose profile than the ion chamber due to minimal volume averaging.

A study by Klein et al. (2010) highlights the excellent spatial resolution of PSDs. In this study, a PSD containing 2 mm of SCSF-3HF scintillating fiber with a diameter of 0.5 mm was used to measure output

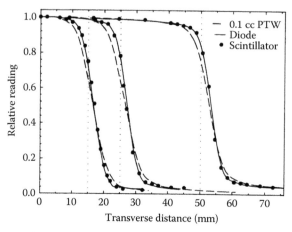

Figure 4.20 Transverse dose profiles of a 6 MV X-ray beam for a 10 × 10, 5 × 5, and 3 × 3 cm² field sizes using the scintillation detector, an ion chamber, and a photon diode. (Reproduced from Beddar, A.S. et al., Water-equivalent plastic scintillation detectors for high-energy beam dosimetry: II. Properties and measurements, *Phys. Med. Biol.*, 37, 1901–1913, 1992b. With permission from IOP Publishing.)

factors of small fields with side lengths falling between 0.5 and 10 cm. The results were compared to output factors measured with a CC01 ion chamber, which has an active volume of 0.01 cm³. The PSD agrees excellently with the CC01 for field sizes larger than 1 × 1 cm². For smaller fields, the PSD reports slightly larger output factors. The reason for this is the small size of the PSD active volume resulting in less volume averaging. For such small fields, there is no *flat region*, and therefore, volume averaging of measurements on the central axis will necessarily result in a lower measured dose (and output factor).

4.3.6 TEMPERATURE DEPENDENCE

Temperature dependence is an important consideration for any detector, and especially for a detector used for *in vivo* dosimetry. Such a detector would likely be calibrated at or near room temperature, and used closer to body temperature. For this reason, Beddar et al. (1992a) investigated the temperature dependence of the original PSD. To do so, the detector probe was immersed in a water-filled beaker and irradiated with a 4 MV beam at temperatures between 0°C and 50°C. A waiting period of 15 min prior to each irradiation ensured that the probe was in thermal equilibrium with the water surrounding it. A negligible increase in response was observed over the temperatures studied, with all measurements falling within ±0.5% of the central value. Furthermore, the response was constant with ±5°C of room temperature (22°C). Flühs et al. (1996) performed a similar experiment with NE102A, a commercial equivalent of BC-400, and also did not observe significant changes between 15°C and 45°C.

After the publication of these results, there were no additional studies initiated by other groups to investigate or independently corroborate the finding. However, although Wootton et al. (2014) were performing a study using PSDs for *in vivo* dosimetry as part of a patient clinical protocol, they observed a consistent systematic under-response of their detectors when used *in vivo* but not otherwise (Beddar 2012). After failing to find any other explanation, they investigated the possibility of temperature dependence. They found that PSDs built with BCF-60 and BCF-12 scintillating fibers under-responded, respectively, 0.5% and 0.1% per °C relative to room temperature (Wootton and Beddar 2013). A plot of the normalized measured dose as a function of temperature is displayed in Figure 4.21. The methodology used was similar to that of Beddar et al. (1992a): PSDs were fixed in a water-filled beaker and irradiated with a 6 MV photon beam at different temperatures. This study also found that the quantity of Cerenkov light produced in PMMA optical fiber was not influenced by temperature, although the transmission of light through cyanoacrylate, an optical coupling agent, was temperature dependent. Finally, the measured spectral distribution of light from these scintillating fibers was found to change with temperature. The results of this study indicate that temperature independence was a specific property of the original PSD, and not a general property of PSDs.

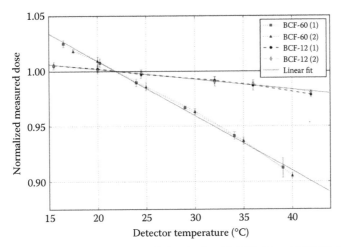

Figure 4.21 Temperature dependence of two pairs of PSDs built with BCF-60 and BCF-12. (Reproduced from Wootton, L. and Beddar, S., Temperature dependence of BCF plastic scintillation detectors, *Phys. Med. Biol.*, 58, 2955–2967, 2013. With permission from IOP Publishing.)

Buranurak et al. (2013) independently investigated the temperature dependence of scintillation detectors as well. Their study used a thermoelectric cooler plate to maintain detectors at various temperatures and irradiated them with a 50 kV X-ray source. BCF-12 and BCF-60 were investigated and found to exhibit temperature dependence in agreement with those found by Wootton et al. They additionally studied two other scintillating fibers produced by Industrial Fiber Optics (model 81-0083, a yellow-emitting scintillator, and model 81-0087, a red-emitting scintillator), and found both to exhibit temperature dependence distinct in magnitude from BCF-60, BCF-12, and each other.

Other groups have since reported temperature dependence for other PSDs, including a commercial PSD, the Exradin W1 (Carrasco et al. 2015, Stefanowicz et al. 2013). Thus, when using PSDs, it is important to note that the detector may have nonnegligible temperature dependence depending on the scintillator material and the implementation of the detector. However, to date the effect has not been observed to be greater than approximately 0.5% per °C, so small variations in temperature are unlikely to result in drastic deviations in measured dose.

4.3.7 PRESSURE DEPENDENCE

Ion chambers exhibit pressure dependence requiring correction because their active volume is gaseous. Therefore, an increase in ambient pressure caused by natural atmospheric conditions will alter the density of the air in the active volume of an unsealed ion chamber, resulting in more or less ionization. Scintillators, however, should not be affected by small variations in pressure as they are solid. Although this has not been explicitly studied for PSDs, there has been to date no data contradicting this assumption.

4.3.8 RADIATION DAMAGE

Exposure of a PSD to ionizing radiation can result in (1) damage to the detector base or associated optical conduits that result in transmission loss and (2) damage to the scintillating molecules embedded in the base rendering them temporarily or permanently incapable of producing light via scintillation. The overall effect manifests as a reduction in scintillation efficiency and a darkening of the plastic scintillator and optical fibers. Measuring the damage caused as a function of absorbed dose is therefore useful for establishing expectations of how often detectors may need to be recalibrated and/or replaced in order to maintain an acceptable level of accuracy.

The first inquiry into the effect of radiation damage on PSDs was performed by Beddar et al. (1992a) by exposing the original PSD to a cumulative dose of 10 kGy using a cesium irradiator at a rate of 7.31 Gy/min. The detector response was found to have decreased by 2.8% at the conclusion of the

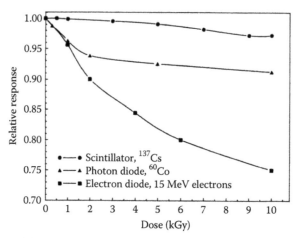

Figure 4.22 Radiation-induced damage to the PSD outlined by Beddar et al. By comparison, the response of two other detectors is plotted as well. The figure legend indicates the source of radiation for each detector. (Reproduced from Beddar, A.S. et al., Water-equivalent plastic scintillation detectors for high-energy beam dosimetry: I. Physical characteristics and theoretical considerations, *Phys. Med. Biol.*, 37, 1883–1900, 1992a. With permission from IOP Publishing.)

irradiation, indicating that PSDs are relatively resistant to radiation damage in comparison with solid-state detectors such as diodes. The detector response is plotted as a function of dose alongside the responses of two diodes in Figure 4.22.

Another study used a PSD constructed with BC-430 and a polycarbonate light pipe rather than silica fiber (Beddar 1994). Approximately 1 kGy was delivered during the course of testing the stability, linearity, and daily and weekly reproducibility of the detector. An additional 9 kGy was then delivered using a ^{60}Co unit to reach a total exposure of 10 kGy for comparison with the results generated with the original PSD. After the first 1 kGy, the response was found to have decreased by 2.1%. The response fell by 7.6% after the full 10 kGy had been delivered. Several possible explanations are proposed for the greater change in response in this PSD relative to the original PSD. The first is that the polycarbonate light pipe is more susceptible to radiation damage than the radiation-resistant optical fiber used in the previous study. Therefore, more of the light produced by the scintillator was attenuated by the light pipe, reducing the system response. A second cause of the discrepancy may have been the use of a silicon photodiode to quantify scintillation light; this photodiode was in close proximity to the radiation field, and therefore may have suffered some radiation damage, causing a loss of sensitivity. The authors also noted that the last 9 kGy was delivered in one session, allowing no time for the detector to recover from temporary damage. Thus, these results are more severe than would be expected for a detector exposed to 10 kGy over the course of an extended period in smaller daily fractions.

Radiation damage effects were also assessed for three commercial PSD systems (all three were DCT-444s, where DCT stands for *daily constancy tool*) designed to monitor daily output constancy of medical linear accelerators (Das et al. 1996). Long-term stability studies on DCT-444 conducted over a period of 25 months showed visible reduction in the detector signal output suggesting radiation damage. The response of one DCT-444 device irradiated with a 6 MV beam exhibited a signal loss of approximately 1% per month when exposed to 100 Gy per month. The other two devices exhibited a much smaller loss of signal but were exposed to a correspondingly smaller monthly dose.

Finally, Carrasco et al. (2015) studied the effect of radiation damage on a commercial PSD (the Exradin W1) as part of a comprehensive characterization of the detector. They found that exposing the detector to 10 kGy resulted in a 2.8% sensitivity loss, in exact agreement with the results from Beddar et al.'s 1992a study. However, they continued to irradiate the detector to approximately 125 kGy. The rate of sensitivity loss decreased as the cumulative dose increased, indicating that PSDs gain radiation resistance with increasing cumulative exposure, similarly to diodes. The commercial PSD exhibited a total 8% loss of sensitivity after irradiation by 125 kGy.

These studies collectively demonstrate that PSDs are relatively radiation resistant in comparison with other detectors such as diodes and metal–oxide–semiconductor field–effect transistors (Jornet et al. 2004, Mijnheer et al. 2013, Yorke et al. 2005). The exact radiation damage characteristics will depend on the design of the PSD, in particular the materials used to construct it and the usage pattern.

4.3.9 MAGNETIC FIELD EFFECTS ON PSDs

Literature from the field of high-energy physics indicates that many scintillators will experience a response-altering effect in the presence of strong magnetic fields. Although PSDs are not typically exposed to strong magnetic fields, if the combined magnetic resonance imaging and radiotherapy units now under development become common, this effect will be a relevant concern.

Balalykin et al. (1997) studied the response of NE-110 (a blue plastic scintillator with a polyvinyltoluene base) to alpha and electron irradiation under a magnetic field that varied between 0 and 3.8 T. They found a nonnegligible loss of signal for fields greater than 1 T. Blömker et al. (1990) studied the response of three plastic scintillators, including NE-102A (a blue plastic scintillator that is a commercial equivalent of BC-400) under proton, low-energy X-ray, and ultraviolet (UV) light irradiation under a magnetic field that varied between 0 and 0.45 T. The scintillator response increased a few percentages as the magnetic field increased in strength for both forms of ionizing radiation. However, they found that the response to UV light, which bypasses the base scintillator and excites secondary fluors directly, did not exhibit a magnetic field-dependent result. This is in agreement with the findings of Veloso et al. (2001) and suggests that excitation/de-excitation or transfer of energy from the base molecule is responsible for the effect. Finally, Green et al. (1995) studied the response of polystyrene-based scintillators in magnetic fields up to 1.5 T irradiated with electrons, and found an increase of a few percentages in the response of the scintillator with increasing field strength.

Unfortunately, these studies are not directly applicable to PSDs for several reasons. For one thing, most used large volumes of scintillator (blocks on the order of 1 to tens of centimeters in each dimension) with no buildup. Furthermore, the irradiating particles were not representative of those encountered in radiotherapy, being mostly electrons with less than 1 MeV of energy. The only photon irradiation used an isotope producing 5.9 and 6.5 keV X-rays. To date, there has been only one study of PSDs irradiated with therapeutic radiation under a magnetic field (Stefanowicz et al. 2013). In that study, PSDs were constructed with 5 mm of BCF-12 or BCF-60 scintillating fiber coupled to PMTs with long optical fibers. The detectors were placed in a PMMA phantom between the poles of an electromagnet and irradiated with a ^{60}Co source. The group found an increase in the PSD response with increasing magnetic field strength, resulting in a 5%–10% increased signal at 1 T. In light of this result, further studies are warranted to conclusively confirm and thoroughly characterize the effect of magnetic fields on PSDs.

4.4 CONCLUSION

PSDs possess a unique set of characteristics that have been outlined in this chapter, along with some of the groundwork establishing these characteristics. The exact nature of some attributes, such as reproducibility, radiation damage effects, and temperature dependence, will depend on the specifics of a given PSD (i.e., what materials and photodetector were used) but will adhere generally to what was presented in this chapter. These characteristics make PSDs suitable for unique applications, many of which are outlined in the following chapters.

REFERENCES

Archambault, L, Beddar, AS, Gingras, L, Roy, R, and Beaulieu, L. 2005. Measurement accuracy and Cerenkov removal for high performance, high spatial resolution scintillation dosimetry. *Medical Physics*. 33: 128–135.

Archambault, L, Briere, T, Pönisch, F et al. 2010. Toward a real-time *in vivo* dosimetry system using plastic scintillation detectors. *International Journal of Radiation Oncology, Biology, Physics*. 78: 280–287.

Attix, FH. 1986. *Introduction to Radiological Physics and Radiation Physics*. New York: Wiley.

Ayotte, G, Archambault, L, Gingras, L, Lacroix, F, Beddar, AS, and Beaulieu, L. 2006. Surface preparation and coupling in plastic scintillator dosimetry. *Medical Physics*. 33: 3519–3525.

Balalykin, NI, Golutvin, IA, Kochetov, OI et al. 1997. Measurement of the plastic scintillator response in the magnetic field. *Nuclear Instruments and Methods in Physics Research A*. 390: 286–292.

Beddar, AS. 1994. A new scintillator detector system for the quality assurance of ^{60}Co and high-energy therapy machines. *Physics in Medicine and Biology*. 39: 253–263.

Beddar, AS. 2012. On possible temperature dependence of plastic scintillator response. *Medical Physics*. 39: 6522.

Beddar, AS, Mackie, TR, and Attix, FH. 1992a. Water-equivalent plastic scintillation detectors for high-energy beam dosimetry: I. Physical characteristics and theoretical considerations. *Physics in Medicine and Biology*. 37: 1883–1900.

Beddar, AS, Mackie, TR, and Attix, FH. 1992b. Water-equivalent plastic scintillation detectors for high-energy beam dosimetry: II. Properties and measurements. *Physics in Medicine and Biology*. 37: 1901–1913.

Beierholm, AR, Behrens, CF, and Andersen, CE. 2014. Dosimetric characterization of the Exradin W1 plastic scintillator detector through comparison with an in-house developed scintillator system. *Radiation Measurements*. 69: 50–56.

Berger, MJ and Seltzer, SM. 1983. Stopping powers and ranges of electrons and positrons. National Bureau of Standards Information Report 82-2550-A. Washington, DC: National Bureau of Standards.

Blömker, D, Holm, U, Klanner, R, and Krebs, B. 1990. Response of plastic scintillators in magnetic fields. *IEEE Transactions on Nuclear Science*. 37: 220–224.

Buranurak, S, Andersen, CE, Beierholm, AR, and Lindvold, LR. 2013. Temperature variations as a source of uncertainty in medical fiber-coupled organic plastic scintillator dosimetry. *Radiation Measurements*. 56: 307–311.

Burlin, TE. 1966. A general theory of cavity ionization. *The British Journal of Radiology*. 39: 727–734.

Carrasco, P, Jornet, N, Jordi, M et al. 2015. Characterization of the Exradin W1 scintillator for use in radiotherapy. *Medical Physics*. 42: 297–304.

Das, IJ, Gazda, MJ, and Beddar, AS. 1996. Characteristics of a scintillator-based daily quality assurance device for radiation oncology beams. *Medical Physics*. 23: 2061–2067.

Fontbonne, JM, Iltis, G, Ban, G et al. 2002. Scintillating fiber dosimeter for radiation therapy accelerator. *IEEE Transactions on Nuclear Science*. 49: 2223–2227.

Francescon, P, Beddar, S, Satariano, N, and Das, IJ. 2014. Variations of $k_{Q_{clin}Q_{msr}}^{f_{clin}f_{msr}}$ for the small-field dosimetric parameters percentage depth dose, tissue-maximum ratio, and off axis ratio. *Medical Physics*. 41: 101708 1–14.

Flühs, D, Heintz, M, Indenkämpen, F, Wieczorek, C, Kolanoski, H, and Quast, U. 1996. Direct reading measurement of absorbed dose with plastic scintillators—The general concept and applications to opthalmic plaque dosimetry. *Medical Physics*. 23: 427–434.

Frelin, AM, Fontbonne, JM, Ban, G et al. 2005. Spectral discrimination of Čerenkov radiation in scintillating dosimeters. *Medical Physics*. 32: 3000–3006.

Green, D, Ronzhin, A, and Hagopian, V. 1995. Magnetic fields and scintillator performance. Fermilab-TM-1937.

Jornet, N, Carrasco, P, Jurado, D, Ruiz, A, Eudaldo, T, and Ribas, M. 2004. Comparison study of MOSFET detectors and diodes for entrance in vivo dosimetry in 18 MV x-ray beams. *Medical Physics*. 31: 2534–2542.

Klein, DM, Tailor, RC, Archambault, L, Wang, L, Therriault-Proulx, F, and Beddar, AS. 2010. Measuring output factors of small fields formed by collimator jaws and multileaf collimator using plastic scintillation detectors. *Medical Physics*. 37: 5541–5549.

LaCroix, F, Beaulieu, L, Archambault, L, and Beddar, AS. 2010. Simulation of the precision limits of plastic scintillation detectors using optimal component selection. *Medical Physics*. 37: 412–418.

Mijnheer, B, Beddar, S, Izewska, J, and Reft, C. 2013. *In vivo* dosimetry in external beam radiotherapy. *Medical Physics*. 40: 070903 1–19.

Stefanowicz, S, Latzel, H, Lindvold, LR, Andersen, CE, Jäkel, O, and Greilich, S. 2013. Dosimetry in clinical static magnetic fields using plastic scintillation detectors. *Radiation Measurements*. 56: 357–360.

Veloso, JFCA, dos Santos, JMF, Conde, CAN et al. 2001. A driftless gas proportional scintillation counter for muonic hydrogen X-ray spectroscopy under strong magnetic fields. *Nuclear Instruments and Methods in Physics Research A*. 460: 297–305.

Wootton, L and Beddar, S. 2013. Temperature dependence of BCF plastic scintillation detectors. *Physics in Medicine and Biology*. 58: 2955–2967.

Wootton, L, Kudchadker, R, Lee, A, and Beddar, S. 2014. Real-time *in vivo* rectal wall dosimetry using plastic scintillation detectors for patients with prostate cancer. *Physics in Medicine and Biology*. 59: 647–660.

Yorke, E, Alecu, R, Ding, L et al. 2005. Diode in vivo dosimetry for patients receiving external beam radiation therapy. Report of Task Group 62 of the Radiation Therapy Committee. AAPM Report No. 87. Madison, WI: Medical Physics Publishing.

5

Čerenkov and its solutions

Luc Beaulieu, Jamil Lambert, Bongsoo Lee, and Wook Jae Yoo

Contents

5.1 INTRODUCTION

When the absorbed dose for high-energy radiation beams is measured by using a plastic scintillation detector (PSD), most of the light reaching the light-measuring device is scintillating light emitted from the scintillator. However, a portion of the light is a result of the stem effect within the optical fiber itself and it has an adverse influence on the real-time measurement of the pure scintillation signal as a noise component.[1–5] The main components causing the stem effect in the optical fiber during beam irradiation are fluorescence and Čerenkov radiation or a combination of these two noise components, as described previously in Chapter 1. First, fluorescence is the emission of light stimulated in an optical fiber simultaneously with the absorption of incident radiation, and it has an average lifetime of nanoseconds range.[6] Next, Čerenkov radiation or light is produced by the direct action of charged particles passing through a transparent medium at a velocity greater than that of light in the same medium.[3,7,8] This phenomenon is akin of a supersonic airplane reaching Mach speed.

5.2 PHYSICS OF ČERENKOV RADIATION

Čerenkov radiation can be generated in the core of any fiber with a core refractive index greater than 1 and it is the main contributor to the stem effect. When an optical fiber is irradiated with external high-energy radiation beams, the charged particles (e.g., secondary electron) are produced by the interactions between the radiation and the core material. In case of therapeutic photon beams, the dominant interaction is Compton scattering and thus, the Čerenkov radiation is mainly produced by Compton electron.

If the energy of Compton electrons exceeds a specific minimal energy, Čerenkov radiation can be produced in the optical fiber and then be detected as unwanted noise signals. Here, this specific energy is called the *Čerenkov threshold energy* (CTE, E_{Th}), and it is the minimum electron energy that is necessary to generate Čerenkov radiation.

5.2.1 ČERENKOV THERSHOLD ENERGY

The above-mentioned threshold condition is related to the relative phase velocity of a charged particle (β) and the refractive index of the fiber (n) and is given as

$$\beta = \frac{v}{c} = \frac{1}{n} \tag{5.1}$$

For relativistic equations, β is also related to the kinetic energy of the electron and can be obtained as

$$\beta = \sqrt{1 - \left(\frac{m_0 c^2}{E + m_0 c^2}\right)^2} \tag{5.2}$$

where:
  m_0 is the rest mass of an electron
  c is the velocity of light in a vacuum

Accordingly, $m_0 c^2$ is the rest mass energy of an electron and equal to 511 keV.[7,9–13] Thus, the minimum particle energy necessary to produce Čerenkov radiation can then be obtained from Equation 5.2

$$E_{Th} = m_0 c^2 \left[\frac{1}{\sqrt{(1-\beta^2)}} - 1\right] \tag{5.3}$$

or in term of the refractive index n using Equation 5.1, Equation 5.3 is now expressed as

$$E_{Th} = m_0 c^2 \left[\frac{1}{\sqrt{\left(1 - (1/n^2)\right)}} - 1\right] \tag{5.4}$$

From Equation 5.4, one can relate the production threshold of various media ranging from about 100 keV for a refractive index of 1.82–400 keV for an index of 1.21, as depicted in Figure 5.1. As a comparison, plastic clear fiber refractive indices are usually between 1.49 and 1.59 leading to threshold energies between 178 keV (1.49) and 146 keV (1.59). For water ($n = 1.33$), E_{min} is higher at 260 keV. This further means that all megavoltage beams used in external beam radiation therapy and proton therapy will produce Čerenkov radiation from high-energy electrons. To a lesser extent, Čerenkov radiation will also be produced in ^{192}Ir-based brachytherapy and at the high-energy range of orthovoltage treatments (at least in plastics).

5.2.2 ČERENKOV EMISSION ANGLE

Considering an electron traveling from A to B as in Figure 5.2, the Čerenkov radiation is emitted at an angle (θ) with respect to the direction of the electron in the transparent medium that is related to the ratio of the vectors $\overrightarrow{AC}$ to $\overrightarrow{AB}$. From Equations 5.1 and 5.2, the Čerenkov emission angle is given as

$$\cos\theta = \frac{\overrightarrow{AC}}{\overrightarrow{AB}} = \frac{(C/n)\Delta t}{\beta C \Delta t} = \frac{(C/n)\Delta t}{v \Delta t} = \frac{1}{\beta n} = \frac{c}{nv} \tag{5.5}$$

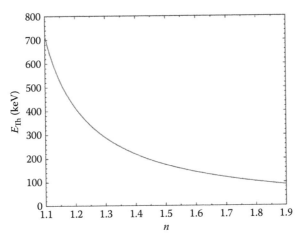

Figure 5.1 Threshold energy (E_{Th}) for Čerenkov light emission as a function of the refraction index n using Equation 5.4.

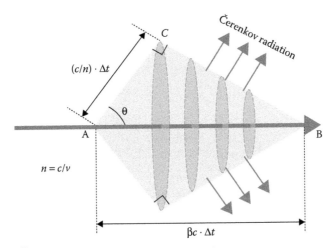

Figure 5.2 Production of Čerenkov radiation in a transparent medium.

$$\theta = \cos^{-1}\left[\frac{1}{n\sqrt{1-\left((1/((E/m_0c^2)+1))\right)^2}}\right] \quad (5.6)$$

where v is velocity of light in the medium and can be obtained using relativistic equations as follows:

$$v = \beta c = c\sqrt{1-\left[\frac{1}{(E/m_0c^2)+1}\right]^2} \quad (5.7)$$

Chapter 21 further describes the energy thresholds and emission angles in various mediums. Because an optical fiber has a critical angle for light transmission (*cf.* Chapter 3), the Čerenkov radiation generated in an optical fiber is therefore dependent on the irradiation angle of the high-energy radiation beam over the CTE. Using Equation 5.6, the angle of maximum Čerenkov emission can be theoretically determined as about 47.8° when the energy of an incident electron beam and the refractive index of poly(methyl methacrylate) as a core material at 6 MeV and 1.49, respectively.[10,14] The effect of fiber-beam angle has been well characterized experimentally early on[3] and measurements close to the critical angle as well as at 90° have been used to separate fluorescence emission from that of Čerenkov[15] (Figure 5.3). An in-depth study of these two modes of emission from clear plastic fibers is also available in the literature.[16]

More recently, Yoo et al. used the same approach to describe the angular dependence of the pure scintillation and Čerenkov signals obtained by using an integrated PSD system based on a subtraction method (Figure 5.4).[17] By rotating the treatment head of clinical Linac (Clinac® 21EX, Varian Medical Systems, Palo Alto, CA) from 0° to 90°, they obtained two light signals: a mixed light signal, including both scintillating light and Čerenkov radiation, and a Čerenkov signal, which are changed with angular variation. In their experiment, the peak value of the Čerenkov signal curve is found at an angle of about 47°, close to the theoretical value of 47.8°. However, the pure scintillation signals obtained by subtraction method have almost uniform values regardless of the irradiation angle.

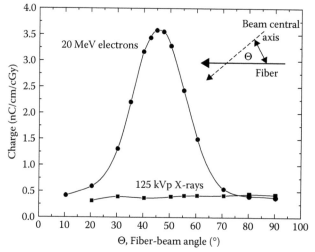

Figure 5.3 Light measurements in a clear optical fiber as a function of the fiber-beam angle for a 20 MeV electron beam and 125 kVp X-ray beam. (From de Boer, S.F. et al., *Phys. Med. Biol.*, 38, 945, 1993. With permission.)

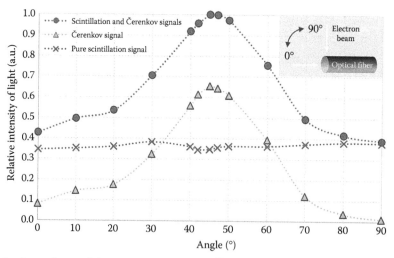

Figure 5.4 Angular dependence of the pure scintillation and Čerenkov signals.

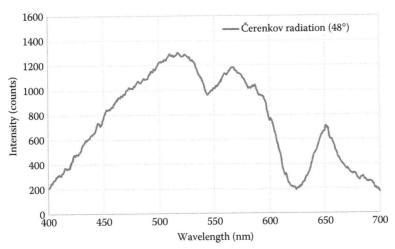

Figure 5.5 Spectrum of Čerenkov radiation transmitted through a plastic optical fiber induced by a 6-MV photon beam.

5.2.3 ČERENKOV WAVELENGTH DISTRIBUTION

The last important point about Čerenkov radiation is its spectral distribution. The number of emitted photon of a given wavelength per unit path length is given by the Frank–Tram equation[18] as

$$\frac{dN}{dx} = 2\pi\alpha z^2 \left(1 - \frac{1}{\beta^2 n^2}\right)\frac{1}{\lambda^2}\,d\lambda \tag{5.8}$$

where:
   z is the particle charge
   α is the fine structure constant

Therefore, the wavelength distribution of Čerenkov radiation is very broad and while it encompassed the entire visible light spectrum, it is predominantly in the ultraviolet (UV)-blue spectral region, decreasing as λ^{-2} as the wavelength increases going into the red portion of the spectrum. If one is interested in the light energy produced instead of the number of photons, then the intensity of Čerenkov radiation becomes proportional to λ^{-3}, enhancing even more the contribution in the UV-blue region.

Čerenkov production is further a function of the irradiated length and the square of the irradiated radius of the fiber.[19] Figure 5.5 shows the spectrum of Čerenkov radiation transmitted via a SI MMF (GH4001, Mitsubishi Rayon, New York) induced by a 6-MV photon beam. The peak wavelength is found at about 515 nm and the original wavelength of the Čerenkov radiation might be affected according to the inherent transmission properties of the fiber. Čerenkov radiation can cause problems as an unwanted light signal and interfere with a dose measurement using the scintillating light. Therefore, various methods have been reported to measure the pure scintillation signal without fluorescence or Čerenkov radiation signal. These will be discussed in the next section.

5.3 ČERENKOV REMOVAL TECHNIQUES

While the amount of Čerenkov light produced and transmitted by mm of clear fiber (a few tens of photons) is small relative to that of scintillators (thousands per MeV of energy deposited), the total length of collecting fiber used is large compared to the scintillation probes, usually a few mm at most. If irradiation conditions were exactly the same at all time, Čerenkov light would simply be a constant signal added to the scintillation light that would be taken into account in the calibration process. However, irradiation conditions will depend on applications, and the amount of clear fiber exposed to the beam will vary from a few cm to as much as 40 cm.[3,20] Thus, Čerenkov light will also vary and can account for up to 25%–30%

of the total light in certain configurations.[21] It is therefore crucial that the Čerenkov signal, which can be considered the equivalent of a stem effect for the purpose of scintillation dosimetry, be removed for accurate measurements. The following subsections will address the most common approaches implemented for the removal of the Čerenkov stem-effect.

5.3.1 TWO-FIBERS METHOD

The very first approach proposed for accurate scintillation dosimetry has been a subtraction method built around a balanced light detection system composed of two photomultiplier tubes: one detecting the signal produced by a scintillation element and the stem signal from the transport fiber, and the other measuring only the signal from a second transport fiber that is exactly the same as the first one but without the scintillating element (Figure 5.6). Thus, in this approach set forth in the seminal twin papers of Beddar et al.[1,2], the Čerenkov signal is effectively subtracted as long as both fibers see the same irradiation conditions; this is the case for most standard field configurations. This design has been used with no or little change by multiple authors over the years. It has further been used for small field output factor measurements using a configuration in which both fibers are on the top of each other (instead of side by side).[22] However, in the presence of large dose gradients, such as those seen in the smallest field radio surgery beams or penumbra measurements, a differential reading from both fibers are expected and constitute the main limitation of the method.[21]

5.3.2 SIMPLE OPTICAL FILTERING

Since the Čerenkov light is mainly emitted in the violet–blue region of the visible spectrum, de Boer et al.[15] tested the possibility of filtering out the Čerenkov light from the scintillation light. To that end, numerous scintillator-filter combinations were tested. Reductions of 50% and more of the Čerenkov light contamination were observed when scintillators with emission in the green or orange wavelength part of the spectrum are used.[15] A similar approach was taken by Clift et al.[23] In both the cases, significant residual contribution of Čerenkov light remains un-subtracted. As such, the simple optical filtering approach is inferior to the two-fiber method for accurate dose measurements.

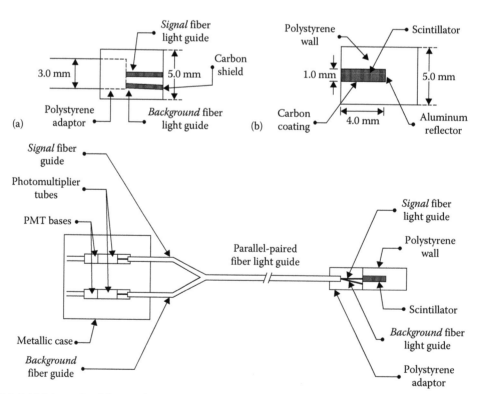

Figure 5.6 (a,b) Schematic of the two-fiber subtraction technique first proposed by Beddar et al. (From Beddar, A.S. et al., *Phys. Med. Biol.*, 37, 1883, 1992; Beddar, A.S. et al., *Phys. Med. Biol.*, 37, 1901, 1992. With permission.)

5.3.3 TIME FILTERING

In the search for a measurement technique that involve only a single fiber, bypassing the key limitation of the two-fiber method, Clift et al. proposed a second method called *time filtering* or *timing*.[4] In that work, a scintillator (BG-444G) with a long decay constant, about 20 times longer than standard plastic scintillator (264 ns instead of 10–15 ns), was used. Since Čerenkov light production is correlated to the beam pulses and possesses a decay time that is faster than the chosen scintillator, a reading window can be appropriately timed, so that the Čerenkov stem signal is decayed-out leaving only the scintillation signal. The results of this method applied to the measurements of a 16 MeV electron beam are shown in Figure 5.7 in term of a depth dose measurement (CFR = Čerenkov and fluorescence radiation), which is compared to the two-fiber method (BC-428-CFR). This method was shown to remove 99.9% of the stem effect while scarifying 44% of the scintillation signal.[4] Time filtering is thus a major improvement over the simple optical filtering. However, the technical aspect involved in the timing circuit was seen as a drawback at the time and made the overall system more complex than the two-fiber method.

5.3.4 CHROMATIC REMOVAL

In 2002, Fontbonne et al. recognized that in making a measurement with a PSD, one effectively get a linear superposition of at least two signals, that is, that of the scintillator and the Čerenkov light (neglecting fluorescence).[24] As such, from two experimental measurement conditions providing the same dose but in two different conditions for Čerenkov light production, it should be possible to *deconvolve* the two signals, provided that measurements can be made in two different spectral bands. Fontbonne et al. expressed this in term of the following matrix system

$$\begin{pmatrix} G_1 & B_1 \\ G_2 & B_2 \end{pmatrix} * \begin{pmatrix} k_g \\ k_b \end{pmatrix} = \begin{pmatrix} D_1 \\ D_2 \end{pmatrix} \tag{5.9}$$

where G_X and B_X correspond to measurements in the green and blue wavelength regions, respectively k are calibration factors D_X are the expected doses for measurement conditions $X = 1$ and 2, with the expectation that $D_1 = D_2$[21,25,26].

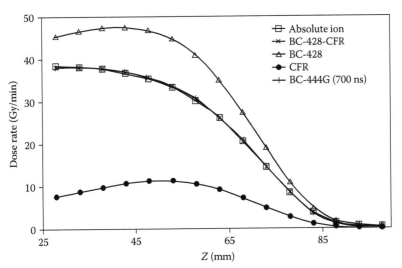

Figure 5.7 Depth dose measurements of a 16 MeV elecron beam. Comparison is made of a reference ion chamber, the two-fiber method (BC-428-CFR), and the time-filtering method (BC-444G). (From Clift, M.A. et al., *Phys. Med. Biol.*, 47, 1421, 2002. With permission.)

Once the calibration factors are obtained, dose can be computed from the green and blue channel measurements using this simple relation

$$D = k_g * G + k_b * B \qquad (5.10)$$

While blue and green spectral bands have been used in the early prototypes, the method is not limited to these two specific spectral bands. This approach eliminates the need for a second fiber or a complex timing circuit to achieve accurate dose measurements (output, depth dose curves, and profiles) of both photon and electron beams.[21,25–27] The burden of obtaining two simultaneous measurements into two wavelength windows is moved to the photodetector apparatus. This can be accomplished with the use of commonly available filters and dichroic mirrors[25] and using RGB detectors such as PIN diodes[24] or charge-coupled device cameras.[21]

The chromatic removal method is quite efficient and has been applied to extract experimentally effective point of measurements of ion chambers for electron beams,[28] silicon diodes,[29] and diode calibration factors for small field dosimetry.[30]

Potential limitation of this approach is related to the calibration conditions chosen to extract the calibration factors k. In fact, as long as the measurements conditions are bounded by two calibration conditions, the technique is expected to achieve a high level of accuracy. Chapters 6 and 10 will discuss this topic in greater details. In general, calibration for small field dosimetry (less than 5×5 cm^2) will be different than for larger fields.[30]

Finally, it was recognized recently that the chromatic removal technique could be embedded in a more general mathematical formulation that allows for both calibration condition involving more spectral bands or multiple scintillators measurement points read by a single coupling fiber, mPSD.[31] The latter is a significant development in the field of scintillation dosimetry that will be addressed in Chapter 10.

5.3.5 AIR CORE FIBERS

The use of an air core light guide to transmit the scintillation signal out of the radiation field can eliminate the generation of Čerenkov light itself. Air has a refractive index close to 1 and therefore there is no Čerenkov light generated by electrons in the MeV energy range.

In a solid core optical fiber, the light is guided through the core by total internal reflection at the boundary between the core and cladding, where the cladding has a lower refractive index than the core. Using total internal reflection for an air core light guide would require a cladding with a refractive index of less than 1, which is not possible under most situations for visible light. Therefore, to transmit light through an air core, a highly reflective surface inside the air core is required. This can be achieved with a multilayered dielectric coating, a microstructured layer or a metal surface with a usefully low attenuation coefficient.

5.3.5.1 Microstructured plastic optical fibers

Light can be guided through an air core by using a microstructured photonic bandgap material surrounding an air core. Air core light guides have been constructed using a pattern of air holes extending along the fiber parallel to a 5–100 μm air core.[32] The loss of these fibers is in the order of 60 dB/m for light with a wavelength of 500 nm. The attenuation loss is higher for larger core fibers and the current range of fibers have a limited maximum core size. This limited core size of the current range of microstructured fibers and the high loss in coupling to a solid core fiber make them unsuitable for scintillation dosimetry. Larger core sizes are currently being studied and may be an option should they show a sufficiently high transmission efficiency for visible light.

5.3.5.2 Silvered silica tube light guides

Another form of an air core light guide is a silica tube with the inner surface coated with a reflective metal or dielectric material. Silver-coated silica tubes have been developed for the transmission of high-power infrared laser light that cannot be transmitted through a silica or plastic fiber due to high attenuations at these wavelengths causing excessive heating.

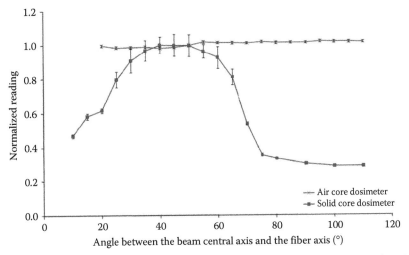

Figure 5.8 Light measurements from a solid core optical PSD (square) and air core PSD as a function of the fiber-beam angle for a 9-MeV electron beam. Measurement geometry is similar to that of Figures 5.3 and 5.4. (From Lambert, J. et al., *Phys. Med. Biol.*, 53, 3071, 2008. With permission.)

Marcatili and Schmeltzer[33] showed that the attenuation coefficient is proportional to $1/R$ and $1/r^3$, where R is the bend radius and r is the internal radius of the cylindrical air core light guide. Due to this dependence of the transmission loss on the bend radius, the hollow light guides used in scintillation dosimetry must be kept at a constant large bend radius to maintain a high and constant transmission efficiency.

The material that is used to provide a reflective coating on the inside of the tube is an important factor in the transmission properties of an air core light guide. The transmission properties of a gold-, silver-, and dielectric-coated hollow waveguide were studied by Matsuura et al.[34] A gold-coated waveguide was developed for use in oxidizing environments that has a lower loss than a silver-coated waveguide for infrared light. But the attenuation loss of the gold-coated waveguide increases for wavelengths shorter than 600 nm, and silver provides a better transmission efficiency. Matsuura and Miyagi[35] studied the transmission properties of a hollow waveguide with a silver or aluminum cladding and compared them with uncoated SiO_2 tubing. For the wavelength region between 400 nm and 700 nm, a cladding made from silver had the lowest loss per meter.

Lambert et al.[36,37] developed a scintillation detector based on a silvered silica tube with a 1 mm internal diameter and a 1×4 mm BC-400 scintillator. They showed that no Čerenkov or florescence was produced in the air core light guide, and the detector provided accurate depth dose measurements in external beam therapy. Figure 5.8 shows a comparison of a regular solid core PSD to an air core one under similar measurements condition presented in Figures 5.3 and 5.4. The detector readings agreed with ionization chamber measurements to within 1.6% in a 6 MV and a 18 MV photon beam. The detector also provided accurate depth dose measurements in 9 and 6 MeV electron beams, agreeing with the ionization chamber measurement to within 1% in most cases with a maximum deviation of 3.6%. The depth dose measurements in the 20 MeV electron beam under estimated the dose in the peak region with a deviation of up to 4.5% from the ionization chamber measurements, but was more accurate for depths larger than 20 mm. The detector was also shown to have potential in small field dosimetry with the measurements agreeing with a diamond detector for all field sizes measured, down to a minimum field size of 1×1 cm^2.

5.4 DOSIMETRY USING ČERENKOV RADIATION

In recent years, a Čerenkov fiber-optic dosimeter (CFOD) has been developed using only Čerenkov radiation generated in two identical optical fibers with different lengths to measure the depth dose distribution for therapeutic proton, photon, and electron beams, because the Čerenkov radiation is also one of the light signals generated from an optical fiber by radiation interactions.[10,17,38] In particular, it is expected that a CFOD can be used as a useful dosimeter to measure relative depth dose in proton therapy dosimetry, because it can measure real-time dose without a quenching effect at high stopping power.[3] Chapters 20 and 21 are covering in details more applications harnessing Čerenkov light for dosimetry purposes.

5.5 CONCLUSION

This chapter has presented the main physical aspect of Čerenkov light emission, in particular the threshold energy, emission angle, and emission wavelength. The issue related to Čerenkov as a stem-effect in PSD has been discussed, and an overview of the various techniques proposed over the years to correct or attenuate this effect is presented. While the two-fiber method is one of the simplest and most accurate approaches, it is not adapted to modern dosimetry needs, in particular for small field dosimetry and *in vivo* applications. The air core approach possesses all the necessary elements for small field dosimetry, it has however limitation in terms of physical flexibility for use in *in vivo* applications. The chromatic removal technique is, to date, the single approach that can cover all dosimetry applications.

In conclusion, it is important to state that Čerenkov light is not only an undesirable stem-effect, but can also be used for specific dosimetry purposes.

REFERENCES

1. Beddar, A. S., Mackie, T. R., and Attix, F. H. Water-equivalent plastic scintillation detectors for high-energy beam dosimetry: I. Physical characteristics and theoretical consideration. *Phys Med Biol* **37**, 1883–1900 (1992).
2. Beddar, A. S., Mackie, T. R., and Attix, F. H. Water-equivalent plastic scintillation detectors for high-energy beam dosimetry: II. Properties and measurements. *Phys Med Biol* **37**, 1901–1913 (1992).
3. Beddar, A. S., Mackie, T. R., and Attix, F. H. Čerenkov light generated in optical fibres and other light pipes irradiated by electron beams. *Phys Med Biol* **37**, 925–935 (1992).
4. Clift, M. A., Johnston, P. N., and Webb, D. V. A temporal method of avoiding the Čerenkov radiation generated in organic scintillator dosimeters by pulsed mega-voltage electron and photon beams. *Phys Med Biol* **47**, 1421–1433 (2002).
5. Hyer, D. E., Fisher, R. F., and Hintenlang, D. E. Characterization of a water-equivalent fiber-optic coupled dosimeter for use in diagnostic radiology. *Med Phys* **36**, 1711–1716 (2009).
6. Marckmann, C. J., Aznar, M. C., Andersen, C. E., and Bøtter-Jensen, L. Influence of the stem effect on radioluminescence signals from optical fibre Al2O3:C dosemeters. *Radiat Prot Dosimetry* **119**, 363–367 (2006).
7. Jelley, J. V. Čerenkov radiation and its applications. *British Journal of Applied Physics* **6**, 227 (1955).
8. Law, S. H., Suchowerska, N., McKenzie, D. R., Fleming, S. C., and Lin, T. Transmission of Čerenkov radiation in optical fibers. *Opt. Lett* **32**, 1205–1207 (2007).
9. Bell, Z. W. and Boatner, L. A. Neutron detection via the Cherenkov effect. *Nucl Sci* **57**, 3800–3806 (2010).
10. Jang, K. W. et al. Application of Čerenkov radiation generated in plastic optical fibers for therapeutic photon beam dosimetry. *J Biomed Opt* **18**, 027001 (2013).
11. Knoll, G. F. *Radiation Detection and Measurement*, 3rd edn. New York: John Wiley & Sons (1999).
12. Ross, H. H. Measurement of β-emitting nuclides using Čerenkov radiation. *Anal Chem* **41**, 1260–1265 (1969).
13. Yoo, W. J. et al. Development of a Čerenkov radiation sensor to detect low-energy beta-particles. *Appl Radiat Isoto* **81**, 196–200 (2013).
14. Lee, B. et al. Measurements and elimination of Cherenkov light in fiber-optic scintillating detector for electron beam therapy dosimetry. *Nucl Instrum Methods Phys Res A* **579**, 344–348 (2007).
15. de Boer, S. F., Beddar, A. S., and Rawlinson, J. A. Optical filtering and spectral measurements of radiation-induced light in plastic scintillation dosimetry. *Phys Med Biol* **38**, 945–958 (1993).
16. Therriault-Proulx, F., Beaulieu, L., Archambault, L., and Beddar, S. On the nature of the light produced within PMMA optical light guides in scintillation fiber-optic dosimetry. *Phys Med Biol* **58**, 2073–2084 (2013).
17. Yoo, W. J. et al. Simultaneous measurements of pure scintillation and Čerenkov signals in an integrated fiber-optic dosimeter for electron beam therapy dosimetry. *Opt Express* **21**, 27770–27779 (2013).
18. Tamm, I. E. and Frank, I. M. Coherent radiation of fast electrons in a medium. *Doklady Akad. Nauk SSSR* **14**, 107–112 (1937).
19. Jang, K. et al. Development of a fiber-optic Čerenkov radiation sensor to verify spent fuel: Characterization of the Čerenkov radiation generated from an optical fiber. *J Korean Phys Soc* **61**, 1704–1708 (2012).
20. Beddar, A. S., Suchowerska, N., and Law, S. H. Plastic scintillation dosimetry for radiation therapy: Minimizing capture of Čerenkov radiation noise. *Phys Med Biol* **49**, 783–790 (2004).

21. Archambault, L., Beddar, A. S., Gingras, L., Roy, R., and Beaulieu, L. Measurement accuracy and Čerenkov removal for high performance, high spatial resolution scintillation dosimetry. *Med Phys* **33**, 128 (2006).
22. Létourneau, D., Pouliot, J., and Roy, R. Miniature scintillating detector for small field radiation therapy. *Med Phys* **26**, 2555–2561 (1999).
23. Clift, M. A., Sutton, R. A., and Webb, D. V. Dealing with Čerenkov radiation generated in organic scintillator dosimeters by bremsstrahlung beams. *Phys Med Biol* **45**, 1165–1182 (2000).
24. Fontbonne, J. M. et al. Scintillating fiber dosimeter for radiation therapy accelerator. *IEEE Trans Nucl Sci* **49**, 2223–2227 (2002).
25. Frelin, A. M. et al. Spectral discrimination of Čerenkov radiation in scintillating dosimeters. *Med Phys* **32**, 3000 (2005).
26. Guillot, M., Gingras, L., Archambault, L., Beddar, S., and Beaulieu, L. Spectral method for the correction of the Čerenkov light effect in plastic scintillation detectors: A comparison study of calibration procedures and validation in Čerenkov light-dominated situations. *Med Phys* **38**, 2140–2150 (2011).
27. Lacroix, F. et al. Clinical prototype of a plastic water-equivalent scintillating fiber dosimeter array for QA applications. *Med Phys* **35**, 3682 (2008).
28. Lacroix, F. et al. Extraction of depth-dependent perturbation factors for parallel-plate chambers in electron beams using a plastic scintillation detector. *Med Phys* **37**, 4331–4342 (2010).
29. Lacroix, F., Guillot, M., McEwen, M., Gingras, L., and Beaulieu, L. Extraction of depth-dependent perturbation factors for silicon diodes using a plastic scintillation detector. *Med Phys* **38**, 5441 (2011).
30. Morin, J. et al. A comparative study of small field total scatter factors and dose profiles using plastic scintillation detectors and other stereotactic dosimeters: The case of the CyberKnife. *Med Phys* **40**, 011719 (2013).
31. Archambault, L., Therriault-Proulx, F., Beddar, S., and Beaulieu, L. A mathematical formalism for hyperspectral, multipoint plastic scintillation detectors. *Phys Med Biol* **57**, 7133–7145 (2012).
32. Argyros, A., van Eijkelenborg, M. A., Large, M. C. J., and Bassett, I. M. Hollow-core microstructured polymer optical fiber. *Opt Lett* **31**, 172–174 (2006).
33. Marcatili, E. A. J. and Schmeltzer, R. A. Hollow metallic and dielectric waveguides for long distance optical transmission and lasers. *Bell Syst Tech J* **43**, 1783–1809 (1964).
34. Matsuura, K., Matsuura, Y., and Harrington, J. A. Evaluation of gold, silver, and dielectric-coated hollow glass waveguides. *Opt Eng* **35**, 3418–3421 (1996).
35. Matsuura, Y. and Miyagi, M. Hollow optical fibers for ultraviolet and vacuum ultraviolet light. *J Quantum Electron* **10**, 1430–1434 (2004).
36. Lambert, J., Yin, Y., McKenzie, D. R., Law, S., and Suchowerska, N. Čerenkov-free scintillation dosimetry in external beam radiotherapy with an air core light guide. *Phys Med Biol* **53**, 3071–3080 (2008).
37. Lambert, J. et al. A prototype scintillation dosimeter customized for small and dynamic megavoltage radiation fields. *Phys Med Biol* **55**, 1115–1126 (2010).
38. Jang, K. W., Yoo, W. J., Shin, S. H., Shin, D., and Lee, B. Fiber-optic Čerenkov radiation sensor for proton therapy dosimetry. *Opt Express* **20**, 13907–13914 (2012).

Part II

Clinical applications using small PSDs

6 Basic quality assurance: Profiles and depth dose curves

Louis Archambault and Madison Rilling

Contents

6.1 RATIONALE FOR USING PSDs IN BEAM PROFILING

Medical physicists have access to a wide range of tools for dose measurements: from Farmer-type ion chambers (ICs) to thermoluminescent detectors (TLDs) and from films to complex arrays of diodes. Each detector type used clinically is based on a different physical principle. Because some assumptions or approximations are almost always used when converting the raw signal from a radiation detector into a dose in Gy, having detectors based on different principles, and thus, different assumptions and approximations are desirable. However, given the already large spectrum of clinical tools available, it is worthwhile to ask if one more should be added to the list, especially for basic measurements such as profiles and depth dose curves.

The physical principles behind plastic scintillation detectors (PSDs) are described in depth in Part I of this book (Chapters 1 through 5). Comparisons between PSDs and other detectors for basic quality assurance (QA) go back to the early 1990s where percentage depth dose (PDD) curves and lateral profile measurements were accurately measured by Beddar et al. (1992). Since then, these types of basic QA measurements have been repeated for a wide range of PSD prototypes. Except for one instance (Beddar 1994), basic QA measurements are rarely the focus of published work. These measurements are rather used mainly as a way to demonstrate that the behavior of a given PSD prototype matches that of a more conventional detector such as an IC. Just by looking at the data published for several PSD prototypes for profile and depth dose measurement, one could argue that they rightfully have their place in the toolbox of the medical physicist for basic measurements. In addition, PSDs have been shown to be valid in a wide range of applications from radiosurgery to intensity modulated radiation therapy, thus illustrating the versatility of this type of instruments.

The rising presence of nonstandard fields in the clinic put a strain on most dose-measuring tools. Highly modulated fields, small fields, rotational delivery, high dose rate (e.g., flattening filter-free irradiations), and irradiation in the presence of strong magnetic fields are increasingly common dose delivery methods that may affect the response of a detector. Dose measurement in nonstandard fields is a subject of active investigations by numerous groups.

Alfonso et al. (2008) presented a formalism to express dose in a nonstandard field as a series of linear factors to account for the various perturbations introduced by a dosimeter on its surroundings, the possible lack of equilibrium, and the fact that the quality of the beam can change for different field sizes. Furthermore, some detectors such as optically stimulated luminescence dosimeters and TLDs even require nonlinear corrections. A more recent study by Azangwe et al. (2014) focusing only on small fields, but comparing a large group of detectors, illustrates well the sometime large corrections that must be applied to most detectors as a function of the field size or beam quality.

Comparison between multiple types of detectors is a good way to safeguard against errors especially when measuring nonstandard fields. Thus, keeping a diverse inventory of detectors is always good practice. Furthermore, to avoid misusing a detector, it is best if the person using it is familiar with its operation. It should be easier to make a nonstandard measurement for commissioning a special technique or a new delivery modality with a detector that is used every month or week for basic measurements than to do the same measurement with a special detector that is kept in its box all year round. However, to be used for both basic measurement and a wide range of nonstandard field, a detector needs versatility. Using the formalism of Alfonso (Alfonso et al. 2008), one could define the versatility of a detector as having a $k_{Q_{clin},Q_{msr}}^{f_{clin},f_{msr}}$ that is relatively constant and only weakly affected by the beam quality. This is the case of PSD, and its versatility therefore makes it a good candidate for basic QA as well as complex treatment delivery.

ICs are unquestionably the tool of choice for commissioning and performing the absolute calibration of a medical linac. Once a machine has been commissioned and calibrated ICs are still commonly used for basic measurements to monitor its performance, but other types of detectors can be used as well. The potential advantages of using PSDs for such measurements will be first presented in Sections 6.1.1 and 6.1.2. Then Sections 6.2 through 6.4 will illustrate how PSDs can be used for such measurements.

6.1.1 THE CASE FOR PHOTON BEAMS

The properties of PSDs are presented and demonstrated in Chapter 4. Several of these properties are especially useful for basic QA and beam profiling:

- *Real-time capability*: It is essential for any scanning system and especially for beam tuning. It must be possible to visualize a beam profile or depth dose curve as soon as it is delivered.
- *Absence of energy and dose rate dependence*: Because of scattered photons, the penumbra and out-of-field regions have a different energy composition than the center of the beam. Furthermore, in these regions, the dose per machine pulse is much lower. Therefore, if the detector was subject to a strong energy or dose rate dependence, measurements in the penumbra or out-of-field regions could be erroneous.
- *Minimal fluence perturbation/water equivalence*: Because most basic QA measurements are made in water or in a water-equivalent plastic phantom, using a water-equivalent detector minimizes the need to apply corrections to the readings. A water-equivalent detector will also be less affected by nonequilibrium regions such as the dose buildup or on field edges.
- *High spatial resolution*: Although not as important as for other applications such as radiosurgery or stereotactic body radiation therapy (SBRT), using a detector with a high spatial resolution is important for an accurate evaluation of the edges of any radiation fields.

Each of these properties can also be found in other detectors, but not at the same time. For example, ICs have essentially no energy or dose rate dependence, but often lack a high spatial resolution; Diodes have an excellent spatial resolution, but can cause fluence perturbation and overrespond to low energies. Therefore, PSDs are the only detector type that regroups all of these properties in a single device.

In the literature, the focus is often on small fields rather than large fields. There are ample data to support the role of PSDs in that area (Beddar et al. 2001, Gagnon et al. 2012, Klein et al. 2012, Morin et al. 2013) (see also Chapter 7). However, larger field sizes (i.e., 20 × 20 cm² and larger) are less commonly discussed. They nonetheless represent an important part of basic QA measurements. Large fields are

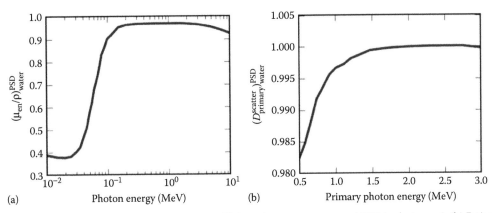

Figure 6.1 (a) Ratio of mass energy attenuation coefficients between water and PSD (polystyrene). (b) Estimation of the ratio between scattered and primary doses for PSD compared to water as a function of the primary photon energy (normalized at 2 MeV).

usually easier to measure than small fields, but suffer from an increased contribution of scatter dose, which can have an impact on the beam quality. A common PSD material is polystyrene, and mass attenuation coefficients of polystyrene are slightly lower than water at low photon energies (see Figure 6.1a). To estimate the possible under-response of PSD in water due to the material difference, a naive, worst-case scenario model was developed. In this simplistic scenario, all primary photons undergo a Compton interaction where the maximum energy is transferred to the electrons so that the scattered photon has the smallest possible energy. The dose received by the PSD comes from either a high-energy electron produced via Compton interaction of primary photons (primary dose) or a scattered photon undergoing photoelectric effect with local dose deposition (scattered dose). Results of this naive model are shown in Figure 6.1b as a function of the primary photon energy. It can be seen that a PSD responds similarly to primary and scattered doses for a wide range of primary photon energies, thus illustrating that no correction should be necessary when using a PSD in large fields or even out-of-field regions where scattered dose dominate.

Another advantage of the dose rate independence of PSDs is that they do not require correction in high dose rate environments such as those produced by flattening filter-free (FFF) linacs. As an example, Table 6.1 shows the ratio of dose measured with two IC models (the Exradin A12 from Standard Imaging and the CC13 from IBA dosimetry) and the Exradin W1 PSD from Standard Imaging in the center of a high dose rate field (2400 MU/min). All detectors were calibrated at a lower dose rate. If the ICs are not corrected for the ion recombination (P_{ion}) specific to this dose rate, there is a discrepancy of about 1% between both types of detectors. However, if the ICs are corrected, then both detectors agree within 0.5% or less. This indicates that a PSD would be a suitable detector for basic QA of FFF beams because no additional corrections are necessary.

6.1.2 THE CASE FOR ELECTRON BEAMS

Most of the advantages of PSDs for photon measurements also apply to electrons. However, in several ways, the physics governing the interactions of an electron beam is more complex than the physics governing the interactions of a photon beam. Electrons continuously interact with the medium, gradually lose energy and eventually stop. They also produce bremsstrahlung photons during their slowing down: at shallow depth, an electron beam consists mostly of electrons, but once most electrons stopped, the photons dominate (i.e., the bremsstrahlung tail). Furthermore, the energy range of electron beams used clinically is much larger than that of photons. Electron beams between 4 and 20 MeV are commonly seen, and although a linac will typically have access to only one or two photon energies, they often have five or six electron

Table 6.1 **Comparison of ion chambers and PSD readings in a high-dose rate environment**

DETECTORS	UNCORRECTED RATIO	AFTER IC CORRECTION FOR P_{ion}
A12/W1	0.989	0.998
CC13/W1	0.991	1.005

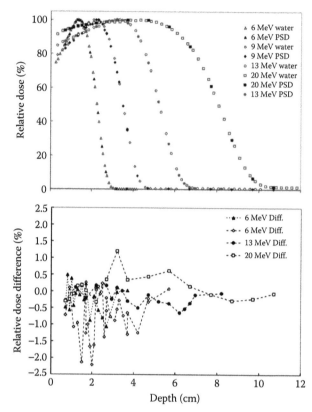

Figure 6.2 Electron depth dose curves simulated by Monte Carlo for four different beam energies and comparing the dose to water and the dose to PSD. The upper figure shows the simulated PDDs and the bottom figure shows the difference between the dose to water and the dose to PSD. (Adapted from Lacroix, F. et al., *Med. Phys.*, 37, 4331, 2010b, doi:10.1118/1.3463383. With permission.)

energies commissioned. To accurately measure an electron beam, a detector must therefore offer a uniform response to both photons and electrons and be insensitive to electron energies. ICs possess these properties, but unfortunately they also perturbation the incident radiation. These perturbations must be corrected by applying various factors or by using a *perturbation-free* detector (Lacroix et al. 2010b). Diamond detectors, radiochromic films, and PSDs can all be considered perturbation free. However, diamond detectors suffer from dose rate dependence and radiochromic film often have precision issues (Hartmann et al. 2010). PSDs exhibit neither disadvantages and have been shown to behave in a manner extremely close to water for a wide range of electron beams using Monte Carlo simulations (see Figure 6.2).

6.2 ACQUISITION METHODS

PSDs can come in many forms. If we exclude 3D detectors (Goulet et al. 2014) and 2D scintillator sheets (Frelin et al. 2008, Petric et al. 2006) there are essentially two methods to measure basic beam profiles: either use a scanning system or an array of detectors. Traditionally, basic beam QA measurements have been performed using a single IC mounted on a motorized arm inside a large water phantom. This process although highly accurate is also quite time consuming. Because the scintillation process is essentially instantaneous, a scanning acquisition setup can also be used with PSDs. It has been used in the past when performing a comparison between an IC and a PSD (Archambault et al. 2006). The precision of a PSD however depends on the total amount of light collected by the system and on the sensitivity of the photodetector used (Lacroix et al. 2010a). Therefore, a PSD that is to be used on a scanning system should be designed to maximize light collection and use a high-sensitivity photodetector in order to have a high precision. If the precision is too low, it will be necessary to repeat and average each scan or use a *step-and-shoot* method where the scintillator stops for integration at every point on the scan path.

As an example, an early PSD prototype required 5 s integration time for a precision of 0.3% (Archambault et al. 2006), this is not ideal for use in a scanning system. Since then, multiple prototypes have however been developed with far better precision (see Archambault et al. (2010), for an example), but even with the best possible precision, using a scanning methods remains time consuming with any detector because it is necessary to move the detector over the water phantom for every acquisition.

One possible way to accelerate basic beam QA measurements is by using a detector array. An array of detectors can sample the dose delivered to a whole plan (or even to a whole volume for an hypothetical 3D array) in a single acquisition, thus at least partially removing the need to move a detector around at every scan. Several commercial types of detector arrays are available. These systems usually rely on IC or diodes and can easily comprise several hundreds of detectors. However, commercial arrays are more often sold and used for patient specific QA rather than basic beam profiling. The reason is twofold. First, having a large number of detectors requires having a large number of electrodes and electrical wiring that often imply high Z materials. Thus, a detector array can perturb the incident beam in a significant manner or, at the very least, perturb the beam for a subset of incident angles (Feygelman et al. 2010, Guillot et al. 2011a, Wolfsberger et al. 2010). The second reason commercial detectors arrays are not often used for basic beam QA is that the measurement frequency is not sufficiently high. For example, the IBA MatriXX uses 5 mm IC every 7 mm; this is not sufficient for most applications. Even diode arrays such as Sun Nuclear's MapCheck or Scandidose's Delta4 cannot have sampling frequency of more than once every 4–5 mm. Even if each individual diode has a high spatial resolution, multiple detectors cannot be placed side by side because of their wiring and connections.

Arrays of PSDs suffer from neither of these limitations. Because of the intrinsic water equivalence of these detectors, it is possible to embed a large number of detectors in a 2D or even a 3D water-equivalent volume without introducing any fluence perturbation. No metal electrodes are required in a PSD. One could have a polystyrene sensitive volume coupled to a polystyrene optical fiber for collecting the light and insert several of these detectors inside a polystyrene phantom. Thus, there would be absolutely no change in density or atomic composition. The detectors and their surrounding materials would be 100% polystyrene. A similar argument goes for stacking detector close to each other: Because no electrodes or electrical wiring are required, it is possible to place PSDs directly side by side. The only requirement is to have some form of optical insulation between them to avoid cross talk, but such insulation can be only hundreds of nanometers thick. Some examples of closely packed scintillator array can be found in the literature (Archambault et al. 2007, Gagnon et al. 2012, Guillot et al. 2010, 2011a, Lacroix et al. 2008). The goal here is not to describe PSD array design and usage in depth (see Chapter 13 in that regard), but rather to discuss how these arrays can serve for basic beam QA. One potential limitation of PSD arrays is that the optical fibers required to collect and guide the scintillation light can take lots of space and make the device cumbersome and clinically impractical (see Figure 6.3). This limitation can be countered by using smaller

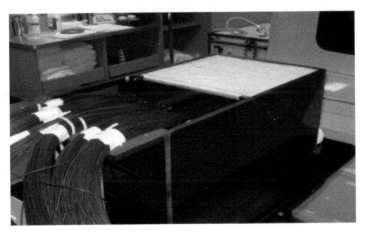

Figure 6.3 Picture of the 781 PSD array. The large bundles to the left of the figures are the optical fibers required for reading out the device. From Guillot, M. et al., *Med. Phys.*, 38, 2140, 2011a, doi:10.1118/1.3562896. With permission.)

optical fiber at the cost of lower sensitivity (i.e., because less scintillation light can reach the photodetector) or by coupling multiple scintillating elements to a single optical fiber as described in Chapter 10.

6.3 ACQUISITION SETUPS

6.3.1 DETECTOR SIZE AND ORIENTATION

Some consideration must be taken when using point-like PSD either with a scanning method or in an array. Most of the time, the sensitive volume of a PSD is cylindrical with diameters between 0.5 and 2 mm, and lengths typically between 1 and 10 mm, although any shape and size is possible (Archambault et al. 2005). Therefore, PSDs are typically longer in one direction than in the others, and this can have an impact on basic QA measurements because the spatial resolution will depend on the orientation of the PSD with respect to the beam. For a cylindrical PSD, there are two possible configurations: parallel and perpendicular. In the parallel configuration, the long axis of the detector is parallel to the beam direction. This configuration is also referred to as the *standup* configuration because the PSD would stand up in a water tank so that the beam can enter through the distal tip of the scintillator. The advantage of this setup is that the cross section of the PSD in the beam plane is smallest and thus maximizes the spatial resolution. In the perpendicular configuration, the long axis of the detector is perpendicular to the beam direction. In this setup, the PSD lie flat on the treatment couch; this is a typical configuration for use in a solid water phantom. The spatial resolution is not as high as with the parallel setup, but manipulations are easier. Both setups are illustrated in Figure 6.4.

6.3.2 CALIBRATION OF A PSD FOR BASIC QA

It has been mentioned on multiple occasions in this book that the calibration of a PSD is different from most detectors. The goal of the calibration process is twofold: (1) it is used to decouple the scintillation signal from the Cerenkov noise and (2) it is used to determine the relation between the dose in Gy and the amount of scintillation produced in the detector. For more details regarding the Cerenkov light and its removal, see Chapter 5. Because of the sensitive nature of basic QA measurements, it is important to have a robust and ideally simple calibration method.

Assuming a superposition of scintillation and Cerenkov, it is possible to get the dose by measuring the signal under two distinct waveband channels of the visible spectra (e.g., in blue and green regions). By doing at least two measurements where the dose delivered to the PSD is accurately known, it is possible

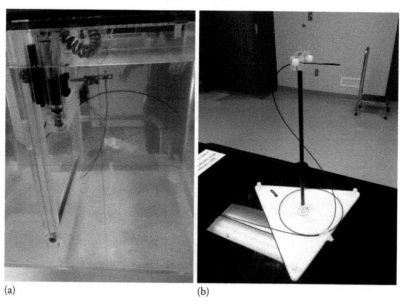

(a)   (b)

Figure 6.4 (a) A PSD mounted in a water tank using a parallel setup. (b) A PSD mounted in a support for measurement in air using a perpendicular setup.

to solve a linear system of equations to extract calibration factors (Archambault et al. 2006, Fontbonne et al. 2002; see Chapter 10 for a more thorough discussion on this method). This general method can be implemented in numerous ways. However, one practical approach to PSD calibration has been described by Guillot et al. (2011b). For this method, the PSD is placed at the center of a relatively large field (i.e., 30×30 cm^2 or larger) with a minimal amount of stem exposed to the radiation field. A first measurement is taken in that configuration, and then the amount of stem exposed to the beam is increased to a maximum (e.g., by rolling the cable inside the field) for a second measurement. Except for the amount of stem exposed, all parameters must remain constant between both measurements, including the dose, the field size, and the position of the PSD. This way, we can assume that the sensitive volume of the PSD received exactly the same dose for both measurements. Hence, the only parameter that changed is the amount of Cerenkov light produced within the PSD. It is then possible to compute the Cerenkov light ratio (CLR):

$$\text{CLR} = \frac{M_1^{\max} - M_1^{\min}}{M_2^{\max} - M_2^{\min}} \tag{6.1}$$

where:

M_i represents the measurement in channel 1 or 2

The *min* and *max* superscripts represent, respectively, the condition where a minimum and a maximum of stem are exposed

Once the CLR is known, a reference measurement of known dose D_{ref} must be done in order to get the calibration factor a in Gy per raw unit of the PSD:

$$a = \frac{D_{\text{ref}}}{M_1^{\text{ref}} - \text{CLR} \cdot M_2^{\text{ref}}} \tag{6.2}$$

Then any dose measured with the PSD can be expressed using the CLR and a:

$$D = a(M_1 - \text{CLR} \cdot M_2) \tag{6.3}$$

This calibration method is simple and advantageous because the dose must be accurately known only for the determination of a. Determination of the CLR only requires that the dose be constant between measurements, but it does not have to be known accurately. If a is not determined, it is still possible to do relative measurement.

As explained in Chapter 10, calibration of a PSD is essentially a linear regression. As for any type of fitting operation, it is better to perform the PSD calibration using conditions close to the ones planned for the measurements. The method described earlier makes use of a large field size to determine the CLR. Because of this, it might not be the best calibration approach for measuring doses from small fields. It is however perfectly appropriate for basic QA as such measurements often use large field sizes. The same reasoning applies to the measurement setup and detector orientation: it is better to calibrate the detector in a setup similar to the one that is to be used for the actual measurements. Calibrating a PSD in a perpendicular setup can be done easily by rolling a large quantity of optical fibers in the field. The commercial W1 PSD from Standard Imaging even comes with a phantom specially designed for this task (see Figure 6.5a). When using a PSD in a parallel setup, calibration may be more challenging. Two possible calibration configurations with a detector in a parallel setup are illustrated in Figure 6.5b and c.

6.3.3 CONSISTENCY MONITORING

Basic QA measurements, especially those that require scanning in a large water tank, can be performed over several days. Although PSDs offer precise and repeatable readings (see Section 6.4), they remain relatively fragile compared to other detector. A short bend radius or other mechanical stress can easily alter the properties of the optical cable carrying the signal from the scintillator to the photodetector. For this reason, tools for integrity or consistency monitoring have sometime been built in PSDs (Yin et al. 2008). The hypothesis behind the whole calibration process is that the shape of the scintillation and Cerenkov

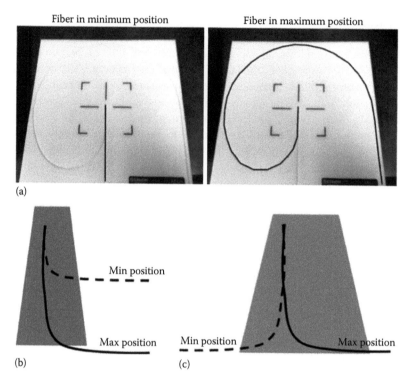

Fiber in minimum position   Fiber in maximum position

(a)

(b)   Min position   Max position   Min position   (c)   Max position

Figure 6.5 Calibration setups for PSDs. (a) Calibration with the detector in a perpendicular configuration using the calibration phantom from Standard Imaging. (b, c) Two possible calibration approaches to use with a parallel configuration. In both case, the sensitive region of the detector (i.e., the distal tip) remains at a fixed position, whereas the amount of stem is varied.

spectra reaching the photodetector remains constant at all time (see Chapter 10). Mechanical stress to the detector can invalidate this hypothesis. Furthermore, depending on the choice of components in the design of a PSD system, it is possible that the spectrum reaching the photodetector changes for a different setup (e.g., if the transmission properties of the optic cable are altered for different bending configurations), making it even more important to calibrate such PSD in a setup similar the one used for measurements.

One simple method to guarantee that the behavior of PSD remains constant from day to day or from setup to setup is to periodically monitor the CLR. The quantity computed in Equation 6.1 is the ratio of the Cerenkov signal measured in each waveband channel. If the underlying hypothesis of the calibration is respected, this ratio should be constant. However, if something changed in the PSD causing the Cerenkov and scintillation spectra reaching the photodetector to change, the CLR will also change. The CLR can therefore be used as simple consistency check without requiring any additional equipment. As an example, Table 6.2 shows some CLR values taken in various configurations using the W1 PSD.

Table 6.2 **CLR measured with the W1 for a large range of configurations**

SETUP	FIELD SIZE (CM²)	DEPTH (CM²)	SURROUNDING MEDIUM	CLR
Perpendicular	30 × 30	1.5	Solid water	0.751
Perpendicular	30 × 30	0	Air	0.747
Perpendicular	40 × 40	10	Solid water	0.746
Parallel	10 × 10	10	Solid water	0.715
Parallel	30 × 30	10	Solid water	0.714

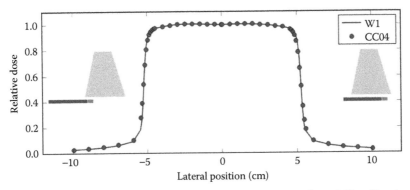

Figure 6.6 A profile measured to test that no residual stem effect is present after a PSD calibration. Comparison is made between a PSD (W1) and an IC (CC04).

In Table 6.2, all measurements made using a perpendicular configuration show a CLR constant to within 0.5%. A similar constancy is also seen by repeating measurement over several days. This illustrates that even for a large range of experimental conditions (measurement at different depth, in different materials, using different field sizes), the behavior of the PSD remains constant. However, when we switch from a perpendicular to a parallel setup, an important shift in the CLR is seen. This is not a problem per se (as long as this effect is reproducible), but only indicates that for that particular detector, the change in setup affects the way the light is guided in the system. In this case, using a calibration for each of the two setups can fix the problem.

After calibrating a PSD, a profile measurement of a large field can be acquired to test that the dose measured is completely free from Cerenkov-induced stem effect. The most strenuous test is to make the acquisition in a perpendicular configuration with the detector entering the field through its distal tip. If the profile measured this way is more asymmetric than the same profile measured with another dosimeter, it means that the PSD signal is still contaminated by Cerenkov. An example of such profile is shown in Figure 6.6 along with a schematic of the PSD position with respect to the radiation field. In Figure 6.6, the rightmost part of the profile contains nearly no Cerenkov, whereas the leftmost part contains an important amount of Cerenkov. In fact, when the tip of the detector is out of the primary field, but the optical fiber cable is entirely covered by it, the Cerenkov signal largely dominates over the scintillation signal. Nevertheless, the excellent agreement between IC and PSD indicates that the calibration could efficiently decouple these two sources of light.

6.4 ACCURACY AND PRECISION

6.4.1 PRECISION OF PSD SYSTEMS

As mentioned in the beginning of Section 6.2, the precision of a PSD directly depends on the number of scintillation photon that can reach the photodetector. The number of scintillation photons produced is relatively high, typically around 6000 photons per MeV of energy absorbed for a plastic scintillating fiber. However, there are numerous sources of loss (e.g., coupling losses, attenuation, limited quantum efficiency) and only a small fraction of the total number of scintillation photons are quantified by the photodetector. Furthermore, there is a compromise to be made between the spatial resolution and the precision of the detector. Small sensitive volume will have high spatial resolution but will produce less scintillation, and hence have a poorer precision. PSD can be divided in two general classes according to the type of photodetector used: (1) camera-based systems and (2) non-camera-based systems. Camera-based systems are often used for making PSD arrays because they can easily measure the light produced by hundreds or even thousands of PSD simultaneously. However, only a small fraction of the light exiting a fiber-optic cable can be captured by the objective lens of a camera. Therefore, there are important losses of signal in any camera system, and this affects the precision of such device. This problem has been studied in depth

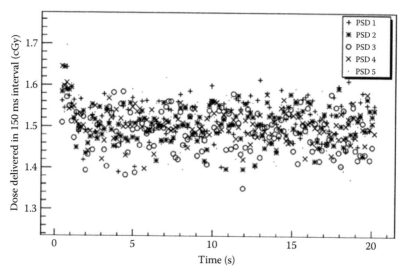

Figure 6.7 Example of the precision of a 5 PSD array read with an electron-multiplying CCD camera at a frequency of about 7 Hz. (From Archambault, L. et al., *Int. J. Radiat. Oncol. Biol. Phys.*, 78, 280, 2010, doi:10.1016/j.ijrobp.2009.11.025. With permission.)

(Lacroix et al. 2009, 2010a), but typically a camera system will require integration time at least on the order of the second to get a precision within 2%–3%. By using amplified cameras such as electron-multiplying charge-coupled device (CCD), it is possible to build a PSD array capable of real-time measurement (Archambault et al. 2010) (see Figure 6.7 for an example). With an amplified CCD, it was possible to get a precision of 3% for measurements made every 150 ms.

PSDs that use non-camera photodetector can typically exhibit better precision because the output of the fiber-optic cable can be directly coupled to a photodetector such as a photodiode or a photomultiplier tube, thus greatly reducing the losses (Beddar et al. 1992, Letourneau et al. 1999). However, with these systems large arrays are more challenging to make. The commercial W1 PSD is a good representative of this type of detector. Figure 6.8 shows the precision and daily reproducibility of the W1. For doses of 200 cGy, the average precision was 0.17% and the daily reproducibility was 0.14%.

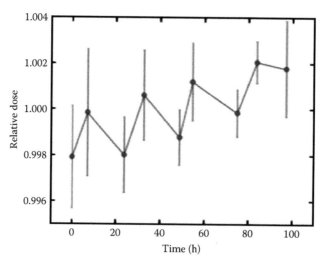

Figure 6.8 Dose measured with the W1. Each point represents the average of 10 irradiations of 200 cGy.

6.4.2 DEPTH DOSE CURVES

Percent depth dose (PDD) curve is a central part of any basic QA as it serves as an indicator of the beam energy and quality. An example of both photon and electron PDDs measured with the W1 PSD and compared to an IC (CC04) is shown in Figure 6.9. On average, the accuracy is 0.5% for photons and 0.6% for electrons using the IC as reference. As of now, the W1 cannot be used in a continuous scanning mode. The measurements presented in Figure 6.9 were therefore acquired using a *step-and-shoot* method, but a continuous scanning acquisition is theoretically and technically feasible. As previously mentioned, PSDs are ideal for measurement in the buildup region due to their water equivalence and spatial resolution.

PSD arrays have also been used to measure PDD as shown in Figure 6.10. It is important to note that PSD arrays are one of the only type of detector arrays capable of such measurements because they are extremely homogeneous in their material composition (i.e., both the detector themselves and the surrounding phantom

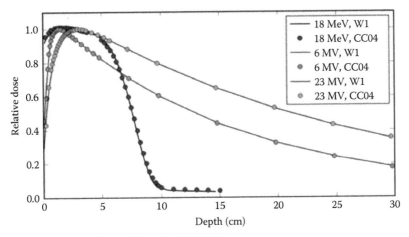

Figure 6.9 PDD measurements with the W1 and an IC (CC04).

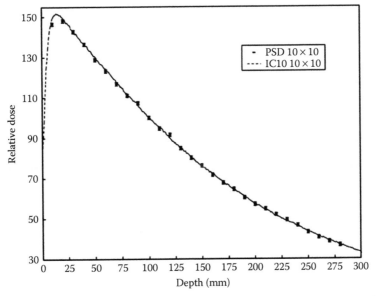

Figure 6.10 A PDD measured with an array of PSD compared to IC measurements. (From Lacroix, F. et al., *Med. Phys.*, 35, 3682, 2008. With permission.)

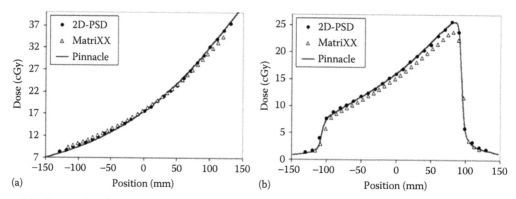

Figure 6.11 Comparison from a PSD array, an IC array (IBA MatriXX), and a dose calculation in a homogeneous water phantom (Pinnacle): (a) using a 90° beam angle; (b) using a 120° beam angle. (From Guillot, M. et al., *J. Phys. Conf. Ser.*, 2010. With permission.)

are made of water-equivalent materials). To illustrate that, Guillot et al. have compared measurements made with a PSD array and the IBA MatriXX. Both arrays were placed on the surface of the treatment couch and irradiated at 90° to simulate a PDD and also at 120°. Results were compared to the calculation from a treatment planning system for a beam incident on a water phantom. As it can be seen from Figure 6.11, measurement in the array is essentially indistinguishable from dose calculation in homogeneous water.

6.4.3 LATERAL DOSE PROFILE

Lateral beam profile is another important basic QA measurement. It serves to quantify precisely the beam flatness and symmetry, and it is an essential part in the commissioning of any treatment planning system. Although PDDs are usually acquired once per beam energy, it is often customary to measure multiple beam profiles. At the very least profiles in the two cardinal directions must be measured (i.e., inplane and crossplane), but it is also common to measure them at different depth or diagonally. On top of that, lateral dose profiles must also be measured for wedges. These measurements can represent a sizeable amount of work. Therefore, 2D or 3D PSD arrays are extremely attractive for lateral profile measurement. However, to be of any use, an array must have a sufficiently high measurement frequency. The most critical regions are the beam edges where the dose decreases rapidly. For these regions, measuring the dose every few millimeters is not sufficient. As discussed earlier, PSD arrays with a large number of points and a high sampling frequency are feasible. However, to limit the complexity of such device, it would be possible to design an array with nonuniform detector distribution. For example, it could be possible to increase the number of PSDs per millimeter every 5 cm so that the edges of field sizes that are multiples of 5 cm could be measured accurately. Such a nonuniform PSD array could be perfect for basic QA. However, nonuniform arrays could be less useful for applications other than basic QA. As for other types of detectors, the most versatile array uses a uniform distribution of detectors. Several examples of lateral beam profile measured with PSD arrays can be found in the literature. Arrays ranging from 8 (Archambault et al. 2007) to 781 detectors (Guillot et al. 2011a) have been used in the past. One interesting example is the array designed by Lacroix et al. (2008) because it was designed specifically for basic QA measurements. Great care was taken in the development of this detector to make it rugged, robust, and easy to use in the clinic. Lateral profiles measured with this tool are shown in Figure 6.12. Finally, even if arrays are better for profile measurements, detector scanning is always possible as it can be seen from Figure 6.6.

6.5 A WORD ON 3D SYSTEMS

While 2D and 3D arrays of PSDs are great tools for beam profiling and basic QA, scintillator-based detectors can be pushed even further and form a truly 3D system without any gap in measurements. The 3D systems based on liquid scintillators are discussed in Chapter 15 and systems designed specifically for brachytherapy are described in Chapter 14. Plastic scintillators can also be used in 3D systems and have a great potential for rapid, but thorough characterization of a given beam or patient plan.

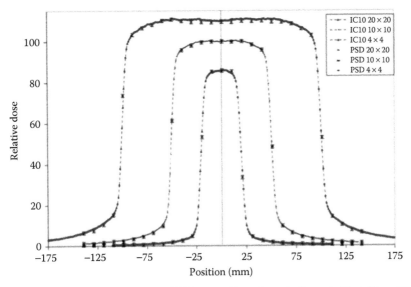

Figure 6.12 Profile measurements using a rugged PSD array. (From Lacroix, F. et al., *Med. Phys.*, 35, 3682, 2008. With permission.)

A variety of 3D dosimetry systems have been developed to allow accurate volumetric measurement of delivered dose distributions. Among typical 3D dosimeters can be found stacks of radiochromic films, and polyacrylamide and Fricke gels, notably for their high spatial resolution. These dosimeters, each composed of a radiosensitive material, are based on the measurement of a radiation-induced change in the material's physical properties, proportionally related to the absorbed dose (e.g., the darkening of a radiochromic film, the degree of polymerization of a polymer gel, or the magnetic relaxation time of a Fricke gel sample) (Baldock et al. 2010; Gore and Kang 1984; Guo et al. 2006; Kirov et al. 2005). The use of film and gel-based dosimetry systems for practical treatment QA is mainly limited by their time-consuming read-out procedures. Moreover, dosimetric gel systems require tedious preparation methods, whereby introducing possible variations between the chemical properties of different batches. Furthermore, these dosimetric systems are intrinsically integrating dosimeters, yielding only total dose information with no real-time information on the dose delivery.

Plastic scintillation detectors are increasingly considered for application in 3D dosimetry. A nonperturbing scintillator volume, which can be either solid or liquid, serves simultaneously as the phantom material and the detector (Beddar 2015). Furthermore, the instantaneous light emission of irradiated scintillator volumes motivates their promising use for real-time 3D dosimetry.

Several scintillator-based dosimeters use tomographic principles to reconstruct 3D dose distributions based on acquisitions or projections of the scintillator's emitted light field (Glaser et al. 2013; Goulet et al. 2013; Kirov et al. 2005). Indeed, Kirov et al. marked the beginning of 3D scintillation dosimetry by using a bracytherapy Ru-106 eye plaque applicator in a liquid scintillator (see Chapter 14 for more details). Though considerable accuracy has been achieved with such scintillation dosimetry systems, application to dynamic treatment deliveries can be challenging: in such cases, a proper balance has to be found between a system's spatial resolution, temporal resolution, and light collection efficiency to provide sufficient precision and accuracy, while accommodating the complexities of the delivered dose distribution.

Hence, 3D dosimetry would be ideal for truly comprehensive treatment QA (Nelms et al. 2011). However, to result ultimately in a usable clinical tool, a dosimetry system must consist of a cost-effective detector providing submillimeter resolution, while being an easy-to-use and, preferably, reusable device. Moreover, patient-specific QAs would greatly benefit from systems providing real-time dose measurements in order to verify the accuracy of a treatment plan throughout the whole delivery (Goulet et al. 2014).

6.5.1 3D DOSIMETRY USING LONG SCINTILLATING FIBERS

Scintillating fibers have commonly been studied for point and planar dose measurements. Goulet et al. (2013) extended the application of long scintillating fibers to absolute 3D dosimetry by simulating an array of fibers along concentric cylindrical planes, as shown in Figure 6.13a. As the cylindrical phantom rotates about its central axis during an irradiation, 1D dose projections are obtained by acquiring scintillation light data using a CCD camera at the opposite ends of the phantom (s1 and s2); tomographic algorithms are then applied to these projections to reconstruct the incident 3D dose distribution. The 3D dose reconstructions, at a resolution of $1 \times 1 \times 1$ mm^3, provided better than 1% accuracy and 1% precision in high dose, low gradient regions of the simulated incident distribution. Examples of three perpendicular planes of a 3D dose reconstruction of a seven-field IMRT prostate treatment are shown in Figure 6.13b, along with their absolute dose differences with respect to the planned distribution calculated by the treatment planning system.

Certain challenges arise when using scintillator fiber-based dosimetry systems for dynamic treatment techniques. Indeed, Goulet et al. (2013) highlighted the presence of dose reconstruction artifacts due to the

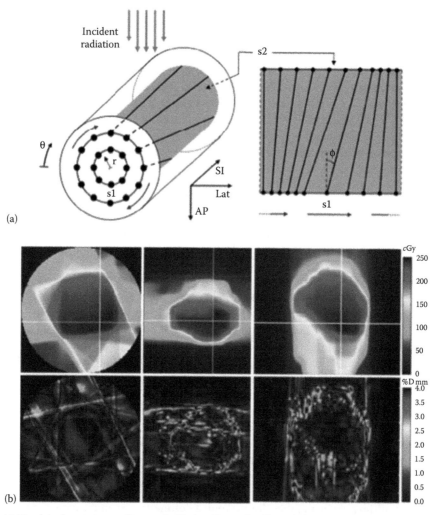

(a)

(b)

Figure 6.13 (a) Simulated prototype of long scintillating fibers distributed along two or more concentric cylindrical planes. Each fiber can be inclined at a different angle φ. (b) Examples of mutually perpendicular planes of a 3D dose reconstruction of a seven-field IMRT prostate treatment (top row), along with their absolute dose differences with respect to the planned distribution (bottom row). (Data from Goulet et al., 3D tomodosimetry using long scintillating fibers: A feasibility study. *Med. Phys.*, 40, 101703, 2013. doi: 10.1118/1.4819937).

continuously varying incident radiation field for such treatment modalities. Though it would be possible to minimize these dose inaccuracies, there remains an important trade-off to consider in such cases between real-time acquisitions, loss in spatial resolution, and increase in reconstruction time (Goulet et al. 2013; Lamanna et al. 2013). Moreover, to obtain optimal accuracy with such scintillating fiber configurations, the optical attenuation of the fibers has to be taken into account to model properly the weighted dose integrals based on measured scintillating light projections.

6.5.2 3D DOSIMETRY USING 3D SCINTILLATOR VOLUME

The tomographic reconstruction approach was also adopted by Kroll et al. (2013) to measure 3D dose distributions using multiple digital cameras to image a scintillator volume from different viewing angles. The authors built a portable detector prototype, shown in Figure 6.14, which uses four digital cameras (C0, C1, C2, C3) to acquire 2D light distribution projections of an irradiated plastic scintillator block (S). Because of the proposed configuration, the detector system is currently limited to unidirectional measurements. Though their system can rapidly and accurately reconstruct 3D dose distributions for electron and proton beams, its use cannot yet be extended to high-energy photon beams due to the positioning of the rear camera (C0 seen in Figure 6.14), which would become quickly saturated and damaged by incident photons.

Recently, Goulet et al. (2014) presented the first 3D dosimetry system capable of high precision, real-time volumetric dose measurements of intensity-modulated treatment plans. The proposed prototype uses a static plenoptic camera, also known as a light-field imager, to acquire images of the light emitted from a cubic scintillator volume of $10\times10\times10$ cm^3 contained in an acrylic phantom. The plenoptic camera is identical to any typical CCD camera, but an array of microlenses is added in front of its active sensor: each pixel integrates incident light over a limited and known range of angles, thus allowing simultaneous recording of both spatial and angular information of an imaged light field (Georgiev and Lumsdaine 2012). Using tomographic reconstruction, algorithms similar to those of previous work (Goulet et al. 2013), the scintillating light distribution is reconstructed by backprojecting the light intensity measured by each pixel into the scintillator volume, whereby the 3D dose distribution is obtained. In this first prototype, continuous acquisitions using the linac's electronic portal imaging device are used to constrain the shape

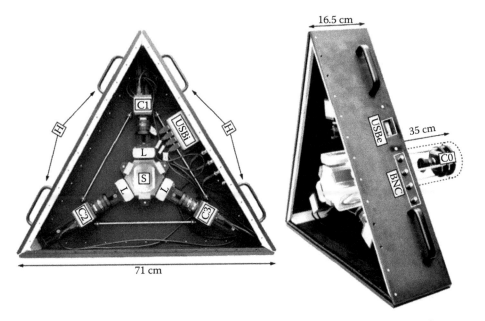

Figure 6.14 Portable detector prototype (shown without the front cover) with handles (H) using four CCD cameras (C0–C3) to image a plastic scintillator block (S) from different viewing angles through aplanatic lenses (L). (From Kroll et al. Preliminary investigations on the determination of three-dimensional dose distributions using scintillator blocks and optical tomography. *Med. Phys.*, 40, 082104, 2013. doi: 10.1118/1.4813898.)

Clinical applications using small PSDs

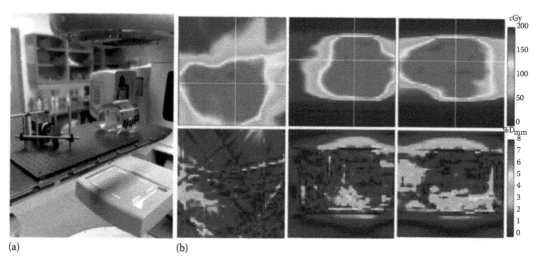

(a)   (b)

Figure 6.15 (a) 3D dosimetry system using a plenoptic camera to image a cubic plastic scintillator volume. Acquisitions with the linac's portal imager are used to constrain the shape of the reconstructed 3D dose distribution; (b) mutually perpendicular planes of the 3D dose reconstruction of an IMRT brain treatment (top row), with their absolute dose differences with respect to the distribution calculated in Pinnacle3 (bottom row). (From Goulet M et al. Novel, full 3D scintillation dosimetry using a static plenoptic camera. *Med. Phys.* 41, 082101, 2014. doi: 10.1118/1.4884036.)

of the resulting 3D distribution by a beam's eye view projection of the incident dose. The experimental prototype is presented in Figure 6.15a.

On average, for IMRT and VMAT dose distributions reconstructed at a 2 mm resolution, the absolute dose difference between the measured 3D dose and the planned distribution calculated in a treatment planning system was less than 1.5%; in the case of a seven-field step-and-shoot IMRT brain treatment, the absolute dose difference remained below 3% for each beam incidence (Goulet et al. 2014). Examples of mutually perpendicular planes of the 3D dose reconstruction of the IMRT brain treatment are shown in Figure 6.15b, with their absolute dose differences with respect to the distribution calculated in Pinnacle3. This real-time 3D dosimetry system offers promising results toward providing truly comprehensive QA of dynamic radiotherapy treatments.

The systems proposed by Kroll et al. (2013) and Goulet et al. both show promising real-time capabilities for 3D dose measurements. However, it is important to note that using tomographic reconstruction algorithms requires a non-negligible amount of time, which increases with the number of reconstructed volume voxels and the number of projections used as input data or camera pixels used for backprojection (Goulet et al. 2014). Hence, implementing the reconstruction process on graphic processing units (GPUs) would distribute the processing load between separate cores, thus significantly reducing the time needed to reconstruct each 3D dose distribution (Cui et al. 2011; Goulet et al. 2014; Jang et al. 2009; Kroll et al. 2013).

6.6 CONCLUSION

Despite being present in the literature for several years, PSDs are only just beginning to appear in clinic throughout the world due, in part, to the recent availability of a first commercial miniature PSD. Thus, PSDs represent a rather new tool in the medical physicist toolbox. Although other chapters of this book discuss how PSD can offer a benefit over other detectors for specialized applications (e.g., SBRT and radiosurgery), the goal of this chapter was essentially to show that PSDs could do as good a job as other detectors for basic QA measurements such as depth dose curves and lateral profiles. The fact that PSDs can be used as well for small fields and large fields, for electrons and photons, and for high dose rate and low dose rate makes them one of the most versatile detectors available in radiation therapy. When looking at basic QA measurements, one strength of PSDs over other types of detectors is their capability of forming closely packed, 3D arrays that do not perturb the incidence fluence because all materials (the phantom, the dose-sensitive volumes, and the optical fibers for light collection) are homogeneous and water equivalent. No other detectors capable of real-time measurement possess this property. Unfortunately, there is currently no such array available commercially, but as the clinical interest for PSD grows, we are likely to see such product in the future.

One last word of caution is that the mechanism of calibration of a PSD is significantly different than most detectors. In a PSD, the goal of the calibration is twofold: (1) to decouple the scintillation component of the signal from the Cerenkov component and (2) to establish the relation between the amount of scintillation light collected and the dose delivered. This type of calibration is unique to PSD and this should be kept in mind when using such detectors. One should not think of a PSD in the same way as an IC or a diode. Nevertheless, once properly calibrated, PSDs are highly accurate dosimeters.

REFERENCES

Alfonso R, Andreo P, Capote R et al. (2008) A new formalism for reference dosimetry of small and nonstandard fields. *Medical Physics* 35:5179–5186.

Archambault L, Arsenault J, Gingras L et al. (2005) Plastic scintillation dosimetry: Optimal selection of scintillating fibers and scintillators. *Medical Physics* 32:2271.

Archambault L, Beddar AS, Gingras L et al. (2006) Measurement accuracy and Cerenkov removal for high performance, high spatial resolution scintillation dosimetry. *Medical Physics* 33:128–135.

Archambault L, Beddar AS, Gingras L et al. (2007) Water-equivalent dosimeter array for small-field external beam radiotherapy. *Medical Physics* 34:1583.

Archambault L, Briere TM, Pönisch F et al. (2010) Toward a real-time in vivo dosimetry system using plastic scintillation detectors. *International Journal of Radiation Oncology, Biology, Physics* 78:280–287. doi:10.1016/j. ijrobp.2009.11.025.

Azangwe G, Grochowska P, Georg D et al. (2014) Detector to detector corrections: A comprehensive experimental study of detector specific correction factors for beam output measurements for small radiotherapy beams. *Medical Physics* 41:072103. doi:10.1118/1.4883795.

Baldock C, De Deene Y, Doran S et al. (2010) Polymer gel dosimetry. *Physics in Medicine and Biology* 55:R1–63. doi: 10.1088/0031-9155/55/5/R01.

Beddar AS (1994) A new scintillator detector system for the quality assurance of 60Co and high-energy therapy machines. *Physics in Medicine and Biology* 39:253–263.

Beddar AS, Kinsella K, Ikhlef A, Sibata C (2001) A miniature "scintillator-fiberoptic-PMT" detector system for the dosimetry of small fields in stereotactic radiosurgery. *IEEE Transactions on Nuclear Science* 48:924–928.

Beddar AS, Mackie TR, Attix FH (1992) Water-equivalent plastic scintillation detectors for high-energy beam dosimetry: II. Properties and measurements. *Physics in Medicine and Biology* 37:1901–1913.

Beddar S (2015) Real-time volumetric scintillation dosimetry. *Journal of Physics Conference Series* 573:012005. doi: 10.1088/1742-6596/573/1/012005.

Cui J-Y, Pratx G, Prevrhal S, Levin CS (2011) Fully 3D list-mode time-of-flight PET image reconstruction on GPUs using CUDA. *Medical Physics* 38:6775–6786. doi: 10.1118/1.3661998.

Feygelman V, Opp D, Javedan K et al. (2010) Evaluation of a 3D diode array dosimeter for helical tomotherapy delivery QA. *Medical Dosimetry* 35:324–329. doi:10.1016/j.meddos.2009.10.007.

Fontbonne J, Iltis G, Ban G et al. (2002) Scintillating fiber dosimeter for radiation therapy accelerator. *IEEE Transactions on Nuclear Science* 49:2223–2227.

Frelin A, Fontbonne J, Ban G et al. (2008) The DosiMap, a new 2D scintillating dosimeter for IMRT quality assurance: Characterization of two Čerenkov discrimination methods. *Medical Physics* 35:1651.

Gagnon J-C, Thériault D, Guillot M et al. (2012) Dosimetric performance and array assessment of plastic scintillation detectors for stereotactic radiosurgery quality assurance. *Medical Physics* 39:429. doi:10.1118/1.3666765.

Georgiev T, Lumsdaine A (2012) The multifocus plenoptic camera. In: *Proceedings of SPIE*, pp. 829908–829908–11.

Glaser AK, Davis SC, McClatchy DM et al. (2013) Projection imaging of photon beams by the Čerenkov effect. *Medical Physics* 40:012101. doi: 10.1118/1.4770286.

Gore JC, Kang YS (1984) Measurement of radiation dose distributions by nuclear magnetic resonance (NMR) imaging. *Physics in Medicine and Biology* 29:1189. doi: 10.1088/0031-9155/29/10/002.

Goulet M, Archambault L, Beaulieu L, Gingras L (2013) 3D tomodosimetry using long scintillating fibers: a feasibility study. *Medical Physics* 40:101703. doi: 10.1118/1.4819937.Goulet M, Rilling M, Gingras L et al. (2014) Novel, full 3D scintillation dosimetry using a static plenoptic camera. *Medical Physics* 41:082101. doi:10.1118/1.4884036.

Guillot M, Beaulieu L, Archambault L et al. (2011a) A new water-equivalent 2D plastic scintillation detectors array for the dosimetry of megavoltage energy photon beams in radiation therapy. *Medical Physics* 38:6763. doi:10.1118/1.3664007.

Guillot M, Gingras L, Archambault L et al. (2010) Toward 3D dosimetry of intensity modulated radiation therapy treatments with plastic scintillation detectors. *Journal of Physics Conference Series* 250:012006.

Guillot M, Gingras L, Archambault L et al. (2011b) Spectral method for the correction of the Cerenkov light effect in plastic scintillation detectors: A comparison study of calibration procedures and validation in Cerenkov light-dominated situations. *Medical Physics* 38:2140. doi:10.1118/1.3562896.

Guo PY, Adamovics JA, Oldham M (2006) Characterization of a new radiochromic three-dimensional dosimeter. *Medical Physics* 33:1338–1345.

Hartmann B, Martišíková M, Jäkel O (2010) Technical note: Homogeneity of Gafchromic EBT2 film. *Medical Physics* 37:1753–1756.

Jang B, Kaeli D, Do S, Pien H (2009) Multi GPU implementation of iterative tomographic reconstruction algorithms. In: *IEEE International Symposium on Biomedical Imaging: From Nano to Macro*, pp. 185–188.

Kirov AS, Piao JZ, Mathur NK et al. (2005) The three-dimensional scintillation dosimetry method: Test for a 106Ru eye plaque applicator. *Physics in Medicine and Biology* 50:3063–3081. doi: 10.1088/0031-9155/50/13/007.

Klein D, Briere TM, Kudchadker R et al. (2012) In-phantom dose verification of prostate IMRT and VMAT deliveries using plastic scintillation detectors. *Radiation Measurements* 47:921–929. doi:10.1016/j.radmeas.2012.08.005.

Kroll F, Pawelke J, Karsch L (2013) Preliminary investigations on the determination of three-dimensional dose distributions using scintillator blocks and optical tomography. *Medical Physics* 40:082104. doi: 10.1118/1.4813898.

Lacroix F, Archambault L, Gingras L et al. (2008) Clinical prototype of a plastic water-equivalent scintillating fiber dosimeter array for QA applications. *Medical Physics* 35:3682.

Lacroix F, Beaulieu L, Archambault L, Beddar A (2010a) Simulation of the precision limits of plastic scintillation detectors using optimal component selection. *Medical Physics* 37:412.

Lacroix F, Beddar AS, Guillot M et al. (2009) A design methodology using signal-to-noise ratio for plastic scintillation detectors design and performance optimization. *Medical Physics* 36:5214. doi:10.1118/1.3231947.

Lacroix F, Guillot M, McEwen M, Cojocaru C (2010b) Extraction of depth-dependent perturbation factors for parallel-plate chambers in electron beams using a plastic scintillation detector. *Medical Physics* 37:4331. doi:10.1118/1.3463383.

Lamanna E, Fiorillo AS, Gallo A et al. (2013) Dosimetric study of therapeutic beams using a homogeneous scintillating fiber layer. *IEEE Transactions on Nuclear Science* 60:109–114. doi: 10.1109/TNS.2012.2223231.

Letourneau D, Pouliot J, Roy R (1999) Miniature scintillating detector for small field radiation therapy. *Medical Physics* 26:2555.

Morin J, Béliveau-Nadeau D, Chung E et al. (2013) A comparative study of small field total scatter factors and dose profiles using plastic scintillation detectors and other stereotactic dosimeters: The case of the CyberKnife. *Medical Physics* 40:011719. doi:10.1118/1.4772190.

Nelms BE, Zhen H, Tomé WA (2011) Per-beam, planar IMRT QA passing rates do not predict clinically relevant patient dose errors. *Medical Physics* 38:1037. doi: 10.1118/1.3544657.

Petric M, Robar J, Clark B (2006) Development and characterization of a tissue equivalent plastic scintillator based dosimetry system. *Medical Physics* 33:96.

Wolfsberger LD, Wagar M, Nitsch P et al. (2010) Angular dose dependence of Matrixx TM and its calibration. *Journal of Applied Clinical Medical Physics* 11:3057.

Yin Y, Lambert J, McKenzie DR, Suchowerska N (2008) Real-time monitoring and diagnosis of scintillation dosimeters using an ultraviolet light emitting diode. *Physics in Medicine and Biology* 53:2303–2312. doi:10.1088/0031-9155/53/9/007.

7
Small field and radiosurgery dosimetry

Kamil M. Yenice, David Klein, and Dany Theriault

Contents

7.1 INTRODUCTION

Small photon fields have been used in stereotactic radiosurgery (SRS) for many years. Their use has recently been expanded into other applications of radiation therapy, as technological improvements in mechanical accuracy, stability, and dosimetric control made it possible for linacs to deliver complex intensity modulated radiotherapy (IMRT) and volumetric arc therapy (VMAT) that are composed of small fields.

The impact of dosimetry of small fields on treatment quality also increased due to significant challenges not encountered in standard photon dosimetry before the era of IMRT/VMAT and SRS with small multileaf collimator (MLC) fields. Therefore, all aspects, including dosimetry, detector response, beam modeling, and geometry, of small fields need to be accurately characterized in order, so that treatment planning systems can correctly predict the delivered dose using small photon fields.

Primary challenges of small photon field dosimetry include the lack of lateral charged particle equilibrium, partial geometrical shielding of the primary photon source as seen from the point of measurement, and detector size and composition perturbing the field. Differences in the output factor (OF) measurements for collimated radiosurgery beams of diameters less than 2 cm have been reported to be up to 12% by Das et al. (2000). Inappropriate selection of a detector in the commissioning of small-field OFs resulted in mistreatment of 145 patients in Toulouse, France, in 2006–2007 (Derreumaux et al., 2008) and 152 patients in Springfield, Missouri, from 2004 to 2009 (Solberg, Ph, & Medin, 2011). These incidents have indicated the continuing challenges in the determination of small field parameters for high-energy photon beams used in radiosurgery delivery.

Accurate characterization of small field dosimetry requires measurements to be made with precisely aligned specialized detectors. Significant development in the understanding of small field dosimetry and new detectors including plastic scintillators (PSs) have taken place in the last 5 years since the publication of the report on small field MV photon dosimetry by the Institute of Physics and Engineering in Medicine (IPEM) (Aspradakis et al., 2010). A task group on small fields and nonequilibrium condition photon beam dosimetry has been commissioned by the American Association of Physicists in Medicine (AAPM) and their report is in the publication stage (Das et al., n.d.).

This chapter reviews the physics of small fields and various detectors available for small field dosimetry and photon beam measurements. Available data for small field dosimetry and measurement methodology, comparison among detectors, and the validity of beam data by various detectors for small fields are reviewed and advantages of plastic scintillating detectors are presented under the light of this discussion.

7.2 SMALL FIELD PROBLEM: *HOW SMALL IS SMALL?*

At present, a small field is generally defined as having dimensions smaller than the lateral range of the charged particles that contribute to the dose at a point along the central axis (Alfonso et al., 2008). These conditions imply that field sizes of less than 3×3 cm^2 are considered to be small for a 6 MV photon beam that needs special attention in both dose measurements and beam modeling. As the field size reduces further, the OFs (the ratio of dose for a particular field size to that for a reference field, e.g., 10×10 cm^2, at a reference depth in water on the central axis) vary rapidly with field size and effects of measurement uncertainties become increasingly significant. It is clear that collimation of a photon beam from a source of finite size will partially occlude the source from the detector's point of view when the field size becomes increasingly narrow beyond a certain field size. The conventional approach of classifying field size based on the full width at half maximum of corresponding dose profiles does not apply in this situation due to the reduction in output on the central axis and overlapping penumbrae (Das et al. 2008). Measurements of OFs for such small fields require even more careful approach, including very precise detector alignment and the measurement of dosimetric field size at the same time as OFs for each field size setting (Cranmer-Sargison, Charles, Trapp, & Thwaites, 2013).

Recently, Charles et al. (2014) introduced the concept of a very small field size based on practical and theoretical considerations. They defined a field to be very small if its output changed by ±1.0% as a result of a change in either the field size or detector position of up to ±1 mm. This criterion practically establishes a field size threshold below which the use of the nominal field size rather than the dosimetric field size would cause a significant impact on the accuracy of reported OFs. Their analysis showed that field sizes ≤15 mm were considered to be very small for 6 MV beams for maximal field size uncertainties of 1 mm. Alternatively, they showed that a very small field size also coincided to the field size when lateral electronic disequilibrium would cause a greater change in OFs than any other effects including photon scatter in the phantom and source occlusion. This effect was found to dominate at field sizes ≤12 mm (side of a square field size) for 6 MV beams.

7.2.1 EFFECTS OF THE RADIATION SOURCE SIZE

The X-ray photon fluence generated by an electron linear accelerator comprise a direct-beam radiation generated from a radiation source at the level of target (focal spot) and an indirect or extra-focal component consisting of photons scattered at structures near or below the target including primary collimator, flattening filter, and secondary collimators. For broad beams, the extrafocal radiation can account for about 8% of the beam output for a 6 MV beam and originates primarily from the flattening filter (Jaffray, 1993). The direct beam source or focal spot is characterized by the full width at half maximum of the bremsstrahlung photon fluence distribution exiting the target and is typically represented by a Gaussian distribution. The size and shape of the focal spot also depend on the shape of the profile of the electron beam incident on the target (Wang & Leszczynski, 2007).

The amount of collimation determines the amount of flattening filter visible from the point of measurement. As the field size is made smaller, less of the flattening filter, as the main component of indirect beam source, is exposed and visible, and therefore extrafocal radiation contribution becomes less important to measured or calculated dose in small fields. At smaller yet collimator settings, the direct beam source becomes occluded by the collimation as *seen* from the point of measurement and the number of primary photons reaching the detector or calculation point from the direct radiation source is reduced. The direct-source occlusion starts occurring at larger collimator settings for larger focal spot sizes and therefore extended beam source is an important parameter for accurate beam modeling in small fields (Treuer et al., 1999). For a 6 MV beam from a Varian iX machine, the direct source occlusion was shown to dominate the small field characteristics at field sizes ≤ 8 mm (Charles et al., 2014).

Although the direct photon beam source is an important parameter for treatment planning systems (TPS), its determination is not typically required as part of the clinical commissioning of a TPS for small fields. In the beam modeling, a typical value is entered as a starting point and iteratively modified until calculated and measured penumbra values match for the range of clinical field sizes to be used. Direct determination of the photon source size requires specialized equipment and sophisticated mathematical analysis. Various methods have been published using foil activation (Munro, Rawlinson, & Fenster, 2012), spot camera technique (Caprile & Hartmann, 2009; Jaffray, 1993; Munro et al., 2012) and Monte Carlo (MC) technique (Wang & Leszczynski, 2007). MC dose engines, on the other hand, require the shape and angular spread of the electron source incident on the target as input parameters, and they are iteratively adjusted until the calculated profile penumbrae and OFs agree with the measurements. Typical values of the full width at half maximum (FWHM) of the electron beam intensity distribution varies between 1 and 1.5 mm in MC simulations of broad photon fields and they only affect the calculation of beam profiles (Sheikh-Bagheri & Rogers, 2002). Scott et al. (Scott, Nahum, & Fenwick, 2009) used a BEAMnrc MC linac model to investigate the variation in penumbra widths and small field OFs with electron spot size. They showed that a 0.3 mm uncertainty in the FWHM of the electron focal spot could vary the photon beam output by 5% for a 5 mm × 5 mm field size at 15 MV. Their MC calculations also showed that a FWHM of 0.7 mm incident electron beam corresponding to a FWHM of 2.1 mm for the direct photon source produced the best agreement with measurements.

7.2.2 ELECTRON RANGE AND LOSS OF CPE

Dose deposition from a photon beam is a two-stage process: incident photons first transfer their energy to secondary charged particles (electrons and positrons) in the medium through photon interactions and these secondary charged particles, having finite ranges based on the energy they have received from primary photon interactions, deliver dose away from the interaction point. For a broad photon beam, charge particle equilibrium is said to exist on the central axis where laterally ejected secondary electrons are replaced by laterally displaced electrons from adjacent regions. In narrow photon fields and/or at high energies, this balance is disrupted and lateral charge particle disequilibrium occurs when the beam radius becomes small in comparison to the maximum range of secondary electrons. Those laterally ejected secondary electrons from the beam axis are no longer compensated by equal numbers displaced laterally from elsewhere in the beam. Because the maximum range of secondary electrons is dependent on beam energy and the composition of the interaction material, the charged-particle disequilibrium is exacerbated in higher beam energies and by the presence of tissue heterogeneities in the treatment sites. A linear relationship between the beam quality ($TPR_{20,10}$) and the minimum beam radius required to achieve lateral electron equilibrium (LEE) was derived by evaluating how the ratio of total dose to total kerma varied with beam radius by Li et al. (Li, Soubra, Szanto, & Gerig, 1995) to be

$$r_{LEE} \ [g/cm^2] = 5.973 \ \{TPR_{20,10}\} - 2.688 \tag{7.1}$$

The authors recommended that this relation not be used for high-Z materials and energies higher than 15 MV.

Figure 7.1 shows values for r_{LEE} in water as a function of beam quality as calculated by Li et al. (1995). The minimum radii for a 6 MV ($TPR_{20,10} = 0.670$) and a 10 MV ($TPR_{20,10} = 0.732$) corresponded to 1.3 (g/cm^2) and 1.7 (g/cm^2), respectively, in their analysis. Iwasaki (1996) experimentally determined the minimum square field size for a 10 MV beam for which LEE is established. He took ion chamber measurements in cork and determined the ratio of chamber readout with and without the built-up cap for increasing water equivalent field sizes to deduce at the field size; LEE was established when this ratio became unity. This analysis exploited the theorems of Fano (2010) and O'Connor (1957) and showed that the side of the minimum water equivalent square field for which LEE existed in a 10 MV beam was 35 mm (corresponding to the minimum radius of an equivalent square field for LEE of 19.6 mm). Bjärngard and Petti used a model to represent scatter to primary kerma ratios as a function of beam radius and depth and concluded that for a 6 MV beam LEE was established at beam radii equal to or greater than 10 mm (Bjärngard & Petti, 1988).

7.2.2.1 Detector composition

A measurement of dose-to-water in the absence of the detector for a small field (under lateral charge particle disequilibrium conditions) produces a different response than that for a broad photon field (under lateral electronic equilibrium conditions) with the same detector at the same measurement point in the phantom (Bouchard, Seuntjens, Carrier, & Kawrakow, 2009; Crop et al., 2009; Francescon, Kilby, Satariano, & Cora, 2012). Differences between readings of a very small (0.5 × 0.5 cm²) field and a reference (10 × 10 cm²) field could be as much as several tens of a percent even for small ion chambers, diodes, and

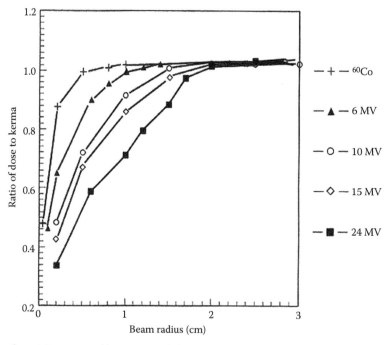

Figure 7.1 Ratios of total dose to total kerma scored along the central axis at 5 cm depth in a water phantom versus the incident narrow beam radii defined at 100 cm SSD for X-ray and at 80 cm SSD for ^{60}Co. (From Li, X. A. et al., *Med. Phys.*, 22, 1167, 1995, http://doi.org/10.1118/1.597508. With permission.)

diamond detectors with detecting volumes of 1–3 mm in diameter (Sánchez-Doblado et al., 2007). The nature of this difference has been found, from MC simulations of detector effects in small fields, to be primarily due to density differences between detector sensitive volumes and water, to dose-averaging over detector sensitive volumes and to a lesser degree of differences in atomic number (Bouchard et al., 2009; Crop et al., 2009; Ding & Ding, 2012; Scott, Kumar, Nahum, & Fenwick, 2012).

The effect of density on detector response is twofold: first, it affects the number of atoms a particle encounters on a given path; second, it also affects electron stopping powers via the polarization effect. It is useful to separate the effect of density of detecting material on detector response from that of mass stopping power to gain better insight into density effects. The combined impact of the density and atomic composition of the sensitive volume on its response is characterized using a ratio, $F_{w,det}$, of doses absorbed by equal volumes of unit density water and detector material co-located within a unit density water phantom (Fenwick, Kumar, Scott, & Nahum, 2013; Scott et al., 2012). Fenwick et al. (2013) isolated the effect of density alone through a similar ratio, P_ρ, of doses absorbed by equal volumes of unit and modified density water (to that of the detector sensitive material). They used a simple cavity theory by splitting the dose absorbed by the sensitive volume into two components, imparted by electrons liberated in photon interactions occurring inside and outside the volume, and showed that P_ρ, was related to the degree of on-axis electronic equilibrium, s_{ee}, a normalized ratio of dose to collision kerma, through a cavity-specific parameter I_{cav} that is determined by the density and geometry of the sensitive volume. The relative variation of s_{ee} in small fields is similar to that of the conventional phantom scatter factor S_p. Following the scheme of Bouchard et al. (2009), $F_{w,det}$ can be written as

$$F_{w,det} = P_\rho P_{fl-} \left[\frac{L_\Delta}{r} \right]_{det}^{w} \qquad (7.2)$$

where:

$[L_\Delta/r]_{det}^{w}$ is the restricted mass stopping power ratio of water to detector material

P_{fl-} is a factor that accounts for any perturbation of the electron fluence in the cavity caused by its non-water equivalent atomic composition

It turns out that both the factor P_{fl-} and the water to detector stopping power ratio, $[L_\Delta/r]_{det}^{w}$ vary minimally for detector materials of silicon, diamond, and pinpoint-like cavity in small fields of width 0.25–10 cm (Fenwick et al., 2013; Scott et al., 2012). Therefore, when the sensitive volume of a detector is irradiated by a field too small to establish lateral electronic equilibrium, the absorbed dose by the sensitive volume depends primarily on its density and the most of field size dependence of $F_{w,det}$ originates from the factor P_ρ. Fenwick et al. (2013) further pointed out that one can take advantage of the fact that P_ρ takes a value of unity at the center of fields wide enough to establish lateral electronic equilibrium to calibrate small field detectors in 3 × 3 or 4 × 4 cm² fields (wide enough for lateral electronic equilibrium for 6 and 15 MV), minimizing the number of correction factors needed for small-field measurements.

7.2.2.2 Energy dependence

Highly collimated narrow photon fields in SRS or IMRT treatments have different energy spectra than that for a broad beam. While the collimation hardens the primary photon beam, shifting the spectrum toward higher energies, this is somewhat offset by the build-up of secondary photons and the increased pair production in the collimation system that increases the lower energy component of the spectrum. On the other hand, if the beam radius is less than the maximum range of primary electrons, the electron energy spectrum at the beam central axis will be deprived of those lower energy electrons originating from points at distances just less than the maximum range, which would have reached the central axis near the end of their travel. The average energy in the electron spectra would then be expected to increase due to the lack of lateral electron equilibrium. In this case, the ratio of these radiation responses such as ionization readings cannot be simply equated with relative dose OF of the narrow beam without consideration for the spectral changes in converting the ion chamber readings to absorbed dose in the medium (Wu, Zwicker, Kalend, & Zheng, 1993).

To study the impact of collimation system on beam spectral changes, Sánchez-Doblado et al. (2003) performed MC simulations for two 6 MV clinical accelerators (Elekta SL-18 and Siemens

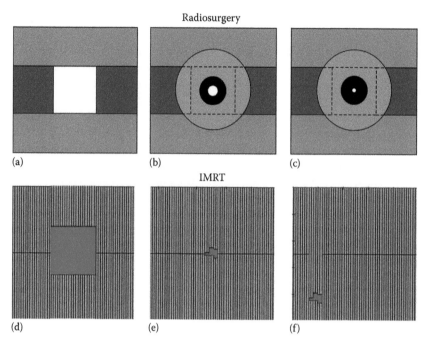

Figure 7.2 The collimator openings for radiosurgery and IMRT delivery: (a) 10 × 10 cm² reference field with collimators; (b) φ = 1.05 cm cone; (c) φ = 0.3 cm cone; (d) 10 × 10 cm² reference field with MLC; (e) on axis: 2 × 2 cm² irregular segment on CAX; and (f) off axis: 2 × 2 cm² irregular segment off CAX. (Adapted from Sánchez-Doblado, F. et al., *Phys. Med. Biol.*, 48, 2081, 2003, http://doi.org/10.1088/0031-9155/48/21/L02. With permission.)

MevatronPrimus), equipped with radiosurgery applicators and MLC. They simulated narrow circular and irregular-shaped on-axis and off-axis fields, as well as broad IMRT configured beams together with reference 10 × 10 cm² beams, as seen in Figure 7.2.

Figure 7.3 shows 6 MV photon fluence spectra in air at the position of the photon surface generated by MC simulations corresponding to these small fields shaped with MLC and radiosurgery cone apertures as shown in Figure 7.2. The upper panels correspond to the relative fluence, normalized to the integral fluence. There is a large spectral difference between the open 10 × 10 cm² beams and the narrow irregular or circular fields with the small-field spectra being shifted toward higher energies, even if the mean energies are somewhat softened by the pair production peak.

In water, the mean energy of photon fluence distribution increases with decreasing field size due to reduced photon scatter with decreasing field size. Figure 7.4 shows the beam spectra calculated by Eklund & Ahnesjö (2008) by integration of the fluence pencil kernels for five square field sizes. The spectra have been calculated at 5 cm depth at the central axis for the incident 6 MV beam and separated into particle categories of primary electrons, scattered electrons, total electron spectrum, and scattered photons (including bremsstrahlung and annihilation photons, and photons produced from primary and scatter electron ionizations). The total electron fluence at the measurement point varies with field size and depth in water influencing the calculated ratios of restricted mass collision stopping powers for conversion of detector reading into absorbed dose in medium (Wu et al., 1993).

For small sensitive volumes for which the Bragg-Gray conditions hold, energy response is dominated by stopping power ratios, while for large volumes the response is dominated by the mass energy absorption ratio and dose is predominantly arisen from photon interactions within the detector sensitive volume. Wu et al. (1993) calculated the mean restricted stopping power ratio of water to air of a 6, 10, and 15 MV photon beam for 10 × 10 cm² and 0.5 × 0.5 cm² field sizes from MC-generated electron energy spectra at 8 cm depth. They found that the ratio of broad beam and narrow beam values, $[(L/\rho)_{w,a}]_{10\times10}/[(L/\rho)_{w,a}]_{0.5\times0.5}$, are 0.997, 0.991, and 0.990 for 6, 10, and 15 MV X-rays, respectively. Sánchez-Doblado et al. (2003) showed that energy dependence of ion chamber response leads to a variation with field size of around 0.5% in the ratio of the measured ionization to dose to

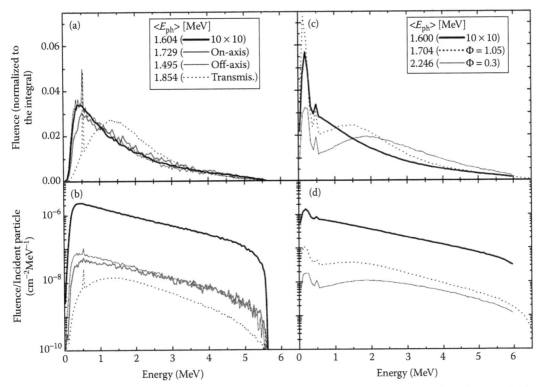

Figure 7.3 Monte Carlo calculated photon fluence in air, at the position of the phantom surface, for 6 MV fields produced with the MLC of a Siemens Mevatron Primus (a) and (b) and the radiosurgery applicators of an Elekta SL-18 linac (c) and (d). The upper panels correspond to the relative fluence, normalized to the integral fluence in each case. The mean energy of the photon spectra is indicated in each case. (From Sánchez-Doblado, F. et al., *Phys. Med. Biol.*, 48, 2081, 2003, http://doi.org/10.1088/0031-9155/48/21/L02. With permission.)

water for a 6 MV beam. For solid state detectors such as a silicon diode, the variation of both mass energy absorption ratio and stopping power ratio of water to silicon with either field size or depth is small, less than 1.1%, within the calculation uncertainties (Scott, Nahum, & Fenwick, 2008).

7.2.3 DETECTOR SIZE

7.2.3.1 Volume averaging

It is known that the penumbral width of measured cross-beam profiles becomes artificially flattened when a detector of finite size is used due to substantial variations of dose over the detector dimensions in the penumbral dose gradient region (García-Vicente, Delgado, & Peraza, 1998; Johns & Darby, 1950; Metcalfe, 1993). While dose averaging over the finite volume of a detector leads to apparent penumbra broadening, this effect is confounded by factors related to detector composition and type and it is offset to some extent by penumbral *sharpening* and profile alteration by solid-state dosimeters such as silicon diodes and diamond detectors (Beddar, Mason, & O'Brien, 1994; Pappas et al., 2006). Mathematically, the effect of the finite size of any detector on profile measurement can be described as a convolution of a kernel ($K(x)$) representative of a measuring system with the real profile

$$D_m(x) = \int_{-\infty}^{+\infty} D(u) K(u - x) \, du \tag{7.3}$$

where:

$D(u)$ represents the real profile of a beam
$D_m(x)$ is the measured profile

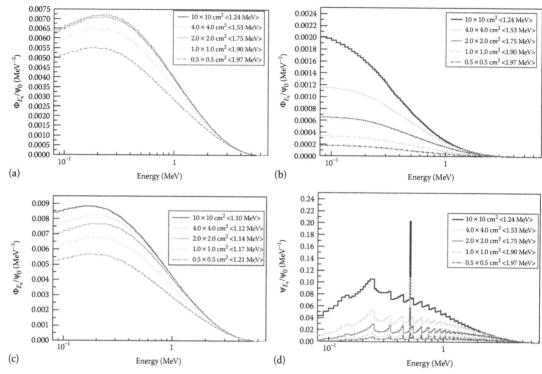

Figure 7.4 (a) Primary, (b) scatter, (c) total electron spectra, and (d) scatter photon spectra from five different square fields (0.5 × 0.5 cm², 1.0 × 1.0 cm², 2.0 × 2.0 cm², 4.0 × 4.0 cm², and 10 × 10 cm²) at 5 cm depth on the central axis for the 6 MV incident beam. The spectra are normalized using the total incident energy. (From Eklund, K. and Ahnesjö, A., *Phys. Med. Biol.*, 53, 4231, 2008, http://doi.org/10.1088/0031-9155/53/16/002. With permission.)

It is possible to correct for the spatial response of finite-sized detectors by deconvolution of the measured profiles with the known kernel of a measuring system (e.g., densitometer for film measurements) and to obtain the *real* penumbra width of cross-beam profiles from ion chamber or other detectors. Alternatively, an experimental approach could be taken and detectors of various sizes are used to measure cross-beam profiles and by extrapolation of measurements to zero detector size, the *true* penumbra width could be assessed (Dawson, 1984, 1986; Rice, Hansen, Svensson, & Siddon, 1987). However, conventional detectors (i.e., air ion chambers and diodes) present some additional challenges for this kind of analysis since a range of detectors with different sizes is needed, and non-tissue equivalency of many detector types makes it difficult to assess response kernels of these systems. More advanced experimental techniques such as gel dosimetry and MC methods have especially become valuable to study the effect of detector size on measurements in small photon beams (Pappas et al., 2006). Pappas et al. measured profiles of 5 mm diameter 6 MV stereotactic beam using polymer gel and magnetic resonance imaging dosimetry and showed that in addition to detector size, its composition and water equivalency were significant factors for correct narrow beam profile measurements. Other studies have shown that a detector with an active area of 1 mm² or less has negligible volume averaging at a typical linear accelerator field size of 5 mm (Cranmer-Sargison, Weston, Evans, Sidhu, & Thwaites, 2012; Scott et al., 2012).

Detector size also becomes an important issue when measuring the OFs for small radiosurgery fields. The measurement is meaningful only if the dose is uniform over the dimensions of the detector. However, for narrow beams, the entire field may be covered by the penumbrae (of opposing field edges) and even detectors that are normally thought of as small, may be large enough that a nonuniform dose is averaged over the detector active volume, making the OF appear to be smaller than it is. This is a particularly serious problem that could potentially result in overdosing the patient. Correcting the signal of a detector under nonstandard conditions is not trivial since response of detectors that are used for small-field measurements can vary with field size (Alfonso et al., 2008; Ding & Ding, 2012). Alfonso et al. (2008) introduced a new formalism for reference dosimetry, where a detector response variation with field size could be corrected for by using the factor $k_{Q_{clin},Q_{msr}}^{f_{clin},f_{msr}}$. An increasing number of studies using experimental (Bassinet et al., 2013; Pantelis et al., 2012; Ralston, Liu, Warrener, McKenzie, & Suchowerska, 2012) and MC techniques (Benmakhlouf, Sempau, & Andreo, 2014; Charles et al., 2013; Cranmer-Sargison et al., 2012; Cranmer-Sargison, Weston, Sidhu, & Thwaites, 2011; Czarnecki & Zink, 2013; Francescon et al., 2012) published these correction factors for various detectors used in small field dosimetry. It has been shown that detectors with the smallest active region dimensions are not always the most suitable detectors for OF determination in small fields, as MC calculations show that correction factors for some silicon diodes with small dimensions exceed those for intermediate-sized diode (Benmakhlouf et al., 2014). It is recommended that detector output correction factors (that account for detector volume averaging effects) be given for specific linac models and field sizes, rather than for a beam quality specifier that necessarily varies with the accelerator type and field size due to the different electron spot dimensions and photon collimation systems used by each accelerator model.

7.2.3.2 Field-perturbation: Departure from Bragg-Gray cavity

In conventional dosimetry, the calculation of absorbed dose from the response of detector relies on the existence of an appropriate cavity theory. Cavity theories such as Bragg-Gray, Spencer-Attix, and Burlin cavity theory are traditionally applied to gas-filled ionization chambers (ICs), where the mass of gas is so small that the presence of a discontinuity in density does not significantly perturb the radiation fluence. The absorbed dose in the medium can then be determined from the ionization in the cavity. On the other hand, with solid-state detectors the mass of material in the detector is generally much greater, so that cavity theories have to be applied with caution. For example, for a silicon diode of an active diameter of 2 mm and thickness of 0.06 mm, the sensitive volume is no longer small compared to the range of Compton electrons. Because of the different densities of air ($\rho = 0.001205$ g/cm^3) and silicon ($\rho = 2.329$ g/cm^3), the range of electrons corresponds to 4 m in air versus 2 mm in silicon, respectively.

Megavoltage photon beams incident on a low-Z absorbing medium generate Compton scatter that is predominantly below 1 MeV. This scatter component influences the dose response, particularly for diode detectors as the field size and depth increase. The diode readout to absorbed dose conversion then requires a general cavity theory with the detector partly behaving as a photon detector (large cavities as the source of the entire electron fluence generation) and as an electron detector (small non-perturbing Bragg-Gray cavities for which all the electron fluence is generated outside the cavity) (Yin, Hugtenburg, & Beddoe, 2004). Burlin (1966) first characterized a general cavity theory and included a parameter $(1 - d)$ representing the average electron fluence generated by photon interactions occurring within the cavity, relative to the equilibrium electron fluence generated in an infinitely large volume of cavity material. The relative diode detector response at depth z_{ref} for a field size r with respect to that at depth z_{ref} for a reference field size r_{ref} could be written according to Burlin theory as

$$\frac{D(r,z_{ref})}{D(r_{ref},z_{ref})} = \frac{M(r,z_{ref})}{M(r_{ref},z_{ref})} k_{det\,r_{ref}}^{r} k_{p\,r_{ref}}^{r} \tag{7.4}$$

where:

$M(r,z_{ref})/M(r_{ref},z_{ref})$ is the relative detector signal

the factor $k_{det\,r_{ref}}^{r}$ determines corrections for large and small cavity effects and is given by

$$k_{det\,r_{ref}}^{r} = \left[d \left(\frac{\overline{S_\Delta}}{\rho} \right)_{det}^{med} + (1-d) \left(\frac{\overline{\mu}_{en}}{\rho} \right)_{det}^{med} \right]_{r_{ref}}^{r} \tag{7.5}$$

In the large cavity approximation, where all electron fluence is assumed to be generated within the detector cavity, the parameter $d = 0$, whereas in the small cavity situation, where all the electron fluence is assumed to be generated outside the cavity, $d = 1$, and the value of d is estimated from the electron fluence generated in the cavity of detector and the electron fluence entering the cavity. As mentioned in Section 7.2.2.1, the values of both the mass stopping power ratios and mass energy absorption coefficients (air to water and silicon to water) vary slowly with energy and field size (Eklund & Ahnesjö, 2010; Scott et al., 2008). But for silicon, the mass energy absorption coefficient ratio $(\bar{\mu}_{en}/\rho)_{det}^{med}$ increases almost by a factor of 8 at low photon energies. Thus, detector signal increases substantially with increasing field size and increasing number of low-energy photons. As a result, the ratio of mass energy coefficient decreases with respect to that for larger fields. The last term in Equation 7.4, $k'_{p_{r_{ref}}}$ is the ratio of detector perturbation factors for field sizes r and r_{ref}, which account for perturbation of the electron fluence due to the presence of the cavity and volume averaging. For solid-state detectors, $k'_{p_{r_{ref}}}$ can be assumed to be unity since the detector size is small.

7.2.3.3 Signal-to-noise ratio

For small nonequilibrium field dosimetry, the requirement of a small detector size conflicts with the high signal-to-noise ratio requirement, especially with stem effect becoming a significant factor in relative dose measurements with respect to a broad reference field. Although higher Z elements as detecting material facilitates higher detector signal and high-resolution measurements, they confound the dosimetry measurements when the sensitive volume of a detector is irradiated by a field too small to establish lateral electronic equilibrium since the absorbed dose by the sensitive volume depends on its density (Scott et al., 2008). Fenwick et al. suggested some practical ways of limiting the size of small-field density-dependent correction by constructing a detector whose sensitive volume has a density close to that of water (e.g., a liquid ion chamber) or reducing the thickness of the cavity in the direction of the radiation beam to minimize the impact of cavity effects from their cavity theory (Fenwick et al., 2013). They also suggested building other regions of different density into the detector to offset the non-water equivalent density of the sensitive volume (e.g., adding a high-density structure to compensate a low-density cavity).

7.3 DETECTORS FOR SMALL FIELD DOSIMETRY

As detailed in Section 7.2, the nature of small fields pose significant problems for the direct measurement of the dose deposited by narrow SRS beams and highly modulated IMRT and VMAT fields composed of many small features. MC calculations provide an excellent avenue for ascertaining doses deposited in complex beam geometries, but any validation scheme attempting to confirm the veracity of something like a radiotherapy treatment plan is somewhat deficient without direct dose measurement. Unfortunately, the rapid development of technology capable of delivering small field radiotherapy is outpacing the development of suitable detector technology, as well as standards for measurement technique. This is not to say that we are currently unable to characterize small fields. Many contemporary detector designs are capable of measuring certain aspects of small radiation fields with a fairly good understanding of their limitations.

In this section, several of the most common detectors that have been used for small field dosimetry will be described. For each detector type, a brief summary of the principle(s) behind their operation will be given, followed by an examination of the relative advantages and disadvantages inherent to the detectors relating to small field dosimetry in particular. The section ends with a list of less commonly used detectors and their associated strengths and weaknesses, and a discussion of the selection of detector(s) appropriate to various common cases in which small radiation fields are encountered.

7.3.1 IONIZATION CHAMBERS

7.3.1.1 Air ionization chambers

Air ICs have been long considered the gold standard in radiation detection and dosimetry. Simply put, they collect the charge generated by ionizing radiation passing through a volume of air between two electrodes across which a voltage potential has been applied. The amount of charge collected is proportional to the dose deposited in the sensitive volume, making ICs suitable for absolute dosimetry. This operating principle lends a great deal of flexibility in their design—allowing for a large assortment of chambers, well suited for different tasks. Naturally, this assortment includes miniaturized detectors having collection volumes in the order of 0.01 cm³ (and less) for the sake of combining the characteristic accuracy and precision of ICs with the high spatial resolution of a small detector.

However, volume-averaging effects are apparent in small field dosimetry even at volumes as small 0.01 cc. The performance of standard air ICs is closely tied to their construction geometry—their sensitivity decreases with decreasing volume of air (see Figure 7.5). Thus, air ICs for small-field measurements require trading off between sensitivity and volume averaging. Losses in sensitivity can be mitigated through the use of higher-sensitivity electrometers, but such devices may prove to be prohibitively expensive for routine use. Additional hurdles faced by air ICs in small fields are dose perturbations caused by the air volume; dependence of the response on the orientation of the detector and associated cabling (with differing detector geometries having different dependencies); and the myriad corrections that may need to be applied to account for environmental conditions, level of applied voltage, beam quality, and so on. (Low, Moran, Dempsey, Dong, & Oldham, 2011).

7.3.1.2 Liquid ionization chambers

Using liquid instead of air as the sensitive medium inside an IC grants a much higher ionization density, and therefore much higher sensitivity. The densities of liquids commonly used in liquid ionization chambers (LICs)—for example, isooctane and tetramethylsilane—are roughly 1000 times that of air, and have ionization densities of hundreds of times that of air (Wickman, Johansson, Bahar-Gogani, Holmström, & Grindborg, 1998). Consequently, LICs can be made much smaller than air ICs while maintaining similar sensitivities. Small liquid volumes also lessen the effects of the much shorter diffusion range and higher rate of recombination of the directly ionized charges that are characteristics of these liquid media. Parallel-plate designs with sub-millimeter plate separations are particularly well suited for LICs. For instance, the microLion by PTW (Freiburg, Germany) uses a parallel-plate design with 1.25 mm radius electrodes separated by a mere 0.35 mm. This and similar designs provide a much higher spatial resolution than that provided by air ICs, and show little dependence on incident beam angle (Dasu et al., 1998). In addition, LICs have high long-term stability (Wickman et al., 1998) and the liquids used are nearly water equivalent, and so the collection volumes are much less perturbing than their air-filled counterparts.

The main disadvantage of LICs is the higher rates of recombination of charges within the sensitive volume prior to collection at the electrodes. The high ionization density that allows for submillimeter plate

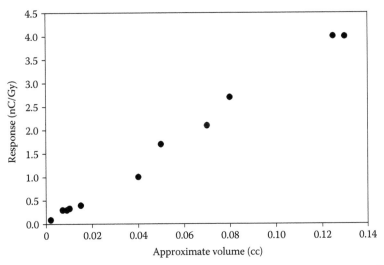

Figure 7.5 Sensitivity in nanocoulomb per gray versus approximate collection volume (cc or cm²) of several commercially available ionization chambers designed for use in small radiation fields. (Data From Low, D.A. et al., Med. Phys., 38, 1313, 2011, http://doi.org/10.1118/1.3514120. With permission.)

separations is also responsible for greater rates of recombination at higher dose rates, when more and more ionization events are occurring within the chamber. Thus, collection efficiencies decrease as dose rate is increased (Andersson, Johansson, & Tölli, 2012; Chung, Davis, & Seuntjens, 2013). Sensitivity loss at higher dose rates can be reduced by using higher bias voltages across the electrodes, or by reducing the plate separation. The higher biases (upward of 800 V or more; Benítez et al., 2013) used for this purpose may not be obtainable with typical electrometers found in clinical settings. Though the sensitive volume is very small, the overall construction of the detector is relatively large. Additional potential weakness of LICs include a temperature dependence of 0.5%/°C (Wickman et al., 1998) and sensitivity to ambient electric fields that increases with bias voltage (Andersson et al., 2012).

7.3.2 FILM

7.3.2.1 Radiographic film

Radiographic film based on silver halide represents a venerable stalwart of radiation measurement. The film itself is made up of a thin paper or plastic sheet that is coated on one or both sides with emulsion consisting of gelatin infused with silver halide. The silver halide crystals also contain defects, which are thought to be responsible for the imaging properties of the emulsion. This mechanism can be briefly described as follows: photons incident to the crystal liberate electrons from halide anions to form neutral halide atoms and free electrons. The free electrons get trapped at the crystal defects and combine with silver cations to form atomic silver. If enough silver atoms accumulate by this process (four or more), a latent-image center is formed (Herz, 1969). The film development process promotes the growth of the silver atom deposits at the sites of crystal defects so that macroscopic silver aggregates are formed (Bushberg, Seibert, Leidholdt Jr., & Boone, 2002). The differing optical densities (i.e., atomic silver concentrations) throughout the developed film form the manifest image.

Film has advantages that make it a strong candidate for small field dosimetry. It is a fully two-dimensional detector that is available in a variety of sizes and has very high inherent spatial resolution. This makes it an ideal tool for characterizing complex dose distributions comprising small fields and/or steep dose gradients. In fact, the spatial resolution of radiographic film is only practically limited by the method used to measure the radiation-induced changes in optical density. Traditional densitometers measure single points using apertures with diameters that are in the order of a millimeter, whereas modern film scanners can provide full two-dimensional readout of the entire film, with resolutions of up to thousands of dots per inch (DPI) or higher.

Nevertheless, performing accurate dosimetry film can be difficult. Task Group 69 of the AAPM describes the many challenges that one must overcome in order to perform accurate dosimetry with radiographic film (Pai et al., 2007). These challenges include the dependence of the film's optical density on energy; dose rate; film orientation with respect to beam direction; inconsistency of silver halide emulsion across film batches, films in the same batch, and even across a single film (making calibration procedures relatively laborious); and the processing and densitometry/scanning conditions. The energy dependence of radiographic film is greatest at energies below the therapeutic range. This is most detrimental at large field sizes and depths due to the greater amount of scattered low-energy photons. However, energy dependence is less significant for small field dosimetry using film (Pai et al., 2007).

7.3.2.2 Radiochromic film

Radiochromic detectors change color at the site of interaction with radiation. The first radiochromic dosimeters were pioneered by McLaughlin in the 1960s (McLaughlin & Chalkley, 1965), and were mainly used in industrial applications. Current radiochromic films, such as the GafChromic™ EBT series of films produced by Ashland (Kentucky), take advantage of the photopolymerization of diacetylene molecules to alter the color characteristics of the film. The degree of color change due to polymerization is proportional to the dose of ionizing radiation received.

Similar to radiographic film, radiochromic film is a thin, flexible detector that exhibits very high spatial resolution in two dimensions. Unlike radiographic film, however, radiochromic film is relatively insensitive to normal room light, eliminating the need for a dark room; and it is self-developing, removing the complications of chemical processing. Other advantages of radiochromic film include water equivalence, has low energy and dose rate dependence in the therapeutic range, and newer versions of the film have a greatly increased dynamic range appropriate for radiotherapy and radiosurgery (Lewis, 2010). Additionally, radiochromic films can be cut to nearly any shape and size for easier integration into a wide range of measurement conditions. These features make the film a very capable detector for small field dosimetry.

The convenience of self-development and easier handling is somewhat offset by some inherent disadvantages, however. Similar to radiographic film, radiochromic film suffers from inconsistency in response between production batches, from film to film, and across the same film (Hartmann, Martišíková, & Jäkel, 2010). The photopolymerization mechanism is temperature and humidity dependent (Rink, Lewis, Varma, Vitkin, & Jaffray, 2008), and relatively slow—the film may take upwards of days or weeks to fully develop. Furthermore, performing reliable dosimetry with radiochromic film can be relatively complicated (Niroomand-Rad et al., 1998). The precise orientation of the film during irradiation, period of time elapsed between irradiation and scanning, positioning of the film in the scanner (e.g., portrait versus landscape and centered versus offset), the spectral characteristics (i.e., center wavelength and width) of the red, blue, and green channels detected by the scanner, all influence the accuracy of radiochromic film dosimetry to varying degrees (Micke, Lewis, & Yu, 2011). However, even considering the limitations mentioned above, future advances in photopolymer technology may reduce or eliminate the weaknesses of current versions of radiochromic film.

7.3.3 DIODE DETECTORS

Diode detectors take advantage of the electron–hole pairs produced by ionizing radiation traversing a semiconducting material, typically silicon. The energy needed to produce an electron–hole pair in silicon is approximately 3 eV, which is about 10 times less than that needed to produce an ion pair in an air-filled IC (Knoll, 2010). This, coupled with the much greater density of semiconductor materials, grants diodes with very high sensitivity, and thus diode detectors can be made very small. Commercially available diode detectors have sensitive volumes in the order of 0.02 mm^3. These diodes provide much greater spatial resolution compared to even the smallest ICs, and are among the most commonly used active detectors for measuring small radiation fields.

Single diodes (SDs) exhibit an inherent dependence on orientation, due to their construction (Beddar et al., 1994). Diode manufacturers have addressed this problem by creating detectors that have two diodes placed immediately adjacent to each other, facing in opposite directions. This significantly reduces the directional dependence of the SD, albeit a slightly asymmetric response has been observed (Westermark, Arndt, Nilsson, & Brahme, 2000). The higher atomic number of the materials commonly used in semiconductor detectors, silicon and germanium, makes diode detectors more sensitive to low-energy photons than water. This could introduce errors when using diodes under large fields where considerable scatter is present. Energy-compensated, or shielded, diodes have been developed to overcome this dependence, but the shielding itself can increase directional dependence for SDs (Duggan & Coffey, 1998). Therefore, unshielded diodes are recommended over shielded diodes when measuring small fields, since the amount of low-energy scatter is minimal (Scott et al., 2008). Finally, the crystal lattice of a semiconductor detector is prone to damage, resulting in decreasing performance with accumulated dose, and diodes exhibit temperature dependence (Yorke, Alecu, Ding, & Fontenla, 2005).

7.3.4 DIAMOND DETECTORS

Diamond detectors have been developed and tested since early 1990's. They show promising results when tested under small radiation fields. Earlier, diamond detectors were based on natural diamond (ND) chips and functioned more or less like ICs, using applied potentials of around 100 V. However, ND chips have to be carefully selected and cut from whatever mined specimens are available, and this leads to small variations in shape and size across individual detectors (De Angelis et al., 2002). Synthetic alternatives to ND, in the form of chemical vapor deposition or high-pressure, high-temperature diamond have been tested as radiation dosimeters. Synthetic diamond's dosimetric performance is similar to that of ND, but can be produced more consistently, with more controllable levels of impurities, and at lower cost than ND.

More recently, single crystal diamond detectors have been developed that feature ohmic contacts that allow the diamond to be function as a Schottky diode, operating in photovoltaic mode and needing no bias voltage.

With respect to small field dosimetry, diamond detectors have several advantages. Because of their high sensitivity (and high cost), they are typically made very small, giving them a high spatial resolution similar to diodes. However, their very small sensitive volumes are typically contained in housings that are relatively large, having dimensions in the order of centimeters. Rustgi found that the response of a ND detector response was directionally independent under ^{60}Co, 6 MV, and 18 MV beams (Rustgi, 1995), whereas Ciancaglioni et al. observed a difference of approximately 3% depending on the angle between the beam and the detector stem (Ciancaglioni et al., 2012). Diamond is also more water equivalent than silicon, and is therefore effectively energy-independent in the therapeutic range (Rustgi, 1995). Several authors have observed a slight decrease in the sensitivity of diamond detectors with increasing dose rate (Hoban, Heydarian, Beckham, & Beddoe, 1994; Laub, Kaulich, & Nüsslin, 1997), while others have found the dose rate dependence to be negligible (Di Venanzio et al., 2013; Scott et al., 2008). Diamond detectors also exhibit good linearity and stability, although they require a 5–10 Gy pre-irradiation in order to ensure detector stability.

7.3.5 METAL–OXIDE–SEMICONDUCTOR FIELD EFFECT TRANSISTORS

Metal–oxide–semiconductor field effect transistor (MOSFET) detectors measure the change in the threshold voltage necessary for current to flow between two separate terminals (a *source* and a *drain*) imbedded into a semiconductor substrate. The threshold voltage changes as free electrons and holes are produced by ionizing radiation within the oxide layer, drift under the influence of a bias placed across the metal layer (the gate) and the semiconductor layer (the body), and then collected at the gate or trapped at the oxide–body interface. Thus, the change in threshold voltage is proportional to absorbed dose (Soubra et al., 1994). The sensitive volume of a MOSFET detector is typically very small, in the order of 0.01 mm^3 or smaller, which makes them very attractive for measuring small radiation fields. Although MOSFETs have been tested under small radiation fields by several authors (Amin, Heaton, Norrlinger, & Islam, 2010; Kaplan et al., 2000; Kurjewicz & Berndt, 2007), the majority of investigations involving MOSFETs have focused on their usefulness for *in vivo* patient dosimetry. In this case, MOSFETs can be used as surface dosimeters, placed in small catheters for internal measurements, or packaged in small, implantable capsules along with miniaturized electronics and an induction coil in order for the detector to be powered and read out wirelessly.

MOSFETs have been shown to have a linear response to dose, and are reproducible to approximately 1% when irradiated with doses of 1 Gy or larger. However, MOSFETs suffer from temperature dependence, energy dependence under lower energies (e.g., scattered radiation), directional dependence, and has a limited lifetime based on accumulated dose. Some of these disadvantages can be overcome by using a dual-MOSFET detector. These detectors measure the difference between the readings from two MOSFETs that operate with different gate biases but are otherwise identical. This difference should be insensitive to changes in the response of the MOSFETs due to measurement conditions. Dual-MOSFET detectors are not temperature dependent, and they exhibit improved linearity, sensitivity, and reproducibility over single MOSFET detectors. However, even dual detectors show variations in reproducibility that depend on the dose measured, and the proper calibration of MOSFETs for high doses (e.g., stereotactic radiosurgery fractions) may not be practical given their limitations on accumulated dose (Tanyi, Krafft, Hagio, Fuss, & Salter, 2008).

7.3.6 THERMOLUMINESCENT DETECTORS

When a thermoluminescent detector (TLD) is irradiated, free electrons and holes are created and conduct through the crystal until getting trapped at crystal defect sites. In this fashion, TLDs are passive, integrating dosimeters. When heat is later applied to the TLD, the electrons are excited out of their traps into the conduction band and then recombine with holes trapped in the valence band. The relaxation of the excited recombination center produces visible light, and the amount of this light produced is proportional to deposited dose (McKeever, 1985). The most widely used TLD material for medical dosimetry is lithium fluoride doped with magnesium and titanium (LiF:Mg,Ti—also known as TLD-100).

TLDs have been used as point dose reference detectors for measuring IMRT and SRS fields with other detectors (Linthout, Verellen, Van Acker, De Cock, & Storme, 2003; Massillon-J L, Cueva-Prócel, Díaz-Aguirre, Rodríguez-Ponce, & Herrera-Martínez, 2013; Pantelis et al., 2010; Richardson, Tomé, Orton, McNutt, & Paliwal, 2003). They have certain advantages in small fields including an effective atomic number close to that of tissue, and availability in a variety of small forms, including 1 mm^3 cubes and square or circular chips having sub-millimeter thicknesses. Furthermore, single-crystal TLDs can be reset by high temperature annealing, and are thus reusable. However, routine clinical use of TLDs can be cumbersome. Great care may be required to achieve consistent post-irradiation readout (Wood & Mayles, 1995). If the highest dosimetric accuracy is desired, a series of relatively burdensome measurements of each TLD may be necessary to fully characterize the detectors response prior to their intended use. Additionally, TLDs exhibit—to varying degrees—energy dependence, thermal fading of the luminescence signal, sensitivity to heating (and cooling) rates, and a nonlinear dose response at higher doses (Horowitz, 1981; Kirby, 1992).

7.3.7 PLASTIC SCINTILLATORS

Plastic scintillation detectors (PSDs) emit light with an intensity proportional to the radiation dose being deposited in the plastic base. A rigorous explanation of the physics behind the scintillation process is given in Chapter 1 of this book, and Chapter 4 provides an extensive description of the basic properties of PSDs. PSDs are very capable candidates with respect to small field radiation dosimetry. They are water equivalent, with mass energy-absorption coefficients that closely match water above 100 keV. They can be manufactured into various shapes to fit the needs of small measurement geometries while maintaining mechanical robustness, and are also readily available in the form of small, 0.5 mm diameter fibers. They are highly sensitive even at small sizes and exhibit fast luminescence with relaxation times in the order of nanoseconds—making them attractive for real-time measurements as well. Furthermore, their response is dose rate, temperature, and energy independent and linear with dose (Beddar, Mackie, & Attix, 1992a,b).

Scintillation is an immediate effect. As such, PSDs cannot function as integrating dosimeters and must be incorporated into devices that can actively measure their luminescence. Typically, this is accomplished by coupling a PS to a light guide (e.g., fiber-optic cable), which transports the scintillation light to a detector that converts the light into a measurable electronic signal. Earlier PSDs used photomultiplier tubes (PMTs) to do this conversion, while more recent investigations have used charge-coupled device (CCD) cameras and highly sensitive photodiodes. The inclusion of a fiber-optic light guide makes PSD dosimetry systems vulnerable to a significant stem effect—specifically, contamination of the dose-proportional scintillation light with Cerenkov light. The reader is directed to Chapter 5 for an in-depth explanation of the Cerenkov effect, and how it can be managed for accurate PSD dosimetry.

Although PSDs can be made much smaller than gas-filled ICs, they are sensitive volumes and are not as small as those found in solid-state detectors, such as diodes and MOSFETs. However, PSDs need only to be shielded from ambient light—which can be easily accomplished with very thin opaque plastic (read: water equivalent) jacketing—and mechanically coupled to a thin light guide. Thus, the overall dimensions of even a 1 mm diameter cylindrical PSD are often much smaller than solid-state detectors and diamond detectors, which have very small active volumes but require additional space for electrodes and wires, guards, build-up layers, and so on. The main drawback of using PSDs lies in the additional complexity necessary for the conversion of dose to electrical signal (PS → light guide → photon detector/imager → electrometer/ computer, as opposed to ion chamber/diode/diamond → electrometer) and the need for careful accounting of the necessary corrections for Cerenkov, PMT voltage stability, CCD camera uniformity, and so on. Once more, the reader is referred to Part I of this book for more information on PSDs.

7.3.8 OTHER DETECTORS

Investigations of small fields have also been performed using detectors other than those described in Sections 7.3.1 through 7.3.7 previous sections. These detectors are either not typically found in clinical settings, or not typically used specifically for small field dosimetry. They include polymer gel detectors, alanine detectors, electronic portal imaging detectors, optically stimulated luminescence detectors (OSLDs), and liquid scintillation detectors (LSDs).

Polymer gel detectors exploit the mechanism by which short monomers (e.g., acrylamide) are transformed into polymers by ionizing radiation at the site of ionization. The concentration of large polymers is proportional to the absorbed dose. Gel dosimeters produce very high-resolution dose reconstructions in three dimensions, are tissue equivalent, energy independent, and their read out is nondestructive (Baldock et al., 2010). They have been shown to perform reasonably well under small fields (Pappas et al., 2006; Wong et al., 2009). However, their preparation is labor intensive and long wait periods (in the order of day[s]) may be necessary after preparation as well as after irradiation. Gels are best evaluated with magnetic resonance imaging, which may not be readily available, though they can also be read out with computed tomography (X-ray or optical).

Alanine is an amino acid that is de-aminated when exposed to ionizing radiation. This results in the production of free radicals that are stabilized by the crystalline structure of the alanine and therefore do not recombine. Accumulated free radicals produce a measurable absorption signal in the electron spin resonance (ESR) spectrum of alanine, and this signal increases with number of free radicals, that is, with absorbed dose (McLaughlin, 1993). Characteristics of alanine that are attractive for small field dosimetry include close water equivalence, good dose linearity, dose rate dependence, and can be manufactured in a wide variety of shapes and sizes. The relative insensitivity of alanine is one important weakness, and limits the size of alanine pellets to ≥ 5 mm diameters to get appreciable signals under clinically relevant doses (Mack et al., 2002). Another disadvantage is the need for an ESR spectrograph for read out.

In the last 10 years, interest has been increasing in using electronic portal imaging devices (EPIDs) for more than just patient positioning verification (van Elmpt et al., 2008). The EPIDs that are being tested include both custom-built detectors and standard gantry-mounted imagers that are common on modern linac therapy machines. They contribute zero-dose perturbation (as they measure the exit dose) and provide two-dimensional data acquisition, as well as the possibility of three-dimensional reconstructions of measured dose. Gantry-mounted EPIDs are conveniently and tightly integrated into existing linac-based SRS machines. Because EPIDs are primarily tailored for imaging, they must be separately and rigorously characterized for performing dosimetry. Many studies have been performed with EPIDs for IMRT-related dosimetry that include fields smaller than 3×3 cm^2 (Low et al., 2011), and studies using high-resolution EPIDs specifically for sub-centimeter fields are beginning to appear (Han et al., 2013).

OSLDs and LSDs represent two other possible candidates for small field dosimeters. OSLDs are very similar to TLDs in that they are solid-state detectors that exploit the stable trapping of electrons freed by incident radiation. However, the mechanism for releasing the trapped charges relies on optical (instead of thermal) stimulation. LSD devices work in a similar fashion as PSDs, only the luminescent fluors are suspended in liquid instead of plastic.

7.3.9 DETECTOR SELECTION

Much of the published literature on small field dosimetry has focused on comparisons of multiple detectors and their relative performance when measuring OFs, depth dose profiles, beam profiles, dosimetric accuracy, and other properties of small fields (Duggan & Coffey, 1998; Heydarian, Hoban, & Beddoe, 1996; Laub & Wong, 2003; Low et al., 2011; Mack et al., 2002; Sánchez-Doblado et al., 2007; Sauer & Wilbert, 2007; Scott et al., 2008; Westermark et al., 2000; Wong et al., 2009). These studies had one common conclusion that there was no single *best* detector for small field dosimetry at present. Because no detector possesses the ideal combination of characteristics, measurements should be cross-referenced between at least two different detector types and compared against values published in the literature when possible (Dieterich & Sherouse, 2011).

Looking at how all of the different detectors can be placed into different categories can make an appropriate selection much easier. One simple category is the desired measurement geometry. Will a point measurement suffice, or is it necessary to obtain simultaneous measurement in one or multiple dimensions? The traditional tool for two-dimensional dosimetry is film, and multiple point detectors can be combined to make linear and two- or three-dimensional arrays. Diode and IC arrays have been commercially available for some time, and though they lack the resolution of film, they can measure in real time, they are more consistent, and they lack the chemical processing and/or scanning equipment and procedures necessary for film dosimetry. PSDs have also been shown to perform very well when arrayed—they can be placed immediately adjacent to each other for high spatial resolution and do not suffer from the cross-talk and dose perturbation effects that limit the array proximity of other point detectors. Orthogonal films and polymer gels can provide high-resolution dosimetry in three dimensions.

Another basic category is active versus passive (or integrating) detection. Detectors such as ICs, diamond detectors, diodes, and PSDs exploit short-lived events, and so these detectors must be connected to measurement devices that are on constant watch while accumulating signal. Other detectors, such as films, gels, and TLDs, take advantage of crystal modifications or polymerizations that accumulate with radiation dose and are stable over the long term. These detectors must be read out after the irradiation is finished. Additionally, if the read out process is nondestructive (as is the case with films and gels), the detectors can act as archival dose information storage devices. Some detectors can be used in either active or passive mode. The threshold voltage changes in MOSFETs can be read out in real time or post-irradiation, and OSLDs produce radioluminescence that can be read in real time while the crystal is collecting trapped electrons to be read out later as OSL signal.

The selection of the proper small field dosimeter depends, of course, on the application. Different applications suffer more or less from the different disadvantages inherent to the various detector types. For routine quality assurance of radiotherapy machines that produce small fields, effects such as local dose perturbation and dependencies on dose rate, energy, and detector orientation may not be very important inasmuch as the characteristics are well understood and can be carefully accounted for under reference conditions. Under nonreference conditions such as patient-specific IMRT QA, the potential complexity of different beam geometries and detector orientations may present a significant departure from reference conditions. Here, the convolution of different detector dependencies may not be well understood, and dose perturbations due to detector construction may prove to be problematic when comparing measured results with calculated expected values obtained, for example, from a radiotherapy treatment planning system.

When moving to an even more demanding application such as *in vivo* patient dosimetry, the choice of detector becomes critical. Ideally, the presence of the chosen detector(s) would have zero influence on the efficacy of the radiation fields being delivered. This is particularly important when using small radiation fields, as dose perturbations from dosimeters would not be *washed out* as they would be in larger fields with higher amounts of scattered radiation. Even small dose perturbations could compromise the efficacy of, say, a radiation therapy treatment in the immediate vicinity of the detector. Detectors capable of real-time measurement could also prove very valuable in patient dosimetry. With these properties in mind, the best detectors for *in vivo* patient dosimetry would be PSDs, LICs, and diamond detectors, as they all share close water equivalence, real-time readout, and small detector size. PSDs and diamond detectors are also very mechanically robust, which may prove to be valuable when a detector is needed for dose delivery verification near or inside a tumor volume and patient motion may be considerable.

7.4 PLASTIC SCINTILLATION DOSIMETRY IN SMALL FIELDS AND THEIR COMPARISON WITH OTHER DETECTORS

Typically, SRS uses radiation field diameters of less than 4 cm and slightly larger for fractionated delivery. Treatment planning data for SRS require measurements of tissue-maximum ratios (TMRs) or percentage depth doses (PDDs), off-axis ratios (OARs) and the associated beam profiles, and OFs (also referred to as *total scatter factors*, Scp) for this range of collimator or field sizes. Lateral electronic disequilibrium and steep dose gradients that exist in a large portion of these fields require the use of high-resolution measurement techniques (Aspradakis et al., 2010). Guidelines for the measurement of PDD curves in standard conditions have been addressed in AAPM reports: TG-106 (Das et al., 2008) and TG-155 (Das et al., 2015). Measurement of PDDs in small fields requires an appropriate detector with a small active volume and excellent spatial resolution. Using a detector that has a large active volume can result in underestimation of the dose and a steeper falloff for the PDD curve. Suitable detectors for measuring depth–dose curves are mini ICs, diodes, and radiochromic film (Aspradakis et al., 2010). Proper setup of the detector should be verified. This includes confirming that the detector is aligned with the central axis, that the scan arm travels parallel with the central axis, and that a minimal amount of stem and cable is exposed to the beam throughout scanning. Whenever possible, PDD curves should be measured and cross-referenced using two different detector types. The following discussion will primarily focus on measurements with PSs and their comparison with data obtained by other detectors used in small field dosimetry.

7.4.1 DEPTH–DOSE MEASUREMENTS

The main factor affecting depth–dose or TMR measurements of small fields is the dose rate dependence of the detector. Changes in energy composition with depth that could make energy dependence a significant factor in standard field sizes is less relevant in small fields because of the low contribution of scattered photons. Since depth–dose or TMR measurements are normalized with respect to a fixed field size, detector composition is also less critical.

Because of their dose rate and energy independence in clinical irradiation situations, PSs are considered reference detectors for standard field sizes depth–dose measurements (Lacroix et al., 2010). There is no reason *a priori* that would make them worse in small fields. Unfortunately, energy composition of MC simulations being often adjusted by fitting experimental depth–dose or TPR curves (Araki, 2006; Francescon, Cora, & Cavedon, 2008; Francescon, Cora, Cavedon, & Scalchi, 2009), reliable and independent simulations of depth–dose or TPR for small fields are missing in literature.

Diodes can be subjected to dose rate dependence due to many factors affecting each diode individually: accumulated dose, doping level, capture cross section, defects and impurities (Araki, 2006; Francescon et al., 2008, 2009; Grusell & Rikner, 1993; Lacroix et al., 2010). As shown in Figure 7.6, dose rate dependence of diodes will typically manifest itself in an under-response at low dose rate compared to higher dose rate. Signal drift not linked to changes in dose rate have also been observed for specific diode models (Derreumaux et al., 2011; Morin et al., 2013). Figure 7.6 also shows recombination effects in liquid

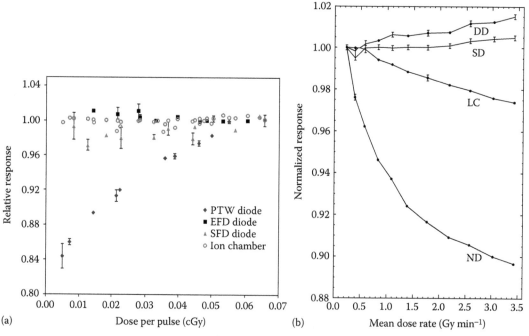

Figure 7.6 (a) Relative response for EFD and SFD diodes (IBA dosimetry), PTW 60012 diode, and PTW 233642 ion chamber (Physikalisch-Technische Werkstatten) when varying dose per pulse. (Reproduced from Lacroix, F. et al., *Med. Phys.*, 37, 4331, 2010. http://doi.org/10.1118/1.3463383.) (b) Response for double diode (DD), single diode (SD), natural diamond (ND), and liquid ionization chamber (LIC) detectors for different dose rates in a 6 MV photon beam. (Reproduced from Westermark, M. et al., *Phys. Med. Biol.*, 45, 685, 2000, http://doi.org/10.1088/0031-9155/45/3/308. With permission.)

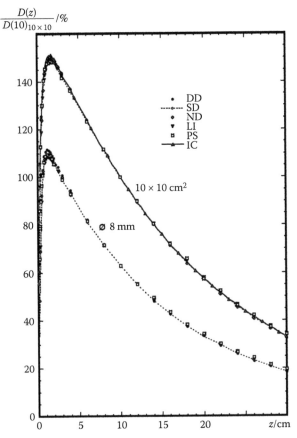

$$\frac{D(z)}{D(10)_{10\times10}}/\%$$

Figure 7.7 Depth doses for a 10 × 10 cm² field and for a 8 mm diameter collimator in a 6 MV photon beam. The data have been normalized to a 10 × 10 cm² field at a depth of 10 cm. (Reproduced from Westermark, M. et al., *Phys. Med. Biol.*, 45, 685, 2000, http://doi.org/10.1088/0031-9155/45/3/308. With permission.)

ionization chamber (LIC) as the dose rate increases. Diamond detector also exhibit considerable signal reduction as a function of dose rate.

Figure 7.7, adapted from Westermark et al. (2000), shows depth–dose measured curves for different detectors: DD, SD, ND, LIC, and IC. Depth–dose curves for a 10 × 10 cm² field and an 8 mm circular collimator are normalized to a 10 × 10 cm² field at a depth of 10 cm. For the 8 mm circular collimator, diode dose rate dependence manifests itself in a slightly steeper curve than that of PS. LI recombination effect makes its depth–dose curve gentler. For diodes in the 10 × 10 cm², increase in signal from scattered photons with depth probably counterbalance to some extent the lost of signal with depth caused by the dose rate dependence. Figure 7.8 shows a sample of TMR measurements for small fields collimated using the 2.5 mm width leaves from the Varian HD120 MLC multileaf collimator at CHU de Québec radiotherapy center. This figure shows a good agreement between TMR measured with a small field IC (CC01, IBA Dosimetry America, Memphis, TN) and the W1 PSD detector (Standard Imaging, Inc., Middleton, WI).

7.4.2 PROFILES (PENUMBRAS)

The main factor influencing measurements of small field penumbras is the spatial resolution of the detector. To a lesser extent, dose rate dependence, nonwater equivalent material embedding the active material, and angular dependence are factors to take into account. Figure 7.9, reproduced from Westermark et al. (2000), shows values of 80%/20% circular collimator penumbras measured for the same detectors as in Figure 7.7. Overall, the penumbras values are roughly ordered according to the spatial resolution of the detectors. MC calculated values in water are shown. For the two water equivalent

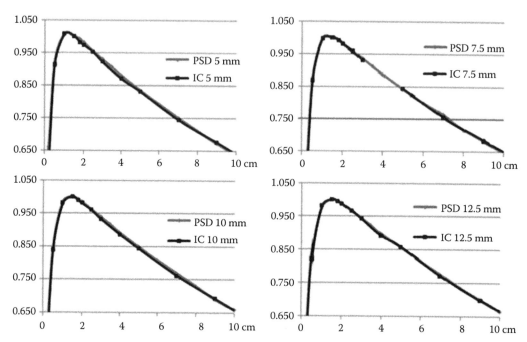

Figure 7.8 TMR measurements for MLC square fields of 5, 7.5, 10, and 12.5 mm side. PSD is compared to the CC01 ionization chamber. (Unpublished data from CHU de Quebec Radiotherapy Center, Quebec, Canada)

detectors (LI and PS), the measurements were corrected for the finite size of the detectors, and they agree within 0.2 mm with the MC simulations. The low penumbra values obtained with the DD is a consequence of the silicon shielding around the sensitive volume reducing the range of secondary electrons. Diamond measurements were corrected for dose rate dependence. Figure 7.10 (Gagnon et al., 2012) shows dose profiles measurements for a 4 mm radiosurgery circular collimator with an IC (Exradin A16, Standard Imaging, Inc.), a shielded diode (SD, PTW 60008, Physikalisch-Technische Werkstatten), an unshielded stereotactic diode (SFD, IBA Dosimetry America, Memphis, TN), Gafchromic EBT2 film (EBT2, ISP, NJ), and a PSD crosshair array (PSDCA). The spatial averaging of the IC is clearly visible on the graph as it has the largest sensitive area facing the beam. Shielded diode is the second largest detector and measures a smaller penumbra. The PSDCA and the SFD gives measurements that are consistent with the EBT2 film, the latter being considered as a reference because of its good resolution (0.17 mm pixels used) and water equivalence.

7.4.3 OUTPUTS FACTORS

Spatial resolution, energy dependence, dose rate dependence, and water equivalence are the main detector characteristics influencing OFs measurements of small fields. Figure 7.11 (Gagnon et al., 2012) shows radiosurgery cones outputs factors obtained with different detectors. IC (Exradin A16, Standard Imaging, Inc.), a shielded diode (SD, PTW 60008, Physikalisch-Technische Werkstatten), an unshielded stereotactic diode (SFD, IBA Dosimetry America, Memphis, TN), a Gafchromic EBT2 film (EBT2, ISP, NJ), and a PSD were used. When normalized to a standard 10×10 cm^2, the UD response is significantly lower than the others detectors for all cone sizes. This is due to the energy spectrum change between the 10×10 cm^2 normalization field and the radiosurgery cones. The high-density UD overresponds by means of photoelectric effect in the normalization field having a larger fraction of scattered low-energy photons. The SD is being shielded to filter low-energy photons and adjust its response to that of an IC in standard field sizes; its response is close to that of IC, EBT2, and PSD for larger cones but clearly overresponds in smaller cones due to its high-density shielding material. For the

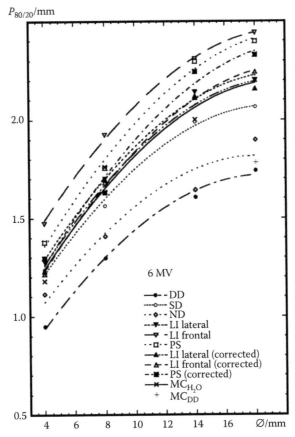

$P_{80/20}$/mm

6 MV
- —●- - DD
- ----○---- SD
- ·· -◇- ·· ND
- ---▼--- LI lateral
- —▽ - LI frontal
- ·· -□- ·· PS
- —▲--- LI lateral (corrected)
- — ▵ - LI frontal (corrected)
- - -■- - PS (corrected)
- —✕— MC_{H_2O}
- + MC_{DD}

Figure 7.9 Penumbra widths (80%–20%) measured with different detectors: double diode (DD), single diode (SD), natural diamond (ND), liquid ionization chamber (LIC), plastic scintillator (PS), and ionization chamber (IC). For the LIC and PS, penumbra widths corrected for the volume averaging effect are included. Monte Carlo simulation results in water are shown for all collimators and result for the double diode is graphed for the 18 mm collimator. (Reproduced from Westermark, M. et al., *Phys. Med. Biol.*, 45, 685, 2000, http://doi.org/10.1088/0031-9155/45/3/308. With permission.)

smallest cones, the over response of the SD seems to diminish as the output factors are closer to those of other detectors. This is probably due to a counterbalance of the nonwater equivalence effect by the dose rate dependence lowering the response of this type of detector (Araki, 2006). The EBT2 and the PSD give very close OFs having a mean difference of 1.3% over all cones. EBT2 has excellent resolution and water equivalence but has the drawback of needing offline processing that make real-time measurement impossible. When normalized to a 35 mm cone, to avoid large energy composition changes, the UD joins the PSD results having a mean difference of 1% over all cones. The IC OFs for the small cones are underestimated because of the volume averaging of this detector. IC OFs lower than those of PSD in subcentimeter MLC fields have also been measured with a different IC model (CC01, IBA Dosimetry America, Memphis, TN) (Klein et al., 2010).

Figure 7.12, reproduced from (Morin et al., 2013), shows OF measurements for the three smallest circular collimators of a CyberKnife unit. The measurements are compared to MC simulation of Francescon et al. (2008). The electron spot size and the nominal electron energy of the MC simulation were adjusted to the specific Cyberknife unit using the procedure described by Francescon et al. (2008, 2009). The simulation uncertainty graphed in Figure 7.12 arose from three sources: statistical uncertainties, nominal electron energy, and spot size. The detectors used are a shielded diode (PTW 60008, Physikalisch-Technische Werkstatten), two unshielded stereotactic diodes (PTW 60012, Physikalisch-Technische

Clinical applications using small PSDs

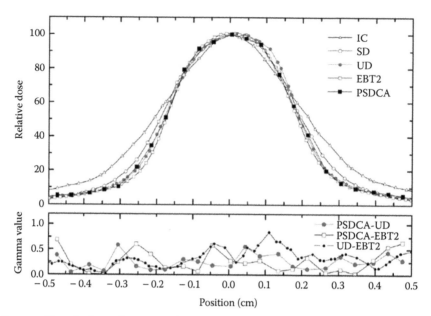

Figure 7.10 Dose profiles of a 4 mm diameter radiosurgery cone measured at 5 cm depth for ionization chamber (IC), shielded diode (SD), unshielded diode (UD), Gafchromic EBT2 film (EBT2), and PSD crosshair array (PSDCA). (Reproduced from Gagnon, J.-C. et al., *Med. Phys.*, 39, 429, 2012, http://doi.org/10.1118/1.3666765. With permission.)

Werkstatten and SFD, IBA Dosimetry America, Memphis, TN), a Gafchromic EBT2 film (ISP, NJ), a LIC (MicroLion, Physikalisch-Technische Werkstatten), and two PSD (1 and 0.5 mm diameter). MicroLion OFs are lower than the simulation result for the 5 mm cone but are a little higher for the 7.5 mm one. If only the volume averaging effect of this detector is considered (2.5 mm diameter), an under-response of about 6% would be expected for the 5 mm cone and one of about 1.6% for the 7.5 mm cone (Morin et al., 2013). Some MC simulations suggested that the volume averaging effect of the microLion in small fields is compensated by an over-responses due to the high-density material surrounding the sensitive volume (Francescon et al., 2012; Morin et al., 2013; Underwood, Winter, Hill, & Fenwick, 2013). All diodes show an over-response for the smallest cones. This over response is a consequence of the high-density sensitive volume. On top of that, the shielded diode (PTW 60008) exhibit an additional over-response caused by its high-density shielding material. The PTW 60008 OF for the 5 mm cone is lowered closer to the simulation result and the SFD because of a mix of volume averaging and dose rate dependence going against the signal increase caused by the high-density composition. Overall, PSD measurements provided the best agreement with MC simulation among all the detectors investigated. The slight under response for the 5 mm cone using the 1.0 mm PSD is probably caused by some volume-averaging.

7.4.4 COMPLETE THREE-DIMENSIONAL PHANTOM QA

Complete three-dimensional phantom quality assurance (QA) is a must for an end-to-end verification of any radiotherapy procedure, especially for small field techniques such as radiosurgery. In the case of cranial radiosurgery, a spherical or anthropomorphic phantom is usually used in which gafchromic films and different detectors can be inserted, provided that well-adjusted insertion cavities are drilled in the phantom. Detector response perturbations discussed in this chapter for depth–dose and OFs measurements will directly affect the result of a complete phantom QA. Another effect will be relevant in the context of a typical 3D non-coplanar radiosurgery QA: angular dependence of the detector. Figure 7.13 shows the setup for the measurements of the complete treatment plan of stereotactic radiosurgery on the phantom containing a dosimeter (Gagnon et al., 2012). The results of two complete treatment plans of stereotactic radiosurgery done with circular collimators of 5 and 35 mm diameter were shown in Table 7.1, adapted from (Gagnon et al., 2012). Since the treatment planning system

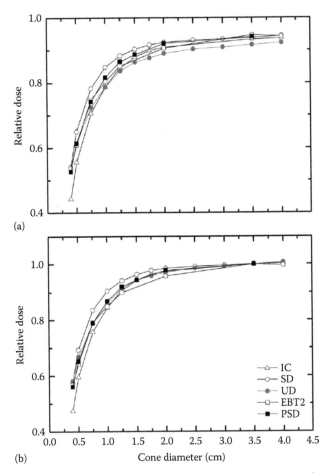

(a)

(b)

Figure 7.11 Output factors of radiosurgery cones measured at 1.8 cm depth. (a) normalized to a 1010 cm² field and (b) normalized to the 35 mm diameter cone. Detectors used are: ionization chamber (IC), shielded diode (SD), unshielded diode (UD), Gafchromic EBT2 film (EBT2), and plastic scintillator detector (PSD). Reproduced from Gagnon, J.-C. et al., *Med. Phys.*, 39, 429, 2012, http://doi.org/10.1118/1.3666765. With permission.)

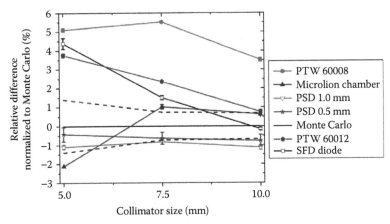

Figure 7.12 Relative total scatter factors difference normalized to Monte Carlo calculations. The continuous black line represents the Monte Carlo results and the dotted lines represent the uncertainties on those results. (Reproduced from Morin, J. et al., *Med. Phys.*, 40, 011719, 2013, http://doi.org/10.1118/1.4772190. With permission.)

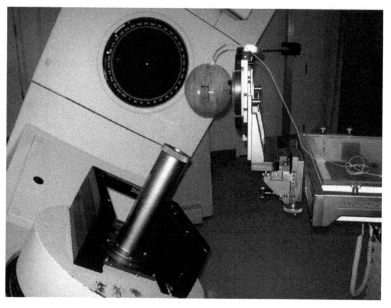

Figure 7.13 Setup for the measurements of the complete treatment plan of stereotactic radiosurgery on the phantom containing a dosimeter. (From Gagnon, J.-C. et al., *Med. Phys.*, 39, 429, 2012, http://doi.org/10.1118/1.3666765. With permission.)

Table 7.1 **Results obtained after performing a complete stereotactic radiosurgery treatment composed of four non-coplanar arcs and a prescription dose of 15 Gy at the isocenter where one of the above detectors was placed**

	5 MM CONE		35 MM CONE	
	$\dfrac{D_{measured}}{D_{prescribed}}$	$\dfrac{D_{measured} \cdot S_{c,p}^{X}}{D_{prescribed} \cdot S_{c,p}^{SD}}$	$\dfrac{D_{measured}}{D_{prescribed}}$	$\dfrac{D_{measured} \cdot S_{c,p}^{X}}{D_{prescribed} \cdot S_{c,p}^{SD}}$
IC	0.847	0.990	0.992	0.997
PSD	0.932	0.986	1.010	1.007
UD	0.913	0.970	0.955	0.978

was commissioned using a shielded diode, the measurements compensated for total OFs bias are also shown. Despite the fact that this procedure wash out all OF effects, the unshielded diode (SFD, IBA Dosimetry America, Memphis, TN) measurements are still about 2% lower than the IC (Exradin A16, Standard Imaging, Inc.) and PSD measurements. Both the IC and the PSD are within 1.4% of the prescription dose. The lower dose for the SFD diode is explained by its angular dependence response. Detectors were inserted in the phantom obliquely from a quasi-anterior position, so that irradiation arcs could hit detectors in multiple orientations. Figure 7.14, reproduced from (Westermark et al., 2000), shows measurements of angular dependency of a *SD* (Scanditronix) having similar properties to the SFD diode in a 18 MV photon beam. Under dosages of up to about 5% can be seen for the oblique backward irradiation angles. The DD, built to overcome the directional dependence of conventional diodes, is much stable with an under dosage of about 2% limited to the most backward angles.

Figure 7.15, reproduced from (Wang, Klein, & Beddar, 2010), shows results of MC simulation for the angular dependency of a PSD (BC-400) plastic scintillator. Results for a bare PSD ant two added tip thickness are shown. For the bare PSD configuration, closest to reference (Gagnon et al., 2012) PSD configuration, a signal variation of at most 0.8% is seen over all irradiation angles.

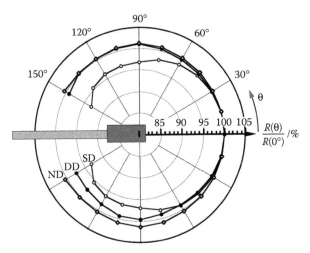

Figure 7.14 Directional response of a single diode (SD), double diode (DD), and a natural diamond detector (ND) at depth of dose maximum in a 18 MV photon beam. (Reproduced from Westermark, M. et al., *Phys. Med. Biol.*, 45, 685, 2000, http://doi.org/10.1088/0031-9155/45/3/308. With permission.)

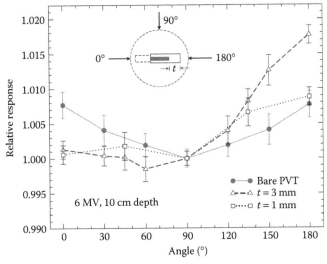

Figure 7.15 Monte Carlo simulation of angular dependence of a bare BC-400 scintillator at a depth of 10 cm in water in a 6 MV photon beam. Two added tip thickness are also studied. (Reproduced from Wang, L.L.W. et al., *Med. Phys.*, 37, 5279, 2010, http://doi.org/10.1118/1.3488904. With permission.)

7.5 CONCLUSION

There is an increasing awareness of the challenges of small field dosimetry in the radiation therapy community with increasing number of publications by the international collaborative groups, task groups, and other professional societies. Both experimental and MC techniques have produced respectable literature in understanding the small field issues in terms of detector properties, small field measurements, beam modeling, and dose calculation problems for the past 5 years. There are now published correction factors available for various detectors with the use of specific machine beams from several vendors. Plastic scintillating detectors, as more widely commercialized, are quickly becoming the reliable detector of choice with their demonstrated water-equivalence, reproducibility, energy independence, and linear response to dose. In spite of all these developments, the user is still cautioned to take careful measurements with different detector choices and to understand the limitations inherently existent in all dosimetry tools.

REFERENCES

Alfonso, R., Andreo, P., Capote, R., Huq, M. S., Kilby, W., Kjäll, P., ... Vatnitsky, S. (2008). A new formalism for reference dosimetry of small and nonstandard fields. *Medical Physics, 35*(November), 5179–5186. http://doi.org/10.1118/1.3005481.

Amin, M. N., Heaton, R., Norrlinger, B., & Islam, M. K. (2010, October 4). Small field electron beam dosimetry using MOSFET detector. *Journal of Applied Clinical Medical Physics.* 12(1), 50–57. http://doi.org/10.1120/jacmp.v12i1.3267.

Andersson, J., Johansson, E., & Tölli, H. (2012). On the property of measurements with the PTW microLion chamber in continuous beams. *Medical Physics, 39*(8), 4775. http://doi.org/10.1118/1.4736804.

Araki, F. (2006). Monte Carlo study of a Cyberknife stereotactic radiosurgery system. *Medical Physics, 33*(8), 2955. http://doi.org/10.1118/1.2219774.

Aspradakis, M. M., Byrne, J. P., Palmans, H., Duane, S., Conway, J., Warrington, A. P., & Rosser, K. (2010). IPEM report 103: Small field MV photon dosimetry. Retrieved from http://www.iaea.org/inis/collection/NCLCollectionStore/_Public/42/026/42026419.pdf.

Baldock, C., De Deene, Y., Doran, S., Ibbott, G., Jirasek, A., Lepage, M., ... Schreiner, L. J. (2010). Polymer gel dosimetry. *Physics in Medicine and Biology, 55*(5), R1–R63. http://doi.org/10.1088/0031-9155/55/5/R01.

Bassinet, C., Huet, C., Derreumaux, S., Brunet, G., Chéa, M., Baumann, M., ... Clairand, I. (2013). Small fields output factors measurements and correction factors determination for several detectors for a CyberKnife® and linear accelerators equipped with microMLC and circular cones. *Medical Physics, 40*(July), 071725. http://doi.org/10.1118/1.4811139.

Beddar, A. S., Mackie, T. R., & Attix, F. H. (1992a). Water-equivalent plastic scintillation detectors for high-energy beam dosimetry: I. Physical characteristics and theoretical considerations. *Physics in Medicine and Biology, 37*(10), 1883–1900. http://doi.org/10.1088/0031-9155/37/10/006.

Beddar, A. S., Mackie, T. R., & Attix, F. H. (1992b). Water-equivalent plastic scintillation detectors for high-energy beam dosimetry: II. Properties and measurements. *Physics in Medicine and Biology, 37*(10), 1901–1913. http://doi.org/10.1088/0031-9155/37/10/007.

Beddar, A. S., Mason, D. J., & O'Brien, P. F. (1994). Absorbed dose perturbation caused by diodes for small field photon dosimetry. *Medical Physics.* 21(7), 1075–1079. http://doi.org/10.1118/1.597350.

Benítez, E. M., Casado, F. J., García-Pareja, S., Martín-Viera, J. A., Moreno, C., & Parra, V. (2013). Evaluation of a liquid ionization chamber for relative dosimetry in small and large fields of radiotherapy photon beams. *Radiation Measurements, 58,* 79–86. http://doi.org/10.1016/j.radmeas.2013.08.009.

Benmakhlouf, H., Sempau, J., & Andreo, P. (2014). Output correction factors for nine small field detectors in 6 MV radiation therapy photon beams: A PENELOPE Monte Carlo study. *Medical Physics, 41,* 041711. http://doi.org/10.1118/1.4868695.

Bjärngard, B. E., & Petti, P. L. (1988). Description of the scatter component in photon-beam data. *Physics in Medicine and Biology, 33*(1), 21–32. http://doi.org/10.1088/0031-9155/33/1/002.

Bouchard, H., Seuntjens, J., Carrier, J.-F., & Kawrakow, I. (2009). Ionization chamber gradient effects in nonstandard beam configurations. *Medical Physics, 36*(2009), 4654–4663. http://doi.org/10.1118/1.3213518.

Burlin, T. E. (1966). A general theory of cavity ionisation. *The British Journal of Radiology, 39*(466), 727–734. http://doi.org/10.1259/0007-1285-39-466-727.

Bushberg, J., Seibert, J., Leidholdt Jr., E., & Boone, J. (2002). *The essential physics of medical imaging* (2nd ed.). Philadelphia, PA: Lippincott Williams & Wilkins.

Caprile, P., & Hartmann, G. H. (2009). A beam model applicable to small fields: Development and validation. *IFMBE Proceedings, 25,* 56–59. http://doi.org/10.1007/978-3-642-03474-9-15.

Charles, P. H., Cranmer-Sargison, G., Thwaites, D. I., Crowe, S. B., Kairn, T., Knight, R. T., ... Trapp, J. V. (2014). A practical and theoretical definition of very small field size for radiotherapy output factor measurements. *Medical Physics, 41,* 041707. http://doi.org/10.1118/1.4868461.

Charles, P. H., Crowe, S. B., Kairn, T., Knight, R. T., Hill, B., Kenny, J., ... Trapp, J. V. (2013). Monte Carlo-based diode design for correction-less small field dosimetry. *Physics in Medicine and Biology, 58,* 4501–4512. http://doi.org/10.1088/0031-9155/58/13/4501.

Chung, E., Davis, S., & Seuntjens, J. (2013). Experimental analysis of general ion recombination in a liquid-filled ionization chamber in high-energy photon beams. *Medical Physics, 40*(6), 062104. http://doi.org/10.1118/1.4805109.

Ciancaglioni, I., Marinelli, M., Milani, E., Prestopino, G., Verona, C., Verona-Rinati, G., ... De Notaristefani, F. (2012). Dosimetric characterization of a synthetic single crystal diamond detector in clinical radiation therapy small photon beams. *Medical Physics, 39*(7), 4493–4501. http://doi.org/10.1118/1.4729739.

Cranmer-Sargison, G., Charles, P. H., Trapp, J. V., & Thwaites, D. I. (2013). A methodological approach to reporting corrected small field relative outputs. *Radiotherapy and Oncology, 109*(3), 350–355. http://doi.org/10.1016/j.radonc.2013.10.002.

Cranmer-Sargison, G., Weston, S., Evans, J. A., Sidhu, N. P., & Thwaites, D. I. (2012). Monte Carlo modelling of diode detectors for small field MV photon dosimetry: Detector model simplification and the sensitivity of correction factors to source parameterization. *Physics in Medicine and Biology, 57*(16), 5141–5153. http://doi.org/10.1088/0031-9155/57/16/5141.

Cranmer-Sargison, G., Weston, S., Sidhu, N. P., & Thwaites, D. I. (2011). Experimental small field 6 MV output ratio analysis for various diode detector and accelerator combinations. *Radiotherapy and Oncology: Journal of the European Society for Therapeutic Radiology and Oncology, 100*(3), 429–435. http://doi.org/10.1016/j.radonc.2011.09.002.

Crop, F., Reynaert, N., Pittomvils, G., Paelinck, L., De Wagter, C., Vakaet, L., & Thierens, H. (2009). The influence of small field sizes, penumbra, spot size and measurement depth on perturbation factors for microionization chambers. *Physics in Medicine and Biology, 54*, 2951–2969. http://doi.org/10.1088/0031-9155/54/9/024.

Czarnecki, D., & Zink, K. (2013). Monte Carlo calculated correction factors for diodes and ion chambers in small photon fields. *Physics in Medicine and Biology, 58*, 2431–2444. http://doi.org/10.1088/0031-9155/58/8/2431

Das, I., Francescon, P., Ahnesjö, A., Aspradakis, M., Cheng, C., Ding, G., ... Sauer, O. (2015). Small fields and non-equilibrium condition photon beam dosimetry: AAPM Task Group Report 155. *Medical Physics*.

Das, I., Paolo, F., Vicenza, O. Di., Rodolfi, V., Ahnesjö, A., Physics, M. R., ... Sauer, O. A. (n.d.). Task Group 155 Report: Small fields and non-equilibrium condition photon beam dosimetry.

Das, I. J., Cheng, C.-W., Watts, R. J., Ahnesjö, A., Gibbons, J., Li, X. A., ... Zhu, T. C. (2008). Accelerator beam data commissioning equipment and procedures: Report of the TG-106 of the Therapy Physics Committee of the AAPM. *Medical Physics, 35*, 4186–4215. http://doi.org/10.1118/1.2969070.

Das, I. J., Downes, M. B., Kassaee, A., & Tochner, Z. (2000). Choice of radiation detector in dosimetry of stereotactic radiosurgery-radiotherapy. *Journal of Radiosurgery, 3*, 177–185.

Dasu A., Löfroth P. O., & Wickman, G. (1998). Liquid ionization chamber measurements of dose distributions in small 6 MV photon beams. *Physics in Medicine and Biology, 43*, 21–36. http://doi.org/10.1088/0031-9155/43/1/002

Dawson, D. J. (1984). Analysis of physical parameters associated with the measurement of high-energy x-ray penumbra. *Medical Physics, 11*(4), 491. http://doi.org/10.1118/1.595542.

Dawson, D. J. (1986). Penumbral measurements in water for high-energy x rays. *Medical Physics, 13*(1), 101. http://doi.org/10.1118/1.595963.

De Angelis, C., Onori, S., Pacilio, M., Cirrone, G. A. P., Cuttone, G., Raffaele, L., ... Mazzocchi, S. (2002). An investigation of the operating characteristics of two PTW diamond detectors in photon and electron beams. *Medical Physics, 29*(2), 248. http://doi.org/10.1118/1.1446101.

Derreumaux, S., Bassinet, C., Huet, C., Chea, M., Boisserie, G., Brunet, G., ... Clairand, I. (2011). SU-E-T-163: Characterization of the response of active detectors and passive dosemeters used for dose measurement in small photon beams. *Medical Physics, 38*(6), 3523. http://doi.org/10.1118/1.3612113.

Derreumaux, S., Etard, C., Huet, C., Trompier, F., Clairand, I., Bottollier-depois, J. F., ... Gourmelon, P. (2008). Lessons from recent accidents in radiation therapy in France. *Radiation Protection Dosimetry, 131*(1), 130–135. http://doi.org/10.1093/rpd/ncn235.

Di Venanzio, C., Marinelli, M., Milani, E., Prestopino, G., Verona, C., Verona-Rinati, G., ... Pimpinella, M. (2013). Characterization of a synthetic single crystal diamond Schottky diode for radiotherapy electron beam dosimetry. *Medical Physics, 40*(2), 021712. http://doi.org/10.1118/1.4774360.

Dieterich, S., & Sherouse, G. W. (2011). Experimental comparison of seven commercial dosimetry diodes for measurement of stereotactic radiosurgery cone factors. *Medical Physics, 38*(7), 4166–4173. http://doi.org/10.1118/1.3592647.

Ding, G. X., & Ding, F. (2012). Beam characteristics and stopping-power ratios of small radiosurgery photon beams. *Physics in Medicine and Biology, 57*(17), 5509–5521. http://doi.org/10.1088/0031-9155/57/17/5509.

Duggan, D. M., & Coffey, C. W. (1998). Small photon field dosimetry for stereotactic radiosurgery. *Medical Dosimetry, 23*(3), 153–159. http://doi.org/10.1016/S0958-3947(98)00013-2.

Eklund, K., & Ahnesjö, A. (2008). Fast modelling of spectra and stopping-power ratios using differentiated fluence pencil kernels. *Physics in Medicine and Biology, 53*, 4231–4247. http://doi.org/10.1088/0031-9155/53/16/002.

Eklund, K., & Ahnesjö, A. (2010). Modeling silicon diode dose response factors for small photon fields. *Physics in Medicine and Biology, 55*, 7411–7423. http://doi.org/10.1088/0031-9155/55/24/002.

Fano, U. (2010). Note on the Bragg-Gray cavity principle for measuring energy dissipation. Retrieved from http://www.rrjournal.org/doi/abs/10.2307/3570368.

Fenwick, J. D., Kumar, S., Scott, A. J. D., & Nahum, A. E. (2013). Using cavity theory to describe the dependence on detector density of dosimeter response in non-equilibrium small fields. *Physics in Medicine and Biology, 58*(9), 2901–2923. http://doi.org/10.1088/0031-9155/58/9/2901.

Francescon, P., Cora, S., & Cavedon, C. (2008). Total scatter factors of small beams: A multidetector and Monte Carlo study. *Medical Physics, 35*(December 2007), 504–513. http://doi.org/10.1118/1.2828195.

Francescon, P., Cora, S., Cavedon, C., & Scalchi, P. (2009). Application of a Monte Carlo-based method for total scatter factors of small beams to new solid state micro-detectors. *Journal of Applied Clinical Medical Physics, 10*(1), 147–152. http://doi.org/10.1120/jacmp.v10i1.2939.

Francescon, P., Kilby, W., Satariano, N., & Cora, S. (2012). Monte Carlo simulated correction factors for machine specific reference field dose calibration and output factor measurement using fixed and iris collimators on the CyberKnife system. *Physics in Medicine and Biology, 57*(12), 3741–3758. http://doi.org/10.1088/0031-9155/57/12/3741.

Gagnon, J.-C., Thériault, D., Guillot, M., Archambault, L., Beddar, S., Gingras, L., & Beaulieu, L. (2012). Dosimetric performance and array assessment of plastic scintillation detectors for stereotactic radiosurgery quality assurance. *Medical Physics, 39*(1), 429. http://doi.org/10.1118/1.3666765.

García-Vicente, F., Delgado, J. M., & Peraza, C. (1998). Experimental determination of the convolution kernel for the study of the spatial response of a detector. *Medical Physics, 25*(2), 202. http://doi.org/10.1118/1.598182.

Grusell, E., & Rikner, G. (1993). Linearity with dose rate of low resistivity p-type silicon semiconductor detectors. *Physics in Medicine and Biology, 38*(6), 785–792. http://doi.org/10.1088/0031-9155/38/6/011.

Han, B., Luxton, G., Yu, S., Lu, M., Wang, L., Mok, E., & Xing, L. (2013). SU-C-105-06: Development of a high resolution EPID solution for small field dosimetry. *Medical Physics, 40*(6), 84. http://doi.org/10.1118/1.4813930.

Hartmann, B., Martišíková, M., & Jäkel, O. (2010). Technical note: Homogeneity of Gafchromic® EBT2 film. *Medical Physics, 37*(4), 1753. http://doi.org/10.1118/1.3368601.

Herz, R. (1969). *The photographic action of ionizing radiation.* New York: John Wiley & Sons.

Heydarian, M., Hoban, P. W., & Beddoe, A. H. (1996). A comparison of dosimetry techniques in stereotactic radiosurgery. *Physics in Medicine and Biology, 41*(1), 93–110. http://doi.org/10.1088/0031-9155/41/1/008.

Hoban, P. W., Heydarian, M., Beckham, W. A., & Beddoe, A. H. (1994). Dose rate dependence of a PTW diamond detector in the dosimetry of a 6 MV photon beam. *Physics in Medicine and Biology, 39*(8), 1219–1229. http://doi.org/10.1088/0031-9155/39/8/003.

Horowitz, Y. S. (1981). The theoretical and microdosimetric basis of thermoluminescence and applications to dosimetry. Retrieved from http://inis.iaea.org/Search/search.aspx?orig_q=RN:13645799.

Iwasaki, A. (1996). 10 MV x-ray SMRs obtained using zero-area correction factors derived by means of the Bjärngard–Petti method. *Physics in Medicine and Biology, 41*(4), 625–636. http://doi.org/10.1088/0031-9155/41/4/004.

Jaffray, D. A. (1993). X-ray sources of medical linear accelerators: Focal and extra-focal radiation. *Medical Physics, 20*(5), 1417. http://doi.org/10.1118/1.597106.

Johns, H. E., & Darby, E. K. (1950). The distribution of radiation near the geometrical edge of an X-ray beam. *The British Journal of Radiology, 23*(267), 193–197. http://doi.org/10.1259/0007-1285-23-267-193.

Kaplan, G. I., Rosenfeld, A. B., Allen, B. J., Booth, J. T., Carolan, M. G., & Holmes-Siedle, A. (2000). Improved spatial resolution by MOSFET dosimetry of an x-ray microbeam. *Medical Physics, 27*(1), 239. http://doi.org/10.1118/1.598866.

Kirby, T. H. (1992). Uncertainty analysis of absorbed dose calculations from thermoluminescence dosimeters. *Medical Physics, 19*(6), 1427. http://doi.org/10.1118/1.596797.

Klein, D. M., Tailor, R. C., Archambault, L., Wang, L., Therriault-Proulx, F., & Beddar, A. S. (2010). Measuring output factors of small fields formed by collimator jaws and multileaf collimator using plastic scintillation detectors. *Medical Physics, 37*(2010), 5541–5549. http://doi.org/10.1118/1.3488981.

Knoll, G. (2010). *Radiation detection and measurement* (4th ed.). Hoboken, NJ: John Wiley & Sons, Inc.

Kurjewicz, L., & Berndt, A. (2007). Measurement of Gamma Knife® helmet factors using MOSFETs. *Medical Physics, 34*(3), 1007. http://doi.org/10.1118/1.2437282.

Lacroix, F., Guillot, M., McEwen, M., Cojocaru, C., Gingras, L., Beddar, A. S., & Beaulieu, L. (2010). Extraction of depth-dependent perturbation factors for parallel-plate chambers in electron beams using a plastic scintillation detector. *Medical Physics, 37*(2010), 4331–4342. http://doi.org/10.1118/1.3463383.

Laub, W. U., Kaulich, T. W., & Nüsslin, F. (1997). Energy and dose rate dependence of a diamond detector in the dosimetry of 4–25 MV photon beams. *Medical Physics, 24*(4), 535. http://doi.org/10.1118/1.597902.

Laub, W. U., & Wong, T. (2003). The volume effect of detectors in the dosimetry of small fields used in IMRT. *Medical Physics, 30*, 341–347. http://doi.org/10.1118/1.1544678.

Lewis, D. (2010). Radiochromic film. Retrieved from http://www.filmqapro.com/Documents/Lewis_Radiochromic_Film_20101020.pdf.

Li, X. A., Soubra, M., Szanto, J., & Gerig, L. H. (1995). Lateral electron equilibrium and electron contamination in measurements of head-scatter factors using miniphantoms and brass caps. *Medical Physics, 22*(1995), 1167–1170. http://doi.org/10.1118/1.597508.

Linthout, N., Verellen, D., Van Acker, S., De Cock, M., & Storme, G. (2003). Dosimetric evaluation of partially overlapping intensity modulated beams using dynamic mini-multileaf collimation. *Medical Physics, 30*(5), 846. http://doi.org/10.1118/1.1562170.

Low, D. A., Moran, J. M., Dempsey, J. F., Dong, L., & Oldham, M. (2011). Dosimetry tools and techniques for IMRT. *Medical Physics, 38*(3), 1313–1338. http://doi.org/10.1118/1.3514120.

Mack, A., Scheib, S. G., Major, J., Gianolini, S., Pazmandi, G., Feist, H., … Kreiner, H.-J. (2002). Precision dosimetry for narrow photon beams used in radiosurgery—Determination of Gamma Knife® output factors. *Medical Physics, 29*(9), 2080. http://doi.org/10.1118/1.1501138.

Massillon-J. L., G., Cueva-Prócel, D., Díaz-Aguirre, P., Rodríguez-Ponce, M., & Herrera-Martínez, F. (2013). Dosimetry for small fields in stereotactic radiosurgery using Gafchromic MD-V2-55 film, TLD-100 and alanine dosimeters. *PLoS One, 8*(5), e63418. http://doi.org/10.1371/journal.pone.0063418.

McKeever, S. (1985). *Thermoluminescence of Solids.* New York: Cambridge University Press.

McLaughlin, W. L. (1993). ESR Dosimetry. *Radiation Protection Dosimetry, 47*(1–4), 255–262. Retrieved from http://rpd.oxfordjournals.org/content/47/1-4/255.short.

McLaughlin, W. L., & Chalkley, L. (1965). Low atomic number dye systems for ionizing radiation measurement. *Photographic Science and Engineering, 9.* Retrieved from http://www.osti.gov/scitech/biblio/4600181.

Metcalfe, P. (1993). Dosimetry of 6-MV x-ray beam penumbra. *Medical Physics, 20*(5), 1439. http://doi.org/10.1118/1.597107.

Micke, A., Lewis, D. F., & Yu, X. (2011). Multichannel film dosimetry with nonuniformity correction. *Medical Physics, 38*(5), 2523. http://doi.org/10.1118/1.3576105.

Morin, J., Beliveau-Nadeau, D., Chung, E., Seuntjens, J., Theriault, D., Archambault, L., … Beaulieu, L. (2013). A comparative study of small field total scatter factors and dose profiles using plastic scintillation detectors and other stereotactic dosimeters: The case of the CyberKnife. *Medical Physics, 40*(1), 011719. http://doi.org/10.1118/1.4772190.

Munro, P., Rawlinson, J. A., & Fenster, A. (2012). Therapy imaging: Source sizes of radiotherapy beams. *Medical Physics, 15*(1988), 517–524. http://doi.org/10.1118/1.596295.

Niroomand-Rad, A., Blackwell, C. R., Coursey, B. M., Gall, K. P., Galvin, J. M., McLaughlin, W. L., … Soares, C. G. (1998). Radiochromic film dosimetry: Recommendations of AAPM Radiation Therapy Committee Task Group 55. *Medical Physics, 25*(11), 2093. http://doi.org/10.1118/1.598407.

O'Connor, J. E. (1957). The variation of scattered x-rays with density in an irradiated body. *Physics in Medicine and Biology, 1*(4), 352–369. http://doi.org/10.1088/0031-9155/1/4/305.

Pai, S., Das, I. J., Dempsey, J. F., Lam, K. L., LoSasso, T. J., Olch, A. J., … Wilcox, E. E. (2007). TG-69: Radiographic film for megavoltage beam dosimetry. *Medical Physics, 34*(6), 2228. http://doi.org/10.1118/1.2736779.

Pantelis, E., Moutsatsos, A., Zourari, K., Kilby, W., Antypas, C., Papagiannis, P., Sakelliou, L. (2010). On the implementation of a recently proposed dosimetric formalism to a robotic radiosurgery system. *Medical Physics, 37*(5), 2369. http://doi.org/10.1118/1.3404289.

Pantelis, E., Moutsatsos, A., Zourari, K., Petrokokkinos, L., Sakelliou, L., Kilby, W., … Seimenis, I. (2012). On the output factor measurements of the CyberKnife iris collimator small fields: Experimental determination of the k(Q(clin),Q(msr)) (f(clin),f(msr)) correction factors for microchamber and diode detectors. *Medical Physics, 39*(8), 4875–4885. http://doi.org/10.1118/1.4736810.

Pappas, E., Maris, T. G., Papadakis, A., Zacharopoulou, F., Damilakis, J., Papanikolaou, N., & Gourtsoyiannis, N. (2006). Experimental determination of the effect of detector size on profile measurements in narrow photon beams. *Medical Physics, 33*, 3700–3710. http://doi.org/10.1118/1.2349691.

Ralston, A., Liu, P., Warrener, K., McKenzie, D., & Suchowerska, N. (2012). Small field diode correction factors derived using an air core fibre optic scintillation dosimeter and EBT2 film. *Physics in Medicine and Biology, 57*, 2587–2602. http://doi.org/10.1088/0031-9155/57/9/2587.

Rice, R. K., Hansen, J. L., Svensson, G. K., & Siddon, R. L. (1987). Measurements of dose distributions in small beams of 6 MV x-rays. *Physics in Medicine and Biology, 32*, 1087–1099. http://doi.org/10.1088/0031-9155/32/9/002.

Richardson, S. L., Tomé, W. A., Orton, N. P., McNutt, T. R., & Paliwal, B. R. (2003). IMRT delivery verification using a spiral phantom. *Medical Physics, 30*(9), 2553. http://doi.org/10.1118/1.1603965.

Rink, A., Lewis, D. F., Varma, S., Vitkin, I. A., & Jaffray, D. A. (2008). Temperature and hydration effects on absorbance spectra and radiation sensitivity of a radiochromic medium. *Medical Physics, 35*(10), 4545. http://doi.org/10.1118/1.2975483.

Rustgi, S. N. (1995). Evaluation of the dosimetric characteristics of a diamond detector for photon beam measurements. *Medical Physics, 22*(5), 567. http://doi.org/10.1118/1.597543.

Sánchez-Doblado, F., Andreo, P., Capote, R., Leal, A., Perucha, M., Arráns, R., … Carrasco, E. (2003). Ionization chamber dosimetry of small photon fields: A Monte Carlo study on stopping-power ratios for radiosurgery and IMRT beams. *Physics in Medicine and Biology, 48,* 2081–2099. http://doi.org/10.1088/0031-9155/48/21/L02.

Sánchez-Doblado, F., Hartmann, G. H., Pena, J., Roselló, J. V., Russiello, G., & Gonzalez-Castaño, D. M. (2007). A new method for output factor determination in MLC shaped narrow beams. *Physica Medica, 23*(2), 58–66. http://doi.org/10.1016/j.ejmp.2007.03.002.

Sauer, O. A., & Wilbert, J. (2007). Measurement of output factors for small photon beams. *Medical Physics, 34*(December 2006), 1983–1988. http://doi.org/10.1118/1.2734383.

Scott, A. J. D., Kumar, S., Nahum, A. E., & Fenwick, J. D. (2012). Characterizing the influence of detector density on dosimeter response in non-equilibrium small photon fields. *Physics in Medicine and Biology, 57,* 4461–4476. http://doi.org/10.1088/0031-9155/57/14/4461.

Scott, A. J. D., Nahum, A. E., & Fenwick, J. D. (2008). Using a Monte Carlo model to predict dosimetric properties of small radiotherapy photon fields. *Medical Physics, 35,* 4671–4684. http://doi.org/10.1118/1.2975223

Scott, A. J. D., Nahum, A. E., & Fenwick, J. D. (2009). Monte Carlo modeling of small photon fields: Quantifying the impact of focal spot size on source occlusion and output factors, and exploring miniphantom design for small-field measurements. *Medical Physics, 36*(2009), 3132–3144. http://doi.org/10.1118/1.3152866.

Sheikh-Bagheri, D., & Rogers, D. W. O. (2002). Sensitivity of megavoltage photon beam Monte Carlo simulations to electron beam and other parameters. *Medical Physics, 29*(2002), 379–390. http://doi.org/10.1118/1.1446109.

Solberg, T. D., Ph, D., & Medin, P. M. (2011). Quality and safety in stereotactic radiosurgery and stereotactic body radiation therapy: Can more be done?, *Journal of Radiosurgery and SBRT, 1,* 13–19.

Soubra, M., Cygler, J., & Mackay, G. (1994). Evaluation of a dual bias dual metal oxide-silicon semiconductor field effect transistor detector as radiation dosimeter. *Medical Physics, 21*(4), 567–572

Tanyi, J. A., Krafft, S. P., Hagio, T., Fuss, M., & Salter, B. J. (2008). MOSFET sensitivity dependence on integrated dose from high-energy photon beams. *Medical Physics, 35*(1), 39. http://doi.org/10.1118/1.2815626.

Treuer, H., Boesecke, R., Schlegel, W., Hartmann, G. H., Muller, R. P., & Sturm, V. (1999). The source-density function: determination from measured lateral dose distributions and use for convolution dosimetry. *Physics in Medicine and Biology, 38,* 1895–1909. http://doi.org/10.1088/0031-9155/38/12/013.

Underwood, T. S. A., Winter, H. C., Hill, M. A., & Fenwick, J. D. (2013). Mass-density compensation can improve the performance of a range of different detectors under non-equilibrium conditions. *Physics in Medicine and Biology, 58*(23), 8295–8310. http://doi.org/10.1088/0031-9155/58/23/8295.

van Elmpt, W., McDermott, L., Nijsten, S., Wendling, M., Lambin, P., & Mijnheer, B. (2008). A literature review of electronic portal imaging for radiotherapy dosimetry. *Radiotherapy and Oncology: Journal of the European Society for Therapeutic Radiology and Oncology, 88*(3), 289–309. http://doi.org/10.1016/j.radonc.2008.07.008.

Wang, L. L. W., Klein, D., & Beddar, A. S. (2010). Monte Carlo study of the energy and angular dependence of the response of plastic scintillation detectors in photon beams. *Medical Physics, 37*(10), 5279. http://doi.org/10.1118/1.3488904.

Wang, L. L. W., & Leszczynski, K. (2007). Estimation of the focal spot size and shape for a medical linear accelerator by Monte Carlo simulation. *Medical Physics, 34*(2007), 485–488. http://doi.org/10.1118/1.2426407.

Westermark, M., Arndt, J., Nilsson, B., & Brahme, A. (2000). Comparative dosimetry in narrow high-energy photon beams. *Physics in Medicine and Biology, 45,* 685–702. http://doi.org/10.1088/0031-9155/45/3/308.

Wickman, G., Johansson, B., Bahar-Gogani, J., Holmström, T., & Grindborg, J. E. (1998). Liquid ionization chambers for absorbed dose measurements in water at low dose rates and intermediate photon energies. *Medical Physics, 25*(1998), 900. http://doi.org/10.1118/1.598268.

Wong, C. J., Ackerly, T., He, C., Patterson, W., Powell, C. E., Qiao, G., … Geso, M. (2009). Small field size dose-profile measurements using gel dosimeters, gafchromic films and micro-thermoluminescent dosimeters. *Radiation Measurements, 44*(3), 249–256. http://doi.org/10.1016/j.radmeas.2009.03.012.

Wood, J. J., & Mayles, W. P. M. (1995). Factors affecting the precision of TLD dose measurements using an automatic TLD reader. *Physics in Medicine and Biology, 40*(2), 309–313. http://doi.org/10.1088/0031-9155/40/2/009.

Wu, A., Zwicker, R. D., Kalend, A. M., & Zheng, Z. (1993). Comments on dose measurements for a narrow beam in radiosurgery. *Medical Physics, 20*(3), 777–779. http://doi.org/10.1118/1.597032.

Yin, Z., Hugtenburg, R. P., & Beddoe, A. H. (2004). Response corrections for solid-state detectors in megavoltage photon dosimetry. *Physics in Medicine and Biology, 49,* 3691–3702. http://doi.org/10.1088/0031-9155/49/16/015.

Yorke, E., Alecu, R., Ding, L., & Fontenla, D. (2005). Diode in vivo dosimetry for patients receiving external beam radiation therapy. Report of Task Group. Retrieved from https://scholar.google.com/scholar?hl=en&q=Diode+in+vivo+dosimetry+for+patients+receiving+external+beam+radiation+therapy&btnG=&as_sdt=1,14&as_sdtp=&search_plus_one=form#0.

In vivo dosimetry I: External beam radiation therapy

Landon Wootton and Sam Beddar

Contents

8.1 INTRODUCTION

8.1.1 BACKGROUND

Safety is of the highest importance in radiotherapy treatment due to the potential for debilitating and even fatal outcomes in the event of treatment errors. Numerous measures are practiced to ensure safe treatment: machine interlocks, secondary dose calculations, chart checks, machine quality assurance (daily, monthly, and annually), patient-specific quality assurance, and other measures all exist to detect and prevent errors that could compromise treatment quality or result in patient injury. *In vivo* dosimetry plays an increasingly important role in radiotherapy safety as a final check at the end of the radiotherapy chain. The purpose of *in vivo* dosimetry is to verify the treatment at the point of delivery, thus detecting any errors that pass undetected through prior safety checks or are introduced after the final pretreatment safety check.

At the time of writing, *in vivo* dosimetry is largely confined to skin dose measurements by thermoluminescent dosimeters (TLDs) and diodes. Furthermore, it is used primarily under special circumstances such as total body irradiation to identify areas receiving too little radiation for local boosts, treatment of pregnant patients to document fetal dose, or treatment of patients with implantable electronic devices that are prone to malfunction if overexposed to radiation. Occasionally, skin dosimetry may also be used to verify the delivery of very complex treatments. However, a large majority of radiation therapy patients do not receive routine *in vivo* dosimetry.

8.1.2 MOTIVATION FOR *IN VIVO* DOSIMETRY

Motivation to adopt *in vivo* practices stems from a number of sources: multiple high-profile radiotherapy accidents have been reported recently that might have been prevented with *in vivo* dosimetry [1–4]; there is a possibility of new governmental regulations (already a reality in parts of Europe) in response to such accidents looms on the horizon; and there is a growing consensus by experts on the importance of *in vivo* dosimetry [5]. The rapidly increasing complexity of treatment delivery underscores the need for *in vivo* dosimetry. An increasing number of systems have to interface flawlessly, and as the complexity of these systems grows more intricate, the likelihood of errors (user errors in particular) increases. Thus, a final safety check at the point of delivery becomes a more valuable measure for ensuring patient safety.

Another motivation for expanded *in vivo* dosimetry is the increased feasibility of implementing such a system. Detectors that are capable of real-time measurement and small enough to perform measurements internally rather at the skin surface are becoming commercially available. There is more research documenting appropriate methods to use these detectors. Some of these detectors are relatively inexpensive, offsetting concerns about cost-effectiveness. As detector technology continues to improve, the weight of these advantages will continue to grow.

In vivo dosimetry also offers the possibility to improve the patient experience as well as enhance their safety. As mentioned earlier, several accidents with grievous outcomes have been profiled in the media, drawing public and government attention to safety risks of radiotherapy. Many patients will have heard or read about these incidents, and being unfamiliar with radiation will be apprehensive about their treatment. *In vivo* dosimetry can be used to reassure frightened patients that the errors they have heard about would be caught and corrected before the consequences became significant. As such, it will improve the actual and perceived quality of care of patients.

8.2 PLASTIC SCINTILLATION DETECTORS FOR *IN VIVO* DOSIMETRY

Plastic scintillation detectors (PSDs) are of particular interest for *in vivo* dosimetry due to a number of favorable characteristics that are discussed below. The goal of this chapter is to highlight the advantages and provide guidance for addressing the challenges of implementing a PSD-based *in vivo* dosimetry system in a clinical setting, drawing on published experience to date.

8.2.1 PROPERTIES

As described elsewhere in this book, PSDs possess an array of unique characteristics distinguishing them from other detectors [6–8]. A number of these characteristics suit the PSD particularly well for *in vivo* applications: specifically, water equivalence; a response independent of energy, dose rate, and other factors; a high sensitivity; and a response time on the order of nanoseconds.

Water equivalence is a highly desirable characteristic for *in vivo* dosimetry. In therapeutic energy ranges, the dose measured by an appropriately calibrated water-equivalent detector is the same dose that would be absorbed by an equivalent volume of tissue (water and soft tissue being approximately equivalent in terms of radiation interactions), obviating the need for a conversion factor. Additionally, water equivalence ensures that detectors do not perturb a radiation field when present. Accordingly, a PSD can be placed in the center of a treatment field in a patient without compromising the treatment field.

The independence of PSD response from factors such as radiation energy and angle of incidence facilitates accurate measurements. The effect of such factors will sometimes be difficult to determine and/or correct *in vivo*, even if they are well characterized *ex vivo*. For example, the orientation of an angularly dependent detector would be difficult to determine *in vivo* if the detector were inserted via nasal catheter to measure the dose to a patient's nasopharyngeal tissue. Even if the effect of angle is well known for this detector, the imprecision in determining its orientation will introduce some uncertainty into the correction. Similarly, the use of other correction factors will often introduce uncertainty in a similar manner. PSDs do not require correction factors at therapeutic energies, excepting temperature dependence (discussed below) in some cases.

Finally, the high signal and the nanosecond response time of PSDs make real-time dose monitoring feasible. The fast response time allows fine temporal resolution. The high signal allows a PSD system to maintain a reasonable signal-to-noise ratio (SNR) even when measuring in the short time intervals with commensurately small doses. Real-time dosimetry is essential if one wishes to be able to detect errors as they happen in order to interrupt treatment and rectify the situation. Temporal information can also be useful when investigating errors. For example, treatment can be investigated on a beam-by-beam basis.

It is important to note that some PSDs exhibit temperature dependence [9,10], a characteristic undesirable for *in vivo* dosimetry. Detectors are typically calibrated near room temperature but when used *in vivo* will come to a thermal equilibrium near body temperature (approximately 37°C). The magnitude of the effect observed in studies ranges from a 10% decrease in light output at body temperature relative to room temperature for BCF-60 to a negligible change in light output for BC-400. A linear correction factor obtained by measuring the relative response at calibration temperature and body temperature is sufficient to account for this effect. The detector can also be calibrated at body temperature, eliminating the need for correction altogether.

8.2.2 COMPARISON TO OTHER DETECTORS

A variety of detectors have been used for external beam *in vivo* dosimetry [11], so for the purpose of comparison, metal–oxide–semiconductor field-effect transistors (MOSFETs), silicon diodes, and electronic portal imaging devices (EPIDs) will be considered. These detectors are considered because unlike many other *in vivo* detectors, they are capable of real-time dosimetry and are well represented in the scientific literature.

Silicon diodes are the most widely used real-time *in vivo* dosimeters [12,13] and are very commonly used for entrance *in vivo* dosimetry (i.e., skin dosimetry, sometimes as a proxy for pacemaker/fetal dose, etc.). Diodes are highly sensitive due to the very low ionization energy of silicon (~3 eV compared to ~30 eV in air). As a direct result of this high sensitivity, diodes with volumes on the order of millimeters or less can be used to achieve excellent spatial resolution. Diodes overrespond to low-energy photons and are often encapsulated in a thin layer of metal to filter out such photons. Diodes also require buildup to achieve a charged particle equilibrium. As a result of the buildup and encapsulation, they are energy-dependent detectors and should only be used in beam energies for which the manufacturer has designed them. Correction factors are also necessary to account for angular dependence (a result of the asymmetrical internal geometry of the detector) and dose rate dependence [14,15]. Finally, diodes are damaged by large cumulative exposure to radiation. As they are irradiated, the sensitivity will decrease, requiring periodic recalibration. The loss of sensitivity is rapid at first but slows as dose is accumulated. For this reason, many manufacturers pre-irradiate diodes before commercially distributing them. In spite of these drawbacks, diodes have been successful *in vivo* detectors. They are capable of real-time readout, and in fact, wireless readout has become available recently. This eliminates the need for cables connecting the diode to an electrical source. They have been used extensively for both published research and clinical application.

MOSFET detectors are a more recent development. When irradiated, the threshold voltage required for a specific level of current flow between the MOSFET transistor source and drain increases linearly with absorbed dose. A negative bias may be applied to increase the sensitivity of the detector. The threshold voltage can be read out at predetermined time intervals to obtain real-time dose information and after treatment as a measure of the cumulative delivered dose [16]. MOSFETs are on the order of millimeters in each dimension, providing excellent spatial resolution and resulting in minimal beam attenuation. They are moderately energy dependent, may exhibit a large angular dependence depending on the design, and can exhibit a small dose rate dependence as well, all of which may be corrected for. Single-bias MOSFETs are slightly temperature dependent, but dual-bias MOSFETs (two MOSFETs on a single silicon chip operating at two different gate bias voltages) minimize temperature effects [17]. A main drawback of

MOSFETs is that they are quite sensitive to radiation damage, becoming essentially useless after exposure to a cumulative 50 Gy. For *in vivo* applications where irradiation of 2 Gy for each fraction is typical, this severely limits their reusability. However, lifetime can be increased at the cost of sensitivity by careful design. Additionally, MOSFETs are relatively expensive [18]. They exhibit a lower precision and accuracy than other *in vivo* detectors, though not drastically so. The use of MOSFETs for *in vivo* dosimetry is widely published in [19–22], and they are commercially available. Like diodes, they can be operated wirelessly and have even been implanted in patients for *in vivo* dosimetry [23].

EPIDs are routinely used for positioning verification via megavoltage portal images, and are not commonly thought of as dosimeters. However, by using the same EPID to collect radiation transmitted through a patient, they can be repurposed as completely noninterventional *in vivo* dosimeters capable of providing 2D or even 3D information [24]. EPIDs are dose rate independent and show an approximately linear response with accumulated dose. Because the imaging device is part of the gantry, its orientation relative to the radiation source is constant, and it is thus angularly independent. EPIDs are prone to a phenomenon referred to as *ghosting*, wherein latent dose images are not cleared completely when information is read out, resulting in a small amount of temporal blurring. This can be accounted for, however. The main difficulty of EPID-based *in vivo* dosimetry is the interpretation of the data. There are two primary methods: forward projection through the simulation computed tomography (CT) dataset or back projection to produce a 3D dose distribution. In forward projection, the planar dose profile the EPID is expected to measure is calculated for each beam and compared to the measured dose profile using gamma analysis or a similar method [25,26]. However, the significance and source of discrepancies between the two is complicated to determine. To overcome this, a method to back-project the transmitted dose through a CT dataset to calculate a dose profile has been developed [27,28]. The result is a 3D dose distribution superimposed on the patient simulation CT image, which allows a more straightforward interpretation of the EPID data.

The PSD compares favorably overall with each of these detectors. Its size is equal to or smaller than diodes and MOSFETs. It does not require buildup or correction for factors such as energy, angle, or dose rate. Because PSDs generate light rather than an electronic signal, it must be physically connected via optical fiber and cannot be implanted like a MOSFET or wireless diode. However, because it is nonelectronic and contains no metal, it is compatible with magnetic resonance imaging, a distinct advantage with the advent of magnetic resonance imaging-linear accelerators. It is far more radio resistant than either the MOSFET or the diode, giving it a comparatively longer lifetime, and is less expensive as well. It cannot provide 3D dose information provided by the EPID (at best a large array of PSDs could be used to sample 3D space), but a multipoint PSD can measure dose at multiple points simultaneously with a single optical fiber for light transmission [29,30]. Like diodes and MOSFETs, some PSDs exhibit temperature dependence, but this can be accounted for straightforwardly [9,10].

8.3 REAL-TIME IMPLEMENTATION

Real-time capability is one of the PSD's most advantageous characteristics for *in vivo* dosimetry. Real-time acquisition allows fine temporal resolution: individual beams and even segments of an intensity-modulated radiation therapy (IMRT) treatment can be distinguished allowing beam-by-beam verification. Errors can be detected during the course of treatment rather than after it has concluded. When an error is detected, treatment can be halted immediately and the source of the error determined before continuing. This ability is of great benefit to the patient for obvious reasons.

Implementing real-time *in vivo* dosimetry with a PSD introduces additional considerations beyond those of static dosimetry, which are covered in this section, namely, the increased sensitivity to the system of background subtraction, an SNR that is sensitive to the sampling rate of data acquisition, and the detector dead time. Each of these arises from the serial nature of real-time dosimetry: Dose is sampled at each point of time exactly once and the final result is the summation of all these measurements. How this occurs will be made clear in the rest of this section.

Note to the reader: To avoid confusion when discussing static and real-time dosimetry, the term *measurement* will refer the acquisition of the entire signal of interest, which in the case of real-time measurement is divided into *sub-measurements*. For example, if the dose to a phantom during a delivery of intensity-modulated radiation is the quantity of interest, the total light signal after delivery—in one acquisition in static dosimetry or the sum of many short acquisitions in real-time—constitutes a measurement. In static dosimetry, measurements are often repeated and averaged together. The same can be done in real-time dosimetry, though in practice one will often not have the chance to do so, for example, if dose is being measured *in vivo*.

8.3.1 BACKGROUND SUBTRACTION

Background subtraction is essential in real-time dosimetry just as in static dosimetry, and the process is largely identical. Measurements of background signal are performed in the absence of radiation and averaged together. This average background signal is subtracted from all subsequent measurements to eliminate the systematic error that would arise from a nonzero background signal.

An import caveat is that because the average background signal is a measured quantity, it will be subject to random error and not precisely equal to the true average background signal. Static dosimetry is largely insensitive to random error in the measured background because the signal generated by the detector (when exposed to radiation) is typically much larger than the error in the background. In real-time dosimetry, signal is measured repeatedly in small segments of time with the measured background subtracted from each. The magnitude of the signal in each segment may not be substantially greater than the background. When cumulative dose is calculated by summing the measurements, the background error will accumulate and potentially result in a nonnegligible deviation in the measured signal from the true signal.

The following three methods are recommended to minimize errors introduced by background subtraction: acquisition of a large sample of background measurements to reduce the random error in the average background; the use of an identical data acquisition rate for background measurement and real-time measurement; and finally calculation a mean background rather than a median background if the values are discretized (grayscale levels of charge-coupled device (CCD) pixels being the primary example).

The adequacy of the number of background measurements acquired can be estimated by calculating the standard error of the mean (i.e., dividing the standard deviation of the background measurements by the square root of the number of measurements), a standard statistical calculation. The result is an indication of how much the mean average background would be expected to vary when measured repeatedly [31]. For example, if the background is distributed normally, approximately 95% of measured average backgrounds would be expected to fall within two standard errors of the mean of a single measured average background. For practical purposes, this can be used to estimate how close the measured average background is to the true background. This method also highlights the diminishing returns of increased background measurements: the standard error decreases as the square root of the number of measurements. Therefore, a balance must be struck between the quality of the average background and the time required to obtain it, especially in time-limited situations such as background collection immediately prior to a patient treatment.

When acquiring an average background, the data acquisition rate should be identical to the rate that will be used for real-time measurements. This is to prevent inaccurate background subtraction in the case of dark current accumulation. If a detector acquires some background signal per second, the background will be a function of measurement duration. Thus, a downside of slow data acquisition (corresponding to long sample durations) is the longer time required to acquire sufficient background.

Finally, it is sometimes important to compute the mean rather than median average background if the measurements are a discrete value, such as pixel values on a CCD camera. This is important because the true average will likely lie between discretized values. For example, if the true average background for a CCD pixel was 1.4, one would expect to see approximately 60% of background measurements equal to 1 and 40% equal to 2. Thus, the median would most likely be 1, off by 0.4. Even an equal number of measurements of one and two would result in a value of 1.5, which is still incorrect. When the signal is composed of the summation of many pixels, this error will accumulate and may result in a nonnegligible error in the measured background.

8.3.2 SIGNAL-TO-NOISE RATIO

Every measurement made includes some statistical uncertainty, as discussed in Section 8.3.1. This uncertainty is a result of random, undesirable fluctuations in signals from a variety of (sometimes unknown) sources, which are collectively referred to as noise. In many applications, including radiation detectors, it is common to quantify the ratio of the useful signal to this noise, a quantity known as the SNR. Inversely, the ratio of the noise to the useful signal is a measure of the uncertainty in the measured signal due to the noise. For the purposes of this section, we will consider noise that is approximately normal in distribution and centered around zero, an assumption that is both reasonable and sufficient for analyzing the function of PSDs.

In static dosimetry, multiple measurements are made under the same conditions and averaged together. There is noise associated with each measurement, but when the measurements are averaged together, the noise will tend to cancel itself out because it is distributed around zero. Accordingly, there will be less variation in the average of measurements than in single measurements. That is to say, the SNR of many measurements averaged together is better than that of a single measurement. The SNR improves as more measurements are averaged together, and the improvement is proportional to the square root of the number of measurements averaged together. For example, if the SNR of one measurement made with a PSD is 100, the SNR of four measurements averaged together is 200.

Real-time dosimetry is fundamentally different in that multiple sub-measurements are *added* together to produce one total measurement, rather than *averaged*. In the case of *in vivo* dosimetry, the total measurement cannot be repeated, which eliminates the possibility of averaging to improve SNR (i.e., the patient cannot be retreated to achieve better dosimetric data). The noise associated with each sub-measurement accumulates when the sub-measurements are summed to produce a *total noise* for the measurement. The total noise will be distributed around zero, just like the noise of the individual sub-measurements. However, if the entire measurement could be repeated, the variation in the outcome would be larger than the variation in individual sub-measurements. Similarly to how the SNR of averaged measurements improves with the square root of the number of measurements, the SNR of a real-time measurement decreases as the square root of the number of sub-measurements. For example, if the SNR of a real-time measurement consisting of 100 sub-measurements is 200, the SNR of an otherwise identical (same dose and time span) real-time measurement consisting of 400 sub-measurements will be reduced to 100.

The result of this limitation is that the duration of sub-measurements must be chosen carefully. The sub-measurement noise can be quantified (e.g., by calculating the standard deviation on a large set of background measurements of equal duration as the sub-measurements) and in combination with the expected total signal, the maximum number of sub-measurements allowable to achieve a given SNR or uncertainty can be calculated:

$$n = \left(\frac{SNR_1}{SNR_n} \right)^2$$

where:

SNR_1 is the SNR if the entire signal were acquired in one sub-measurement (i.e., the expected signal divided by the sub-measurement noise)

SNR_n is the minimum acceptable SNR that results from dividing the measurement into n sub-measurements, a user defined quantity

In general, real-time measurement of very long treatments will require longer sub-measurements than that of very fast treatments. An example of the former would be an IMRT treatment involving many beam angles and many *beamlets* per beam angle which may take 10–15 min, although an example of the former would be volumetric modulated arc therapy which can be delivered in 3 min.

8.3.3 DETECTOR DEAD TIME

The final consideration regarding real-time measurement is detector dead time. Dead time is the interval of time between two subsequent periods of data acquisition during which a detector does not collect signal. For PSDs, the dead time is due to the photodetecting element of the system, which is responsible for converting light into a quantifiable electrical signal.

The effect of dead time depends on the duration of the dead time relative to the measurements. For a continuous source of radiation, such as cobalt-60, the calculation of signal lost due to dead time is straightforward. The calculation for pulsed sources of radiation such as linear accelerators is more complicated. The signal lost depends on the number of pulses falling within the dead time, which may vary between subsequent dead times if the detector and linear accelerator are not synced. In practice, if the duration of measurements is significantly longer than the dead time, this effect is negligible for continuous and pulsed radiation. This will often be the case as many modern photodetectors have extremely short dead times on the order of 10–100 μs.

8.4 LOCALIZATION AND EXPECTED DOSE

Determining the position of a detector *in vivo* is necessary to determine the expected dose to the detector, and thus what measured dose constitutes a significant deviation from the planned dose. The simplest method of localizing the detector is with radiographic imaging. Image-based localization is divided into two categories for discussion here: 2D localization and 3D localization, referring to the type of image acquired for localization. Additionally, electronic positioning systems that do not rely on imaging are available and will be discussed briefly as well.

8.4.1 2D LOCALIZATION

Most external beam radiation therapy (EBRT) patients undergo daily portal imaging for the purpose of alignment. Typically, the locations of bony anatomy or implanted fiducials on orthogonal portal images are compared to digitally reconstructed radiographs (DRRs) from simulation, and appropriate shifts are made to align the markers on the daily portal images with their position on the DRRs. An *in vivo* regimen can capitalize on daily alignment by using the portal images for the purpose of detector localization as well, integrating the *in vivo* process seamlessly into the clinical workflow. PSDs present a unique challenge in this regard, however, as the property of water equivalence renders PSDs indistinguishable from tissue on radiographs.

One way to circumvent this problem is the incorporation of radiopaque fiducials into the detector system in a rigid geometry relative to the detector. How this is done will depend on the application, but often it will involve attaching fiducials directly to the detector near to but not directly adjacent to the detector active volume. See Figure 8.1 for an example. The reason for separation between the fiducials and the active volume is to prevent the fiducials from attenuating the dose measured by the detector.

After incorporating fiducials, each portal image acquired during daily setup will capture the location of the fiducials in the two dimensions perpendicular to the imaging axis. Because setup uses two portal images offset by 90°, all three spatial dimensions will be represented and the 3D location of each fiducial can be calculated relative to isocenter, which should be indicated on the portal images. The known location of the detector relative to the fiducials can then be used to calculate its position relative to isocenter. To determine the expected dose to the detector, the dose on the simulation CT at the calculated position of the detector can be determined using the treatment planning system. It is important to note that shifts are usually

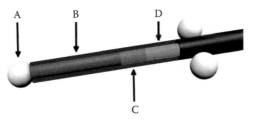

Figure 8.1 Example diagram of an *in vivo* plastic scintillation detector (PSD). Three ceramic fiducials (A) are attached to the PSD. They are separated from the active volume (C) by a spacer (B) and a small length of optical fiber (D) to ensure that the PSD active volume is not in a fiducial generated dose shadow. If the materials used permit the assumption of a rigid geometry, the location of the active volume can be calculated based on the position of the fiducials. (Reproduced from Wootton, L. et al., Real-time *in vivo* rectal wall dosimetry using plastic scintillation detectors for patients with prostate cancer, *Phys. Med. Biol.*, 59, 647–660, 2014. With permission from IOP Publishing.)

applied after portal imaging, and if the shifts are small, additional portal images may not be acquired, depending on the preferences of the treating physician and/or hospital protocol. It is therefore important to note and incorporate any shifts made into the calculated location of the detector.

An example of such detector localization can be found in an *in vivo* study performed by Hsi et al. [32]. Four fiducials were attached to an endorectal balloon in addition to a fiducial already present in the balloon, for a total of five fiducials. One fiducial was located in the tip of the endorectal balloon and another at the base of the balloon on the stem. These fiducials were used to visualize the axis of the endorectal balloon. The remaining fiducials were positioned laterally on the endorectal balloon, one on the left lateral side and two on the right. Finally, detectors (TLDs in this case) were mounted on the anterior surface of the balloon to place them in contact with the anterior rectal wall. When the balloon is appropriately inflated in this setup, the fiducials and detectors are in an approximately rigid geometry. Hsi et al. were able to accurately localize the detector as evidenced by the −2.1% average difference between the measured dose and the expected dose. Additionally, the position of the three lateral fiducials on the lateral portal image served as a proxy for the rotation of the balloon, allowing more consistent balloon placement.

8.4.2 3D LOCALIZATION

For some treatments, particularly for complicated or challenging treatments, CT images acquired on linear accelerators equipped with cone beam CT or CT-on-rails may be used for daily setup instead of portal imaging because more accurate alignment is possible. Once again, it is possible to capitalize on this setup methodology to localize *in vivo* detectors and integrate *in vivo* dosimetry into the clinical workflow.

As in 2D localization, fiducials in a rigid geometry with the detector are required because PSDs are water equivalent, and therefore indistinguishable from tissue on radiographic images. To localize the detector, the fiducials are identified on the CT dataset from patient setup. The coordinates of each fiducial can then be used to calculate the position of the detector on the dataset. Discrepancies between the imaging isocenter and radiation isocenter and shifts made after imaging must be taken into account when determining the position of isocenter on the dataset relative to the detector position. Overall, the method is very similar to 2D localization: the primary advantage of 3D localization is that once the detector position is determined, the user may choose to then extract the expected dose from the treatment plan on simulation CT image set or to calculate it on the daily CT image set with the treatment planning system.

In the latter case, the expected dose will take into account changes in the patient anatomy following simulation and differences in setup, isolating the contribution of the treatment delivery and intrafractional motion between the time the daily image is acquired and the treatment is delivered. Once *in vivo* dosimetry has verified that the treatment was delivered correctly, the dose distribution on the daily image can be examined to verify that anatomic changes have not compromised the target coverage or organ at risk sparing. Compare this to computing an expected dose from the simulation image, wherein discrepancies between the measured and expected doses may be caused by gross anatomic changes following simulation, incorrect delivery (treatment errors), or intrafractional anatomic motion.

Wootton et al. [33] used 3D localization in their *in vivo* protocol. Their PSD system employed a ceramic fiducial at the distal tip of the detector and two fiducials attached laterally to the optical fiber proximal to the active volume of the detector (Figure 8.1). To validate this method, they constructed detectors with radiopaque metal wire in place of scintillating fiber and acquired CT scans of the detectors in a phantom. They compared the position of the wire calculated with fiducials to the position observed in CT scans and found within the axial plane the calculated position's net deviation from the observed position was just 0.1 mm, with all observations deviating less than 0.7 mm. However, only 65% (13 out of 20) of the calculated locations were on the correct CT slice, with the rest being one slice off. This was deemed acceptable for their study as the dose was fairly homogeneous in the craniocaudal direction but highlights an important aspect of 3D localization: the craniocaudal accuracy of localization will be limited by the slice thickness of the scan.

The reason for this limitation is the ambiguity of determining the fiducial location exactly in the craniocaudal direction. The center of the fiducial will rarely lie at the center of a CT slice, but partial volume artifacts will render the task of determining where the fiducial center truly lies difficult if not impossible. If a CT slice is large enough to encompass the entire fiducial, it cannot be determined where in that slice the fiducial is centered. As the slice thickness becomes increasingly fine, the fiducial will begin to span several slices and determining the center can be done much more accurately by comparing subsequent slices.

8.4.3 ELECTRONIC POSITIONING SYSTEMS

Positioning sensors that wirelessly measure and report their position relative to an external reference (a local magnetic field for example) may also be used for localization. The position reported is relative to a point of origin in a transmitter. By determining the position of the transmitter relative to a treatment isocenter and fixing the geometry between the sensor and the detector, the position of the detector can be determined continuously. There is relatively little in the literature about the use of such devices. Cherpak et al. [34] used electronic position sensors in conjunction with MOSFETs to measure skin dose and position during lung IMRT.

Electronic positioning systems offer the potential to combine real-time dose information with real-time position information, as the sensors can report their location multiple times per second (location was recorded at a rate of approximately 20 Hz in the study of Cherpak et al.). This would enable the unique ability to determine the contribution of intrafractional motion to the discrepancy between measured and expected dose.

8.5 APPLICATORS

In addition to being attached directly to the skin, PSDs may be introduced into patient body cavities via applicators designed to conform to the shape of the body cavity for consistent positioning. Ideally, these applicators would cause minimal patient discomfort and provide a barrier between the detector and the tissue to avoid contamination. Often they will additionally perform some function unrelated to the detector such as immobilizing anatomy.

8.5.1 CATHETERS

Catheters are a natural applicator as PSDs are small enough to run through a catheter with relative ease. They are also a very general applicator if relatively invasive, and not limited to any specific portion of the anatomy. However, to date the primary use of catheters for *in vivo* dosimetry has been urethral catheterization of male patients being treated for prostate cancer to enable measurement of dose to the prostate [35–37]. Although this procedure has been performed only for brachytherapy patients, there is no reason that it could not be applied to external beam treatments if desired. This particular setup would have the advantage of placing the detector in the center of the target volume where the dose is homogeneous. Because the dose distribution is homogeneous, it is not necessary to localize the detector with high precision to generate an accurate expected dose.

When using a catheter as a PSD applicator, a single fiducial at the tip of the detector may be sufficient to localize the detector if the catheter is visible on radiographic images. A liquid contrast agent can be used to enhance the visibility of the catheter and aid in the localization of the detector if the catheter is sealed at the interior end to avoid needless administration of contrast to the patient. The length of the detector inserted into the catheter should be carefully noted, perhaps using length indicators on the detector fiber itself, and the detector clamped at the exterior end after being inserted the correct distance to immobilize it within the catheter.

8.5.2 ENDORECTAL BALLOON/CYLINDER

The endorectal balloon is an inflatable device designed for use in prostate cancer patients being treated with external beam radiation. Prior to treatment, the endorectal balloon is inserted into the rectum of the patient deflated, and then inflated with air or water to immobilize the prostate against the pubic symphysis. This limits the intrafractional motion of the prostate and allows the use of tighter margins in the dose distribution. It has the additional benefit of displacing rectal tissue from the high dose regions of the treatment field (Figure 8.2).

A detector embedded in the surface of the balloon will be directly adjacent to the rectal wall, a dose-limiting organ. If the detector is positioned anteriorly, in addition to being in contact with the radiosensitive anterior rectal wall, it will also be in close proximity to the prostate. As such, it can monitor dose to a critical organ at risk and verify the target dose simultaneously. This applicator is used by

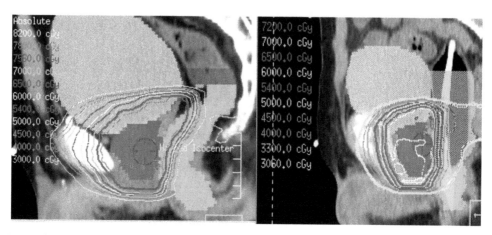

Figure 8.2 Endorectal balloons are a popular applicator for *in vivo* dosimetry of prostate cancer patients. By mounting detectors to the lumen of the balloon, they can be positioned against the rectal wall for dose monitoring. Additionally, the balloon aids in immobilizing the prostate (allowing tighter margins) and displaces rectal tissue from the high dose region of the treatment field. Sagittal computed tomography (CT) slices taken from prostate treatment plans are presented as examples, with the endorectal balloon on the right.

Wootton et al. [33] in their *in vivo* study. Sometimes, a hard cylinder is used instead of an endorectal balloon. Its purpose and usefulness as an applicator are similar to those of the endorectal balloon.

8.5.3 ORAL PLATE

For treatment of cancer of the head and neck, custom mouthpieces designed to fit a patient's teeth and oral cavity are sometimes fabricated [38] for reproducible immobilization and to provide separation between tissues surrounding the oral cavity in order to reduce the dose to normal or healthy tissue. These mouthpieces are fabricated by dentists and sometimes referred to as oral stents. They provide an opportunity to reproducibly position a detector within the oral cavity for *in vivo* measurements.

A PSD embedded within an oral plate would allow dose measurements in the vicinity of the parotid and submandibular glands, which are critical organs at risk. Excessive dose to the parotids severely limits a patient's ability to produce saliva and results in dry mouth and difficulty eating and swallowing. An oral plate applicator could also position detectors close to the hard and soft palates, the mandible, the teeth, and other areas of interest for head and neck treatment.

As an example, Marcié et al. [39] used oral plate applicators in an *in vivo* trial monitoring the treatment of oropharyngeal and nasopharyngeal tumors using MOSFETs. A picture of the applicator used is shown in Figure 8.3.

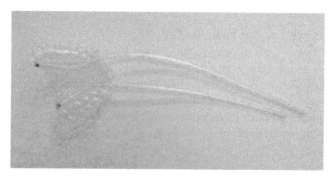

Figure 8.3 Oral plate applicator used by Marcié et al. (Reproduced from *IJROBP*, 61, Marcié, S. et al., *In vivo* measurements with MOSFET detectors in oropharynx and nasopharynx intensity-modulated radiation therapy, 1603–1606. Copyright 2005, with permission from Elsevier.)

8.5.4 NASOESOPHAGEAL TUBE

Nasoesophageal tubes are another type of applicator that can be employed for monitoring of head and neck cancers. As the name suggests, nasoesophageal tubes are inserted through the patient's nasal cavity into the esophagus. PSDs can be inserted into these tubes and positioned to monitor dose in the nasal cavity or esophagus as needed. They have been used for *in vivo* trials by Engström et al. [40] and Gagliardi et al. [41] using TLDs as detectors . In general, nasoesophageal tubes are well tolerated (37 out of 43 patients tolerated the applicator in Gagliardi et al.'s study and 10 out of 10 patients in Engström et al.'s study). Local anesthetic sprayed into the nasal cavity and a light lubrication of the tube can improve the patient experience. Photographs of a nasoesophageal tube alone and that inserted into a patient's nasal cavity are displayed in Figure 8.4.

8.6 OTHER SOURCES OF UNCERTAINTY

There are a variety of uncertainties extrinsic to the detector in *in vivo* dosimetry. It is important to account for these when interpreting measured *in vivo* doses in relation to expected doses. The contribution of each will vary greatly depending on the application, so they are described in general here, and it is left to the reader to determine the applicability to specific situations.

8.6.1 DOSE GRADIENT AND POSITION

Uncertainty in position, whether it stems from intrafractional motion [42–44] or from other sources such as the localization process, will introduce a variable uncertainty into the expected dose that depends on the gradient of the local dose distribution. If a detector is in an area with a negligible dose gradient, uncertainty in position will have a negligible effect on the uncertainty of the measured dose: regardless of where the detector is, the dose is the same. Alternatively, if a detector is in a very steep dose gradient, a small uncertainty in the position can translate into a large uncertainty in the expected dose. A shift of a millimeter by the detector in an extreme dose gradient may change the expected dose by more than 5% of the target dose. This effect is illustrated in Figure 8.5.

Either situation can be advantageous. When a detector is positioned in a flat dose region, such as in the planning target volume, an excellent agreement between the measured and expected doses is possible in spite of positional uncertainty. Deviations from the expected dose in this situation are more likely to be related to the treatment itself—for example, an incorrect machine output or wrong accessory used—because the positional uncertainty does not factor into the deviation. The disadvantage of this is that a detector located in such a region would be insensitive to set up errors such as incorrect alignment because such errors have the effect of shifting the detector in the dose distribution.

A detector placed in a steep dose gradient however will be exquisitely sensitive to set up errors. Depending on the magnitude of the dose gradient, even an acceptably small error in the alignment of the patient may result in a large deviation between the expected dose and the measured dose. For this reason, one should consider using a distance to agreement criterion to evaluate the measured *in vivo* dose. This is done by

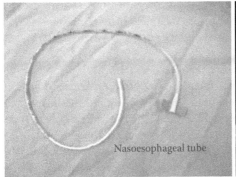

Figure 8.4 Nasoesophageal tube used by Gagliardi et al. (left) and tube inserted through patient nasal cavity prior to treatment (right). (Reproduced from Gagliardi, F. et al., Intra-cavitary dosimetry for IMRT head and neck treatment using thermoluminescent dosimeters in a nasoesophageal tube, *Phys. Med. Biol.*, 54, 3649–3657, 2009. With permission from IOP Publishing.)

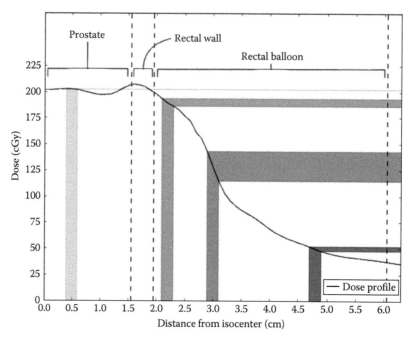

Figure 8.5 A typical dose profile taken from a patient's prostate intensity-modulated radiation therapy (IMRT) treatment plan extending from the prostate (treatment isocenter) to the posterior rectal wall. Anatomical regions are labeled and delineated by dashed lines. The effect of dose gradient on uncertainty in the expected dose as measured by a hypothetical detector is illustrated with vertical shaded bars representing a ±1 mm positional uncertainty and horizontal shaded bars representing the resulting uncertainty in the expected dose. In the prostate (yellow), the expected dose varies by only ±0.1% due to the flat dose gradient. At the anterior rectal wall (green), the same positional uncertainty results in an expected dose uncertainty of ±2%. This increases to ±11% for the lateral rectal wall (red). Finally, at the posterior rectal wall, the relative uncertainty is ±4% because the absolute dose is low. When expressed in absolute terms, the uncertainties, respectively, become ±0.25, ±3.8, ±14.15, and ± 2.15 cGy. (Reproduced from Wootton, L. et al., Real-time *in vivo* rectal wall dosimetry using plastic scintillation detectors for patients with prostate cancer, *Phys. Med. Biol.*, 59, 647–660, 2014. With permission from IOP Publishing.)

calculating the minimum change in the position of the detector required to change the expected dose to the measured dose. If the shift is large, the possibility of a setup error can be investigated, whereas if the shift is small (within treatment tolerance), the treatment can be assumed to have been delivered correctly, even if a fairly significant deviation between the expected and measured doses is observed. Again, what constitutes a large or small shift will depend on the treatment site and is left to the user's discretion.

8.6.2 TREATMENT PLANNING SYSTEM

An additional consideration when interpreting *in vivo* measurements is the accuracy of the treatment planning system. The dose distribution calculated by the treatment planning system is limited by the accuracy of model-based approaches to dose calculation (as opposed to Monte Carlo) and a finite resolution that is generally poorer than the spatial resolution of PSDs.

In order to calculate dose distributions in reasonable times, most treatment planning systems use a model-based approach where machine-specific parameters are used to model the beam output, rather than performing full radiation transport to determine dose distributions. This method is highly useful but suffers from a loss of accuracy at heterogeneous interfaces and outside the primary beams [45]. As a result, PSDs positioned at such an interface may measure a dose that does not agree with the treatment plan because the TPS is incapable of correct calculation. The magnitude of this effect should be estimated by referring to situation-appropriate published Monte Carlo simulations or by performing appropriate Monte Carlo simulations oneself. Depending on the nature of the interface and the algorithm used, the magnitude of the effect will vary; though for modern algorithms, it will generally be less than 5% [46].

An additional approach used to limit calculation times in the planning process is using a limited resolution when calculating dose distributions. The resolution is chosen by the user, typically on the order of a few millimeters (2–4) in each direction. Because dose is calculated in three dimensions, each doubling of resolution requires an eightfold increase in calculation time. As the spatial resolution of PSDs can be 1 mm or even less (well below the resolution of the TPS), the expected dose to the detector calculated by the TPS will often be an interpolated value. It is therefore possible that a PSD may measure a very fine detail in the dose distribution that the TPS is not capable of resolving.

8.7 PERFORMANCE CHARACTERISTICS

At the time of writing, the only published study using PSDs for *in vivo* dosimetry in EBRT is that of Wootton et al. [33]. The results of that study are described here as an indication of what might be expected from an *in vivo* PSD system. The reader can also refer to the next chapter on *in vivo* dosimetry of brachytherapy, a subject on which more clinical experience has been published, for additional insight into the performance of PSDs *in vivo*.

This study included five prostate cancer patients being treated with IMRT. Twice weekly *in vivo* measurements using a PSD duplex (i.e., two PSDs in conjunction) were performed for the duration of treatment, resulting in a total of 142 *in vivo* measurements. The detectors were attached to the anterior surface of endorectal balloons to position them in close proximity to the anterior rectal wall. CT scans were acquired using a CT-on-rails unit immediately prior to each *in vivo* monitored treatment fraction. The CT images were used to determine the location of the detectors and calculate the dose expected to be measured by the detector. The expected dose was compared with the measured dose to assess the performance of the system. Posttreatment validations were also performed in which the detectors were irradiated with a known dose under controlled conditions. This was done to determine whether deviations between the expected and measured *in vivo* doses were caused by a failure of the detector itself or extrinsic factors.

The mean difference between the measured and expected doses for individual patients ranged between –3.3% and 3.3%. The standard deviation ranged between 5.6% and 7.1% for four of the enrolled patients, and was 14.0% for the remaining patient (Figure 8.6). This outlier was determined to be caused by difficulty accurately positioning the rectal balloon with detectors due to the patient's obesity. This caused the detectors to be positioned more frequently in a high-gradient, lower absolute dose region. Posttreatment validation

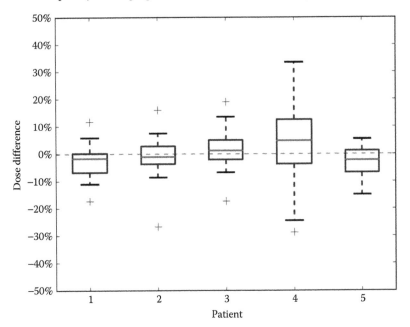

Figure 8.6 Results from a five-patient *in vivo* dosimetry protocol utilizing plastic scintillation detectors (PSDs). The average values range between approximately ±3% for all patients, and the distribution of dose differences was similar for all patients excepting patient 4 due to difficulty reproducibly positioning the endorectal balloon. (Reproduced from Wootton, L. et al., Real-time *in vivo* rectal wall dosimetry using plastic scintillation detectors for patients with prostate cancer, *Phys. Med. Biol.*, 59, 647–660, 2014. With permission from IOP Publishing.)

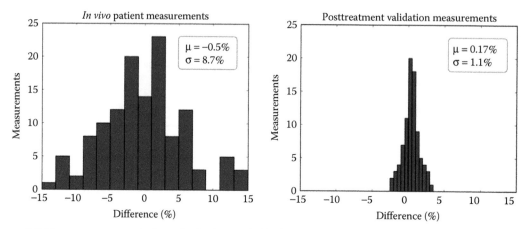

Figure 8.7 The aggregate results from the *in vivo* study displayed as a histogram (left), compared with the aggregate results of the posttreatment validation of the detectors (right). Both distributions exhibit an excellent average agreement, but the spread is considerably smaller for the validation measurements. This suggests that deviations *in vivo* were caused by effects extrinsic to the detector.

demonstrated that the detectors were measuring dose accurately (Figure 8.7); thus, the discrepancies between the measured and expected doses were caused by extrinsic effects such as anatomical motion.

When the measurements were considered in aggregate (i.e., all 142 measurements considered simultaneously), the mean difference between the expected and measured doses was −0.5%, with a standard deviation of 8.7%. The ability of a PSD *in vivo* dosimetry system to produce highly accurate real-time information was demonstrated as well (Figure 8.8). Overall the system was demonstrated to be capable of

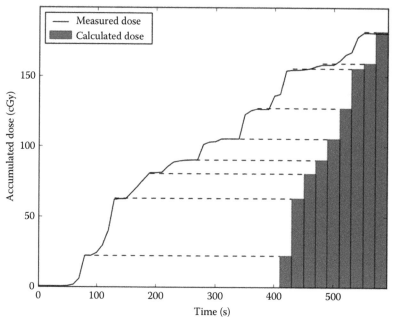

Figure 8.8 Real-time cumulative dose measured with a PSD plotted against expected cumulative per-beam expected dose for IMRT treatment of the prostate. The treatment planning system does not provide real-time cumulative dose for comparison (a therapist can take as much time as necessary between beams), so instead the cumulative dose after each beam is indicated by bars on the right with dotted lines to facilitate comparison to the real-time dose profile. Flat regions in the measured dose indicate zero dose rate, and therefore correspond to the time between beams. These flat regions agree extremely well with the expected beam-by-beam dose, demonstrating highly accurate real-time measurement. (Reproduced from Wootton, L. et al., Real-time *in vivo* rectal wall dosimetry using plastic scintillation detectors for patients with prostate cancer, *Phys. Med. Biol.*, 59, 647–660, 2014. With permission from IOP Publishing.)

Table 8.1 **Results of Wootton et al.'s five-patient *in vivo* dosimetry study**

PATIENT	NUMBER OF MEASUREMENTS	MEAN DIFFERENCE (%)	STANDARD DEVIATION (%)	VALIDATION (%)
1	30	−2.6	5.6	−0.1
2	28	−1.1	7.1	0.5
3	30	1.5	6.7	0.3
4	28	3.3	14.0	0.5
5	21	−3.3	5.8	−0.5

Note: The mean difference between the expected dose (calculated on daily CT images) and the PSD measured dose is displayed, along with the standard deviation of these differences. Validation represents the average agreement between a known delivered dose and a PSD measured dose under controlled conditions after treatment. The excellent validation results suggest that the differences observed *in vivo* are extrinsic to the detector, caused by effects such as anatomical motion.

measuring dose *in vivo* with high accuracy and precision in high dose gradient regions. The results from this system are summarized in Table 8.1.

Because the results across the patients were highly similar (with the exception of one patient due to complications arising from that patient's obesity), the authors posited that these results were typical and could reasonably be expected when using PSDs for rectal wall *in vivo* dosimetry. Given the steep dose gradient and the anatomical motion associated with the prostate, improved performance could be expected in more controlled circumstances. For example, if the detectors were inserted into the urethra (which would be located in the center of the homogeneous target dose for a prostate cancer patient), this system would measure dose *in vivo* with a precision and accuracy similar to those produced by the validation measurements in this study, as localization would be of minor importance in determining the expected dose.

8.8 CONCLUSION

In vivo EBRT dosimetry with PSDs is not yet a mature field, and as such much remains to be done. However, due to its favorable characteristics and a positive initial experience, PSDs remain a strong candidate to make routine, accurate *in vivo* dosimetry practicable. Continuing commercialization of PSDs, improving detector technology, and continued research into their use will bolster this position. It is hoped that this reference will serve as a helpful guide to aid the continued development of PSDs for *in vivo* dosimetry.

REFERENCES

1. International Atomic Energy Agency, Investigation of an accidental exposure of radiotherapy patients in Panama/ Report of a team of experts (International Atomic Energy Agency [IAEA], Vienna, Austria, 2001).
2. Mayles W. The Glasgow incident—A physicist's reflections. *Clin. Oncol.* **19** 4–7 (2007).
3. Derreumaux S, Etard C, Huet C, Trompier F, Clairand I, Bottollier-Depois J, Aubert B, Gourmelon P. Lessons from recent accidents in radiation therapy in France. *Radiat. Prot. Dosim.* **131** 130–135 (2008).
4. Bogdanich W, Rebelo K. A pinpoint beam strays invisibly, harming instead of healing, *New York Times*, December 29, 2010.
5. Edwards C, Mountford P. Characteristics of *in vivo* radiotherapy dosimetry. *Br. J. Radiol.* **82** 881–883 (2009).
6. Beddar S, Mackie T, Attix F. Water-equivalent plastic scintillation detectors for high-energy beam dosimetry: I. Physical characteristics and theoretical considerations. *Phys. Med. Biol.* **37** 1883–1900 (1992).
7. Beddar S, Mackie T, Attix F. Water-equivalent plastic scintillation detectors for high-energy beam dosimetry: II. Properties and measurements. *Phys. Med. Biol.* **37** 1901–1913 (1992).
8. Archambault L, Briere T, Pönisch F, Beaulieu L, Kuban D, Lee A, Beddar S. Toward a real-time *in vivo* dosimetry system using plastic scintillation detectors. *IJROBP.* **78** 280–287 (2010).
9. Wootton L, Beddar S. Temperature dependence of BCF plastic scintillation detectors. *Phys. Med. Biol.* **58** 2955–2967 (2013).

10. Buranurak S, Andersen C, Beierholm A, Lindvold L. Temperature variations as a source of uncertainty in medical fiber-coupled organic plastic scintillator dosimetry. *Rad. Meas.* **56** 307–311 (2013).

11. Mijnheer B, Beddar S, Izewska J, Reft C. *In vivo* dosimetry in external beam radiotherapy. *Med. Phys.* **40** 070903 (2013).

12. Edwards C, Hamer E, Mountford P, Moloney A. An update survey of UK *in vivo* radiotherapy dosimetry practice. *Br. J. Radiol.* **80** 1011–1014 (2007).

13. York E, Alecu R, Ding L, Fontenla D, Kalend A, Kaurin D, Masterson-McGray ME et al. Diode *in vivo* dosimetry for patients receiving external beam radiation therapy: Report of the American Association of Physicists in Medicine (AAPM) Task Group 62. AAPM Report No. 87 (Madison, WI: Medical Physics Publishing, 2005).

14. Saini A, Zhu T. Dose rate and SDD dependence of commercially available diode detectors. *Med. Phys.* **31** 914–924 (2004).

15. Saini A, Zhu T. Energy dependence of commercially available diode detectors for *in-vivo* dosimetry. *Med. Phys.* **34** 1704–1711 (2007).

16. Gladstone D, Chin L. Automated data collection and analysis system for MOSFET radiation detectors. *Med. Phys.* **18** 542–548 (1991).

17. International Atomic Energy Agency, IAEA Human Health Report No. 8. Development of procedures for in vivo dosimetry in radiotherapy (International Atomic Energy Agency [IAEA], Vienna, Austria, 2013).

18. Jornet N, Carrasco P, Jurado D, Ruiz A, Eudaldo T, Ribas M. Comparison study of MOSFET detectors and diodes for entrance *in vivo* dosimetry in 18 MV x-ray beams. *Med. Phys.* **31** 2534–2542 (2004).

19. Ramaseshan R, Kohli K, Zhang T, Lam T, Norlinger B, Hallil A, Islam M. Performance characteristics of a microMOSFET as an *in vivo* dosimeter in radiation therapy. *Phys. Med. Biol.* **49** 4031–4048 (2004).

20. Scalchi P, Francescon P, Rajaguru P. Characterization of a new MOSFET detector configuration for *in vivo* skin dosimetry. *Med. Phys.* **32** 1571–1578 (2005).

21. Bloemen-van Gurp E, du Bois W, Visser P, Bruinvis I, Jalink D, Hermans J, Lambin P. Clinical dosimetry with MOSFET dosimeters to determine the dose along the field junction in a split beam technique. *Radiother. Onc.* **67** 351–357 (2003).

22. Chuang C, Verhey L, Xia P. Investigation of the use of MOSFET for clinical IMRT dosimetric verification. *Med. Phys.* **29** 1109–1115 (2002).

23. Den R, Nowak K, Buzurovic I, Cao J, Harrison A, Lawrence Y, Dicker A, Showalter T. Implanted dosimeters identify radiation overdoses during IMRT for prostate cancer. *IJROBP.* **83** e371–e376 (2012).

24. van Elmpt W, McDermott L, Nijsten S, Wendling M, Lambin P, Mijnheer B. A literature review of electronic portal imaging for radiotherapy dosimetry. *Radiother. Oncol.* **88** 289–309 (2008).

25. Nijsten S, van Elmpt W, Jacobs M, Mijnheer B, Dekker A, Lambin P, Minken A. A global calibration model for a-Si EPIDS used for transit dosimetry. *Med. Phys.* **34** 3872–3884 (2007).

26. Piermattei A, Fidanzio A, Azario L, Grimaldi L, D'Onofrio G, Cilla S, Stimato G et al. Application of a practical method for the isocenter point *in vivo* dosimetry by a transit signal. *Phys. Med. Biol.* **52** 5101–5117 (2007).

27. Hansen V, Evans P, Swindell W. The application of transit dosimetry to precision radiotherapy. *Med. Phys.* **23** 713–721 (1996).

28. Wendling M, McDermott L, Mans A, Sonke J, van Herk M, Mijnheer B. A simple back projection algorithm for 3D *in vivo* EPID dosimetry of IMRT treatments. *Med. Phys.* **36** 3310–3321 (2009).

29. Archambault L, Therriault-Proulx F, Beddar S, Beaulieu L. A mathematical formalism for hyperspectral, multipoint plastic scintillation detectors. *Phys. Med. Biol.* **57** 7133–7145 (2012).

30. Therriault-Proulx F, Archambault L, Beaulieu L, Beddar S. Development of a novel multi-point plastic scintillation detector with a single optical transmission line for radiation dose measurement. *Phys. Med. Biol.* **57** 7147–7159 (2006).

31. Rosner B. *Fundamentals of Biostatistics*, 6th edn. (Belmont, CA: Brooks/Cole Publishing Co. 2005).

32. Hsi W, Fagundes M, Zeidan O, Hug E, Schreuder N. Image-guided method for TLD-based *in vivo* rectal dose verification with endorectal balloon in proton therapy for prostate cancer. *Med. Phys.* **40** 051715 (2013).

33. Wootton L, Kudchadker R, Lee A, Beddar S. Real-time *in vivo* rectal wall dosimetry using plastic scintillation detectors for patients with prostate cancer. *Phys. Med. Biol.* **59** 647–660 (2014).

34. Cherpak A, Cygler J, Andrusyk S, Pantarotto J, MacRae R, Perry G. Clinical use of a novel *in vivo* 4D monitoring system for simultaneous patient motion and dose measurements. *Radiother. Oncol.* **102** 290–296 (2012).

35. Bloemen-van Gurp E, Murrer L, Haanstra B, van Gils F, Dekker A, Mijnheer B, Lambin P. In vivo dosimetry using a linear MOSFET-array dosimeter to determine the urethra dose in ^{125}I permanent prostate implants. *IJROBP.* **73** 314–321 (2009).

36. Brezovich I, Duan J, Pareek P, Fiveash J, Ezekiel M. *In vivo* urethral dose measurements: A method to verify high dose rate prostate treatments. *Med. Phys.* **27** 2297–2301 (2000).

37. Nose T, Koizumi M, Yoshida K, Nishiyama K, Sasaki J, Ohnishi T, Kozuka T et al. *In vivo* dosimetry of high-dose-rate interstitial brachytherapy in the pelvic region: Use of a radiophotoluminescence glass dosimeter for measurement of 1004 points in 66 patients with pelvic malignancy. *IJROBP.* **70** 626–633 (2008).

38. Zhen-Yu Q, Deng X, Huang S, Shiu A, Lerch M, Metcalfe P, Rosenfeld A, Kron T. Real-time *in vivo* dosimetry with MOSFET detectors in serial tomotherapy for head and neck cancer patients. *IJROBP.* **80** 1581–1588 (2011).

39. Marcié S, Charpiot E, Bensadoun R, Ciais G, Hérault J, Costa A, Gérard J. *In vivo* measurements with MOSFET detectors in oropharynx and nasopharynx intensity-modulated radiation therapy. *IJROBP.* **61** 1603–1606 (2005).

40. Engström P, Haraldsson P, Landberg T, Hansen H, Engelholm S, Nyström H. In vivo dose verification of IMRT treated head and neck cancer patients. *Acta Oncol.* **44** 572–578 (2005).

41. Gagliardi F, Roxby K, Engström P, Crosbie J. Intra-cavitary dosimetry for IMRT head and neck treatment using thermoluminescent dosimeters in a naso-oesophageal tube. *Phys. Med. Biol.* **54** 3649–3657 (2009).

42. Litzenberg D, Balter J, Hadley S, Sandler H, Willoughby T, Kupelian P, Levine L. Influence of intrafraction motion on margins for prostate radiotherapy. *IJROBP.* **65** 548–553 (2006).

43. Keal PJ, Mageras GS, Balter JM, Emery RS, Forster KM, Jiang SB, Kapatoes JM et al. The management of respiratory motion in radiation oncology report of AAPM Task Group 76. *Med. Phys.* **33** 3874–3900 (2006).

44. Shirato H, Seppenwoolde Y, Kitamura K, Onimura R, Shimizu S. Intrafractional tumor motion: Lung and liver. *Seminars in Radiation Oncology.* **14** 10–18 (2004).

45. Papanikolaou N, Battista J, Boyer A, Kappas C, Klein E, Mackie T, Sharpe M, Dyk J. Tissue inhomogeneity corrections for megavoltage photon beams: Report of Task Group No. 65 of the Radiation Therapy Committee of the American Association of Physicists in Medicine (Madison, WI: Medical Physics Publishing, 2004).

46. Knöös T, Wieslander E, Cozzi L, Brink C, Fogliata A, Albers D, Nyström H, Lassen S. Comparison of dose calculation algorithms for treatment planning in external photon beam therapy for clinical situations. *Phys. Med. Biol.* **51** 5785–5807 (2006).

9 In vivo dosimetry II: Brachytherapy

Jamil Lambert

Contents

9.1 INTRODUCTION

Brachytherapy is a method of treatment where sealed radioactive sources are used to deliver radiation at a short distance [1]. It allows a high radiation dose to be given to the target volume while giving a lower dose to surrounding normal tissue. Brachytherapy can be broken up into two categories: low and high dose rate (LDR and HDR). In LDR brachytherapy, low activity sources or seeds are implanted into the area being treated. These may be removed after a few days once the required dose is reached, or implanted with a set activity and left in the patient permanently so that the desired dose is delivered in the time taken to decay completely. Three common isotopes used in LDR brachytherapy are ^{137}Cs, ^{125}I, and ^{103}Pd, which have half-lives of 30.17 years, 60 days, and 17 days, respectively, and emit photons of average energy 662, 28, and 21 keV, respectively. The relatively low energy photons of ^{125}I and ^{103}Pd have a short range in tissue and are mostly attenuated by the patient's body, requiring less shielding to protect staff and the general public.

In HDR brachytherapy, sources with activities approximately three orders of magnitude higher are used to treat the patient. The most common source used is ^{192}Ir with a half-life of 74.2 days and an average photon energy of 380 keV. The source is remotely moved out of a lead safe along a transfer tube into a catheter, which has been placed next to or inside the area to be treated [2]. The source stops at set dwell positions along the catheter for a set time to deliver the required dose to the surrounding tissue.

Multiple catheters can be used, especially in treating the prostate, where up to 20 or more catheters are inserted through the perineum into the prostate. Radiopaque markers are placed inside the catheters and two orthogonal radiographs or a CT scan is taken prior to the treatment to determine the position of the catheters. A computer planning system is then used to determine the dwell positions and dwell times of the source along these catheters. These are optimized to give an even dose throughout the target volume and to deliver the minimum possible dose to the surrounding normal tissue, especially sensitive structures. Standard applicators are also used, for example, in gynecological treatment where no computer planning is done. A predetermined set of dwell positions and dwell times are used to deliver a set dose along the applicator, which has previously been determined to be the optimal distribution by clinical studies or other means. The length and thickness of the applicator can be changed to suit the anatomy of the patient. In all of these techniques, there is currently no correction for changes in position of the catheters due to swelling or the patient moving between the imaging and the treatment. These effects can result in a difference between the planned dose distribution and the actual dose delivered to the patient, which will only be quantified if *in vivo* dosimetry is performed.

The dose around a brachytherapy source drops off rapidly with distance, due to the inverse square law and attenuation, allowing a high dose to be given to the target volume and a low dose to surrounding areas. But any movement of the source catheters or internal organs between the CT scan used for planning and the treatment delivery will result in potentially large changes in the dose delivered. Unless *in vivo* dosimetry is performed or a second CT scan is taken immediately prior to the treatment, there is no record of the actual dose delivered. Deviations in the delivered dose distribution from the planned dose distribution will go unnoticed. Taking a second CT scan prior to treatment delivers an extra unwanted dose to the patient and is often not a practical option. There is therefore a need for *in vivo* dosimetry with high spatial resolution in the target volume and surrounding critical organs. A promising dosimeter for this application is the plastic scintillation detector, which has favorable dosimetric and physical characteristics for *in vivo* dosimetry in brachytherapy.

9.2 *IN VIVO* DOSIMETERS USED IN BRACHYTHERAPY

In vivo dosimetry has been performed in brachytherapy over the past few decades with a range of different dosimeters. The most commonly used dosimeters in the clinic are thermo-luminescent dosimeters (TLD), semiconductor diodes, and metal–oxide–semiconductor field effect transistors (MOSFETs). There has also been work done with optically stimulated luminescent (OSL) dosimeters with both aluminum oxide and silver-activated phosphate glass, Alanine, diamond detectors, Fricke gels, and scintillation detectors. Each of these dosimeters has drawbacks and limitations in their use as an *in vivo* dosimeter in brachytherapy [3].

The major difficulties are the high dose gradients, which require a dosimeter with a small sensitive volume, the physical size and flexibility of the dosimeter, and the accuracy in measuring often low photon energies.

Ionization chambers are often considered the gold standard for dosimetry, but the smaller volume ionization chambers do not have the sensitivity to accurately measure the dose from a brachytherapy source at a centimeter distance in water or tissue. Larger volume ionization chambers are well suited for the calibration of brachytherapy sources but cannot be used *in vivo* due to the high dose gradients and impracticalities due to their large physical size.

Diamond detectors have a small sensitive volume and can be used to measure the dose around a brachytherapy source, [4] but they have a relatively large physical size and rigid stem, which prevents their use in some *in vivo* dosimetry applications, such as insertion into the urethra or as part of a rectal dose array.

Lithium fluoride (LiF) TLDs have been used extensively for *in vivo* dosimetry in brachytherapy [5–9]; they can provide integral dose readings at multiple points but do not provide any real-time information. The read out and annealing procedure of TLDs is relatively time consuming compared to other similar dosimeters such as OSL, and the dose precision of individual chips can be problematic and cause large uncertainties in the readings.

Brezovich et al. [5] found that LiF TLDs have a linear dose response down to 2 Gy but then become nonlinear. LiF is also more sensitive to low energy photons and the shift in the photon spectrum to lower energies with distance from the brachytherapy source can cause a change in sensitivity. Meigooni et al. [10]

showed an 8.5% increase in the sensitivity of LiF at a distance of 100 mm from an ^{192}Ir source compared to the sensitivity at 10 mm. In a later study by Pradhan and Quast [11], the depth-dependent sensitivity of LiF was found to be lower, with a 2.5% difference between the sensitivity at 10 and 100 mm.

Diodes are also commonly used, in particular to measure the rectal dose during HDR brachytherapy in the pelvic region. Diodes can have a large angular dependence, depending on their construction. Waldhausl et al. [12] found the angular dependence of diodes in a PTW (Germany) type 9112 rectum probes and type 9113 bladder probes to vary from −2% to +1.8% in the longitudinal direction and −4.9% to +5.4% in the axial direction. The sensitivity of the diodes was also found to change with distance from the source with a difference between measured and calculated dose of −3.5% at 30 mm and −7.3% at 100 mm.

Kirov et al. [13] found the relative uncertainty of diode measurements to be 5% and for TLDs up to 10% in a phantom study comparing Monte Carlo calculations with the readings of TLD and diode dosimeters around an ^{192}Ir source.

MOSFET dosimeters have gained in popularity in recent years, and there are a range of types that have been used for both HDR [14–18] and LDR [19–21] brachytherapy. These dosimeters provide real-time measurements from a small detection volume and have a small physical size. The dosimetric properties reported by the various groups, which may be due to the construction of different MOSFETs, or the calculation and application of the correction factors required. One such correction factor that can affect the accuracy of the readings is an energy dependence that results in a change in sensitivity with distance from the source. For *in vivo* dosimetry, the temperature dependence of a MOSFET needs to be measured and corrected for, and for any measurement, the change in sensitivity with absorbed dose must also be corrected for. The angular dependence of a MOSFET is highly dependent on its construction with some having as much as 10% angular dependence [22] and others being as low as 2% [15].

Zilio et al. [17] compared the performance of RADFET type TOT 500 MOSFETs to an ionization chamber and Monte Carlo calculations using MCNP4C. A thin waterproof plastic holder was used to protect the electrical components since this type of dosimeter is not inherently waterproof. Corrections were applied for the sensitivity dependence with absorbed dose and as a function of distance. The dose was measured at distances of 5–50 mm in water from an ^{192}Ir HDR source and compared to the dose calculated using Monte Carlo. The mean MOSFET dose readings were within 10% of the Monte Carlo calculation for 95% of the measurement points and the net uncertainty was calculated to be 8.6% and 10.8% for R and K type MOSFETs, respectively. This is similar to deviations between ^{192}Ir dose measurements and Monte Carlo of 8.9% and 3.8% for the single MOSkin and dual-MOSkin MOSFET dosimeters developed by the Center for Medical Radiation Physics of the University of Wollongong (Australia) [14]. Qi et al. [15] had deviations of the measurements to those planned of less than 5% for all measurements, with a better than 3% reproducibility, but showed that the energy dependence results in 10% deviation in the response between a distance of 10 and 20 mm from an ^{192}Ir source.

Alanine dosimeters can also be used as pellets with a similar size and shape to TLDs; although their reproducibility and uncertainty may be better than TLDs, they are limited to measuring high doses, in the order of 5–50 Gy [23]. Alanine dosimetry requires expensive electron spin resonance spectrometers and is therefore limited to only a small number of centers, with other centers gaining access through standards laboratories with an alanine dosimetry mailing service.

Fricke gels can be used inside a catheter to give a one-dimensional dose distribution along the path of the catheter. These gels exhibit a temperature dependence, with a difference in sensitivity between body temperature and room temperature of 7%; the gel is also limited to doses over 4 Gy but then saturates at doses higher than 20 Gy and at dose rates higher than 400 cGy/min [24,25].

OSL dosimeters based on Al_2O_3:C can be used for *in vivo* dosimetry in a similar way to TLDs differing only in the readout and annealing methods [26–28]. They can also be used coupled to an optical fiber to give real-time dose measurements [29]. The readout is much faster than with TLDs, and commercial systems can have dosimeters inside light tight plastic cases that are individually barcoded. One issue with the plastic cases is the angular dependence, which can be as high as 20% [27] or even 70% with an asymmetric field on the surface of a phantom [28]. The OSL chips have an energy dependence that depends on the dose history of the chip [28]; this has the effect of changing the correction factor required if the chips are calibrated in ^{60}Co or MV photon beam. The response of the detector also changes by

10% between a distance of 10 and 100 mm in water from a ^{192}Ir brachytherapy source. OSL dosimeters have been shown to exhibit a nonlinear dose response [30], but this can be accounted for if the nonlinear response is adequately characterized.

9.3 OPTIMIZATION OF SCINTILLATION DETECTOR DESIGN FOR BRACHYTHERAPY

The main difference in brachytherapy that will affect the design of the scintillation detector is the relatively lower energies, which for ^{125}I and ^{103}Pd are below the threshold for Cerenkov radiation production. For ^{192}Ir, the photon energies are above the threshold energy for Cerenkov production in a plastic optical fiber, but there are arguments both for and against using a Cerenkov correction technique (see Section 9.4.5), such as those discussed in Chapter 5.

The energy spectrum from a brachytherapy source changes with distance, shifting to lower energies. Inorganic scintillators are less water equivalent than plastic scintillators and potentially have a larger depth-dependent sensitivity, due to their energy dependence. Plastic scintillators are the preferred option for brachytherapy dosimetry.

Increase in the signal-to-noise ratio (SNR) in a scintillation detector is achieved by maximizing the amount of scintillation light generated and transmitted to the photo detector. This can be achieved by increasing the core diameter and numerical aperture of the optical fiber to maximize the amount of scintillation light coupling into the optical fiber. Increasing the diameter of the scintillator will also increase the amount of light coupling into the optical fiber, but for a set scintillator diameter, the optimal solution is an optical fiber with an equal diameter [31–33]. Increasing the length of the scintillator will increase the amount of light produced, but will also increase the sensitive volume, decreasing the spatial resolution of the dosimeter. Adding a reflective coating at the end of the scintillator can effectively double the length of the scintillator in terms of light coupling into the optical fiber, with no change in the sensitive volume [34].

The emission of light from the scintillator is isotropic and it is possible to construct a dosimeter with no angular dependence. A cylindrical scintillator with a length equal to its diameter will have little angular dependence, but the length of the scintillator may not be adequate to give the required sensitivity or to overcome the Cerenkov background signal, if no removal techniques are used. Increasing the length of the scintillator will create an angular dependence due to both the change in light-collection efficiency with distance from the end of the optical fiber and volume averaging effects when measuring in high dose gradients. Therefore, there is a compromise between the sensitivity of the dosimeter and its spatial resolution and angular dependence.

To maximize the amount of scintillator light transmitted, the optical fiber material needs to be highly transparent to light in the wavelength range emitted by the scintillator, commonly in the blue region. Silica is a common material for optical fibers and is transparent to light in this wavelength range, but becomes too rigid with larger diameters needed in this application. Poly(methyl methacrylate) (PMMA) is also transparent to visible light and is more flexible than silica. PMMA fibers with diameters of 1 and 0.5 mm are readily available and are suitable light guides for most applications. The higher the numerical aperture of the optical fiber, the more light from the scintillator will couple into it. The numerical aperture of a PMMA fiber is typically close to 0.5, for example with a refractive index of 1.492 in the core and 1.405 in the cladding.

The use of a photomultiplier tube will increase the sensitivity for a single-point detector over other light detectors such as charge-coupled devices (CCDs) and photodiodes [35]. For large arrays, a CCD may be preferable for both cost and practicality.

Accurate location of the dosimeter is required in brachytherapy due to the high dose gradients. Plastic scintillators and optical fibers are difficult to see on a CT or X-ray image due to their near water equivalence. A radiopaque marker can be placed near the scintillator so that it can be located on a CT or X-ray image (Figure 9.1). The radiopaque marker can shadow the scintillator in certain orientations reducing the signal by up to 20% when the marker is between the source and the scintillator [36]. If the scintillation detectors are placed inside an applicator, the markers can be placed in locations

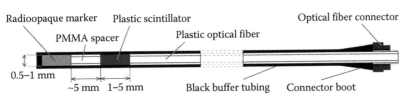

Figure 9.1 An example of a plastic scintillation detector designed for *in vivo* point dose measurements in brachytherapy.

that will not be between the source and scintillator so that this dose shadowing effect does not occur. For a single dosimeter, the marker should be placed at an adequate distance so that the angle when the dose shadowing occurs is reduced. Cartwright et al. [36] used a separation of 4 mm between the scintillator end and the marker, the reading of the dosimeter was only affected by the marker when the source was above the scintillator and within 10° of the fiber axis. Outside of this region, the dosimeter had less than 2% angular dependence.

In most *in vivo* brachytherapy dosimetry, the scintillation detector will need to be inserted into the organ at risk or target volume that is being measured. This will require insertion into a catheter or applicator (Figure 9.2), the detection volume can be correctly placed by withdrawing the detector to a set distance from the end of the catheter, a radiopaque marker can then be used to verify this position in relation to the patient anatomy. It is important to keep the entire length of the optical fiber and scintillator light tight, but the black plastic sheath that comes on common commercial PMMA fiber may be too thick. Sourcing a fiber with a thinner sheath will allow the dosimeter to be inserted into thinner catheters; another option is removing the section of sheath from the part of the dosimeter that is inserted and replacing it with a thinner one, or inserting it into a light tight applicator or catheter.

If an extension fiber or fibers are used to transmit the light from the scintillation detector outside the bunker to the detector, it is important that they are physically protected. The black protective sheath on a common PMMA fiber will not be very durable and will not provide adequate protection in a clinical environment. Plastic optical fibers with an extra protective sheath are available and will reduce the chances of the fiber being damaged when it is trodden on, passed under a bunker door, or repeatedly rolled out and up again. It is also important to have an appropriate boot on the connector as the bending stresses near the

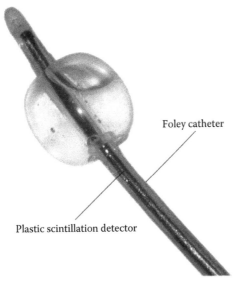

Figure 9.2 A photograph of a Foley catheter with a 1 mm diameter plastic scintillation detector inserted into it.

Clinical applications using small PSDs

connector can cause the fiber to break. The fiber is often weakened at this point due to the removal of the sheath and the attachment of the connector.

9.4 DOSIMETRIC PERFORMANCE OF SCINTILLATION DETECTORS IN BRACHYTHERAPY

9.4.1 DEPTH DOSE MEASUREMENTS

Depth dose curves are an essential tool in assessing the accuracy of a dosimeter in brachytherapy, even though the curves themselves may not be as important as they are in external beam therapy. For *in vivo* dosimetry, the dosimeter will be at a range of different distances from the brachytherapy source or sources; it must therefore provide an accurate dose reading over a relevant range of distances. For small distances from the source, the geometry of the source and its encapsulation will cause a difference in the dose delivered to that calculated or planned. Small changes in the position of the dosimeter will also result in large differences in the measured dose, for example, at a distance of 10 mm from an ^{192}Ir source, a 1 mm change in the distance results in a 20% change in the dose. For the purposes of assessing the quality of a dosimeter, measuring at distances of less than 10 mm from the source may not give any meaningful results.

Lambert et al. [37] measured the depth dose from an ^{192}Ir source in water with a scintillation detector. They found that the dosimeter with a 0.5-mm diameter scintillator and fiber was accurate to within ±2.9% and the 1-mm diameter dosimeter was accurate to within ±2.2% for distances between 10 and 100 mm (Figure 9.3). For distances larger than 100 mm from the source, the dose calculated using TG43 [38] becomes inaccurate. Monte Carlo calculations using EGSnrc [22] are also shown in Figure 9.1, and show that the scintillation detector remains accurate for all distances measured, up to the maximum of 250 mm.

9.4.2 ANGULAR DEPENDENCE

The angular dependence of a scintillation detector is determined by its physical construction. The major contributor is the ratio of the scintillator length to the diameter. It is in theory possible to construct a dosimeter with no angular dependence. The scintillation light is emitted isotropically and the angular dependence is only due to a dose gradient across the scintillation volume.

The angular dependence of a 5-mm long, 1-mm diameter scintillator can be kept below ±1.5% [37], increasing the length-to-diameter ratio to 8:1, with a 4-mm long 0.5-mm diameter scintillator increases the angular dependence to ±2.5%. Therefore, for diameter ratios below 5:1, essentially there is no angular dependence.

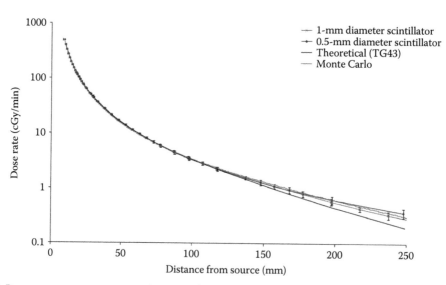

Figure 9.3 Dose rate measurements as a function of distance from an ^{192}Ir HDR brachytherapy source in water.

9.4.3 ENERGY DEPENDENCE

The energy spectrum from a brachytherapy source changes with distance and any energy dependence in the dosimeter may result in a depth-dependent sensitivity. There is a decrease in the response of a plastic scintillation detector with decreasing photon energy below 100 keV [39,40]. This could theoretically result in a change in the sensitivity of the dosimeter with distance, but this is not what is seen in practice [37]. The response of the scintillation detector to ^{192}Ir remains constant and agrees with Monte Carlo calculations over the range in distances of 10–250 mm in water (Figure 9.3).

The energy dependence does result in a difference in response to a brachytherapy source and a ^{60}Co or megavoltage photon beam. A radiation quality correction factor will need to be used if the scintillation detector is calibrated with a radiation source different to that being measured.

9.4.4 TEMPERATURE DEPENDENCE

The temperature dependence of a scintillation detector can be relatively small compared other dosimeters and may not need to be corrected for measurements performed around room temperature. When the dosimeter is inserted into a patient, the dosimeter will increase to near body temperature and the small temperature dependence should be corrected to reduce a potential systematic error. For a plastic scintillator based on polyvinyl toluene doped with anthracene, the temperature dependence can be as low as 0.1%/°C for a Saint-Gobain Bicron BC-400 scintillator [22] (Figure 9.4), and 0.09%–0.15%/°C for BCF-12 [41,42]. But temperature dependence of other anthracene-based plastic scintillators can be higher, a study by Beddar [43] measured an average temperature dependence of 0.6%/°C for a BCF-60 scintillator over the temperature range of 15°C–50°C and the studies by Buranurak et al. [41], and Wootton and Beddar [42] show dependences of 0.5%/°C and 0.55%/°C, respectively.

9.4.5 CERENKOV AND FIBER FLUORESCENCE

The need for Cerenkov removal in brachytherapy is dependent on the type of brachytherapy source and the measurement geometry. Most LDR brachytherapy sources do not have a high enough energy to produce Cerenkov radiation in a plastic optical fiber, for example ^{125}I has a maximum gamma energy of 35.5 keV and ^{103}Pd has 21 keV.

^{192}Ir has emission peaks at 316 and 468 keV. The peak at 316 keV is not far above the threshold of 177 keV for Cerenkov production in PMMA (with a refractive index of 1.492) and will produce only a

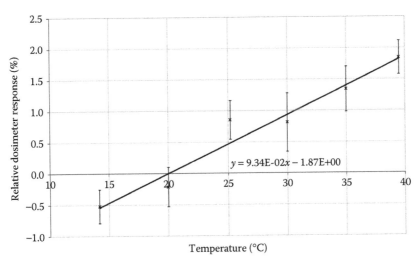

Figure 9.4 The temperature dependence of a BC-400 plastic scintillation detector irradiated with an ^{192}Ir HDR brachytherapy source.

small amount of Cerenkov radiation. The peak at 468 keV will potentially produce a detectable amount of Cerenkov radiation.

Whether or not a Cerenkov removal technique needs to be used for accurate dosimetry of an ^{192}Ir HDR brachytherapy source is a matter of discussion. Therriault-Proulx et al. [44] showed that when the source is within 10 mm of the optical fiber and greater than 25 mm from the scintillator polychromatic, Cerenkov removal is necessary. They show in a subsequent publication [45] that the error can be as large as 25% in a water phantom study of an *in vivo* scintillation dosimetry system. In a similar geometry, another group [37] showed that in the worst case, where the source is 5 mm from the fiber and 50 mm from the scintillator, the signal due to Cerenkov is 5% of the magnitude of the scintillator signal. And, when the distance is comparable, the Cerenkov is less than 0.1%.

It is clear that there is a detectable Cerenkov radiation signal produced from an ^{192}Ir source in a plastic optical fiber. The case where it will potentially cause an incorrectly large dose reading is when the source is close to the optical fiber and far from the scintillator. But the actual measurement in these cases in terms of dose will be small. These cases will only make up a small part of the total measurement and the absolute dose error over the total irradiation will be much smaller. This is supported by the measurements by the first group that showed up to a 25% error for a single-source position, but over the treatment the measured dose difference was 0.6% ± 3% higher than expected with stem effect removal and 0.4% ± 3% low without stem effect removal [45].

9.4.6 REPRODUCIBILITY

The reproducibility of a dosimeter's readings is important in *in vivo* dosimetry since making a repeat measurement under the same conditions is not possible. The purpose of *in vivo* dosimetry is to record the actual dose delivered during a fraction, or part thereof, or to detect any errors in the treatment delivery. If the dosimeters readings are not reproducible then the dose recorded will be incorrect or there may be a false positive or false negative in any error detection.

Scintillation detectors rely on a constant transmission efficiency of the optical path between the scintillator and the detector. Any changes in the transmission efficiency will relate directly to a change in the scintillators sensitivity. There are a number of possible causes of a change in the sensitivity of a scintillation detector:

- Damage to the join between the scintillator and the fiber
- Damage to the extension fiber
- Damage to the scintillator itself
- Reproducibility of the coupling efficiency of any connections in the optical path
- Changes in bending losses in the optical fiber
- Stability of the light detection system

In *in vivo* brachytherapy dosimetry, the scintillation detector may need to be inserted into a catheter. As the detector is inserted, particularly if there are tight bends, stress is placed on the joint between the scintillator and the optical fiber. If the glue holding the scintillator in optical contact to the optical fiber is not strong enough, the join may break, causing a small air gap. This will result in losses due to reflections at the scintillator–air and air–fiber interfaces. The optical fiber can also be accidentally damaged as it may be run across the floor of a room with multiple staff members concentrating on other tasks. Both of these types of damage to the detector will result in a reduction of the signal.

The reproducibility of the coupling efficiency of any optical connections and changes in bending losses can result in both an increase and decrease in the transmission efficiency. These changes will be smaller and more of a random nature, and can be minimized with a well-designed optical system with protected optical fibers and high-quality connectors that are kept clean and protected when not in use.

9.5 PRACTICALITIES OF *IN VIVO* DOSIMETRY

Along with the dosimetric characteristics of the scintillation dosimeter, there must also be considerations made to the practical aspects of routinely using such a dosimeter in the clinic for *in vivo* dosimetry.

9.5.1 POINT DOSE MEASUREMENTS

The majority of publications on *in vivo* brachytherapy dosimetry with a scintillation dosimeter are point dose measurements. One such system has been clinically trialed to measure the dose to the urethra during HDR prostate brachytherapy with ^{192}Ir [46]. Accurate location of the dosimeter is essential due to the high dose gradients in brachytherapy, and the placement of a radiopaque marker near the scintillator allows the dosimeter to be located on a CT or orthogonal X-rays. Single-point dose measurements simplify almost all aspects of *in vivo* dosimetry as compared to arrays, but the information that can be gained is also reduced and any error in the dosimeter cannot be cross-checked with other measurements.

9.5.2 ARRAYS

Arrays of individual dosimeters can be used to increase the number of measurement points and therefore the available information that can be gained from a single measurement. This increases the complexity of the measurement, requiring multiple optical fibers and multiple detectors, such as the pixels on a CCD. Arrays can be made by simply combining multiple identical scintillation dosimeters such as the system described by Liu et al. [35]. Another approach is to combine multiple scintillators with different emission peaks in series on a single fiber [47]. This single-fiber multipoint dosimeter can be used to give three dose measurements, a combination of which can then be used to calculate the location of the source along the catheter to measure the positioning accuracy and potentially detect an error in positioning (Chapter 10).

An array of scintillation dosimeters can be placed inside a specially made applicator and can be used to give a map of the dose to the rectal wall [48]. This method was also used to locate the source using a combination of readings from 16 dosimeters.

Both of the multifiber arrays described by Cartwright et al. [48] and Liu et al. [35] do not use any Cerenkov removal, making the detection of the scintillation light simpler, with the first implementation using a black-and-white CCD, and the second implementation using a multianode PMT. Adding Cerenkov removal to an array can be done using the same methods as used in external beam therapy described in Chapters 5 and 8, but will add cost and complexity to the construction of the system and its calibration.

9.5.3 USE INSIDE A CATHETER OR APPLICATOR

Inserting the scintillation dosimeter into a catheter or applicator places physical stresses on the detector, in particular on the joint between the scintillator and the optical fiber. The feasibility of inserting a scintillation dosimeter into a urethral catheter has been demonstrated in the clinical trial by Suchowerska et al. [46]. They did not report of any breakages of the dosimeter itself due to its insertion into the catheter. In one case, the balloon on the end of the catheter was punctured by a source needle resulting in the incorrect location of the dosimeter, which gave a reading of 67% below what was expected at the correct location.

An applicator can be used to keep the multiple detection points of an array dosimeter in a fixed geometry, such as in the brachytherapy rectal applicator presented by Cartwright et al. [48]. Multiple point dose measurements can then be used in an accurate geometry to locate the brachytherapy source and to give a dose map. Radiopaque markers can be placed in the applicator instead of attaching it to the dosimeter itself. This allows the markers to be placed at a larger distance from the scintillator, and in a location that will not be between the source and scintillator during the treatment.

9.5.4 STERILIZATION

Some sterilization methods used in the clinic may cause damage to the plastic materials of a scintillation dosimeter. High temperatures can deform the optical fiber and scintillator, such as those used in an autoclave, and should not be used for a plastic scintillation dosimeter. Chemical sterilization could damage the plastic materials of the dosimeter, and they must be tested before use in real patient measurements. The surface of the optical fiber at the connector end may be damaged or covered in a thin film that will reduce the coupling efficiency and therefore the sensitivity of the dosimeter.

Whichever sterilization method is chosen, it is important to check that a scintillation dosimeter is not damaged. This should be checked by calibrating a scintillation dosimeter of the same type as to be used in the *in vivo* measurements, both before and after sterilization and ensuring the two factors are within the measurement uncertainty.

9.5.5 DISPOSABLE VERSUS REUSABLE

There are arguments for both a disposable and a reusable design of *in vivo* medical devices [49–51]. For a scintillation dosimeter, the issues of cost, sterilization, and calibration need to be considered. The physical components of a single-point scintillation dosimeter are inexpensive, but there is time required for the construction of the dosimeter and for its calibration. The sterilization issues become more important for a reusable dosimeter and the level of sterilization may need to be increased. Due to the issues surrounding sterilization, it may not be advisable to reuse a dosimeter on more than one patient, but for multiple measurements on the same patient, particularly if the dosimeter is used inside a catheter or other protective sleeve, reusing the dosimeter can save time required for calibration and may save some cost.

Array dosimeters multiply both the effort and cost associated with a disposable design, with each individual element requiring calibration. A reusable array will save time in the recalibration before each use, in particular if a Cerenkov removal technique is used that has a time-consuming calibration procedure.

9.5.6 CALIBRATION

Spectral filtering requires a longer and more complicated calibration procedure (refer to Chapter 5 for the calibration of Cerenkov removal techniques). When there is no Cerenkov correction, the calibration of the system can be done by exposing the dosimeter to a known dose or dose rate. The light output of the scintillator is proportional to the dose rate and a single calibration factor can be used to convert the reading in counts, nC or other unit into dose. The temperature of the scintillator should also be recorded to enable a correction for the temperature dependence. The temperature dependence is linear and should be corrected for in *in vivo* dosimetry measurements to improve the accuracy. Due to the variation of its magnitude in the literature, it is advisable to measure the temperature dependence of the system being used. Assuming the temperature dependence is the same for all identical dosimeters with the same scintillation material, this characterization of the temperature dependence will only need to be done once.

Calibrating the dosimeter with the same type of brachytherapy source that will be measured is recommended. This removes any error in correcting the dosimeter reading for different radiation qualities, which has not yet been systematically characterized for a scintillation dosimeter in brachytherapy. For an array dosimeter with Cerenkov removal, it may be possible to calibrate only a few individual elements with the brachytherapy source, and to perform a uniformity calibration in a uniform field from a linac or ^{60}Co unit. If no Cerenkov removal is used, this technique could cause errors due to different magnitudes of Cerenkov radiation produced in the different elements of the array.

Measurements in water or a water-equivalent phantom are also recommended to remove any dependence on variations in the thickness of the protective sheath around the scintillator and any errors in correcting the reading in air kerma to absorbed dose to water. Accurate measurement of the distance between the source and the scintillator is essential to obtain an accurate calibration factor. Because of the high dose gradient around a brachytherapy source, positioning uncertainties in the calibration of the dosimeter will be one of the largest contributors to the overall uncertainty in the calibration factor. Using a ridged holder in a water tank or a custom-designed solid phantom can produce a precise, reproducible placement of the source and scintillator. Care must be taken when using a solid phantom due to the changes in water equivalent thickness with photon energy. Once a solid phantom has been constructed, the water equivalent thickness between the source and scintillator should be checked by comparing the obtained calibration factor to that obtained in a water tank at multiple distances from the source.

9.6 SUMMARY

A scintillation dosimeter has many desirable characteristics for *in vivo* dosimetry in brachytherapy. The energy dependence is small enough so that there is no noticeable change in sensitivity with distance from the brachytherapy source, as there is with most other dosimeters. The dosimeter can be designed to have no angular dependence and the temperature dependence is small and can easily be corrected for. For LDR brachytherapy with ^{125}I or ^{103}Pd, the photon energies are too low to produce Cerenkov radiation and therefore no Cerenkov correction is required. For ^{192}Ir, there is a detectable amount of Cerenkov produced in the optical fiber, over the course of a fraction this will only contribute a small amount to the total signal and some groups have used no Cerenkov correction at all.

Plastic scintillation detectors can be made from thin flexible plastic optical fibers with no electrical components, which allow insertion into catheters or applicators for dosimetry measurements inside the target volume or organ at risk. The components are inexpensive and a disposable *in vivo* dosimeter is a possibility.

REFERENCES

1. Venselaar, J., A.S. Meigooni, D. Baltas, and P.J. Hoskin. *Comprehensive Brachytherapy: Physical and Clinical Aspects*. Boca Raton, FL: CRC Press (2012).
2. Joslin, F. and E.J. Hall. *Principles and Practice of Brachytherapy: Using Afterloading Systems*. Boca Raton, FL: CRC Press (2001).
3. Tanderup, K., S. Beddar, C.E. Andersen, G. Kertzscher, and J.E. Cygler. In vivo dosimetry in brachytherapy. *Med Phys.* **40** 070902 (2013).
4. Rustgi, S.N. Application of a diamond detector to brachytherapy dosimetry. *Phys Med Biol.* **43** 2085–2094 (1998).
5. Brezovich, I.A., J. Duan, P.N. Pareek, J. Fiveash, and M. Ezekiel. In vivo urethral dose measurements: A method to verify high dose rate prostate treatments. *Med Phys.* **27** 2297–2301 (2000).
6. Anagnostopoulos, G., D. Baltas, A. Geretschlaeger, T. Martin, P. Papagiannis, N. Tselis, and N. Zamboglou. In vivo thermoluminescence dosimetry dose verification of transperineal ^{192}Ir high-dose-rate brachytherapy using CT-based planning for the treatment of prostate cancer. *Int J Radiat Oncol Biol Phys.* **57** 1183–1191 (2003).
7. Gambarini, G., M. Borroni, S. Grisotto, A. Maucione, A. Cerrotta, C. Fallai, and M. Carrara. Solid state TL detectors for in vivo dosimetry in brachytherapy. *Appl Radiat Isot.* **71**(Suppl.) 48–51 (2012).
8. Toye, W., R. Das, T. Kron, R. Franich, P. Johnston, and G. Duchesne. An in vivo investigative protocol for HDR prostate brachytherapy using urethral and rectal thermoluminescence dosimetry. *Radiother Oncol.* **91** 243–248 (2009).
9. Das, R., W. Toye, T. Kron, S. Williams, and G. Duchesne. Thermoluminescence dosimetry for in-vivo verification of high dose rate brachytherapy for prostate cancer. *Australas Phys Eng Sci Med.* **30** 178–184 (2007).
10. Meigooni, A.S., J.A. Meli, and R. Nath. Influence of the variation of energy spectra with depth in the dosimetry of ^{192}Ir using LiF TLD. *Phys Med Biol.* **33** 1159–1170 (1988).
11. Pradhan, A.S. and U. Quast. In-phantom response of LiF TLD-100 for dosimetry of ^{192}Ir HDR source. *Med Phys.* **27** 1025–1029 (2000).
12. Waldhausl, C., A. Wambersie, R. Potter, and D. Georg. In-vivo dosimetry for gynaecological brachytherapy: Physical and clinical considerations. *Radiother Oncol.* **77** 310–317 (2005).
13. Kirov, A., J.F. Williamson, A.S. Meigooni, and Y. Zhu. TLD, diode and Monte Carlo dosimetry of an ^{192}Ir source for high dose-rate brachytherapy. *Phys Med Biol.* **40** 2015–2036 (1995).
14. Gambarini, G. Carrara, M., Tenconi, C., Mantaut, N., Borroni, M., Cutajar, D., Petasecca, M. et al. Online in vivo dosimetry in high dose rate prostate brchytherapy with MOSkin detectors: In phantom feasibility study. *Appl Radiat Isot.* **83** (Pt. C) 222–226 (2014).
15. Qi, Z.Y., X.W. Deng, S.-M. Huang, J. Lu, M. Lerch, D. Cutajar, and A. Rosenfeld. Verification of the plan dosimetry for high dose rate brachytherapy using metal–oxide–semiconductor field effect transistor detectors. *Med Phys.* **34** 2007 (2007).
16. Qi, Z.Y., X.W. Deng, X.P. Cao, S.M. Huang, M. Lerch, and A. Rosenfeld. A real-time in vivo dosimetric verification method for high-dose rate intracavitary brachytherapy of nasopharyngeal carcinoma. *Med Phys.* **39** 6757–6763 (2012).

17. Zilio, V.O., O.P. Joneja, Y. Popowski, A. Rosenfeld, and R. Chawla. Absolute depth-dose-rate measurements for an ^{192}Ir HDR brachytherapy source in water using MOSFET detectors. *Med Phys.* **33** 1532 (2006).

18. Reniers, B., G. Landry, R. Eichner, A. Hallil, and F. Verhaegen. In vivo dosimetry for gynaecological brachytherapy using a novel position sensitive radiation detector: Feasibility study. *Med Phys.* **39** 1925–1935 (2012).

19. Bloemen-van Gurp, E.J., B.K. Haanstra, L.H. Murrer, F.C. van Gils, A.L. Dekker, B.J. Mijnheer, and P. Lambin. In vivo dosimetry with a linear MOSFET array to evaluate the urethra dose during permanent implant brachytherapy using iodine-125. *Int J Radiat Oncol Biol Phys.* **75** 1266–1272 (2009).

20. Bloemen-van Gurp, E.J., L.H. Murrer, B.K. Haanstra, F.C. van Gils, A.L. Dekker, B.J. Mijnheer, and P. Lambin. In vivo dosimetry using a linear Mosfet-array dosimeter to determine the urethra dose in ^{125}I permanent prostate implants. *Int J Radiat Oncol Biol Phys.* **73** 314–321 (2009).

21. Cygler, J.E., A. Saoudi, G. Perry, C. Morash, and E. Choan. Feasibility study of using MOSFET detectors for in vivo dosimetry during permanent low-dose-rate prostate implants. *Radiother Oncol.* **80** 296–301 (2006).

22. Lambert, J., T. Nakano, S. Law, J. Elsey, D.R. McKenzie, and N. Suchowerska. In vivo dosimeters for HDR brachytherapy: A comparison of a diamond detector, MOSFET, TLD, and scintillation detector. *Med Phys.* **34** 1759 (2007).

23. Anton, M., D. Wagner, H.J. Selbach, T. Hackel, R.M. Hermann, C.F. Hess, and H. Vorwerk. In vivo dosimetry in the urethra using alanine/ESR during (192)Ir HDR brachytherapy of prostate cancer—A phantom study. *Phys Med Biol.* **54** 2915–2931 (2009).

24. Carrara, M., G. Gambarini, M. Borroni, A. Cerrotta, C. Fallai, M. Invernizzi, C. Cavatorta, and G. Zonca. Fricke gel dosimetric catheters in high dose rate brachytherapy. In phantom dose distribution measurements of a 5 catheter implant. *Radiat Meas.* **46** 1924–1927 (2011).

25. Carrara, M., G. Gambarini, M. Borroni, S. Tomatis, A. Negri, L. Pirola, A. Cerrotta, C. Fallai, and G. Zonca. Characterisation of a Fricke gel compound adopted to produce dosimetric catheters for in vivo dose measurements in HDR brachytherapy. *Nucl Instrum Methods A.* **652** 888–890 (2011).

26. Kertzscher, G., C.E. Andersen, F.A. Siebert, S.K. Nielsen, J.C. Lindegaard, and K. Tanderup. Identifying afterloading PDR and HDR brachytherapy errors using real-time fiber-coupled Al(2)O(3):C dosimetry and a novel statistical error decision criterion. *Radiother Oncol.* **100** 456–462 (2011).

27. Tien, C.J., R. Ebeling, 3rd, J.R. Hiatt, B. Curran, and E. Sternick. Optically stimulated luminescent dosimetry for high dose rate brachytherapy. *Front Oncol.* **2** 91 (2012).

28. Sharma, R. and P.A. Jursinic. In vivo measurements for high dose rate brachytherapy with optically stimulated luminescent dosimeters. *Med Phys.* **40** 071730 (2013).

29. Andersen, C.E., S.K. Nielsen, J.C. Lindegaard, and K. Tanderup. Time-resolved in vivo luminescence dosimetry for online error detection in pulsed dose-rate brachytherapy. *Med Phys.* **36** 5033 (2009).

30. Andersen, C.E., S.K. Nielsen, S. Greilich, J. Helt-Hansen, J.C. Lindegaard, and K. Tanderup. Characterization of a fiber-coupled Al$_2$O$_3$:C luminescence dosimetry system for online in vivo dose verification during ^{192}Ir brachytherapy. *Med Phys.* **36** 708 (2009).

31. Beddar, A.S., S. Law, N. Suchowerska, and T.R. Mackie. Plastic scintillation dosimetry: Optimization of light collection efficiency. *Phys Med Biol.* **48** 1141–1152 (2003).

32. Archambault, L., J. Arsenault, L. Gingras, A. Sam Beddar, R. Roy, and L. Beaulieu. Plastic scintillation dosimetry: Optimal selection of scintillating fibers and scintillators. *Med Phys.* **32** 2271 (2005).

33. Elsey, J., D.R. McKenzie, J. Lambert, N. Suchowerska, S.L. Law, and S.C. Fleming. Optimal coupling of light from a cylindrical scintillator into an optical fiber. *Appl Optics.* **46** 397–404 (2007).

34. Ayotte, G., L. Archambault, L. Gingras, F. Lacroix, A.S. Beddar, and L. Beaulieu. Surface preparation and coupling in plastic scintillator dosimetry. *Med Phys.* **33** 3519 (2006).

35. Liu, P.Z., N. Suchowerska, P. Abolfathi, and D.R. McKenzie. Real-time scintillation array dosimetry for radiotherapy: The advantages of photomultiplier detectors. *Med Phys.* **39** 1688–1695 (2012).

36. Cartwright, L.E., J. Lambert, D.R. McKenzie, and N. Suchowerska. The angular dependence and effective point of measurement of a cylindrical scintillation dosimeter with and without a radio-opaque marker for brachytherapy. *Phys Med Biol.* **54** 2217–2227 (2009).

37. Lambert, J., D.R. McKenzie, S. Law, J. Elsey, and N. Suchowerska. A plastic scintillation dosimeter for high dose rate brachytherapy. *Phys Med Biol.* **51** 5505–5516 (2006).

38. Rivard, M.J., B.M. Coursey, L.A. DeWerd, W.F. Hanson, M.S. Huq, G.S. Ibbott, M.G. Mitch, R. Nath, and J.F. Williamson. Update of AAPM Task Group No. 43 Report: A revised AAPM protocol for brachytherapy dose calculations. *Med Phys.* **31** 633–674 (2004).

39. Williamson, J.F., J.F. Dempsey, A.S. Kirov, J.I. Monroe, W.R. Binns, and H. Hedtjarn. Plastic scintillator response to low-energy photons. *Phys Med Biol.* **44** 857–871 (1999).

40. Frelin, A.M., J.M. Fontbonne, G. Ban, J. Colin, and M. Labalme. Comparative study of plastic scintillators for dosimetric applications. *IEEE Trans Nucl Sci.* **55** 2749–2756 (2008).

41. Buranurak, S., C.E. Andersen, A.R. Beierholm, and L.R. Lindvold. Temperature variations as a source of uncertainty in medical fiber-coupled organic plastic scintillator dosimetry. *Radiat Meas.* **56** 307–311 (2013).

42. Wootton, L. and S. Beddar. Temperature dependence of BCF plastic scintillation detectors. *Phys Med Biol.* **58** 2955–2967 (2013).

43. Beddar, S. On possible temperature dependence of plastic scintillator response. *Med Phys.* **39** 6522 (2012).

44. Therriault-Proulx, F., S. Beddar, T.M. Briere, L. Archambault, and L. Beaulieu. Technical note: Removing the stem effect when performing Ir-192 HDR brachytherapy in vivo dosimetry using plastic scintillation detectors: A relevant and necessary step. *Med Phys.* **38** 2176 (2011).

45. Therriault-Proulx, F., T.M. Briere, F. Mourtada, S. Aubin, S. Beddar, and L. Beaulieu. A phantom study of an in vivo dosimetry system using plastic scintillation detectors for real-time verification of ^{192}Ir HDR brachytherapy. *Med Phys.* **38** 2542 (2011).

46. Suchowerska, N., M. Jackson, J. Lambert, Y.B. Yin, G. Hruby, and D.R. McKenzie. Clinical trials of a urethral dose measurement system in brachytherapy using scintillation detectors. *Int J Radiat Oncol Biol Phys.* **79** 609–615 (2011).

47. Therriault-Proulx, F., L. Archambault, L. Beaulieu, and S. Beddar. Development of a novel multi-point plastic scintillation detector with a single optical transmission line for radiation dose measurement. *Phys Med Biol.* **57** 7147–7159 (2012).

48. Cartwright, L.E., N. Suchowerska, Y. Yin, J. Lambert, M. Haque, and D.R. McKenzie. Dose mapping of the rectal wall during brachytherapy with an array of scintillation dosimeters. *Med Phys.* **37** 2247 (2010).

49. Johnson, M.T., T.A. Khemees, and B.E. Knudsen. Resilience of disposable endoscope optical fiber properties after repeat sterilization. *J Endourol.* **27** 71–74 (2013).

50. Aissou, M., M. Coroir, C. Debes, T. Camus, N. Hadri, C. Gutton, and M. Beaussier. Cost analysis comparing single-use (Ambu(R) aScope) and conventional reusable fiberoptic flexible scopes for difficult tracheal intubation. *Ann Fr Anesth Reanim.* **32** 291–295 (2013).

51. Lee, R.M., Vida, F., Kozarek, R.A., Raltz, S.L., Ball, T.J., Patterson, D.J., Brandabur, J.J. et al. In vitro and in vivo evaluation of a reusable double-channel sphincterotome. *Gastrointest Endosc.* **49** 477–482 (1999).

Multipoint plastic scintillation detectors

François Therriault-Proulx and Louis Archambault

Contents

10.1 INTRODUCTION

Since their first application to radiation therapy dose measurements, plastic scintillation detectors (PSDs) have been mainly used as single-point detectors. They are well suited for dosimetry in radiation therapy because of their typically high spatial resolution. Applications to radiosurgery and intensity-modulated radiation therapy are good examples (Klein et al. 2010, 2012; Morin et al. 2013). Nevertheless, physicists must often measure complex spatial distributions. As a result, other detectors such as ion chambers and diodes are available in arrays. PSDs also exist in arrays (Guillot et al. 2011a, 2013; Gagnon et al. 2012) but can be cumbersome and delicate to use.

It is well known that the raw signal produced within a PSD has two main sources: the scintillation signal, which is assumed to be proportional to the dose, and the Cerenkov signal, which is produced by charged particles traveling at velocity greater than the speed of light in a given transparent dielectric medium. It was understood quite early in the history of PSDs in medical physics that the Cerenkov signal could represent a significant source of noise (Beddar et al. 1992a), as it is produced both within the scintillator itself and within the clear optical light guide used to transport the scintillation light to a photodetector. For this reason, the first PSD prototypes included a *background* fiber used to subtract the Cerenkov signal from the scintillation signal (Beddar et al. 1992b,c). While this technique works well in most cases, the necessity to have two sensors (i.e., the *real* PSD and the background sensor) side by side makes it difficult to accurately measure high dose gradients (Archambault et al. 2006).

To alleviate the problem of the background fiber, several solutions were proposed from temporal filtration (Clift et al. 2002) to a hollowed light guide (Naseri et al. 2010). Nevertheless, most of these solutions proved to be difficult, costly, or even impossible to implement in a clinical environment. One set of solutions aimed to take advantage of the difference between the emission spectra of plastic scintillators and Cerenkov light by using optical filters to block the Cerenkov signal (De Boer et al. 1993). Unfortunately, because Cerenkov light is emitted over the entire visible spectrum, this filtration technique could not completely isolate the scintillation signal. Nevertheless, the idea of exploiting the spectral differences between both sources of light was further investigated, and it was shown that the scintillation signal can be accurately decoupled from the Cerenkov signal by using two different wavelength filters (Fontbonne et al. 2002; Frelin et al. 2005; Archambault et al. 2006). The extension to a larger number of filters was later proposed to account more accurately for the fact that the Cerenkov effect is not the sole contributor to the stem effect (Therriault-Proulx et al. 2013a).

Going toward spectral-based approaches not only allowed to correct for the presence of Cerenkov in the optical signal, but it also allowed for the development of multipoint plastic scintillation detectors (mPSDs) composed of only one optical guide (Archambault et al. 2012; Therriault-Proulx et al. 2012). The goal of this chapter is thus to present the theoretical development of these spectral-based approaches as well as an experimental proof of feasibility for mPSDs, and discuss the new applications made possible by these techniques.

10.2 MATHEMATICAL FORMALISM

Before going further, it is useful to define a model describing the light produced within a PSD, its propagation within the components of the PSD, and its quantification by a photodetector. Let us assume that irradiating a PSD will produce light coming from a given source n (e.g., scintillator and Cerenkov effect), according to a normalized spectrum S_n, which is a function of wavelength λ. This emitted light will pass through the various components of the detector (e.g., scintillator-light guide interface, light-guide, optical filters, and photodetector). During this propagation, the light will suffer some attenuation. This attenuation is a function of the wavelength and depends on the path l taken by the light. Thus, the attenuated spectrum along l can be described as $S_n W_l$, where W_l is the attenuation function. Finally, after the signal is read by a photodetector, the outcome will be a single value proportional to the total number of photons that reached the photodetector. Therefore, the measurement, q, done by the system for a given light source n that emitted a total of A photons that propagated through a given path l can be described as

$$q_{l,n} = \int_{-\infty}^{\infty} AS_n(\lambda)W_l(\lambda)d\lambda \tag{10.1}$$

The response r can simply be defined as the fraction of initial photons that contributes to the measurement (i.e., $r = q/A$)

$$r_{l,n} = \int_{-\infty}^{\infty} S_n(\lambda)W_l(\lambda)d\lambda \tag{10.2}$$

10.2.1 THEORY OF SINGLE-POINT PSD CALIBRATION

It is now known that sources of spurious signal other than the Cerenkov effect exists in any PSD (Therriault-Proulx et al. 2013a). However, in most cases, it is possible to approximate the spurious signal as originating from a single source. With megavoltage photon beams, the Cerenkov signal dominates the other sources, while in low dose rate brachytherapy or proton therapy, it is usually the opposite. It is often reasonable to approximate the signal produced within a PSD as the superposition of only two sources: the scintillation signal and the spurious signal (S_1, S_2).

The spectral-based calibration of single-point PSD requires dividing the total light produced within the PSD into two distinct paths (W_1, W_2). Each path leads to a different photodetector and must attenuate the signal differently so that $r_{1,n}$ differs significantly from $r_{2,n}$. In practice, a dichroic filter is commonly used to create these two paths. When a measurement is performed, the PSD is irradiated and light is produced by both scintillation and the Cerenkov effect. The light then travels toward the photodetectors, which then digitize the fraction of signal each received. The outcome of this measurement is two values, M_1 and M_2, one for each photodetector. In terms of Equation 10.1, one can write

$$M_1 = q_{1,1} + q_{1,2}$$
$$M_2 = q_{2,1} + q_{2,2}$$

(10.3)

The objective of the calibration process is to find a way to combine M_1 and M_2, so that we can get a value that is solely proportional to the scintillation signal, which is also proportional to the dose, D. In matrix form, we can write

$$\begin{bmatrix} M_1 \\ M_2 \end{bmatrix} = \begin{bmatrix} r_{11} & r_{12} \\ r_{21} & r_{22} \end{bmatrix} \begin{bmatrix} D \\ C \end{bmatrix}$$

(10.4)

where C represents the stimulation of the Cerenkov effect. This is a linear system, which can easily be inverted to obtain D. In most cases, the r_{ij} will be unknown. Nevertheless, inversion of Equation 10.4 states that D must be of the form

$$D = aM_1 + bM_2$$

(10.5)

where a and b are the calibration factors. To determine these calibration factors, two measurements must be made where the dose of each, D_1, D_2 is known. This way, a second set of equations can be built

$$\begin{bmatrix} D_1 \\ D_2 \end{bmatrix} = \begin{bmatrix} M_{D_1,1} & M_{D_1,2} \\ M_{D_2,1} & M_{D_2,2} \end{bmatrix} \begin{bmatrix} a \\ b \end{bmatrix}$$

(10.6)

All parameters in this equation are known except the calibration factors. It is therefore possible to invert the system to obtain the values of a and b assuming that the measurement matrix is not singular. This is the essence of the spectral calibration approach of a single-point PSD. Variants of this method can be found in the literature (Frelin et al. 2005; Guillot et al. 2011b), but the principle is the same as described here.

10.2.2 OVERVIEW OF THE MULTISPECTRAL AND HYPERSPECTRAL FORMALISM

In the previous section, we described a system where two spectra are linearly superposed and we gave a method to decouple one from the other. Problems involving spectral superposition are not common in medical physics, but are frequently encountered in other area of physics. In astrophysics, geoscience and other fields, it is common to observe data points comprising tens of spectra superposed (Keshava 2003). The acquisition of such data and its interpretation is called *multispectral, hyperspectral,* or simply *spectral analysis.* In spectral analysis, a data point (e.g., the pixel of an image) is assumed to be a linear superposition of a limited number of *end members.* The goal of the analysis is thus to decouple these *end members* and determine the contributing fraction of each to a given data point. To achieve this goal, it is necessary to measure the light emission of all data points in several wavebands. In other words, the emission spectrum of the initial data must be known. Once this information is known, decoupling of the *end members* can be performed with a technique called *spectral unmixing* (Keshava 2003) (Figure 10.1).

There is a clear parallel between the task involved in spectral analysis and the calibration of a PSD. The PSD calibration problem is a case of spectral analysis where only two *end members* are present (scintillation and Cerenkov light). It was shown that the formalism described in Section 10.2.1 can be generalized to a N superposed spectra measured over L wavebands, or channel (Archambault et al. 2012). This generalization offers two direct advantages. First, it can be used to improve the noise removal in situations where Cerenkov light is not the only dominant source of noise (Therriault-Proulx et al. 2013a). Second, it paves

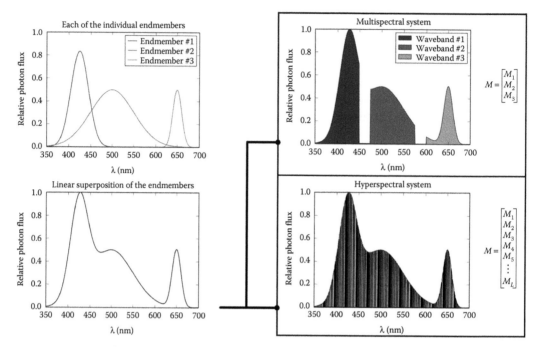

Figure 10.1 Overview of a linear superposition of spectra and illustration of how a measurement is performed in terms of wavelength channels or wavebands for both a multispectral and an hyperspectral system.

the way for multipoint detectors (Archambault et al. 2012; Therriault-Proulx et al. 2012). By coupling several scintillators with different emission spectrum to a single light guide, it is possible to build a detector that can simultaneously measure dose at multiple locations.

In the general PSD signal formalism, performing a measurement means sampling the signal over all of the L wavebands. Thus, Equation 10.4 can be changed to

$$\mathbf{m} = \mathbf{R}\mathbf{x} \tag{10.7}$$

where:

$\mathbf{m}$ represents a vector of L elements replacing $[M_1, M_2]$ in Equation 10.4
$\mathbf{R}$ is a $L \times N$ matrix where the elements represent the response of the system as defined by Equation 10.1

Finally, $\mathbf{x}$ is a vector of N elements representing the photon flux (i.e., the number of photons emitted for a given emission source). Here and throughout the chapter, bold characters in small letters will represent vectors and bold capitalized characters will represent matrices. It is worth pointing out that for $L = 2$ and $N = 2$, Equation 10.7 is identical to Equation 10.4. The goal is thus to determine the photon flux x_i for each of the scintillator because the number of scintillation photons is proportional to the dose received by the scintillator in the absence of quenching. Therefore, $\mathbf{x}$ can be replaced with $\mathbf{d}$ by assuming a linear relationship ($d_i = a_i x_i$). Equation 10.7 represents a single measurement. If a set of K consecutive measurements is performed, it is possible to write

$$\mathbf{M} = \mathbf{R}\mathbf{D} \tag{10.8}$$

where $\mathbf{M}$ and $\mathbf{D}$ are now matrices with dimensions of $L \times K$ and $N \times K$, respectively. The goal of spectral analysis is to get $\mathbf{D}$ from measuring $\mathbf{M}$. If $\mathbf{R}$ is full rank and $L > N$, then the left inverse of $\mathbf{R}$ exists and we can write

$$\mathbf{D} = \left(\mathbf{R}^{\mathsf{T}}\mathbf{R}\right)^{-1}\mathbf{R}^{\mathsf{T}}\mathbf{M} \tag{10.9}$$

Of course, to solve Equation 10.9 directly, **R** must be known, which is not always the case. So far, two approaches were proposed to solve Equation 10.9. The first approach is a multispectral technique that is a direct extension of the process described in Section 10.2.1. The second approach is based on hyperspectral analysis, and it has the potential to be more accurate and more robust than the first, but requires measuring the signal in a large number of wavebands (L typically larger than 50) to be efficient. The distinction between multispectral and hyperspectral techniques essentially lies in the number of wavebands acquired. Multispectral analysis typically uses a limited number of waveband (e.g., less than 30), while hyperspectral analysis uses a large number of narrow wavebands over a continuous spectral range.

10.2.3 APPLICATION OF THE MULTISPECTRAL APPROACH TO mPSD

The minimum requirement to convert a measured signal into dose with Equation 10.9 is that the number of wavebands, L, must be equal or larger than the number of light emitter, N. Thus, for mPSD comprising a limited number of scintillators, a multispectral approach can be used to determine the dose. Experimentally, this can be achieved by using a combination of beam splitter, optical filters, and/or dichroic mirrors (see Section 10.3.2). Throughout this chapter, this kind of system will be referred to as a *multispectral* setup.

In order to directly use Equation 10.9 to determine the dose from a given measurement, it would be necessary to know **R**, which means knowing the response of every light emitter along each of the L possible paths. In a multispectral setup, it might not be practical or even possible to determine **R**. These difficulties arise from two factors:

1. Because of the beam penumbra and scattered dose involved in any irradiation, it can be difficult to stimulate only one light emitter at a given time.
2. Any change in the light collection apparatus will affect **R**, therefore determining **R** prior to assembling the system is not an option.

Nevertheless, it is possible to estimate **R** to an arbitrary degree of precisions by performing a series of measurements of known doses. Equation 10.9 can be rewritten as

$$\mathbf{D}^{\mathrm{T}} = \mathbf{M}^{\mathrm{T}}\mathbf{F} \tag{10.10}$$

where **F** is a $L \times N$ matrix whose elements are functions of the $r_{i,j}$ coefficients. **F** represents the calibration factors of the system that are to be determined experimentally. To do so, it is necessary to perform a series of K_c irradiations where the dose delivered at the location of every scintillating elements of the mPSD is known. After these irradiations, the only unknown in Equation 10.10 is **F**, which can be found with the following equation:

$$\mathbf{F} = \left(\mathbf{MM}^{\mathrm{T}}\right)^{-1}\mathbf{MD}^{\mathrm{T}} \tag{10.11}$$

Once **F** is known, Equation 10.10 can be used to find the dose received by any element of the mPSD from a given measurement. Equation 10.11 can be used to determine **F** as long as the number of calibration measurements (K_c) is equal or larger than the number wavebands, L. In the case, K_c equal L, Equation 10.11 can be rewritten as Equation 10.12.

$$\mathbf{F} = \left(\mathbf{M}^{\mathrm{T}}\right)^{-1}\mathbf{D}^{\mathrm{T}} \tag{10.12}$$

Furthermore, each calibration measurement must use a unique irradiation pattern. If a measurement pattern is repeated (e.g., by irradiating an open 10×10 cm^2 field with 100 cGy then doing the same with 200 cGy in a second measurement), it will be impossible to find a unique solution for **F**, because **M** would not be of full column rank.

10.2.4 APPLICATION OF THE HYPERSPECTRAL APPROACH TO mPSD

In a multispectral setup, the light produced within a mPSD is divided into a limited number of relatively broad optical filters. In theory, Equation 10.10 can always be solved given the minimal requirement that K_c is greater or equal than L and that each calibration irradiation uses a unique pattern. However, in practice, the choice of optical filters will have an impact on the measurement quality. If the wavebands do not match the emission of the scintillating elements or if there are too much overlaps between them, **M** will be poorly conditioned and the calibration will be strongly affected by the input noise.

An alternative to the multispectral setup is the hyperspectral approach, where a large number of narrow wavebands are used. In this case, we have $L \gg N$. The choice of filters therefore ceases to be a potential source of error. A hyperspectral setup can be achieved by using a spectrometer as the photodetector. The main limitations of this setup are the cost of the device and the relatively low signal per waveband (i.e., because we are dividing the total signal in tens or hundreds of small wavebands). Nevertheless, the use of an off-the-shelves spectrometer simplifies the optical path of the system: the output of the mPSD is directly connected to the input of the spectrometer. Furthermore, the spectrometer itself can be calibrated on a regular basis both in intensity and in wavelength, which guarantee consistency between measurements done over a long period of time.

The formalism presented in Section 10.2.3 could theoretically be applied to a hyperspectral setup. However, the requirement that $K_c \geq L$ would make the calibration tedious because tens or hundreds of calibration measurements would be necessary. It is therefore preferable to calibrate a hyperspectral system by finding **R** directly instead of performing K_c calibration measurements of known dose. The robustness of the spectrometer and our ability to monitor its response independently from the mPSD, make the use of Equation 10.9 instead of Equation 10.11 feasible. Several techniques and algorithms have been developed to perform the spectral unmixing of a linear system such as the one described in Equation 10.9. A review of several algorithm can be found in the literature (Keshava 2003). Most algorithms assume that each *end member* contributes positively to the signal (i.e., no negative signal) and that the relative sum of all *end member* contribution to a given data point is 1. Both of these hypotheses can be directly applied to the case of a mPSD. Most unmixing algorithms require *pure* data points, which are points produced by a single light source. In the case of mPSD, pure data points can be obtained either by measuring the emission spectrum of each scintillator independently prior to the detector assembly or by using an extremely well-collimated radiation source to stimulate the luminescence. While these measurements are not necessarily practical, only N of them are required, which is more convenient than performing $K_c \times N$ calibration measurements. Furthermore, once measured, the pure data points do not need to be measured again unless the emission spectrum changes. The practical aspects of developing mPSDs based on the multispectral and hyperspectral approaches are discussed in Sections 10.3 and 10.4.

10.3 DESIGN AND DEVELOPMENT OF EXPERIMENTAL PROTOTYPES

Breaking the device down into its constituent parts is important to understand the development of an experimental mPSD prototype. The dose detector is composed of a multipoint radiosensitive scintillation device optically coupled to a single optical conduct transmitting the light up to a photodetection setup. The main experimental challenge lies in the evaluation of the contribution from each light-emitting component to the total signal while having access only to a single incoming light beam. The photodetection setup must therefore separate light in multiple components (i.e., along multiple wavelength channels) before converting the optical signal in an electrical one. In the following sections, different possibilities are presented for each component of a mPSD device. The best choice depends on the nature of the targeted application.

10.3.1 THE MULTIPOINT SINGLE FIBER DETECTOR

Multiple possibilities exist regarding the construction of the multipoint radiosensitive scintillation device. Figure 10.2 presents some of them. Abutting different scintillators together, with different light emission spectrum, is the first option. Another is to separate each scintillating element by a desired length of optical fiber or another optically conducting element. The same type of scintillator could also be used more than once in a given mPSD, but the use of optical filters then become necessary. For such an approach, the use

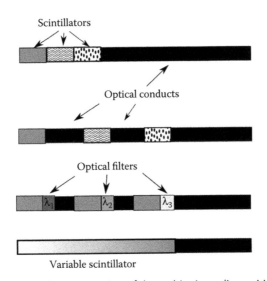

Figure 10.2 Different possibilities for the construction of the multipoint radiosensitive optical probe.

of bandpass filters with increasing bandwidth from the most distal to the most proximal is recommended. Finally, another option would be to use a long scintillating element specially engineered such that its emission spectrum varies depending on the position where the dose is deposited along its length.

The selection of the coupling material is particularly important when building an mPSD. The most distal scintillating component is subject to a loss that grows exponentially with the number of coupling interfaces. It is therefore important to select a coupling agent that is as optically transparent as possible to prevent unwanted filtering of the signal. Some coupling agents may appear transparent at first sight, but take on color when cured. The order in which the different scintillating elements are placed along the mPSD also affects the performance of the detector. It was shown that choosing the order of the different components based on the spectrum as well as on the intensity can lead to an improvement in the performance of the mPSD (Therriault-Proulx et al. 2012). A scintillator with a more intense and distinct (i.e., unique) light emission spectrum should be placed further from the photodetector. The choice of the transmission guide should also be part of the optimization process when selecting the components to include in the construction of the mPSD. Each transmission guide possesses its own attenuation spectrum and will therefore attenuate light of different wavelengths differently.

10.3.2 SEPARATING AND MEASURING THE LIGHT COMPONENTS FROM A SINGLE PROBE

10.3.2.1 The multispectral approach

One approach to obtaining the dose at the different scintillating elements consists of separating the signal into limited number of optical pathways with different spectral windows, which is defined here as the multispectral approach. A network of optical fiber splitters coupled to optical filters (Figure 10.3a) or beam splitters (Figure 10.3b) can be used. It is recommended that the selection of the transmission spectrum characteristics follows an optimization process discussed in Section 10.5. Light collection is performed using one photodetector per optical window or using an array of photodetectors (e.g., CCDs and CMOS cameras).

The collected light signal can be translated to dose using Equation 10.10, as detailed in Section 10.2.3. In order to successfully use Equation 10.10, it is necessary to have a minimum of one wavelength channel per light emitting element. The calibration factors must be obtained from at least as many different irradiation conditions as there are wavelength channels. As observed by Guillot et al. (2011b) for single scintillating element detectors, the choice of calibration conditions (i.e., the amount of dose received by each element) is very important. This process is further complicated by the necessity that the dose to each

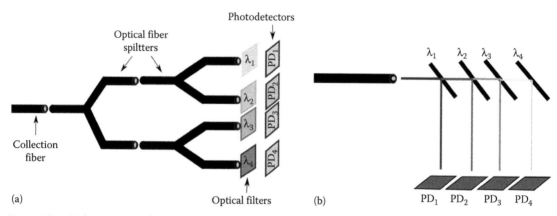

Figure 10.3 Light separation for the multispectral approach using (a) optical fiber splitters and (b) beam splitters.

scintillating element has to be known precisely. This approach was shown to be feasible for mPSDs with only a few scintillating elements and will be discussed in Section 10.4.1. However, for high numbers of scintillating elements, it is recommended to move toward the hyperspectral approach.

10.3.2.2 The hyperspectral approach

Instead of measuring the light output through transmission in some specific optical windows, the entire optical spectral information from the incoming light is acquired in the hyperspectral approach. Using a spectrograph, the incoming light is split or dispersed into its constituent wavelengths using either a prism or a diffraction grating, and measured using an array of photodetectors. In most common spectrographs, light is dispersed along a single dimension, and the number of photodetectors across this dimension determines the number of wavebands available for hyperspectral processing. CCD and CMOS cameras are typically used, as they are composed of hundreds to thousands of pixels along a single dimension and therefore allow for the implementation of the hyperspectral approach.

This approach is based on the assumption that any incoming light spectrum is a linear superposition of its different constituents, whether it is light originating in the scintillating elements or from the stem effect. The main part of the calibration process consists of acquiring the relative emission spectrum from each of the light emitting element composing the mPSD individually. In order to isolate the scintillating light from any Cerenkov effect contamination, an irradiator of sufficiently low energy to avoid producing Cerenkov light is recommended. In the case of a PMMA, one of the most common optical fibers in scintillation dosimetry, the energy threshold for Cerenkov light production is 178 keV. The other component of the stem effect (i.e., the fluorescence) can be limited by shielding the optical transmission conducts using highly attenuating material such as lead.

It was shown that the contribution from fluorescence and Cerenkov effect can vary from one irradiation condition to another (Therriault-Proulx et al. 2013a). Whether to account for the Cerenkov light effect and the fluorescence separately or as a single stem effect component remains the choice of the user and depends on the application. If the stem effect composition is expected to vary significantly with the different irradiation conditions, it would be prudent to account for fluorescence and Cerenkov separately. In this situation, the fluorescence spectrum would have to be acquired by irradiating the optical conduct with the irradiation sub-threshold for Cerenkov production. As it is impossible to completely avoid the fluorescence production, the acquisition of the Cerenkov spectrum should be done under a condition that will maximize the Cerenkov production as well as its collection in the optical fiber. However, if only small fluctuations in the stem effect spectrum are expected across all the irradiation conditions, it may be sufficient to consider the stem effect to be a single light source (i.e., one term in Equation 10.8). With the spectrum of all the light components known, the last remaining step for calibrating the mPSD is to perform at least one irradiation with a known nonzero dose to each of the scintillators. With the calibration process completed, the dose under any irradiation can be obtained from a single spectrum acquisition converted to dose using Equation 10.9.

10.4 EXPERIMENTAL PROOF OF FEASIBILITY

10.4.1 THE 2-POINT DETECTOR USING THE MULTISPECTRAL APPROACH

The feasibility of using the multispectral approach was proven by Therriault-Proulx et al. (2012) for a 2-point detector (Figure 10.4). A network of optical fiber splitters (DieMount GmbH, Wernigerode, Germany) were used to separate the incoming light in different pathways and transmit it through different optical bandpass wavelength filters (Roscolux; Rosco, Stanford, CT). Their output was imaged simultaneously using an EMCCD camera (Luca-R; Andor Technology, Belfast, Northern Ireland). Scintillating light from BCF-12, BCF-60 as well as the light coming from the stem effect were accounted for. Dose deposited by a 10 cm × 10 cm 6 MV photon beam (2100 Clinac; Varian, Palo Alto, CA) was measured in a water tank at different depths with the detector either perpendicular (horizontal) or along (vertical) the beam longitudinal axis. The calibration of the detector was performed by performing three different irradiations: with different dose to each detector, with same dose to both, and with a maximization of the stem effect production by putting more fiber in the radiation field. Figure 10.5 shows

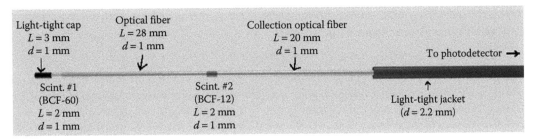

Figure 10.4 The 2-point plastic scintillation detector. (From Therriault-Proulx, F. et al., *Phys. Med. Biol.*, 57, 7147, 2012, doi: 10.1088/0031-9155/57/21/7147.)

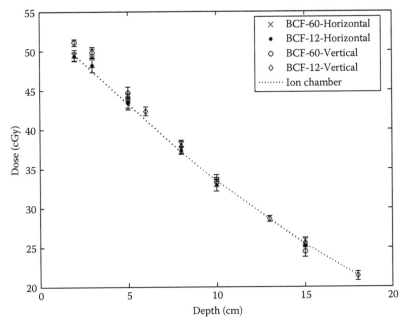

Figure 10.5 Measurement of dose from a 6 MV photon beam at different depths using a 2-point multipoint plastic scintillation detector. (From Therriault-Proulx, F. et al., *Phys. Med. Biol.*, 57, 7147, 2012, doi: 10.1088/0031-9155/57/21/7147.)

the measured doses with the mPSD together with the same measurements using an ion chamber (CC-04). The average difference between the mPSD and ion chamber was 2.4% for BCF-60 and 1.3% for BCF-12.

10.4.2 THE 3-POINT DETECTOR USING THE HYPERSPECTRAL APPROACH

10.4.2.1 Validation for external beam radiation therapy

The feasibility of implementing the hyperspectral approach experimentally was first shown for external beam irradiations. This was demonstrated for a 3-point detector and the details of its construction are shown in Figure 10.6. As discussed in Section 10.3.2, the hyperspectral approach was implemented by connecting the collection fiber to a spectrograph (Shamrock, Andor Technology, Belfast, Northern Ireland) that dispersed the light over the detection chip of a CCD camera (iDus, Andor Technology, Belfast, Northern Ireland). The relative spectrum of each individual scintillating element was obtained using the 125 kVp photon beam from a superficial therapy unit (Philips RT-250, Philips Corp., Eindhoven, Holland). The stem effect spectrum acquisition, combining both fluorescence and Cerenkov effect, and the determination of the calibration factors were performed using a linac (2100 Clinac, Varian).

The mPSD was used to measure a 6 MV photon beam depth-dose curve as well as profiles of a 10 cm × 10 cm field and of a 45° wedge. The maximum and average relative differences to measurements performed with an ionization chamber (CC-04) are shown in Table 10.1. Average relative differences of 2.3% ± 1.1%, 1.6% ± 0.4%, and 0.3% ± 0.2% were obtained over the many measurements for the BCF-60, BCF-12, and BCF-10 scintillating elements, respectively.

10.4.2.2 Validation for high dose rate brachytherapy

With the feasibility of using the hyperspectral approach proven for external beam dosimetry, Therriault-Proulx et al. (2013b) then used a similar 3-point detector for dosimetry of an Ir-192 source used in high dose rate brachytherapy. The order of the scintillating elements was slightly changed to optimize the overall performance of the detector. The BCF-10 and BCF-12 elements were swapped for optimization of the signal decoupling. The length of the BCF-60 scintillating element was chosen to be 1 mm longer to compensate for the losses due to the multiple coupling interfaces. An index-matching epoxy (Epo-Tek 305, AngstromBond; Fiber Optics Center, New Bedford, MA) was also used instead of direct coupling in order to increase the signal transmission efficiency at each interface. The calibration process was similar, with the exception that the Ir-192 source was used instead of a linac to determine the calibration factors and acquire the stem effect spectrum. This was motivated by the fact that the composition of the stem effect depends on the irradiation conditions (Therriault-Proulx et al. 2013a). The measurements using the mPSD were compared to the expected doses around the Ir-192 radioactive source (Daskalov et al. 1998; Rivard et al. 2004).

The 3-point detector was shown to accurately measure the dose at each of his scintillating elements with average relative differences to the expected values of 3.4% ± 2.1%, 3.0% ± 0.7%, and 4.5% ± 1.0% for BCF-60, BCF-10, and BCF-12, respectively. Additional approaches were proposed to improve the overall precision and accuracy of the detector. Another interesting finding is related to the use of the detector to determine source positioning. As the position from one scintillating element to the

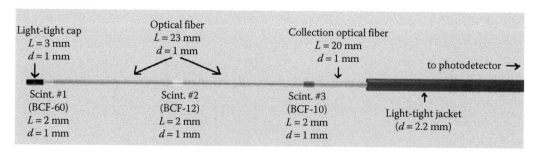

Figure 10.6 A 3-point plastic scintillation detector. (From Therriault-Proulx, F. et al., *Phys. Med. Biol.*, 57, 7147, 2012, doi: 10.1088/0031-9155/57/21/7147.)

Table 10.1 **Average ± standard deviation and maximum relative differences between measurements obtained using the 3-point plastic scintillation detector and those obtained using the ion chamber for 100 MUs irradiations**

SCINTILLATING ELEMENT	RELATIVE DIFFERENCE (%)	
	AVERAGE ± SD	MAX.
BCF-60		
Profile (depth = 10 cm)	2.1 ± 1.4	4.2 ± 0.7
Profile (depth = 5 cm)	2.0 ± 1.5	3.8 ± 0.5
Depth–dose	1.3 ± 1.4	4.6 ± 2.2
45° wedge	3.9 ± 2.4	7.9 ± 3.7
BCF-12		
Profile (depth = 10 cm)	1.6 ± 1.1	3.9 ± 0.4
Profile (depth = 5 cm)	1.7 ± 1.3	2.6 ± 2.8
Depth–dose	1.1 ± 0.5	2.0 ± 0.5
45° wedge	2.1 ± 1.1	4.1 ± 2.1
BCF-10		
Profile (depth = 10 cm)	0.3 ± 0.2	0.9 ± 0.4
Profile (depth = 5 cm)	0.2 ± 0.3	0.6 ± 0.2
Depth–dose	0.2 ± 0.1	0.5 ± 0.2
45° wedge	0.6 ± 0.3	1.3 ± 0.6

Source: Therriault-Proulx, F. et al., *Phys. Med. Biol.*, 57, 7147, 2012, doi: 10.1088/0031-9155/57/21/7147.

other is known from the construction, the dosimetry data can be used to infer the position of the radioactive source. A proposed approach weighs the average based on the dose and standard deviation of the measurement. This approach led to determination of the position within 0.5 mm of the expected position for most of the clinically relevant positions tested.

10.5 THEORETICAL BENEFITS FROM SPECTRAL APPROACHES

Based on a simple linear superposition model, a general expression for the relationship between the dose delivered to scintillating elements of a PSD and the signal measured along a given number of wavelength channels was found (i.e., Equation 10.10). From there two methods to calibrate a mPSD were proposed and tested. The choice of the method depends mainly on the number of wavelength channels available and on the ability to measure precisely the individual signal from each scintillating elements. The first and most obvious advantage of this spectral approach was, of course, the possibility to build a multipoint PSD. Nevertheless, the spectral-based mathematical formalism yields other advantages.

10.5.1 CALIBRATION AS A LINEAR REGRESSION

To determine the dose to the nth scintillating element (d_n) of the mPSD, the following can be written (from Equation 10.10):

$$d_n = \mathbf{M}^\mathrm{T}\mathbf{f}_n \tag{10.13}$$

where:

d_n is a vector representing the dose for each of the K_c calibration measurements
$\mathbf{f}_n$ represents the calibration vector of the same scintillating element

Calibrating the mPSD by solving Equation 10.13 corresponds to performing a linear regression over $\mathbf{M}$ to estimate $\mathbf{f}_n$. This can be achieved using an ordinary least-square (OLS) optimization. Figure 10.7 illustrated this concept for the simple case where $L = N = 2$. In Figure 10.7, the calibration is visualized as a plane defined by $\mathbf{f}_n$.

Understanding the calibration process as a multidimensional linear regression offers new insights especially when dealing with noise. There are two important aspects to consider while selecting a set of calibration measurements. First, the calibration measurement and setup should be somewhat representative of the planned use of the device. If this is not the case, a small error in the calibration measurements can result in large errors when using the mPSD. In other words, if the calibration points (circle in Figure 10.7) are far from the measurements (squares in Figure 10.7), the hyperplane defined by the calibration (the plane in Figure 10.7) might not represent a good fit at the location of the measurements. In practice, this can be achieved by choosing calibration doses that are roughly in the same range as the ones to be measured. Furthermore, a similar setup should be used for both calibration and future measurements. This is

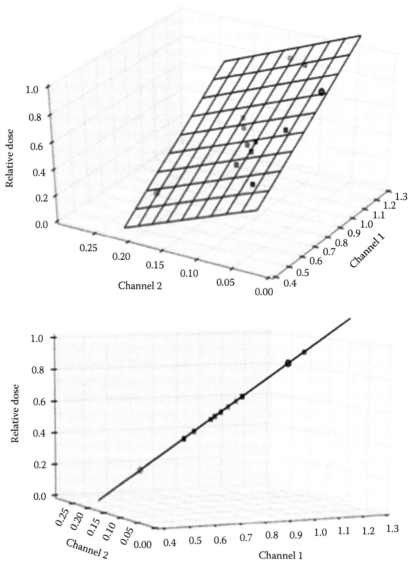

Figure 10.7 Two views of the same data illustrating the calibration of a plastic scintillation detector as a linear regression. The calibration points are the circles, the calibration itself is shown as the plane and some measurements are shown as squares. All units are arbitrary.

important because Cerenkov light can vary strongly depending on the setup. For example, measurements in air with the detector perpendicular to the beam's longitudinal axis will exhibit minimal Cerenkov component, while measurements with the optical fiber immersed in water will exhibit a strong one. Therefore, it would not be ideal to perform a calibration in water and then use the calibrated detector in air.

The second important factor to consider while selecting a set of calibration measurements is to have a well-conditioned system. Calibrating a mPSD involve solving a linear system of equations that can be characterized by a condition number. The condition number is a value commonly used in linear algebra to evaluate how much impact the input noise will have on the solution. The smaller the condition number, the less the input noise will have an effect on the measured dose. In practice, good conditioning will come from a choice of calibration measurements that are significantly different from each other.

In theory, if a calibration measurement can be done with absolutely no noise or error, any choice of calibration setups can be used. In practice, there will always be noise. To minimize its impact, two aspects should be considered in light of the previous paragraphs:

1. Calibration measurements should be performed in a setup similar to the one used for the actual measurements
2. Calibration measurements should be chosen in a way that minimizes the condition number of the linear system of equation

To demonstrate the importance of these two considerations, three different sets of calibrations are applied to the PSD system originally shown in Figure 10.7, but this time a systematic error of 5% is introduced on one of the two calibration points. This unrealistically large error is used only for illustration purpose. The first set of calibration measurements is the one shown in Figure 10.7 (circles); these calibration measurements are considered ideal because they encompass the scopes of the measurement performed with the system (squares) and the condition number is relatively low (7.6). The second calibration setup consists of measurements that possess the same condition number as the ideal calibration (7.6) but does not cover the range of the measurements performed with the system. Finally, the third calibration setup consists of measurements with poor conditioning (condition number of 296), but that are situated roughly in the range of measurements performed with the system. The impact of the 5% error on these three sets of calibration setup is shown in Figure 10.8. In this figure, the plane represents the calibration, the calibration measurements themselves are circles and the measurements are squares. In the case, where no calibration error is made (Figure 10.7) the measurements lie directly on the calibration plane. However, when errors are present, most of the measurements do not lie on the plane. The distance between the plane and the measured data points in Figure 10.8 represents the measurement error caused by the initial calibration error. The average error over 20 measurements for the ideal calibration setup (Figure 10.8a) is 5.7% for the initial 5% error on one calibration point. For the case with the good conditioning, but where the calibration setup did not cover adequately the range of measurement (Figure 10.8b), the average error over 20 measurements is 11.5%. Finally, with the poorly conditioned system (Figure 10.8c), the average error is 90% for the same initial calibration error of 5%.

10.5.2 ACCOUNTING FOR THE STOCHASTIC NOISE ON EACH CALIBRATION MEASUREMENT

One advantage of the *calibration-as-a-regression* paradigm is to better identify the optimal calibration points to use for a given application. Once these points are identified, the calibration measurements are done and the linear system of equation is solved using OLS optimization. However, one hypothesis behind OLS is that all points have the same variance (i.e., the data must be homoscedastic). With PSD in general and mPSD in particular, the random noise for a given measurement is linked to the total signal collected (Lacroix et al. 2010). The lower the signal is, the higher the relative noise will be. For mPSD, the division of the signal over multiple wavebands and the increased optical attenuation that can come from the multiple coupling of the scintillating elements (see Section 10.3) can result in relatively low signals and noise needs to be considered carefully. Thus, it is likely that each calibration measurement will have a unique variance (i.e., the set of calibration measurement is likely heteroscedastic).

These differences in the variance of each calibration measurement can be mitigated by using techniques other than OLS optimization that can account for the variance specific to each data point. The most

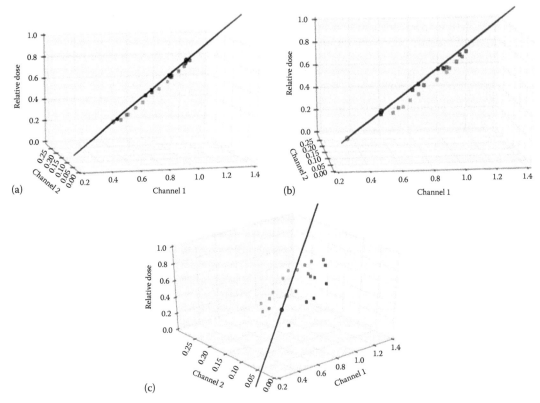

(a)

(b)

(c)

Figure 10.8 Calibration in the presence of a systematic error of 5% on one calibration point. The plane represents the calibration, the calibration measurements themselves are in circles and the measurements are in squares. The distance between the plane and the squares illustrate the errors that would be made by using the erroneous calibration. (a) The choice of calibration point is ideal and the condition number is low. (b) The condition number is the same as for (a), but the calibration points are not representative of the measurements. (c) The calibration setup is poorly conditioned.

obvious technique to achieve this is probably the weighted least square (WLS) optimization. The idea of the WLS approach is to weight each term in a manner inversely proportional to its variance thus making the noisier terms less important in the optimization process. If the variance of each of the K_c calibration measurement is known, Equation 10.13 can be transformed into

$$d_{n,w} = \mathbf{M}_w^T \mathbf{f}_n \tag{10.14}$$

where the dose vector and the measurement matrix have been multiplied by the weight matrix $\mathbf{W}$ which is a $K_c \times K_c$ diagonal matrix where each element is the inverse of the combined variance of all channel for a given measurement:

$$W_{kk}^2 = \frac{1}{\sum_{l=1}^{L} \sigma_{k,l}^2} \tag{10.15}$$

10.5.2.1 From an hyperspectral system to an optimal multispectral system

In a hyperspectral system, the number of wavelength channels is large. If the photodetector used to make the acquisition is a spectrometer, the wavelength channels will typically be uniformly spaced such as the one shown in Figure 10.1. In such cases, a measurement is a vector of L elements. Depending on the type of spectrometer used, L can be several hundreds. Thus, each measurement can be viewed as a point in a L-dimensional space. For a given mPSD, all these wavebands (and therefore all these dimensions) are not

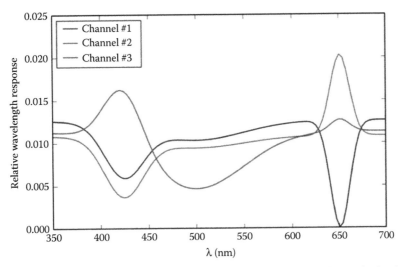

Figure 10.9 Optimal spectral response of the three wavelength filters required to decouple the three spectra shown in Figure 10.1.

equally important for the decoupling of the superposed spectra. Their relative importance will depend on the shape of the N spectra composing an mPSD. Using linear algebraic tools, such as principal component analysis (PCA), it is possible to identify which dimensions are more important when solving Equation 10.9. In practice, this means that starting from a hyperspectral system it is possible to determine the properties of a small subset of optimal wavelength channels that would be ideal for decoupling the N spectra of an mPSD. It is therefore possible to use the data from a large number of measurements performed with a hyperspectral system to conceive the best possible multispectral system. This is interesting from a detector-design point of view, because multispectral systems would typically be more affordable than hyperspectral systems and would involve simpler data analysis because of the reduced dimensionality.

As an example, the same hyperspectral mPSD system as the one presented in Figure 10.1 was used to determine the optimal wavelength response for a 3-channel multispectral system. Five hundred measurements were simulated by assigning a random photon flux to each of the three spectral sources. The simulated measurements matrix **M** was then analyzed by PCA to determine the three best optical filters required for the multispectral system. The result is presented in Figure 10.9 and shows the relative spectral response of the best wavelength channels. Each channel represents one of the principal axes of the hyperspectral system.

10.6 FIELDS OF APPLICATION

The use of a single optical probe to assess the dose at multiple points opens up new possibilities compared to single-point PSDs. The anatomical regions of interest for *in vivo* dose measurement are often spatially constrained and can only accept the insertion of a single optical probe. Using mPSDs multiply the number of measurements that can be performed per optical probe insertion. *In vivo* dosimetry at multiple positions therefore becomes less invasive for the patient. In brachytherapy, a single catheter can be used to assess the dose at multiple points. It can also allow for source position detection (see Section 10.4).

If the main application for mPSDs appears to be their use *in vivo*, they can also open up new possibilities and improve some quality assurance applications. Being able to access dose measurement points stacked right next to each other with the use of a single fiber would be very useful for small field dosimetry and would decrease the uncertainty on the measurements due to positioning. Also, as there is theoretically no limit to the number of scintillating elements composing an mPSD, it could be possible to use a single probe to obtain the profile or depth dose curve of an irradiation as the result of a single measurement. Dose measurements at different positions in a plane or even within a volume could also be performed with a single probe or at least a fewer number of them. The use of mPSDs in medical physics is new, but the number of applications should increase over time.

REFERENCES

Archambault L, Beddar AS, Gingras L et al. (2006) Measurement accuracy and Cerenkov removal for high performance, high spatial resolution scintillation dosimetry. *Med Phys* 33:128–135.

Archambault L, Therriault-Proulx F, Beddar S, Beaulieu L (2012) A mathematical formalism for hyperspectral, multipoint plastic scintillation detectors. *Phys Med Biol* 57:7133–7145. doi: 10.1088/0031-9155/57/21/7133.

Beddar AS, Mackie TR, Attix FH (1992a) Cerenkov light generated in optical fibres and other light pipes irradiated by electron beams. *Phys Med Biol* 37:925–935.

Beddar AS, Mackie TR, Attix FH (1992b) Water-equivalent plastic scintillation detectors for high-energy beam dosimetry: I. Physical characteristics and theoretical consideration. *Phys Med Biol* 37:1883–1900.

Beddar AS, Mackie TR, Attix FH (1992c) Water-equivalent plastic scintillation detectors for high-energy beam dosimetry: II. Properties and measurements. *Phys Med Biol* 37:1901–1913.

Clift MA, Johnston PN, Webb DV (2002) A temporal method of avoiding the Cerenkov radiation generated in organic scintillator dosimeters by pulsed mega-voltage electron and photon beams. *Phys Med Biol* 47:1421–1433.

Daskalov GM, Löffler E, Williamson JF (1998) Monte Carlo-aided dosimetry of a new high dose-rate brachytherapy source. *Med Phys* 25:2200–2208.

De Boer S, Beddar A, Rawlinson J (1993) Optical filtering and spectral measurements of radiation-induced light in plastic scintillation dosimetry. *Phys Med Biol* 38:945–958.

Fontbonne J, Iltis G, Ban G et al. (2002) Scintillating fiber dosimeter for radiation therapy accelerator. *IEEE Trans Nucl Sci* 49:2223–2227.

Frelin A, Fontbonne J, Ban G et al. (2005) Spectral discrimination of Čerenkov radiation in scintillating dosimeters. *Med Phys* 32:3000.

Gagnon J-C, Thériault D, Guillot M et al. (2012) Dosimetric performance and array assessment of plastic scintillation detectors for stereotactic radiosurgery quality assurance. *Med Phys* 39:429. doi: 10.1118/1.3666765.

Guillot M, Beaulieu L, Archambault L et al. (2011a) A new water-equivalent 2D plastic scintillation detectors array for the dosimetry of megavoltage energy photon beams in radiation therapy. *Med Phys* 38:6763. doi: 10.1118/1.3664007.

Guillot M, Gingras L, Archambault L et al. (2011b) Spectral method for the correction of the Cerenkov light effect in plastic scintillation detectors: A comparison study of calibration procedures and validation in Cerenkov light-dominated situations. *Med Phys* 38:2140. doi: 10.1118/1.3562896.

Guillot M, Gingras L, Archambault L et al. (2013) Performance assessment of a 2D array of plastic scintillation detectors for IMRT quality assurance. *Phys Med Biol* 58:4439–4454. doi: 10.1088/0031-9155/58/13/4439.

Keshava N (2003) A survey of spectral unmixing algorithms. *Lincoln Lab J* 14:55–78.

Klein DM, Briere TM, Kudchadker R et al. (2012) In-phantom dose verification of prostate IMRT and VMAT deliveries using plastic scintillation detectors. *Radiat Meas* 47:921–929. doi: 10.1016/j.radmeas.2012.08.005.

Klein DM, Tailor RC, Archambault L et al. (2010) Measuring output factors of small fields formed by collimator jaws and multileaf collimator using plastic scintillation detectors. *Med Phys* 37:5541–5549.

Lacroix F, Beaulieu L, Archambault L, Beddar A (2010) Simulation of the precision limits of plastic scintillation detectors using optimal component selection. *Med Phys* 37:412.

Morin J, Béliveau-Nadeau D, Chung E et al. (2013) A comparative study of small field total scatter factors and dose profiles using plastic scintillation detectors and other stereotactic dosimeters: The case of the CyberKnife. *Med Phys* 40:011719. doi: 10.1118/1.4772190.

Naseri P, Suchowerska N, McKenzie DR (2010) Scintillation dosimeter arrays using air core light guides: Simulation and experiment. *Phys Med Biol* 55:3401–3415. doi: 10.1088/0031-9155/55/12/009.

Rivard MJ, Coursey BM, DeWerd LA et al. (2004) Update of AAPM Task Group No. 43 Report: A revised AAPM protocol for brachytherapy dose calculations. *Med Phys* 31:633–674. doi: 10.1118/1.1646040.

Therriault-Proulx F, Archambault L, Beaulieu L, Beddar S (2012) Development of a novel multi-point plastic scintillation detector with a single optical transmission line for radiation dose measurement. *Phys Med Biol* 57:7147–7159. doi: 10.1088/0031-9155/57/21/7147.

Therriault-Proulx F, Beaulieu L, Archambault L, Beddar S (2013a) On the nature of the light produced within PMMA optical light guides in scintillation fiber-optic dosimetry. *Phys Med Biol* 58:2073–2084. doi: 10.1088/0031-9155/58/7/2073.

Therriault-Proulx F, Beddar S, Beaulieu L (2013b) On the use of a single-fiber multipoint plastic scintillation detector for ^{192}Ir high-dose-rate brachytherapy. *Med Phys* 40:062101. doi: 10.1118/1.4803510.

Applications in radiology

Daniel E. Hyer, Ryan F. Fisher, and Maxime Guillemette

Contents

11.1 INTRODUCTION

Diagnostic radiology is a quickly growing field that has seen a wealth of technological developments in recent decades. These developments, driven largely by the proliferation of digital detectors and increase in computational power, have led to a great increase in the utility, and subsequently the use, of ionizing radiation for diagnostic purposes in modern medicine. This is most evident when viewing the growth of computed tomography (CT), which has seen a 20-fold and 12-fold increase in the United States and the United Kingdom, respectively, over the past two decades [1].

With the growth of diagnostic radiology comes the need to accurately quantify the radiation dose received from each procedure. The driving force behind this desire stems from concerns over stochastic radiation risks, with the most concerning end point being fatal radiation-induced cancers [1,2]. Additionally, recent well-publicized radiation accidents have increased awareness of the potential dangers of acute radiation exposure during diagnostic procedures [3] and have resulted in more stringent oversight and recommendations for radiation-producing equipment [4,5].

The ability to better quantify radiation dose from imaging procedures begins with the development of appropriate dosimeters to perform both *in-vivo* and in-phantom measurements. Measurements at diagnostic photon energies (10s–100s of keV) present special challenges that expose many of the weaknesses of dosimeters currently available today. One of these challenges is the fact that the photoelectric effect becomes the dominant interaction at low photon energies and its cross section is very sensitive to even small changes in atomic number (proportional to Z^3). Many dosimeters, such as diodes or metal–oxide–semiconductor field-effect transistors (MOSFETs), contain high-Z components that make them susceptible to energy dependence issues. This is especially troublesome when measurements are performed at various depths in a medium, resulting in a change to the energy spectrum because of beam hardening and X-ray scattering. Other dosimeters, such as ion chambers, overcome the issues of energy dependency but require large collection volumes that limit their ability to measure the point doses important both for *in vivo* dosimetry and for quantifying CT doses in standard phantoms [6].

The topic of this chapter, which focuses on the use of plastic scintillating dosimeters (PSD) in diagnostic radiology, has great promise to overcome these challenges. The general construction of a PSD, as shown in Figure 11.1, consists of a plastic scintillator that acts as the sensitive element and an optical fiber that transmits the scintillation photons from the sensitive element to a readout device, such as a photomultiplier tube. As discussed in Chapter 4, advantages of PSDs include minimal angular dependence (based on rotations around the assumed cylindrical axis), excellent linearity and reproducibility, instantaneous readout, immunity to electromagnetic interference, and physical sizes as small as 500 µm in diameter and 2 mm in length [7,8]. While PSDs do exhibit an energy dependence, compensation methods that account for the changes in beam spectrum with depth have been shown to yield acceptable results when used in conjunction with PSDs [9]. An additional advantage of using PSD at diagnostic energies is that methods of accounting for Cerenkov light can be ignored, as secondary electrons created at diagnostic energies do not have enough energy to produce Cerenkov light in the optical fiber [10].

The following sections will outline applications as well as the current state of the art in PSD technology in the field of diagnostic radiology.

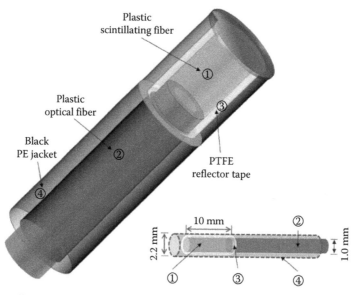

Figure 11.1 Schematic illustrating the main components of a plastic scintillating dosimeter (PSD). The main components include a scintillating fiber (1) which is wrapped in polytetrafluoroethylene (PTFE)-based reflector tape (3) to increase the collection efficiency of the scintillation light. Scintillation photons are transmitted to a readout device by an optical fiber (2) and the entire assembly is wrapped in a black polyethylene jacket (4) to eliminate background light. (From Yoo, W. J. et al., *Optical Review*, 20, 173–177, 2013.)

11.2 REVIEW OF RADIOLOGY APPLICATIONS

Ionizing radiation is utilized for an ever-expanding range of applications in modern diagnostic imaging. These applications include two- and three-dimensional imaging of both static and dynamic anatomy including the visualization of functional processes such as blood flow and digestion. Additionally, beyond diagnosing potential disease and malady, radiological imaging is often used for image guidance in the treatment of disease, replacing open cavity surgeries and turning extended hospital stays into outpatient procedures.

As the use of ionizing radiation for imaging and image-guided intervention expands, the potential for adverse patient effects from radiation increases as well. Studies that attempt to quantify the cancer risks from CT scans [1] are often cited, but cases of acute, and at times severe, skin effects from interventional and cardiac procedures have also been well documented, as seen in the image of skin necrosis in Figure 11.2 [12].

As the use of ionizing radiation increases, so too does the need for accurate methods to determine the amount of radiation administered to the patient for diagnostic cases. Plastic scintillation dosimeters (PSDs) show promise for use in both direct measurement during patient procedures and in phantom measurements that can be used to prospectively predict patient doses from these procedures. The following sections will serve as an overview of the imaging applications of ionizing radiation in which PSDs could be useful for dosimetric measurements.

11.2.1 FLUOROSCOPY

Fluoroscopy is an imaging technique that utilizes X-rays in order to form real-time moving images of patient anatomy, often for the purposes of visualizing dynamic processes. Fluoroscopy is utilized in a variety of applications ranging from simple swallow and gastrointestinal procedures to more complex interventional procedures such as the diagnosis and treatment of brain aneurisms, and the assessment and treatment of cardiac disease. Fluoroscopic procedures can range from several seconds for basic procedures, to multiple hours for more involved interventional procedures. The X-ray output is often pulsed in order to reduce patient doses during procedures.

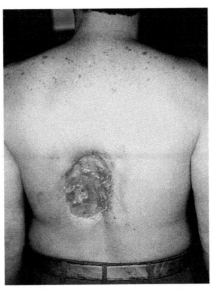

Figure 11.2 Severe tissue necrosis caused by high levels of radiation from multiple coronary procedures. Image was taken approximately 20 weeks post exposure. (Courtesy of Shope, T. B., *Radiographics: A Review Publication of the Radiological Society of North America, Inc.*, 16, 1195–1199, 1996.)

Clinical applications using small PSDs

11.2.1.1 General fluoroscopy

General fluoroscopy refers to exams performed with basic equipment in the form of either a dedicated room located in a radiology department or mobile pieces of equipment that are used in operating and procedure rooms for guidance during procedures. Examples of fixed and mobile general fluoroscopic equipment are shown in Figures 11.3 and 11.4.

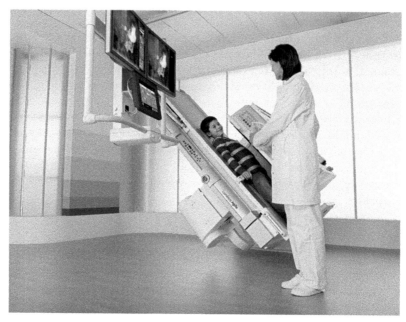

Figure 11.3 A general fluoroscopic system in use for a pediatric study. Copyright Siemens Healthcare 2014. Used with permission.)

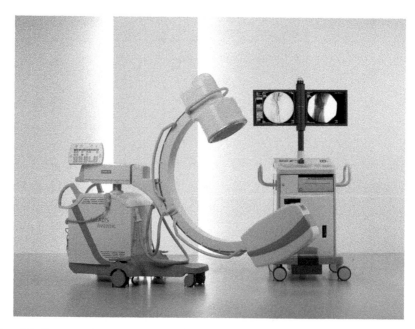

Figure 11.4 A mobile fluoroscopic system, commonly called a *C-arm*, often used for image guidance during surgical procedures. Copyright Siemens Healthcare 2014. Used with permission.)

General fluoroscopic procedures require a total beam on time ranging from several seconds to several minutes, and patients undergoing such procedures are generally not at risk for deterministic skin effects such as hair loss or burns. These procedures commonly utilize a contrast agent such as iodine or barium in order to produce higher contrast between the structures being investigated and background anatomy. The contrast agents can be administered by either injection into a particular cavity via needle or catheter, or ingestion after being mixed with solids or liquids of various viscosities. A brief summary of a variety of general fluoroscopic procedures is provided below.

Gastrointestinal (GI) exams: GI procedures involve diagnosing diseases of the GI system and involve the patient swallowing a barium contrast agent. A barium swallow exam, as seen in Figure 11.5, involves capturing high-frame-rate images of the patient swallowing a contrast agent in order to assess and diagnose potential issues with the oral cavity, pharynx, and esophagus. An upper GI series, as shown in Figure 11.6, is similar but focuses on the esophagus, stomach, and duodenum to assess their functionality.

A small bowel follow through study involves a series of images taken over the course of several hours in order to visualize the small intestines during the digestion process. A barium or air enema, depicted in Figure 11.7, involves the injection of either contrast or air into the rectum for visualizing the anatomy of the colon and large intestines.

Orthopedic exams: Orthopedic procedures involve the utilization of fluoroscopy for the diagnosis of disorders of bones and joints. An arthrogram procedure uses fluoroscopy to guide needle placement into joint spaces in order to inject a contrast agent for the purpose of imaging potential tears in soft connective tissues. Such tears are often difficult to visualize using standard X-ray imaging. Similarly, arthrocentesis procedures involve fluoroscopically guided needle placement for collecting synovial fluid from the joint space in order to determine the cause of joint swelling or arthritis. An image from an arthrocentesis procedure is shown in Figure 11.8.

In addition to needle-guided procedures, fluoroscopy is commonly used during orthopedic surgery to verify the placement of internal or external fixation structures such as pins, screws, plates, and rods, as seen in Figure 11.9.

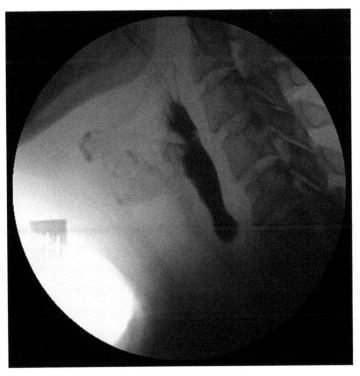

Figure 11.5 Still image taken from a barium swallow procedure. The ingested contrast is visible in the esophagus. (Courtesy of Cleveland Clinic, Cleveland, OH.)

Clinical applications using small PSDs

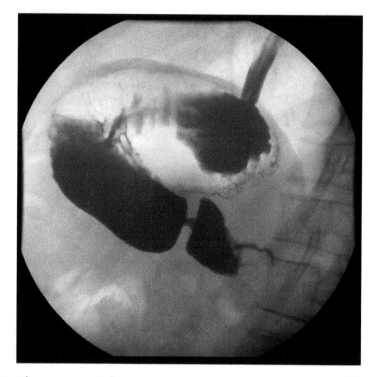

Figure 11.6 An image from an upper GI fluoroscopy study, showing contrast entering and filling the stomach. (Courtesy of Cleveland Clinic.)

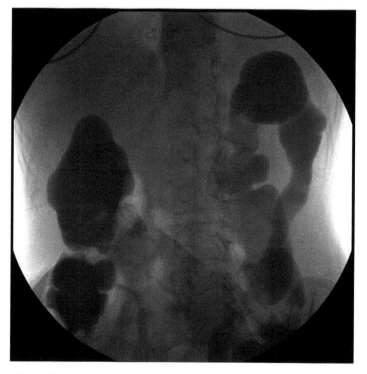

Figure 11.7 An image from a barium enema procedure highlights the folds and structure of the colon. (Courtesy of Cleveland Clinic.)

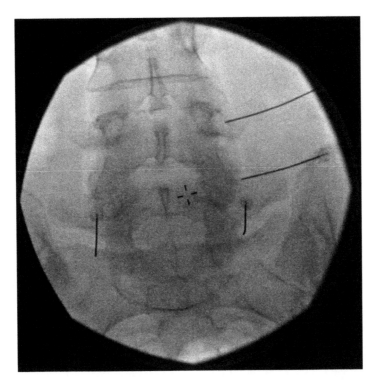

Figure 11.8 Fluoroscopy is used to guide needle placement into the space between vertebrae. (Courtesy of Cleveland Clinic.)

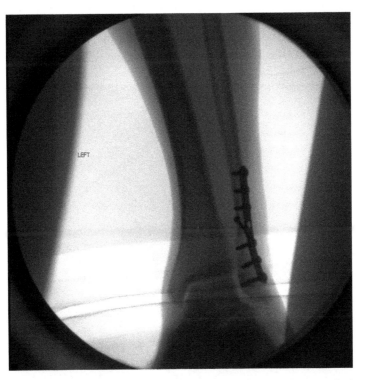

Figure 11.9 Fluoroscopy is utilized during orthopedic surgery procedures in order to verify the position of plates and screws in the ankle. (Courtesy of Cleveland Clinic.)

Genitourinary exams: Genitourinary procedures involve fluoroscopic imaging of the urinary and genital organs in order to diagnose disease or defect. An intravenous pyelogram (IVP) procedure uses fluoroscopy to observe kidney, bladder, and ureter functionality. An IVP involves injecting an iodine-based contrast agent into a vein in the arm, and then taking a series of fluoroscopic images over a period of time to visualize how the dye is filtered by the kidneys, collected in the bladder, and ultimately expelled through the ureters and urethra.

A cystogram procedure involves the injection of contrast via catheter into the bladder, and the use of fluoroscopic imaging to assess the bladder's shape and size. Similarly, a voiding cystourethrogram (VCUG) procedure involves capturing fluoroscopic images of the bladder, ureters, and urethra during urination, in order to visualize the functionality of the entire urinary system, with specific attention to the reflux of fluid up the ureters. A still image from a VCUG procedure is seen in Figure 11.10.

A hysterosalpinogram (HSG) is a fluoroscopic procedure evaluating the female reproductive system that involves the injection of contrast into the uterus and imaging the runoff through the fallopian tubes. An HSG procedure is used to visualize possible polyps and fistulas as well as to identify potentially obstructed fallopian tubes, which can result in fertility problems. An HSG procedure is shown in Figure 11.11.

Neurological exams: Neurological procedures involve the injection of contrast dye into the subarachnoid space between spinal vertebrae. A myelogram, as seen in Figure 11.12, uses fluoroscopic guidance to inject the contrast via needle, and then multiple images of the dye as it spreads through the spinal canal. These exams are used to diagnose spinal pain or inflammation and are effective in locating tumors, infection, herniated discs, or an arrowing of the spinal canal known as *spinal stenosis*.

11.2.1.2 Interventional cardiology

Interventional cardiology is an application of fluoroscopic imaging that deals with catheter-based diagnosis and treatment of cardiac disease. Such procedures are minimally invasive and have replaced open-heart surgery in the treatment of most low-level forms of cardiac disease. The equipment used for these procedures is typically much more advanced than that used for general fluoroscopic procedures. The equipment can include multiple imaging planes and much greater flexibility for angulation in order to allow physicians to obtain the views of the heart needed for these complex procedures. Both single-plane and bi-plane units used for interventional cardiology are shown in Figures 11.13 and 11.14.

Cardiac catheterization procedures involve inserting a catheter into arteries or veins, usually through the femoral artery or vein, and using fluoroscopic guidance to move it through the vascular system to

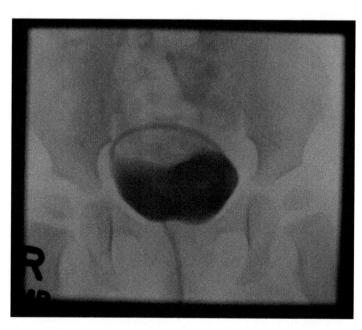

Figure 11.10 Contrast is injected into the bladder during a voiding cystourethrogram. (Courtesy of Cleveland Clinic.)

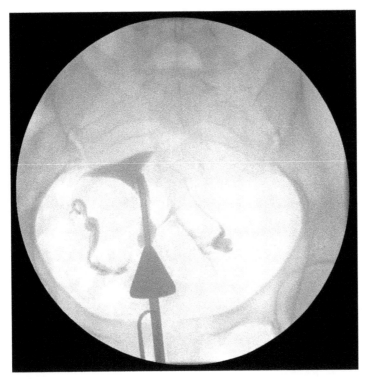

Figure 11.11 Contrast fills the uterus and spills out of the fallopian tubes during a hysterosalpingram procedure. (Courtesy of Cleveland Clinic.)

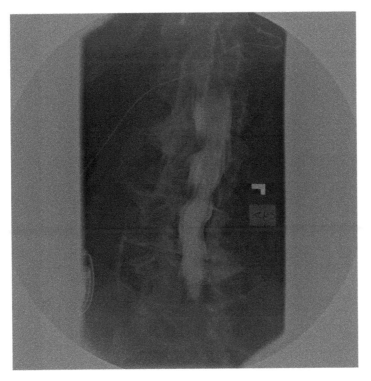

Figure 11.12 Contrast is injected into spinal nerve tracts in a myelogram procedure. (Courtesy of Cleveland Clinic.)

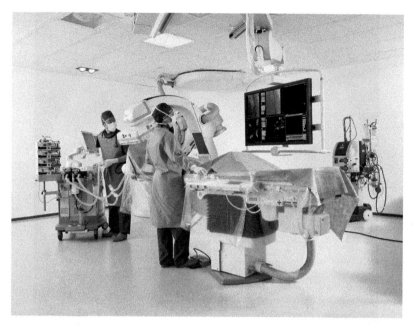

Figure 11.13 A single-plane cardiac system in clinical use. (Copyright Siemens Healthcare 2014. Used with permission.)

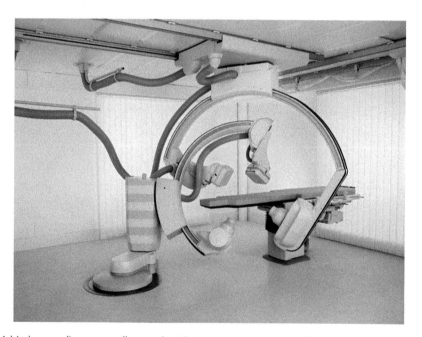

Figure 11.14 A bi-plane cardiac system allows a physician to capture two views of the heart simultaneously. (Copyright Siemens Healthcare 2014. Used with permission.)

a particular position in the body. The catheter can then be used to take pressure measurements in the chambers and vessels of the heart in order to diagnose potential cardiac disease. Contrast dye can also be injected through the catheter in order to visualize the vessels and inspect them for blockages in a process known as *angiography*, depicted in Figure 11.15.

In the case, where angiography reveals blockages in vessels, corrective measures can be completed via catheter as well. Angioplasty procedures involve mechanically widening blocked vessels using pressure from a balloon catheter. Fluoroscopy is again used to guide the balloon to the area of blockage. Once the

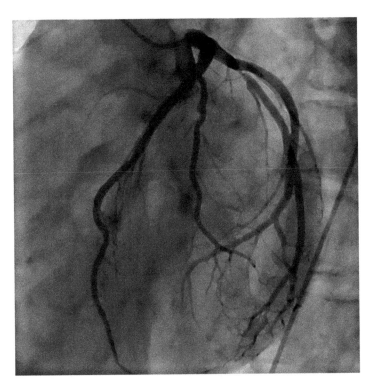

Figure 11.15 Injected contrast is used to visualize the vessels of the heart. (Copyright Siemens Healthcare 2014. Used with permission.)

balloon catheter is in place, it is inflated in order to press the atherosclerotic plaque into the vessel wall in order to restore blood flow. Afterward, the balloon is deflated and removed. A metallic, mesh tube called a *stent* is often placed in the area of the blockage following balloon angioplasty in order to ensure the vessel remains open.

An alternative to clearing blocked vessels using balloon angioplasty procedure is atherectomy. These cases involve using fluoroscopic guidance to place devices with sharp rotating blades at the site of blockages and using these devices to scrape away plaque to restore blood flow.

Cardiac electrophysiology (EP) procedures involve the assessment of cardiac electrical function, and they can be used to diagnose and treat cardiac arrhythmias. These procedures involve the fluoroscopic guidance of catheters to particular areas of the heart, where sensors are used to measure electrical activity in the muscle tissue. Once the areas of cardiac tissue causing the arrhythmia are located, the tissue is destroyed using a variety of methods including radio frequency ablation or cryotherapy. Cardiac catheterization and EP procedures are similar with regard to invasiveness, with both being less invasive than surgical procedures.

11.2.1.3 Interventional radiology

Many of the same techniques previously discussed in the section on interventional cardiology are also utilized in interventional radiographic (IR) procedures to diagnose and treat areas of the body outside of the heart. The equipment used for these procedures is similar to that used in interventional cardiology, though the detectors are often larger in order to provide a wider field of view for imaging areas larger than the heart. IR procedures are commonly performed in the diagnosis and treatment of strokes, aneurysms, many forms of cancer, liver and kidney disease, and even spinal fractures. In all cases, fluoroscopy is used to guide instrumentation through the body in order to either diagnose or treat a particular disease.

As in cardiology, angiography is commonly used in IR cases to visualize arteries and veins in order to find blockages or other issues, as shown in Figures 11.16 and 11.17.

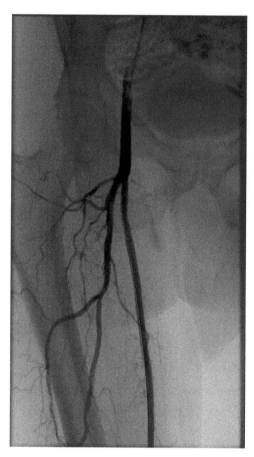

Figure 11.16 Angiography is used to visualize the vasculature of the lower extremities. (Courtesy of Cleveland Clinic.)

If obstructions in the vasculature are found, balloon angioplasty can then be used to clear vessels of the arms, legs, brain, and various other organs. Stents are also commonly used in IR procedures to open blocked blood vessels as well as blocked drainage ducts in organs such as the kidneys and the liver.

Embolization procedures involve the targeted delivery of clotting agents such as coils, gels, or small particles, to a particular area of the body in order to cut off blood flow. This can be done to stop the blood supply to a tumor or aneurysm, or to stop areas of internal bleeding. Similarly, chemoembolization procedures allow radiologists to deliver chemotherapy drugs through a catheter directly to the site of a tumor.

Fluoroscopic guidance is also utilized by radiologists in order to guide needle biopsy procedures and to aid in the placement of drains and feeding tubes.

While general fluoroscopic procedures often employ total fluoroscopy times in the order of a few minutes and relatively low levels of radiation, interventional cardiac and radiological procedures can at times be very complex and involve hours of fluoroscopy time. These extended cases can lead to radiation levels that are high enough to elicit hair loss and potentially cause permanent and dangerous skin damage [12]. As such, physicians must take care to balance the clinical benefit of a given a procedure with the potential risk for radiation effects. Careful attention to factors such as X-ray beam pulse rate and geometry can be used in order to minimize skin dose, and PSD systems could prove valuable in providing real-time peak skin dose monitoring during these procedures.

11.2.2 COMPUTED TOMOGRAPHY

While fluoroscopic procedures make use of two-dimensional moving images, CT exams produce cross-sectional 3D images of a patient, such as those seen in Figure 11.18, which remove the superposition of structures inherent in two-dimensional images, making the visualization of anatomy and disease much easier.

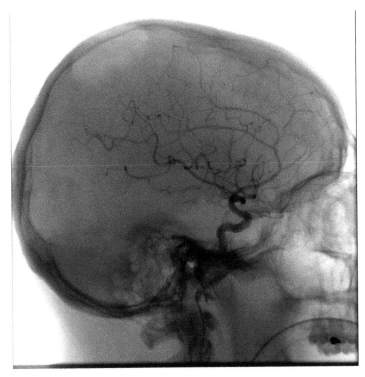

Figure 11.17 Angiography is used to visualize the vasculature of the brain. (Courtesy of Cleveland Clinic.)

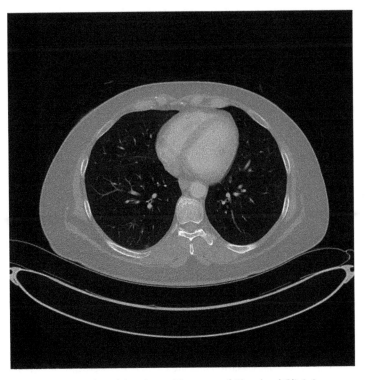

Figure 11.18 A cross-sectional CT image of the chest. (Courtesy of Cleveland Clinic.)

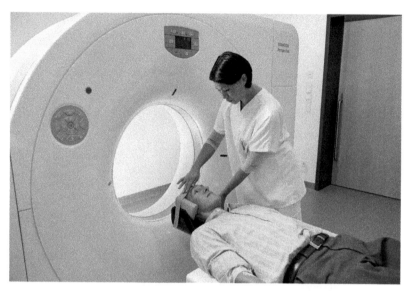

Figure 11.19 A patient is prepared for a head CT scan. (Copyright Siemens Healthcare 2014. Used with permission.)

Because of the clarity of images and advances in scanning technology allowing larger volumes of anatomy to be scanned in even shorter time frames, CT utilization has increased dramatically in recent years.

General CT imaging involves an X-ray tube and detector assembly rotating around an exam table that is translated through the bore of the scanner, as shown in Figure 11.19.

The raw, X-ray transmission data received by the detectors is then reconstructed via complex computer algorithms in order to reconstruct cross-sectional images of the patient anatomy. Iodine-based contrast agents are often intravenously injected into patients in order to better visualize anatomy, and studies involving scans both with and without injected contrast are common.

Recent developments in CT scanning technology have allowed for large volumes of anatomy to be imaged in very short periods of time, which has allowed for drastic improvements in cardiac and brain imaging. It is now possible for an entire heart or brain to be imaged in a single scanner rotation, allowing for 3D angiography and blood perfusion scans. CT angiography cases are similar to those performed in cardiac and interventional radiography, but in three dimensions, which can allow for better vessel visualization while being less invasive than catheter-based angiography studies. Blood perfusion scans of the brain have recently become common in diagnosing potential tissue damage from strokes. This application of CT technology allows for quantitative measurements of blood flow through tissues, allowing for the identification of areas of restricted blood flow and to assess the level of stroke related damage. An example of a three-dimensional CT brain perfusion study is shown in Figure 11.20.

The recent expansion in the use of CT imaging has raised concerns regarding the radiation levels associated with its use. While the radiation dose levels associated with general CT scans are well below the thresholds for hair loss or skin burns, advanced CT applications such as brain and cardiac perfusion imaging can raise doses to potentially dangerous levels. Several cases involving the improper use of advanced CT applications leading to skin burns and hair loss were highlighted by a series of articles published in *The New York Times* [3], which drew national media scrutiny to CT imaging. Beyond these deterministic skin effects, other recent publications have attempted to attribute a range of solid and blood cancers to diagnostic CT scans [14,15]. Media attention brought about by these publications have led to calls for tighter regulation on CT imaging, with the state of California enacting legislation related to CT doses in 2010. This law requires the outside accreditation of all CT scanners in state and mandates reporting of CT dose metrics for all patient cases. Additionally, certain events such as repeat scans, dose metrics above set thresholds, and scans performed without physician approval, must now be immediately reported to the state. Similar legislation has been proposed in several other states. Nationally, the Centers for Medicare and Medicaid Services (CMS) has mandated the accreditation of all non-hospital facilities providing CT imaging. In the wake of the

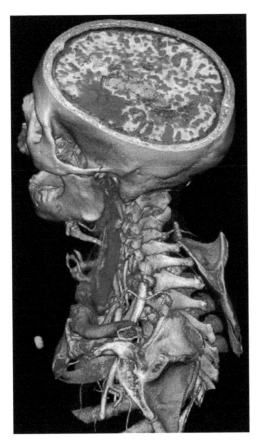

Figure 11.20 Colored areas of the brain reveal blood flow levels in a 3D brain perfusion study. The purple area along the cranial wall indicates areas of restricted blood flow due to stroke. (Copyright Siemens Healthcare 2014. Used with permission.)

enhanced scrutiny on CT imaging, technologies such as fiber-optic plastic scintillation dosimetry can prove useful in future research into the creation of more accurate CT dose metrics, as well as possible use in clinical direct patient dose measurements.

11.3 CURRENT DOSIMETERS

11.3.1 ION CHAMBERS

Gas-filled ionization chambers represent the gold standard of radiation detectors used today. In diagnostic radiology, ionization chambers have found uses ranging from large volume survey detectors to small volume detectors used for dose measurement. In this section, small volume chambers will be discussed, as they are most pertinent to the field of PSDs.

Small volume ionization chambers are available in two primary designs: parallel plate and cylindrical chambers, both of which are illustrated in Figure 11.21. The main components of an ionization chamber include two conducting plates (a collector plate and a ground plate) along with a guard electrode. A large external bias (~300 V) is applied between the collector and ground plates, which creates an electric field between the two electrodes. The gas-filled volume between the plates is referred to as the collection volume of the chamber. When exposed to radiation, air molecules within the collection volume are ionized, and the resulting ions and electrons are swept toward the appropriate collector because of the applied electric field. This induces a current between the two electrodes that can be related to the amount of radiation dose delivered to the detector. The guard electrode keeps the electric field lines uniform and eliminates extraneous signal [16].

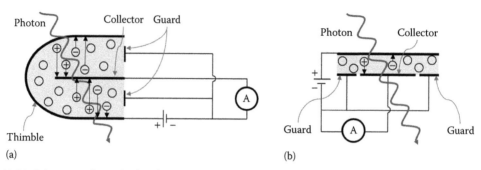

Figure 11.21 Schematic of (a) cylindrical and (b) parallel plate ionization chambers.

Parallel plate chambers are often used for exposure measurements in diagnostic radiology. The parallel plate chamber is meant to be used with the collection plates oriented perpendicular to the X-ray beam axis and the chambers utilize a very thin front window to minimize attenuation of the incident beam. The energy dependence of a thin window parallel plate chamber compared to a thick wall cylindrical chamber can be seen in Figure 11.22. This figure illustrates that a thin window chamber has minimal energy dependence but should still be calibrated over the expected energy range in which it will be used.

In general, ionization chambers utilize a relatively large collection volume, which limits their utility as a point detector. Small volume ion chambers do exist, but the sensitivity of ion chambers decreases with the collection volume. The need for a large collection volume, coupled with the fact that a high voltage must be applied to the chamber during irradiation has also limited the utility of ion chambers as *in vivo* dosimeters.

11.3.2 DIODES

A simple diode dosimeter consists of a silicon *p-n* junction, as shown in Figure 11.23 [17]. The *p*-region is deficient in electrons and is referred to as an *acceptor*, while the *n*-region has an excess of electrons and is referred to as a *donor*. At the interface between the *p*-type and *n*-type material, a small region,

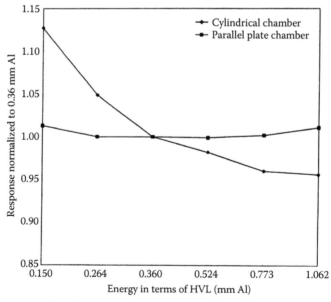

Figure 11.22 Energy response of a cylindrical and parallel plate chamber in low energy region. (From DeWerd, L. A., and L. K. Wagner, *Applied Radiation and Isotopes: Including Data, Instrumentation and Methods for Use in Agriculture, Industry and Medicine*, 50, 125–136, 1999.)

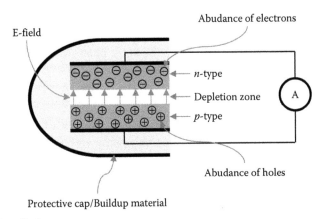

Figure 11.23 *p-n* junction diode.

referred to as the *depletion zone* (or space charge region), is created due to the diffusion of electrons from the *n*-region and holes from the *p*-region. The depletion zone creates an electric field that opposes further diffusion of electrons and holes once equilibrium is achieved. When irradiated, electron–hole pairs are created within the depletion zone, which are subsequently swept away by the electric field present in the depletion zone—giving rise to a radiation-induced current that can then be measured by an electrometer.

One of the main advantages of diode dosimeters is their intrinsic high sensitivity and small size. The increased sensitivity is because the energy required to create an electron–hole pair in silicon is only 3–4 eV, compared to 34 eV required to create an ion-pair in air. Additionally, the density of solid-state diode detectors is approximately 1000 times greater than gas, which increases the probability of X-ray interactions and contributes to making diodes much more sensitive than ion chambers of the same volume [18]. Both of these factors allow diode detectors to be much smaller than ionization chambers, while still producing enough signals to be useful as a measurement device.

The use of photodiodes has been investigated for measuring organ doses in a phantom during diagnostic imaging procedures with acceptable sensitivity, linearity, dose rate dependence, and angular dependence [19]. However, an energy dependence of up to 10%/10 keV was noted (Figure 11.24). This energy dependence is

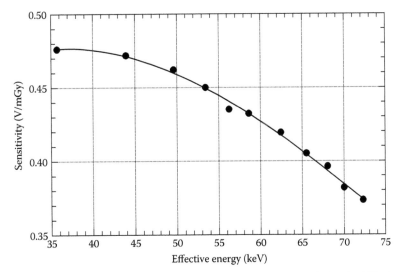

Figure 11.24 X-ray energy dependence of a diode system. (From Aoyama, T. et al., *Medical Physics*, 29, 1504–1510, 2002.)

difficult to avoid, as diodes are primarily manufactured out of silicon, whose atomic number of 14 is higher than the effective atomic number of water, 7.42. For kilovoltage photon beams used in diagnostic radiology, this difference in atomic number becomes very important, as the photoelectric effect, which is prominent at low energies, is highly dependent on atomic number.

11.3.3 MOSFETs

MOSFET have proven to be useful in many different dosimetry applications. For dosimetric use, *p*-channel MOSFETs are common, which consist of a transistor with a source and drain, both made of *p*-type silicon, separated by a gate. This entire assembly is set on an *n*-type silicon substrate (Figure 11.25). In order for current to flow in the *p*-channel, a sufficient voltage must be applied to the gate [20]. When the voltage is applied, the movement of charges (holes) from the source to drain is made possible as the valence band gap is driven away from the Fermi level. The amount of voltage required to enable this current flow is dependent on the accumulated radiation dose.

While MOSFETs have excellent sensitivity in the radiotherapy energy range, these detectors are not as sensitive for applications in the radiological energy range [21]. They also display temperature and dose rate dependencies, which require careful calibrations in order to produce accurate measurements [22,23]. For radiological applications, the temperature dependence could be an important factor due to differences in room temperature and patient body temperature. MOSFET radiosensitivity is a potentially larger issue, as their ability to function as a detector deteriorates with accumulated dose, which requires the detectors to be periodically replaced. Another drawback is the cables used in MOSFET systems are not radio transparent, which produces image artifacts when used during patient imaging, limiting their use primarily to phantom studies [24]. Finally, MOSFETs also exhibit angular dependency [25], which can lead to measurement errors in cases where the incident beam angle changes in relation to the detector, something quite common in fluoroscopic and CT procedures.

11.3.4 THERMOLUMINESCENT DOSIMETERS

Thermoluminescence is the process by which visible light is emitted following the heating of a material. Thermoluminescent dosimeters (TLDs) are impurity-doped crystalline dielectric materials with the ability to function as radiation dosimeters. Incident radiation causes electrons to be excited to higher energy states. These electrons are then trapped in their elevated energy state by impurities in the crystals. Upon heating, the trapped electrons are released back to their normal energy state, giving off light proportional to the incident radiation dose in the process [20]. TLDs function as integrating dosimeters and are widely used as personnel dosimeters since real-time reading is rarely necessary for such applications. TLDs have also been widely used in research applications, including the gluing of an array of detectors to a garment worn by patients during fluoroscopically guided interventions in order to assess peak skin dose and monitor areas of skin at risk for potential radiation induced damage [26]. Unfortunately, TLDs do not offer the possibility of real-time reading, nor are they sufficiently accurate (having an overall precision of 10%–20%) for most radiological applications.

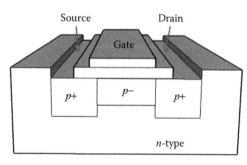

Figure 11.25 Schematic cross section of a MOSFET.

11.4 ADVANTAGES OF PLASTIC SCINTILLATION DOSIMETERS IN RADIOLOGY

11.4.1 ENERGY DEPENDENCE

For photon beams in the megavoltage energy range, Compton scattering is the dominant photon interaction, and PSDs have been shown to be nearly water equivalent. However, for the kilovoltage energy beams used in diagnostic radiology, photoelectric interactions become increasingly important, and the cross section for photoelectric interactions is highly dependent on the atomic number, which can be problematic for PSDs. The atomic number for polystyrene, the primary material used in the sensitive volume of PSDs, is 5.74, compared to 7.42 for water [9]. One simple experiment that can be done to illustrate the energy dependence of a PSD is to measure a depth dose curve with a fixed tube potential using both a calibrated ion chamber and a PSD. The PSD readings can be normalized to the air kerma measured by the ion chamber for easy comparison, as shown in Figure 11.26 for a 120 kV beam. Assuming energy independence of the ion chamber, the energy dependence of the PSD due to the change in beam spectrum as a result of beam hardening and scatter results in a change in response of the PSD of approximately 10% over a range of depths from 0 to 16 cm.

A method to account for the energy dependence of PSDs at low energies was proposed by Lessard et al. [9] and is based on the assumption that energy deposition for low-energy photons occurs at the location of the photon interaction. This is a valid assumption for low-energy photons because the range of secondary particles is negligible. In this approximation, which is known as *large cavity theory* (LCT), the cross-sectional ratio between two media is equivalent to their mass energy absorption coefficient ratio (μ_{en}/ρ). The μ_{en}/ρ ratio between water and polystyrene, the main component of the sensitive element of PSDs, shows a significant variation as a function of energy below 100 keV, as shown in Figure 11.27. Applying LCT to a depth dose curve for an effective photon energy of 48 keV, as shown in Figure 11.28, demonstrates how well LCT technique accounts for the energy dependence of the PSD due to changes in the beam spectrum from beam hardening and scatter. The corrected PSD reading shows less than a 5% variaion from the Monte Carlo simulated depth dose curve while the uncorrected PSD reading varies by nearly 15% over the same depth range. Gafchromic film studies also yielded good results, but separate calibrations must be performed for each batch of film. Data from Jurado et al. [27] is shown for comparison purposes as they performed PDD measurements with the same model of X-ray unit (Therapax SXT 150).

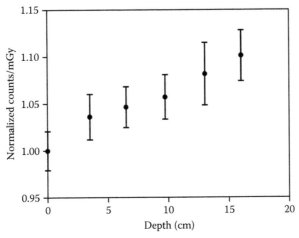

Figure 11.26 Energy dependence of a PSD as a function of depth in soft tissue equivalent material. Data have been normalized to the surface measurement (depth = 0). Error bars correspond to ±1 standard deviation of the mean. (From Hyer, D. E. et al., *Medical Physics*, 36, 1711–1716, 2009.)

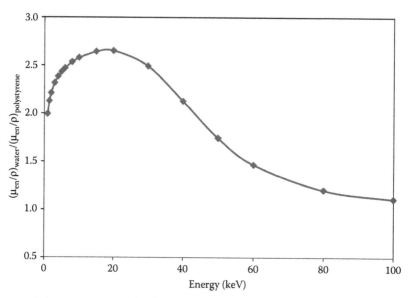

Figure 11.27 μ_{en}/ρ ratio between water and polystyrene in the low-energy range. According to LCT, this ratio is the correction factor to obtain deposited dose in water from a measurement in polystyrene. (Data from Hubbell, J. H., and S. M. Seltzer, *Tables of X-Ray Mass Attenuation Coefficients and Mass Energy-Absorption Coefficients from 1 keV to 20 MeV for Elements Z = 1 to 92 and 48 Additional Substances of Dosimetric Interest*, 2004, National Institute of Standards and Technology.)

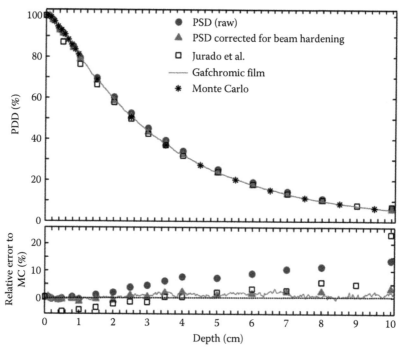

Figure 11.28 (Top) PDD for a 44 keV effective energy beam obtained using various detectors. (Bottom) Relative errors to MC simulation. (From Lessard, F. et al., *Medical Physics*, 39, 5308–5316, 2012.)

11.4.2 ANGULAR DEPENDENCE

Given the cylindrical geometry of most PSDs, the angular dependence when irradiated around the axis of the cylinder is expected to be small. An example of the angular dependence is shown in Figure 11.29, in which a PSD was irradiated around its axis using a CT scanner with a tube potential of 120 kVp. Static shots at fixed tube angles were taken at intervals of 30°. In this configuration, the PSD yielded less than a 5% angular dependence over an entire revolution.

The same PSD was also irradiated in a normal-to-axial configuration, where a large reduction in the number of counts was noted when the beam was incident from a narrow range of angles in the direction of the optical fiber (180° in Figure 11.30). This effect can be explained by the attenuation of X-rays by the optical fiber itself. Additionally, the response of the dosimeter dropped by approximately 12% when irradiated from the distal tip, which is attributable to the smaller solid angle provided by the cylindrical geometry of the sensitive element when viewed head-on.

Overall, the PSD yields expected results in terms of angular dependence, considering its geometry. The angular dependence of other shapes of PSDs would need to be evaluated on an individual basis.

11.4.3 LINEARITY

It has been shown in the seminal papers of Beddar et al. [29,30] that PSDs have a linear output signal in the radiotherapy energy range (300 keV–25 MeV). More recently, it has been shown that PSDs exhibit similar behavior in the lower, diagnostic energy range used in radiology. Hyer et al.'s study [7] elegantly

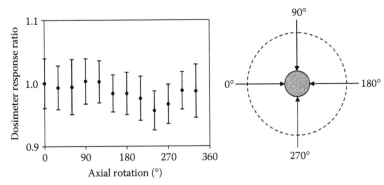

Figure 11.29 Angular dependence of PSD dosimeter to an axial irradiation. Data have been normalized to a 0° axial angular response. Error bars correspond to ±1 standard deviation of the mean. Experimental setup is also shown. (From Hyer, D. E. et al., *Medical Physics*, 36, 1711–1716, 2009.)

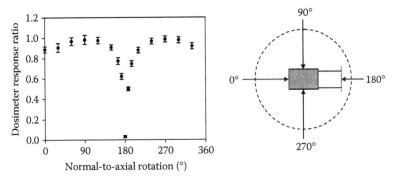

Figure 11.30 Angular dependence of PSD dosimeter to a normal-to-axial irradiation. Data have been normalized to a 0° axial angular response. Error bars correspond to ±1 standard deviation of the mean. Experimental setup is also shown. The shaded area represents the active element. (From Hyer, D. E. et al., *Medical Physics*, 36, 1711–1716, 2009.)

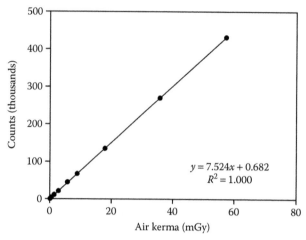

Figure 11.31 Dose linearity of plastic scintillation dosimeter. (From Hyer, D. E. et al., *Medical Physics*, 36, 1711–1716, 2009.)

demonstrated how linear the output signal is at 120 kVp (Figure 11.31), making them ideal candidates for radiological applications where a wide range of exposures may be encountered.

11.4.4 TRANSPARENCY

An important factor in the search for an ideal dosimeter in radiological applications is its transparency or water equivalence. This allows for dose measurements without the need for correction factors, as well as providing the physicians with a clear view of anatomical details, unobstructed by the dosimeter. Guiding a catheter through an artery or finding a small defect in a patient's heart is challenging, thus, the need for dose measurements without creation of artifacts is of great importance. Since plastic scintillation dosimeters have a low effective atomic number, it has a tremendous advantage over other dosimeters with a higher effective Z. Figure 11.32 illustrates the superior transparency and safety of PSDs under diagnostic conditions as compared to a traditional ion chamber.

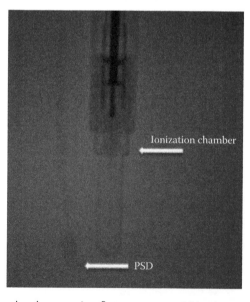

Figure 11.32 PSD and ionization chamber seen in a fluoroscopy acquisition image.

11.4.5 REAL-TIME READING

It is important to monitor peak patient skin dose during a fluoroscopic procedure such that deterministic effects, such as skin damage, can be avoided. However, as we have seen with most dosimeters in this chapter, none is particularly suited to the task of real-time dose measurements at diagnostic energies. With a PSD, doses during procedures can be continuously monitored in order to ensure patient safety, allowing physicians to adjust techniques such as X-ray tube angulation, table height, or beam pulse rate, in order to reduce dose to the patient's skin.

11.5 CLINICAL EXPERIENCE

11.5.1 PHANTOM MEASUREMENTS

As the applications for, and utilization of, X-ray imaging continues to expand, the need for accurate methods to measure and estimate patient doses from these procedures becomes increasingly important. In the face of increasing scrutiny and regulation regarding radiation dose, this dosimetric information is valuable for both the purposes of research as well as clinical implementation. Patient dose measurements can take the form of either direct measurements made during individual patient cases or through phantom experiments in order to better predict the doses received by patients for similar exams and procedures.

Plastic scintillation dosimeters are well suited for the collection of patient and phantom measurements for several reasons. Their water equivalence allows them to be used in the direct X-ray beam during patient procedures without creating image artifacts. The small size of PSDs allows them to be used in combination with catheters for *in-vivo* patient measurements. The small size also makes PSDs well suited for phantom measurements, as they can be inserted into a variety of patient equivalent phantoms without creating large air gaps or otherwise disrupting the material or dose measurement.

Additionally, the real-time measurement capabilities of PSD dosimetry makes them very well suited for applications such as fluoroscopic dose monitoring during complex cases. During these cases, physicians can monitor real-time, cumulative patient skin dose measurements in order to best balance the risks and benefits of a procedure. Several research groups have been involved in the development of plastic scintillation dosimetry systems and the associated collection of phantom measurements for a variety of imaging applications. Similar research into the use of optically stimulated luminescence (OSL) detectors has also been expanded to include cadaver and patient *in-vivo* measurements. While these studies were not conducted using PSD dosimeters, the applications are comparable enough that PSDs could be used, perhaps more easily, as they eliminate the time-consuming postirradiation read out process associated with OSL detectors.

One research group investigating the use of PSDs for diagnostic applications is under the direction of Luc Beaulieu at Laval University in Quebec City, Canada. Their system, as seen in Figure 11.33, utilizes a plastic scintillating element coupled to an optical fiber, which is joined to an avalanche photodiode.

This system makes use of spectral filtration methods in order to remove fluorescence noise signal created in the fiber. In order to use this system in the diagnostic energy range, correction factors based on the effective beam energy and spectrum were developed in order to correct for the energy dependence of the system [9]. Their results showed that PSD systems could be useful for dosimetry in radiographic applications, and further research has expanded into real-time phantom measurements during simulated fluoroscopically guided interventional cases [31].

A second research group, led by David Hintenlang at the University of Florida, has also constructed a PSD dosimetry system [7]. This system differed from that of the Laval group, in that the primary signal collected from the scintillating fiber is collected along with signal from a similar fiber with no scintillator, known as a *dummy fiber*, using two photomultiplier tubes. This dummy fiber accounts for any additional signal or stem effect created within the fiber itself. The signal from the dummy fiber is subtracted from that of the scintillating fiber in order to obtain the overall light output for a given exposure, as discussed in Chapters 4 and 5. This PSD system, as shown in Figures 11.34 and 11.35, was characterized across a range of diagnostic energies and showed good correlation to a small volume ionization chamber.

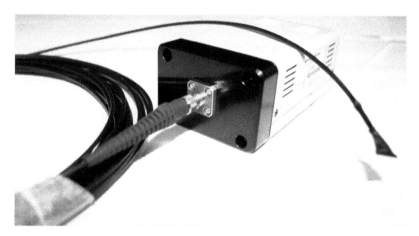

Figure 11.33 The Laval University PSD system utilizing an avalanche photodiode.

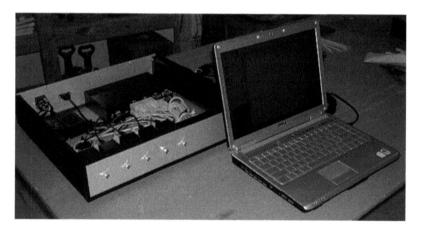

Figure 11.34 University of Florida PSD system with five fiber-optic channels.

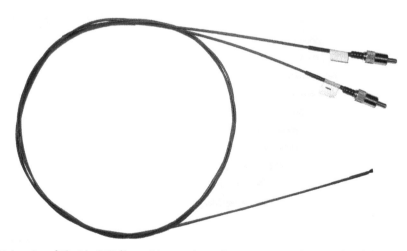

Figure 11.35 University of Florida PSD fiber with two channels representing the signal and dummy fibers.

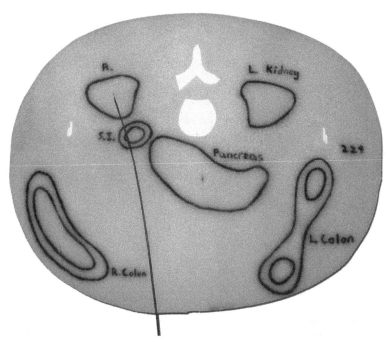

Figure 11.36 A cross-sectional slice of an anthropomorphic phantom used for CT dosimetry. A PSD is in place to measure dose to the right kidney. (From Hyer, D. E. et al., *Journal of Applied Clinical Medical Physics/American College of Medical Physics*, 11, 3183, 2010.)

The University of Florida system has since been utilized along with an anthropometric phantom in order to collect organ doses for a variety of patient procedures. The small size of the PSD allowed for easy insertion into the phantom at predetermined organ locations in order to measure specific organ doses, as depicted in Figure 11.36.

Hyer et al. used the system in order to collect a library of organ doses from various kilo-voltage cone beam CT (CBCT) systems used for patient set-up verification in radiation therapy [32]. This research showed clinically relevant differences in the standard imaging protocols used by several major vendors for CBCT imaging.

The Florida PSD system was also used by Fisher in order to catalog doses to phantoms, as shown in Figure 11.37, representing 50th and 90th percentile-by-weight patients undergoing standard CT imaging of the chest, abdomen, and pelvis [33].

Patient doses from CT scans can be highly variable, as modern scanners employ automatic exposure modulation where the X-ray tube output is changed during the course of the scan based on the habitus of the patient being imaged. In fact, the measured patient doses were found to be higher in the larger phantom as compared to the smaller one for the same scanner settings across all investigated exams. The average increase in measured organ doses in the larger phantom was found to be approximately 10% for a chest exam, and 20% for abdominal and pelvic scans on one CT scanner [33]. Additionally, these increases in dose were accompanied by an 11% increase in image noise. When the same organ dose measurements were repeated on a CT scanner from a different vendor, the larger phantom received dose increases in the order of 30% for chest and abdominal scans, and 15% for a pelvic scan. These increases were accompanied by a 6% increase in image noise. The changes in the magnitude of dose differences seen between the two vendors highlighted the potential deviations in clinical patient doses resulting from differences in the algorithms employed for automatically adapting tube output to patient body habitus.

Additional research using the University of Florida PSD system and phantoms highlighted differences in peripheral organ doses based on variable X-ray tube start location and patient positioning [34]. This research found variations in dose to the eye and thyroid of up to 60% based on these factors. A separate study used the PSD system in order to validate Monte Carlo simulations of organ dose for a variety of CT scan

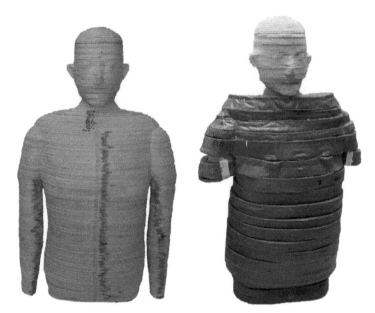

Figure 11.37 50th and 90th percentile by weight phantoms constructed for use with PSD dosimeters to measure patient organ doses during CT procedures.

protocols [35]. Point dose measurements made in adult and pediatric phantoms for multiple CT scans were compared to those simulated using Monte Carlo methods and computational phantoms, with agreements in the order of 10%–15% observed.

Wook Jae Yoo and his team from Konkuk University in South Korea have also constructed a PSD system similar to that developed by the University of Florida group. Their system was successfully utilized in phantom measurements of entrance skin dose (ESD) for general radiographic exams using acrylic phantoms. The light output of the scintillating detectors was calibrated against a commercially available semiconductor dosimeter. They found that for such measurements, there was a good correlation between the measured exposure and PSD light output across a range of imaging techniques including kVp, imaging time, and total tube output [8]. While these studies were performed on general radiographic equipment, they are equally applicable to fluoroscopy cases, where such a system could be placed on patient or phantom surfaces in order to measure ESD in real time.

In addition to PSD systems, OSL dosimeters have also seen expanded use in measuring phantom and patient doses for diagnostic procedures, aided by the off-the-shelf commercial availability of such systems. OSL detectors are usually constructed of high Z materials that are not tissue equivalent and they do not possess the real-time capabilities of PSD systems due to the post-exposure readout required for their use. Despite these drawbacks, their small size makes them well suited for patient and phantom use, and they have been successfully characterized and utilized for both fluoroscopic and CT-related dosimetry applications [36].

Specifically, OSL dosimeters have successfully been used for the measurement of patient ESDs in interventional procedures [37] as well as organ doses in cadavers of various body mass indexes for a range of CT studies [38].While these particular studies did not involve PSD systems, plastic scintillation dosimeters could easily be used for similar applications.

11.5.2 *IN VIVO* DOSIMETRY

The next step for the radiological application of plastic scintillation dosimetry is patient experimentation. Research has shown how suitable these dosimeters are for low-energy applications, and it is clear that they have many advantages over other traditional dosimeters. For fluoroscopically guided interventions, a simple utilization would be to place the dosimeter between the patient's back and the mattress of the examination table. A real-time reading of the patient's dose would then be available for the physician, allowing them to make a clear decision on the impact of dose accumulation during the procedure. Because of their size and shape, PSDs could also be inserted into the patient via a catheter, allowing for true *in vivo* dosimetry, though to date, no such studies have been published.

Recently, work has been published utilizing TLDs to measure internal patient dosimetry during CT colonoscopy procedures for assessing the accuracy of CT dose metrics [39]. PSDs could be used in similar applications and would also add the benefit of real-time dose tracking, eliminating the time-consuming post irradiation readout process of integrating dosimeters such as TLDs and OSLs.

11.6 CONCLUSIONS

Compared to many traditional dosimeters used in diagnostic radiology, PSDs offer many advantages, including a linear output signal, transparency, and real-time capabilities, all of which could be used to improve patient safety. One of the drawbacks of PSD dosimeters is their energy dependence within the low energy range, however, using Lessard et al.'s [9] correction method, the energy dependence can be corrected for radiological applications. Phantom measurements have proven that PSDs offer a suitable alternative to traditional dosimetry methods, and with time, *in-vivo* measurements will further prove their clinical utility.

REFERENCES

1. Hall, E. J., and D. J. Brenner. 2008. Cancer risks from diagnostic radiology. *The British Journal of Radiology* 81:362–378.
2. Linet, M. S., T. L. Slovis, D. L. Miller, R. Kleinerman, C. Lee, P. Rajaraman, and A. Berrington de Gonzalez. 2012. Cancer risks associated with external radiation from diagnostic imaging procedures. *CA: A Cancer Journal for Clinicians* 62:75–100.
3. Bogdanich, W. 2010. After stroke scans, patients face serious health risks. *The New York Times*.
4. U.S. Food and Drug Administration. 2009. Safety investigation of CT brain perfusion scans: Initial notification.
5. National Council on Radiation Protection & Measurement. 2010. Report No. 168—Radiation dose management for fluoroscopically-guided interventional medical procedures. Bethesda, MD.
6. Dixon, R. L. 2006. Restructuring CT dosimetry—A realistic strategy for the future Requiem for the pencil chamber. *Medical Physics* 33:3973–3976.
7. Hyer, D. E., R. F. Fisher, and D. E. Hintenlang. 2009. Characterization of a water-equivalent fiber-optic coupled dosimeter for use in diagnostic radiology. *Medical Physics* 36:1711–1716.
8. Yoo, W. J., D. Jeon, J. K. Seo, S. H. Shin, K.-T. Han, W. S. Youn, S. Cho, and B. Lee. 2013. Development of a scintillating fiber-optic dosimeter for measuring the entrance surface dose in diagnostic radiology. *Radiation Measurements* 48:29–34.
9. Lessard, F., L. Archambault, M. Plamondon, P. Despres, F. Therriault-Proulx, S. Beddar, and L. Beaulieu. 2012. Validating plastic scintillation detectors for photon dosimetry in the radiologic energy range. *Medical Physics* 39:5308–5316.
10. Arnfield, M. R., H. E. Gaballa, R. D. Zwicker, Q. Islam, and R. Schmidt-Ullrich. 1996. Radiation-induced light in optical fibers and plastic scintillators: Application to brachytherapy dosimetry. *IEEE Transactions on Nuclear Science* 43:2077–2084.
11. Yoo, W. J., J. K. Seo, S. H. Shin, K.-T. Han, D. Jeon, K. W. Jang, H. I. Sim, B. Lee, and J.-Y. Park. 2013. Measurements of entrance surface dose using a fiber-optic dosimeter in diagnostic radiology. *Optical Review* 20:173–177.
12. Balter, S., J. W. Hopewell, D. L. Miller, L. K. Wagner, and M. J. Zelefsky. 2010. Fluoroscopically guided interventional procedures: A review of radiation effects on patients' skin and hair. *Radiology* 254:326–341.
13. Shope, T. B. 1996. Radiation-induced skin injuries from fluoroscopy. *Radiographics: A Review Publication of the Radiological Society of North America, Inc* 16:1195–1199.
14. Brenner, D. J., and E. J. Hall. 2007. Computed tomography—An increasing source of radiation exposure. *The New England Journal of Medicine* 357:2277–2284.
15. Pearce, M. S., J. A. Salotti, M. P. Little, K. McHugh, C. Lee, K. P. Kim, N. L. Howe et al. 2012. Radiation exposure from CT scans in childhood and subsequent risk of leukaemia and brain tumours: A retrospective cohort study. *Lancet* 380:499–505.
16. DeWerd, L. A., and L. K. Wagner. 1999. Characteristics of radiation detectors for diagnostic radiology. *Applied Radiation and Isotopes: Including Data, Instrumentation and Methods for Use in Agriculture, Industry and Medicine* 50:125–136.
17. Khan, F. M. 2010. *The Physics of Radiation Therapy*. Lippincott Williams & Wilkins, Baltimore, MD.

18. Knoll, G. F. 2000. *Radiation Detection and Measurement.* John Wiley & Sons, Hoboken, NJ.
19. Aoyama, T., S. Koyama, and C. Kawaura. 2002. An in-phantom dosimetry system using pin silicon photodiode radiation sensors for measuring organ doses in x-ray CT and other diagnostic radiology. *Medical Physics* 29:1504–1510.
20. DeWerd, L. A., L. J. Bartol, and S. D. Davis. 2009. Thermoluminescent dosimetry. In D. W. O. Rogers and J. E. Cygle (eds.), *Clinical Dosimetry Measurements in Radiotherapy.* Madison, WI.
21. Lavallee, M. C., L. Gingras, and L. Beaulieu. 2006. Energy and integrated dose dependence of MOSFET dosimeter sensitivity for irradiation energies between 30 kV and 60Co. *Medical Physics* 33:3683–3689.
22. Falco, M. D., M. D'Andrea, L. Strigari, D. D'Alessio, F. Quagliani, R. Santoni, and A. L. Bosco. 2012. Characterization of a cable-free system based on p-type MOSFET detectors for "in vivo" entrance skin dose measurements in interventional radiology. *Medical Physics* 39:4866–4874.
23. Glennie, D., B. L. Connolly, and C. Gordon. 2008. Entrance skin dose measured with MOSFETs in children undergoing interventional radiology procedures. *Pediatric Radiology* 38:1180–1187.
24. Miksys, N., C. L. Gordon, K. Thomas, and B. L. Connolly. 2010. Estimating effective dose to pediatric patients undergoing interventional radiology procedures using anthropomorphic phantoms and MOSFET dosimeters. *AJR. American Journal of Roentgenology* 194:1315–1322.
25. Chuang, C. F., L. J. Verhey, and P. Xia. 2002. Investigation of the use of MOSFET for clinical IMRT dosimetric verification. *Medical Physics* 29:1109–1115.
26. Lickfett, L., M. Mahesh, C. Vasamreddy, D. Bradley, V. Jayam, Z. Eldadah, T. Dickfeld et al. 2004. Radiation exposure during catheter ablation of atrial fibrillation. *Circulation* 110:3003–3010.
27. Jurado, D., T. Eudaldo, P. Carrasco, N. Jornet, A. Ruiz, and M. Ribas. 2005. Pantak Therapax SXT 150: Performance assessment and dose determination using IAEA TRS-398 protocol. *The British Journal of Radiology* 78:721–732.
28. Hubbell, J. H., and S. M. Seltzer. 2004. *Tables of X-Ray Mass Attenuation Coefficients and Mass Energy-Absorption Coefficients from 1 keV to 20 MeV for Elements Z = 1 to 92 and 48 Additional Substances of Dosimetric Interest.* National Institute of Standards and Technology, Gaithersburg, MD.
29. Beddar, A. S., T. R. Mackie, and F. H. Attix. 1992. Water-equivalent plastic scintillation detectors for high-energy beam dosimetry: I. Physical characteristics and theoretical consideration. *Physics in Medicine and Biology* 37:1883–1900.
30. Beddar, A. S., T. R. Mackie, and F. H. Attix. 1992. Water-equivalent plastic scintillation detectors for high-energy beam dosimetry: II. Properties and measurements. *Physics in Medicine and Biology* 37:1901–1913.
31. Boivin, J., M. Guillemette, and L. Beaulieu. 2013. Real-time plastic scintillation dosimeters for fluroscopically-guided interventional procedures. *Medical Physics* 40:427.
32. Hyer, D. E., C. F. Serago, S. Kim, J. G. Li, and D. E. Hintenlang. 2010. An organ and effective dose study of XVI and OBI cone-beam CT systems. *Journal of Applied Clinical Medical Physics/American College of Medical Physics* 11:3183.
33. Fisher, R. F. 2010. *Dose Assessment and Prediction in Tube-current Modulated Computed Tomography.* Nuclear and Radiological Engineering, University of Florida, Gainesville, FL.
34. Winslow, J. F., C. J. Tien, and D. E. Hintenlang. 2011. Organ dose and inherent uncertainty in helical CT dosimetry due to quasiperiodic dose distributions. *Medical Physics* 38:3177–3185.
35. Long, D. J., C. Lee, C. Tien, R. Fisher, M. R. Hoerner, D. Hintenlang, and W. E. Bolch. 2013. Monte Carlo simulations of adult and pediatric computed tomography exams: Validation studies of organ doses with physical phantoms. *Medical Physics* 40:013901.
36. Lavoie, L., M. Ghita, L. Brateman, and M. Arreola. 2011. Characterization of a commercially-available, optically-stimulated luminescent dosimetry system for use in computed tomography. *Health Physics* 101:299–310.
37. Lekovic, G. P., L. J. Kim, L. F. Gonzalez, A. Bice, F. C. Albuquerque, and C. G. McDougall. 2008. Radiation exposure during endovascular procedures. *Neurosurgery* 63:ONS81–ONS85; discussion ONS85–ONS86.
38. Griglock, T. 2012. *Determining Organ Doses from Computed Tomography Scanners using Cadaveric Subjects.* Biomedical Engineering, University of Florida, Gainesville, FL.
39. Mueller, J. W., D. J. Vining, A. K. Jones, D. Followill, V. E. Johnson, P. Bhosale, J. Rong, and D. D. Cody. 2014. JOURNAL CLUB: In vivo CT dosimetry during CT colonography. *AJR. American Journal of Roentgenology* 202:703–710.

Part III

Applications for 1D, 2D, and 3D dosimetry

1D plastic scintillation dosimeters for photons and electrons

Bongsoo Lee and Kyoung Won Jang

Contents

12.1 INTRODUCTION

Modern radiotherapy dosimetry depends on the accuracy of radiation delivery to the target volume in the field of large dose gradients and requires high-precision dosimeters in its dose measurements. Dosimeters should therefore be small enough for high spatial resolution in their dose measurements and have tissue-equivalent characteristics for accurate measurement without the need for complex calibration processes [1–3]. Also, radiotherapy dosimetry requires multidimensional dose measurement in real time [3]. In these respects, the plastic scintillation dosimeter (PSD) has a sufficiently small sensitive volume to obtain high measurement resolution in regions of high dose gradients; it also has water- or tissue-equivalent characteristics to avoid complex conversions arising from material differences [4,5]. Therefore, PSD can offer reproducibility, linear responses to dose and dose rate, energy independence, absolute dose measurements, relative dose measurements, and resistance to radiation damage [6–8]. Additionally, a multidimensional PSD consisting of PSD arrays can provide multidimensional dose distributions of therapeutic radiation beams in real time [9].

For the past two decades, various studies for one-dimensional (1D) PSDs have been conducted. Flühs et al. developed a 16-channel PSD array system based on plastic scintillators, optical fiber light guides and a multichannel photomultiplier tube (PMT) for eye plaque dosimetry [10]. In the research, they reported that the PSD array system is successfully employed for dosimetric treatment optimization, and this system allows an online measurement unlike film or TLD dosimetry. The multichannel PMT system is well described in the work of Liu et al. [11]. According to their research, this kind of PMT provides a response to meet the demands of future developments in treatment delivery. Archambault et al. [5] introduced a water-equivalent 10-channel PSD array system constructed with scintillating fibers coupled to clear optical fibers and a charge coupled device (CCD) camera for dose measurement in external beam radiotherapy. They addressed that the array system allows precise, rapid dose evaluation of small photon fields. As a photodetector, the CCD camera can be used for measuring multidimensional scintillation signals simultaneously due to its imaging ability. In this research, they reported that the actual field of view of the camera used in their study could accept more than 4000 scintillating fiber detectors simultaneously. The enhanced PSD array-CCD system was fabricated by Lacroix et al. [12] In their research, the array consists of a linear array of 29 PSDs embedded in a water-equivalent plastic sheet coupled to optical fibers used to guide optical photons to a CCD camera. Also, they reported that the array system provides excellent dose measurement reproducibility (0.8%) in-field and good accuracy (1.6% maximum deviation) relative to the dose measured with an IC10 ionization chamber.

In Chapter 12, three types of 1D PSDs will be treated in details except for the dosimeters described above. One is a 1D PSD for electron beam therapy dosimetry, and the other is an integration-type 1D PSD to measure percentage depth doses (PDDs) for therapeutic photon beams. Finally, the third is a T-shaped 1D PSD for small field radiation therapy dosimetry.

12.2 1D PSD FOR ELECTRON BEAM THERAPY DOSIMETRY

In this section, we will introduce a 1D PSD that consists of 10 PSDs and a polymethyl methacrylate (PMMA) phantom for electron beam therapy dosimetry. Here, each PSD is made up of an organic scintillator and a plastic optical fiber (POF). In order to convert light signals to electrical signals, photodiode arrays are employed. In general, this type of photodetector is compact, cheap, and easy to make in array form. Using the 1D PSD, the intensities of scintillating light are measured at two field sizes and two levels of electron beam energy. In addition, isodose and PDD curves are obtained for two different energy beams of a clinical linear accelerator (CLINAC).

12.2.1 CONSTRUCTION OF 1D PSD

The POFs used for 1D PSD are step-index multimode fibers (GH4001, Mitsubishi Rayon, New York). The outer diameter and the cladding thickness of these fibers are 1.0 and 0.01 mm, respectively. The refractive indices of the core and the cladding are 1.49 and 1.402, respectively, and the numerical aperture (NA) is 0.504. NA denotes the light-gathering power; more light can be captured using a POF with a higher NA.

An organic scintillator synthesized with polyvinyl-toluene and wavelength-shifting fluors was used as a detector probe of the PSD. In this research, a commercially available organic scintillator (BCF-20, Saint-Gobain Ceramic & Plastics, Malvern, PA) was exploited as the sensor probe of the PDS. The BCF-20 emits green light with a peak wavelength of 492 nm. The decay time and the amount of emitted photons of BCF-20 are 2.7 ns and about 8000 photons/MeV, respectively.

As a light-measuring device, a photodiode (S1336-18BK, Hamamatsu Photonics, New Jersey)-amplifier system was used. The measurable wavelength range of the system is from 320 to 1100 nm, with a peak wavelength of 960 nm. The quantum efficiency (QE) of a photodiode is commonly expressed in percent (%) and has the following relationship with the photosensitivity (S) at a given wavelength (λ) [13]:

$$QE\ (\%) = \frac{S \times 1240}{\lambda} \times 100 \qquad (12.1)$$

where:
S is the photosensitivity in A/W at a given wavelength
λ is the wavelength in nm

The S of the photodiode used in this study was about 0.26 A/W at the emission peak of BCF-20 (492 nm) and the QE of the photodiode was about 65.5%. To obtain a large signal-to-noise ratio with simple electronics, a homemade photodiode amplifier system was used. The amplification ratio was variable with the resistors that are used in the amplifier circuit; the maximum amplification ratio was about 300 times.

In this research, the 1D PSD system was fabricated via the following processes. First, the surfaces of both a 10 mm length organic scintillator and a 10 m length POF were polished with various kinds of polishing pads in a regular sequence. Second, the scintillator was glued with an optical epoxy to the 10 m length POF. Third, the surface of the PSD probe was surrounded by reflector paint (BC-620, Saint-Gobain Ceramic & Plastics, Malvern, PA) based on titanium oxide (TiO_2) to increase the scintillating light-collection efficiency and to intercept light noise from outside. Finally, PSD probes were embedded in PMMA to make a 1D detector array; these probes were connected to the photodiode array using POFs.

12.2.2 EXPERIMENTAL SETUP USING 1D PSD

Figure 12.1 shows the experimental setup for measuring scintillating light using the 1D PSD with the photodiode-amplifier system, and for measuring Cerenkov radiation (or light) using the PSD and the POF as a dummy fiber irradiated by the high-energy electron beam from the CLINAC. To measure Cerenkov radiation, the detector and the POF were aligned along the CLINAC's axis of rotation so as to locate them at the center of the electron beam. Depending on the design, a CLINAC provides multiple numbers of electron energies with a typical range of 4–20 MeV. In this study, two electron beams (6 and 12 MeV) from the Varian CLINAC® 2100C/D were used. The field sizes of the electron beams were 3 cm × 3 cm and 6 cm × 6 cm.

12.2.3 EXPERIMENTAL RESULTS OBTAINED BY 1D PSD

Generally, the intensity of Cerenkov radiation generated in an optical fiber depends on the field size of the irradiation beam, the refractive index of the optical fiber material, and the incident angle between the electron beam and the optical fiber. The dominant factor for deciding the intensity of the Cerenkov radiation is the incident angle. Theoretically, the maximum Cerenkov transmission in PMMA occurs at an angle of 47.7° for 6 MeV electron beams [14]. The angle of maximum Cerenkov transmission increases only slightly as the electron energy increases above 4 MeV. It also increases as the refractive index of the medium increases.

The measured intensity of Cerenkov radiation was highest at an incident angle of 50° and lowest at an incident angle of 90°, as shown in Figure 12.2. In general, the intensity of the Cerenkov radiation is

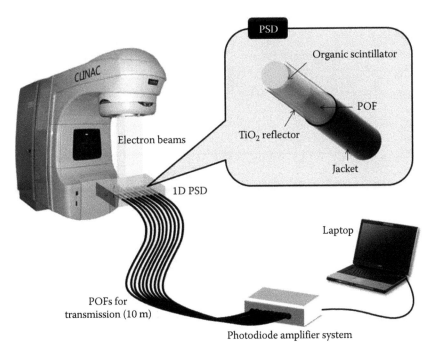

Figure 12.1 Experimental setup for measuring dose distributions of electron beams.

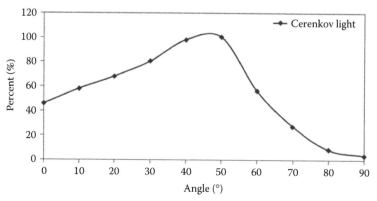

Figure 12.2 Angular dependences of Cerenkov radiation measured from a POF with 20 cm × 20 cm field size of 6 MeV electron beam.

insignificant when the angle between the incident electron beam and the dosimeter is perpendicular [15]. In addition, since the influence of Cerenkov radiation is reduced in small-size fields, the influence of Cerenkov radiation on the process of detecting scintillating light with 1D PSD was ignored throughout this study.

Figure 12.3 shows the measured scintillating light signals from the 1D PSD array with different electron energies and irradiation field sizes. In this experiment, PSDs were placed at a depth of 1 cm of the PMMA phantom. The measured light signals show almost identical values and are distributed uniformly over a field size of 6 cm × 6 cm for 6 and 12 MeV electron-beam energies. The results for the 3 cm × 3 cm field sizes, which can be seen in Figure 12.3a and b, show that due to the reduced field size, only six detectors can measure distinct signals.

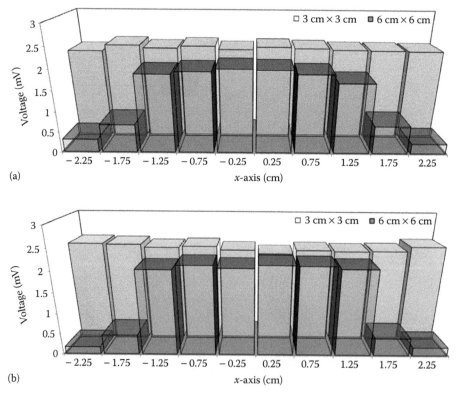

Figure 12.3 Measured scintillating light of a 1D PSD with 3 cm × 3 cm and 6 cm × 6 cm field sizes of two different energy electron beams: (a) 6 MeV and (b) 12 MeV electron beams.

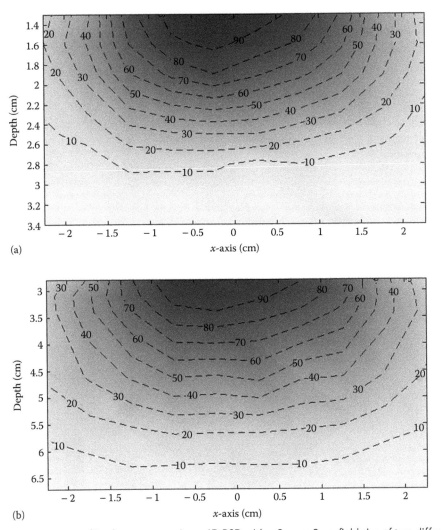

Figure 12.4 Measurements of isodose curves using a 1D PSD with a 3 cm × 3 cm field size of two different energy electron beams: (a) 6 MeV and (b) 12 MeV electron beams.

Figure 12.4 shows measured isodose curves for two different energy beams using the 1D PSD. Generally, isodose curves describe the dose distributions in two dimensions, and are particularly useful tools for representing dose distributions for specific combinations of field size and beam energy. The 10 PSDs were placed with gaps of 5 mm on the PMMA phantom; then, the intensities of scintillating light were measured at different depths of the phantom. The irradiation field size was 3 cm × 3 cm; the source-surface distance (SSD) was 100 cm. For the 6 MeV electron beam, as can be seen in Figure 12.4a, all the isodose curves show some expansion; only the low isodose levels bulge out for the 12 MeV electron beam, as can be seen in Figure 12.4b [15].

Figure 12.4 also shows that the depth of maximum dose increases with increase in the energy of the incident electron beams; the tissue penetration depths are about 3 and 6 cm for the 6 and 12 MeV electron beams, respectively. The penetration depth can be roughly calculated with the following expression [16]:

$$\text{Penetration depth (cm)} = \frac{\text{Electron energy (MeV)}}{2 \text{ (MeV/cm)}} \tag{12.2}$$

Compared with the isodose curves of a photon beam, the field size for an electron beam expands rapidly below the surface of a phantom because electrons scatter rapidly as the electron beam penetrates the medium.

Applications for 1D, 2D, and 3D dosimetry

In clinical practice, the central axis dose distribution is characterized by PDD, which can be defined as the quotient of the absorbed dose at any depth t (d_t) to the absorbed dose at a fixed reference depth to (d_{t0}). PDD can be expressed as

$$\text{PDD (\%)} = \frac{d_t}{d_{t0}} \times 100 \qquad (12.3)$$

For high-energy electron beams, the reference depth is usually taken at the position of the peak absorbed dose. The peak absorbed dose on the central axis is called *maximum depth dose* (d_{max}), and is defined by the following expression:

$$d_{max} = \frac{d_t}{\text{PDD}} \times 100 \qquad (12.4)$$

Figure 12.5 shows two-dimensional dose distributions obtained from profile measurements at multiple depths in the 3 cm × 3 cm field for 6 and 12 MeV electron beams. As given in Equation 12.3, PDD is the

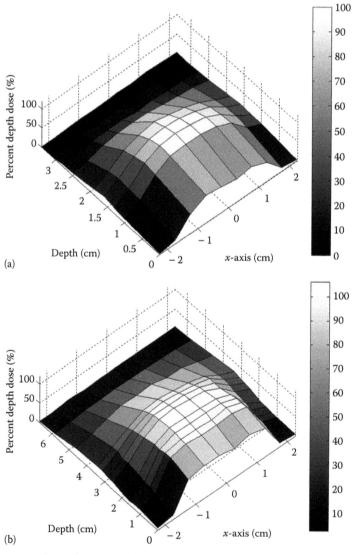

Figure 12.5 Measurements of two-dimensional dose distributions in the 3 cm × 3 cm field for two different energy electron beams: (a) 6 MeV and (b) 12 MeV electron beams.

ratio of the dose at a certain depth to the dose at a fixed reference depth, expressed as a percentage; this value varies according to SSD, energy, and field size. It has been observed that typical depths of maximum dosage along the central axis for a 10 cm × 10 cm field in a water phantom are about 1.3 and 2.8 cm for 6 and 12 MeV electron beams, respectively [16].

In Figure 12.5, the value of d_{max} along the central axis is 1.3 cm for the 6 MeV electron beam. For the 12 MeV electron beam, the value of d_{max} is in the range of 2–3 cm. It is more difficult to find d_{max} for higher electron beam energies, as the dose builds up more slowly and over a longer distance at higher energies.

12.2.4 SUMMARY

A 1D PSD using organic scintillators, POFs, and a PMMA phantom for electron beam therapy dosimetry was presented in this section. The 1D beam distributions in a PMMA phantom with two different energies and field sizes of electron beams were measured using the1D PSD. Also, isodose curves and two-dimensional dose distributions for high-energy electron beams were obtained using the 1D PSD. Through this 1D PSD, it was possible to obtain many reasonable results in terms of penetration depths, isodose curves, and PDDs. These results are consistent with well-known data that have been measured using conventional dosimeters in radiotherapy.

12.3 INTEGRATION-TYPE 1D PSD FOR THERAPEUTIC PHOTON BEAM DOSIMETRY

In this section, an integration-type 1D PSD to measure PDDs for therapeutic photon beams will be introduced. The integration-type 1D PSD consists of a PMMA phantom, square-type organic scintillators, and square-type polystyrene (PS) clear waveguides. And, a complementary metal–oxide–semiconductor (CMOS) is used to measure the scintillating light output of the integration-type 1D PSD. PDDs for 6 and 15 MV photon beams were measured with the integration-type 1D PSD; the results are compared with those obtained using an ionization chamber.

12.3.1 MATERIALS FOR INTEGRATION-TYPE 1D PSD

In this research, nine PSDs were fabricated using square-type organic scintillators and square-type POFs. The square-type organic scintillators (BCF-12, Saint-Gobain Ceramic & Plastics, Malvern, PA), comprising PS with a small amount of wavelength-shifting fluors, were used as the sensor probes of the PSDs. The scintillators have a rectangular parallel-piped shape and 1 mm × 1 mm × 5 mm dimensions, with a core/cladding structure similar to that of optical fiber. The materials of the core and the cladding are PS and PMMA, respectively. It was found that typical organic scintillators, which are based on PS, do not show significant aging for absorbed doses up to 1.0 kGy.

Commercial grade square-type POFs (BCF-98, Saint-Gobain Ceramic & Plastics, Malvern, PA) were used to guide scintillating light from the organic scintillators to the distal end of an integration-type 1D PSD. The outer dimensions of these fibers are 1 mm × 1 mm × 150 mm. The refractive indices of the core and the cladding are 1.60 and 1.49, respectively, and the NA is 0.583. The materials of the core and the cladding are the same as those of BCF-12.

In order to transmit the scintillating light generated from the 1D PSD to a light measuring device, round-type POFs (GH4001, Mitsubishi Rayon, New York) were employed throughout this study. The outer diameter of the POFs is 1.0 mm and the cladding thickness is 0.01 mm. The refractive indices of the core and the cladding are 1.492 and 1.402, respectively, and the NA is 0.510.

As a light measuring device, a CMOS (EO-5012C, Edmund Optics, Barrington, NJ) having 2560 × 1920 pixels was used to measure scintillating light from the 1D PSD. The size of each pixel in the CMOS is 2.2 μm × 2.2 μm; the size of the total sensing area is 5.6 mm × 4.2 mm.

12.3.2 FABRICATION OF INTEGRATION-TYPE 1D PSD AND EXPERIMENTAL SETUP

Figure 12.6 shows the structure of the PSD array. In order to minimize scintillating light interference between the PSDs, a black polyvinyl chloride (PVC) film having 0.1 mm thickness was employed as a septum. Each PSD was composed of an organic scintillator (BCF-12) and a clear waveguide (BCF-98). The BCF-12 was glued with an optical epoxy onto a BCF-98; the surfaces of both the BCF-12 and the BCF-98 were polished with various kinds of polishing pads in a regular sequence.

Figure 12.7 shows the structure of the integration-type 1D PSD; also shown is the experimental setup. The integration-type 1D PSD is composed of nine PSDs in a PMMA block. The size of the PMMA block is 20 cm × 20 cm and the thickness is 10 mm. Also, subminiature-type A 905 (SMA 905) connectors are used to connect the PSDs and the 20-m length of POFs, which deliver light signals from the integration-type 1D PSD to the CMOS. As radiation sources, throughout this study, 6 and 15 MV photon beams were provided by a CLINAC (Varian CLINAC® 2100C/D). The field size of the 6 and 15 MV photon beams used in these experiments was 10 cm × 10 cm.

12.3.3 EXPERIMENTAL RESULTS OBTAINED BY INTEGRATION-TYPE 1D PSD

Figure 12.8 shows the reproducibility of the integration-type1D PSD. In this experiment, the standard deviations of scintillation light signals were measured five times from the nine channels of the 1D PSD. The average and maximum standard deviations of all measurements were 0.22% and ±0.79%, respectively.

Figure 12.9 shows PDD curves for 6 and 15 MV photon beams using the integration-type 1D PSD and an ionization chamber. Generally, the PDD curve for a high-energy photon beam increases sharply to the maximum dose (D_{max}) depth, which is referred to as the build-up region due to the increasing amount of scattered radiation. The curve then decreases slowly beyond the build-up region with increased depth of PMMA phantom.

In this study, it was impossible to measure the PDD in the region of 0–9 mm using the ionization chamber due to that chamber's relatively large volume. In general, the D_{max} depths of 6 and 15 MV photon beams are 15 and 30 mm, respectively, in a 10 cm × 10 cm field size [15]. The PDDs obtained using the integration-type 1D PSD showed good agreement with those obtained using the ionization chamber.

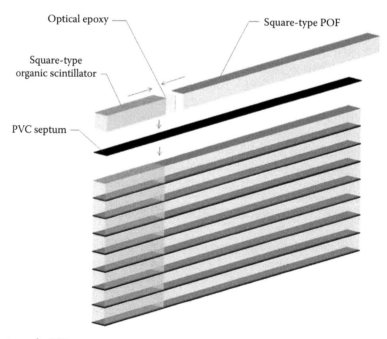

Figure 12.6 Structure of a PSD array.

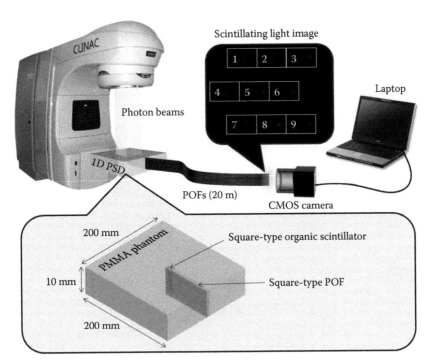

Figure 12.7 Structure of an integration-type 1D PSD and experimental setup.

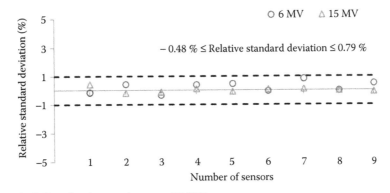

Figure 12.8 Reproducibility of an integration-type 1D PSD.

The mean differences between the results of the integration-type 1D PSD and those of the ionization chamber for 6 and 15 MV photon beams were about 0.49% and 0.74%, respectively.

12.3.4 SUMMARY

In this section, we have reported an integration-type 1D PSD that consists of nine PSDs, and septa and PMMA blocks to measure relative depth doses for therapeutic photon beams. The nine PSDs were composed of square-type organic scintillators and square-type POFs and were arrayed vertically in a PMMA block; the rectangular structure of these PSDs enables a minimization of the air-gap among the closely packed optical fibers, thereby avoiding dose measurement errors arising from air-gaps. Black PVC films were used as septa to minimize the cross-talk between the PSDs. Using the integration-type 1D PSD, PDDs for 6 and 15 MV photon beams were measured and the results were compared with those obtained using an ionization chamber. As for the results, the PDDs obtained using the integration-type 1D PSD showed good agreement with those formed using the ionization chamber.

Applications for 1D, 2D, and 3D dosimetry

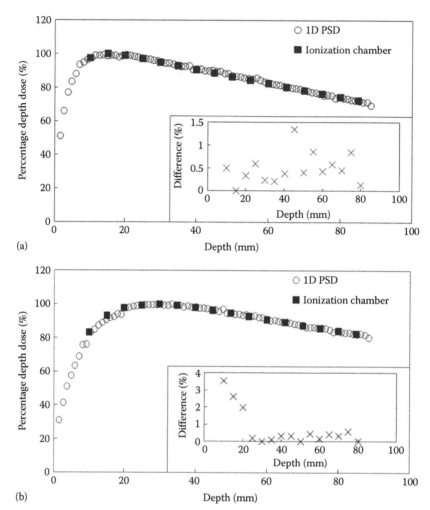

Figure 12.9 Measured PDDs for 6 and 15 MV photon beams using an integration-type 1D PSD and an ionization chamber: (a) 6 MV and (b) 15 MV.

12.4 T-SHAPED 1D PSD FOR SMALL FIELD RADIATION THERAPY DOSIMETRY

In this section, we will consider a T-shaped 1D PSD that is constructed by arraying PSDs with a T-shape in a PMMA phantom. In order to obtain light signals generated from the T-shaped 1D PSD, a CMOS camera with a lens system was employed. Using the T-shaped 1D PSD, the transverse and longitudinal dose distributions were simultaneously measured with small field sizes.

12.4.1 FABRICATION OF T-SHAPED 1D PSD

Throughout this study, square-type organic scintillators and square-type POFs were selected as sensing elements and waveguides of the T-shaped 1D PSD. The physical properties of the scintillators and the POFs were the same as those described in Subsection 12.3.1.

Figure 12.10 illustrates the structure of the T-shaped 1D PSD. To fabricate a PSD, an organic scintillator with a length of 5 mm was polished and connected with optical epoxy to the distal end of a POF whose length was 100 mm. PSDs (50 ea.) were then constructed in a T-shape on PMMA blocks. The gaps between the PSDs were coated using black PVC thin films, as septa, to minimize the scintillating light interference among the closely packed square-fibers due to the cross-talk effect. The outer surface of the T-shaped 1D PSD was covered with a black PVC film to intercept the ambient light noise from the

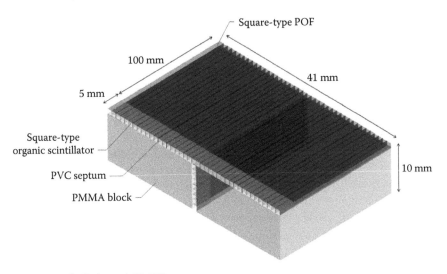

Figure 12.10 Structure of a T-shaped 1D PSD.

measuring environment. In addition, extra PMMA blocks were also used as a stopping material; these blocks were placed at both sides and the bottom of the 1D PSD to provide charge particle equilibrium with respect to the surface of the scintillating fibers.

12.4.2 EXPERIMENTAL SETUP AND DATA ACQUISITION

The experimental setup employed for measuring the scintillating light distribution using the T-shaped 1D PSD is shown in Figure 12.11. In order to guide the scintillating light image from the 1D PSD to a CMOS image camera (EO-5012C, Edmund Optics, Barrington, NJ), a lens system was constructed using a double convex lens (DCX 0504, Daeduk Optical Instruments, Daejeon, South Korea) and a telephoto lens (smc-PENTAX-M 1:2 85 mm, Asahi Optics, Tokyo, Japan). The material of the double convex lens was BK7 grade A optical glass. The diameter and the thickness of a double convex lens were 50 and 5.4 mm, respectively, and the focus length of this lens was 250 mm. In the case of the 85 mm telephoto lens, it was mounted with a CMOS camera using a PK-C camera adapter-ring (KIPON K6517, Shanghai Transvision Photographic Equipment, Shanghai, China); the aperture of the telephoto lens was set to F 4.0 when taking pictures. As can be seen in Figure 12.11, the double convex lens and the CMOS camera having a telephoto lens were equipped on carriers on an optical rail (DOI-POR-1000, Daeduk Optical Instruments, Daejeon, South Korea). In this lens system, the distance L_1 between the back end of the fiber-optic phantom and

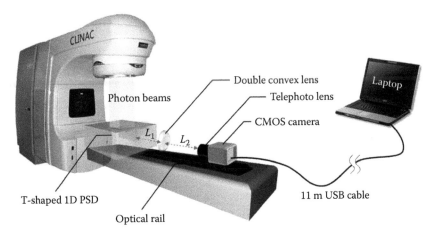

Figure 12.11 Experimental setup using a T-shaped 1D PSD.

the curved surface of double convex lens was 10.5 cm and the distance L_2 between the curved surface of the double convex lens and the CMOS element of the camera was fixed at 90 cm. To generate scintillating light, the center of the T-shaped 1D PSD was accurately located at the isocenter of a CLINAC (CLINAC® 1800, Varian Medical Systems, Palo Alto, CA) system using a laser alignment system. Therefore, the total distance between the distal end of the organic scintillator and the CMOS element of the camera was about 111 cm. In this experiment, a geminated shielding box, which was made up of aluminum (outside) and lead (inside), was used to protect the CMOS camera from the effects of scattered radiation.

In this research, high-resolution and real-time scintillating light images were obtained using the T-shaped 1D PSD according to the field size and different depths of a PMMA phantom irradiated by a 6 MV photon beam of a CLINAC. The SSD was set to 100 cm as the condition for calibrating the CLINAC; the beam field-sizes were set to 1 cm × 1 cm and 3 cm × 3 cm, similar to those used in stereotactic radiosurgery. The scintillating light image generated by interactions between a 6 MV photon beam and an organic scintillator array in a 1D PSD was magnified using a double convex and telephoto lens system; the image was converted to an electric signal using a CMOS camera. The output image signal generated from the CMOS camera was transmitted using an 11 m universal serial bus (USB) cable; this was connected to a desktop computer in a control room. Finally, the intensities of the scintillating lights were analyzed to obtain the dose distributions and the beam profile. A video file having many frames is read and converted to red–green–blue (RGB) color image files. Second, the RGB scale of the image files is changed to gray scale. Third, the positions of each scintillating light generated from the organic scintillator array are selected and the light intensities are calculated. Finally, the light intensity values obtained from many images are calculated to find the average.

In this experiment, the beam profile was measured on one side from the center of the T array because the cross-beam profile generally has a symmetrical distribution over a given beam field due to the radially symmetric conical flattening filter in the CLINAC. However, the entire cross-beam profile, which includes both sides from the center of the T-shaped 1D PSD, is also measurable by changing the lens array. The images of the scintillating light distribution obtained by the T-shaped 1D PSD can be easily and effectively used for calibrating the conditions of a high-dose photon beam in radiosurgery. Furthermore, the transverse dose distributions at specific depths, including that of a skin layer, can be obtained using these scintillating light images. In this study, PDD according to the depth variation was also obtained using the average value of the light intensity.

12.4.3 EXPERIMENTAL RESULTS OBTAINED BY T-SHAPED 1D PSD

The beam profile and the beam edge according to the field size are shown in Figure 12.12. When the PMMA-stacked phantom with 15 mm thickness was placed on the T-shaped 1D PSD, scintillating light images of the transverse fiber-optic array about the beam profile were obtained according to two beam field-sizes of 1 cm × 1 cm and 3 cm × 3 cm. Generally, cross-beam characteristics are affected by the radially symmetric conical flattening filter; a beam transmitted through the flattening filter in the treatment head is flattened by differentially absorbing photons. Therefore, the measured light signals are distributed uniformly over the given field-sizes as shown in Figure 12.12.

For accurate cross-beam measurements, the detector size is very important. In the off-axis region, the measured light signal shows a fairly sharp falloff, but it does not show a perfectly perpendicular beam edge because it is affected by the penumbra, the scattered radiation, and the size of the detector. If the detector size is smaller than the proposed dosimeter, the slope of the measured beam edge is steeper than that shown in Figure 12.12 [17,18].

Figure 12.13 shows the field flatness and the transverse dose distribution of 3 cm × 3 cm field size according to the depth. The beam field flatness changes with depth. At 15 mm depth, the measured light signals show similar output values. However, the transverse dose distributions gradually become uneven with increasing depth of the PMMA phantom, as shown in Figure 12.13. This phenomenon is attributed to either an increase in the scattering of the primary dose ratio in accordance with increasing depth of the human body or phantom and decreasing incident photon energy off-axis. Also, the measured scintillating light signals around the beam edge also have a gentle slope with increasing depth due to their location in the penumbra region.

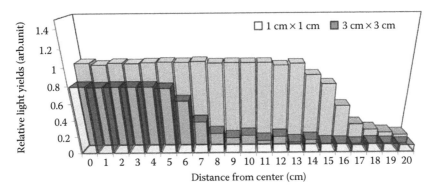

Figure 12.12 Beam profile and beam edge according to the field size variation at 15 mm depth.

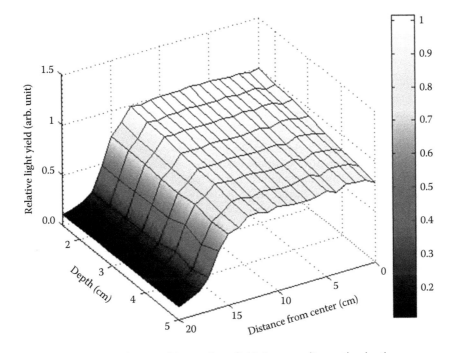

Figure 12.13 Transverse dose distribution of 3 cm × 3 cm field size according to the depth.

The PDDs that were measured using the T-shaped 1D PSD with a 6 MV photon beam are shown in Figure 12.14. In this test, the maximum PDD values were found to be at a depth of approximately 15 mm for the 6 MV photon beam in both cases [4,7,19]. However, the PDDs for the 1 cm × 1 cm beam field have relatively high fluctuation due to the unstable beam conditions of the CLINAC.

In order to maximize the therapeutic gain and to minimize radiation hazard, the skin dose is an important factor to consider during radiosurgery treatment. Therefore, the dose distribution near the skin surface should be measured precisely prior to radiosurgery treatments. To measure the skin dose, the dosimeter size is very important and the output signals of the dosimeter are normally affected by the thickness of the dosimeter. In general, the skin dose measured using radiochromic film is lower than that measured using a scintillating fiber-based dosimeter or a small ionization chamber due to the relatively small sensing volume of radiochromic film. It is known that a fiber-optic dosimeter using an organic scintillator with a diameter of 1 mm leads to a skin dose for a 6 MV photon beam that is higher than 40% [4,19,20].

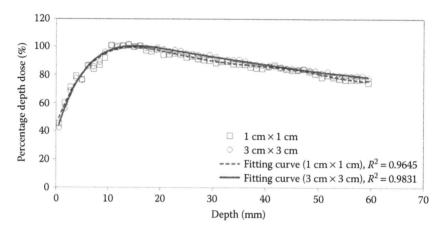

Figure 12.14 Measurement of percentage depth doses with 1 cm × 1 cm and 3 cm × 3 cm field sizes.

12.4.4 SUMMARY

In this section, a novel T-shaped 1D PSD system using square-type PSDs, a lens system, and a CMOS image camera was reported. In order to test the fabricated T-shaped 1D PSD, the transverse and longitudinal dose distributions were simultaneously measured with small field sizes such as 1 cm × 1 cm and 3 cm × 3 cm. In terms of experimental results, the measured beam profiles were found to gradually become uneven with increasing depth of the PMMA phantom; also, the beam edge had a gentle slope with increasing depth. In addition, the maximum dose values for a 6 MV photon beam with field sizes of 1 cm × 1 cm and 3 cm × 3 cm were found to occur at a depth of approximately 15 mm; the PDDs for both field sizes were nearly in agreement at the skin dose level.

12.5 CONCLUSION

In this chapter, we reported three types of 1D PSDs. These types of dosimeters have many advantages over conventional dosimeters in radiation therapy. First, water-equivalent organic scintillators, POFs, and PMMA phantoms make it possible to measure the dose distribution with minimal perturbation. Second, the small sensitive volume of these dosimeters can contribute to dose distribution measurements with high resolution. Third, 1D multichannel measurements systems can be used for quality assurance purposes for electron beam therapy with reduced time. Especially, the 1D dose measurement systems reported in this chapter allow us to obtain dose distributions in a very short time. Fourth, real-time measurements not only help in saving time but also provide chances for real-time quality assurance feedback. Fifth, there should be no corrections of the values of temperature, pressure, or humidity for accurate dose measurements in the PSD array. The 1D PSDs proposed in this chapter can be expected to be effective for measuring dose distributions (i.e., isodose curves) and PDDs in radiotherapy dosimetry.

REFERENCES

1. Frelin A-M, Fontbonne J-M, Ban G, Colin J, Labalme M, Batalla A, Isambert A, Vela A, Leroux T. Spectral discrimination of Cerenkov radiation in scintillating dosimeters. *Med. Phys.* **32**: 3000–3006;2005.
2. White T O. Scintillating fibers. *Nucl.Instr. and Meth. Phys. Res. A* **273**: 820–825;1988.
3. Beddar AS. Plastic scintillation dosimetry and its application to radiotherapy. *Radiat. Meas.* **41**: S124–S133;2007.
4. Beddar A S, Kinsella T J, Ikhlef A, Sibata C H. A miniature "Scintillator-Fiberoptic-PMT" detector system for the dosimetry of small fields in stereotactic radiosurgery. *IEEE Trans. Nucl. Sci.* **48**: 924–928;2001.
5. Archambault L, Beddar A S, Gingras L, Lacroix F, Roy R, Beaulieu L. Water-equivalent dosimeter array for small-field external beam radiotherapy. *Med. Phys.* **34**: 1583–1592;2007.
6. Beddar A S, Mackie T R, Attix F H. Water-equivalent plastic scintillation detectors for high-energy beam dosimetry: I. Physical characteristics and theoretical considerations. *Phys. Med. Biol.* **37**: 1883–1900;1992.

7. Beddar A S, Mackie T R, Attix F H. Water-equivalent plastic scintillation detectors for high-energy beam dosimetry: II. Properties and measurements. *Phys. Med. Biol.* **37**: 1901–1913;1992.

8. Jang K W, Yoo W J, Moon J, Han K T, Park J-Y, Lee B. Measurements of relative depth doses and Cerenkov light using a scintillating fiber-optic dosimeter with Co-60 radiotherapy source. *Appl. Radiat. Isot.* **70**: 274–277;2012.

9. Lee B, Jang K W, Cho D H, Yoo W J, Kim H S, Chung S-C, Yi J H. Development of one-dimensional fiber-optic radiation sensor for measuring dose distributions of high energy photon beams. *Opt. Rev.* **14**: 351–354;2007.

10. Flühs D, Heintz M, Indenkämpen F, Wieczorek C, Kolanoski H, Quast U. Direct reading measurement of absorbed dose with plastic scintillators—The general concept and applications to ophthalmic plaque dosimetry. *Med. Phys.* **23**: 427–434;1996.

11. Liu P Z Y, Suchowerska N, Abolfathi P, McKenzie D R. Real-time scintillation array dosimetry for radiotherapy: The advantages of photomultiplier detectors. *Med. Phys.* **39**: 1688–1695;2012.

12. Lacroix F, Archambault L, Gingras L, Guillot M, Beddar A S, Beaulieu L. Clinical prototype of a plastic water-equivalent scintillating fiber dosimeter array for QA applications. *Med. Phys.* **35**: 3682–3690;2008.

13. Hamamatsu Photonics Co. *Selection guide of Si photodiodes*. Hamamatsu, Japan: Hamamatsu Photonics Co., 2006.

14. Lee B, Cho D H, Jang K W, Chung S-C, Lee J-W, Kim S, Cho H. Measurements and characterizations of Cerenkov light in fiber-optic radiation sensor irradiated by high energy electron beam. *Jpn. J. Appl. Phys.* **45**: 7980–7982;2006.

15. Hendee W R, Ibbott G S, Hendee E G. *Radiation therapy physics*. 3rd ed. Hoboken, NJ: John Wiley & Sons, 2005.

16. Khan F M. *The physics of radiation therapy*. 2nd ed. Baltimore, MD: Williams & Wilkins, 1994.

17. Lee B, Jang K W, Cho D H, Yoo W J, Shin S H, Kim H S, Yi J H et al. Measurement of two-dimensional photon beam distributions using a fiber-optic radiation sensor for small field radiation therapy. *IEEE Trans. Nucl. Sci.* **55**: 2632–2636;2008.

18. Petric M P, Robar J L, Clark B G. Development and characterization of a tissue equivalent plastic scintillator based dosimetry system. *Med. Phys.* **33**: 96–105;2006.

19. Fontbonne JM, Iltis G, Ban G, Battala A, Vernhes J C, Tillier J, Bellaize N et al. Scintillating fiber dosimeter for radiation therapy accelerator. *IEEE Trans. Nucl. Sci.* **49**: 2223–2227;2002.

20. Letourneau D, Pouliot J, Roy R. Miniature scintillating detector for small field radiation therapy. *Med. Phys.* **26**: 2555–2561;1999.

2D plastic scintillation dosimetry for photons

Mathieu Guillot and Anne-Marie Frelin

Contents

13.1 MOTIVATION FOR THE USE OF PLASTIC SCINTILLATORS FOR 2D DOSIMETRY

Two-dimensional (2D) detectors are of great importance for megavoltage photon radiation therapy because they provide dosimetric data that are closely correlated to the 2D fluence of the incident beams. These detectors can be used to measure basic characteristics of beams such as symmetry, flatness, positions and speeds of multileaf collimator (MLC) leaves, dose output of irregular fields, or to assess the quality of treatment plans through comparative analysis of dose maps [1–4]. They therefore play a role in a wide range of situations, including the commissioning of treatment-planning systems, the periodic quality assurances (QAs) of treatment machines, and the patient-specific QAs.

Four types of 2D detectors are currently available in radiation therapy clinics: ion chamber arrays, silicon diode arrays, electronic portal imaging devices (EPIDs), and radiochromic films [3]. Ion chamber arrays are composed of hundreds of air-filled or liquid-filled ion chambers uniformly distributed into a plane of a water-equivalent phantom [5]. Arrays of air-filled ion chambers are characterized by volume averaging effects over a length of about 5 mm and are commonly used for measuring simple open fields and patients' treatment plans in intensity modulated radiation therapy (IMRT). Higher spatial resolution can be obtained with ion chambers filled with liquid isooctane [6]. In this case, the sensitive volumes have sizes of about 2 mm and are nearly water equivalent, making the liquid ion chambers more suited for small field dosimetry. As ion chamber arrays, diode arrays are also composed of hundreds of detectors providing online readout, which is advantageous for verifying the synchronicity of treatment parameters in dynamic fields. The width of the active area of silicon diodes typically varies between 0.6 and 2 mm, which is well adapted for obtaining high spatial resolution measurements in small fields and high dose gradients regions [7]. Diode arrays are however characterized by relatively large dead regions between the individual detectors where radiation dose is not directly detected. EPIDs are primarily designed for imaging purposes but several methods have been developed for using them for 2D and 3D dosimetry [8,9]. They typically consist of an inorganic scintillator sheet of high atomic number that is coupled to an array of silicon amorphous photodiodes. EPIDs provide online measurements with submillimetric spatial resolution, but their response depends on the energy spectrum of the primary and scattered photons, which changes with the field size and the off-axis distance. Because of their water equivalence and submillimetric spatial resolution, radiochromic films are well suited for dosimetry of small fields and high dose gradients. They consist of radiosensitive monomers dissolved in a transparent polymer sheet [10]. The interaction of ionizing radiation with films triggers the polymerization of the monomers, which changes the absorption spectrum of the films according to a nonlinear relation with dose.

Despite their many advantages, current 2D radiation detectors also possess limitations. The most important one being that they contain materials that have densities and interaction cross sections strongly different of those of water and organic tissues. The path length of the particles and the energy transferred to the medium are therefore perturbed when the particles pass through many detectors. This limits the number of detectors that can be placed in a given volume and can cause inaccuracies in situations of charged particle disequilibrium. Films, on the other hand, cannot provide online readout and strict calibration procedures, which are time consuming, are necessary to obtain accurate dose measurements [11]. Overall, 2D detectors would be better adapted to advanced external beams radiation therapy techniques if real-time, water-equivalent, and angular independent systems were available.

In this context, the use of plastic scintillators represents a promising avenue because of their unique dosimetric properties (see Chapter 4). Different types of 2D plastic scintillation detectors (PSDs) have been developed up to date and they can be classified into three categories: arrays of PSDs, arrays of long scintillating fibers, and plastic scintillation sheet detectors. The purpose of this chapter is to discuss the conception and applications of these three types of implementations.

13.2 ARRAYS OF PSDs

Arrays of PSDs are an extension of the technology of PSDs where many detectors are imbedded into a water-equivalent phantom. The development of such systems allows increasing the range of applications of scintillation dosimetry but also brings new challenges of design and fabrication. Three prototypes of 2D arrays of PSDs have been presented in the literature to date. Guillot et al. developed, in 2011, a detector array for IMRT dosimetry, and Gagnon et al. and Yoo et al. both developed, in 2012, detector arrays adapted to small field dosimetry [12–14]. The Sections 13.2.1 through 13.2.3 will discuss these three prototype systems.

13.2.1 ACQUISITION SETUPS

13.2.1.1 Description of 2D arrays

The T-shaped array developed by Yoo et al. was designed to measure simultaneously the dose profile and the percentage depth dose of radiation fields smaller than 40 mm width [14]. The array was composed of 50 PSDs placed side-by-side to form a vertical line of detection of 10 mm and a horizontal line of detection of 41 mm. The PSDs were oriented horizontally and two acrylic blocks were used as supports for the fibers. The sensitive volumes of the PSDs consisted of square-type scintillating fibers, 5 mm in length and 1 mm in width, emitting in the blue part of the optical spectrum. Square optical fibers of a length of 10 cm were coupled to the scintillating fibers to guide the scintillation outside the phantom. The motivation for using square-type fibers instead of round-type fibers in the T-shaped array was to avoid the presence of air gaps between the closely packed PSDs. To isolate the PSDs from each other and from external light sources, the sides of each fiber were covered with opaque polyvinyl chloride thin films. The PSDs were overall highly water equivalent for photon beams of megavoltage energy: the core of the fibers being in polystyrene and the optical cladding being in acrylic.

The purpose of the cross-shaped array developed by Gagnon et al. was to enable the simultaneous measurement of the in-line and cross-line dose profiles of small radiation fields used for stereotactic radiosurgery [13]. A photograph of the prototype is shown in Figure 13.1. The detector array was composed of 49 PSDs inserted vertically into a 1 cm thick plastic water slab with a distance of 1.3 mm between the centers of the PSDs. Each PSD consisted of a cylindrical polystyrene scintillating fiber of 1 mm diameter and 1 mm length to which an optical fiber with a core in polymethyl methacrylate (PMMA) was coupled. The emission spectrum of the scintillators was in the green spectral region. The PSDs were covered by a thin layer of black acrylic paint for optical isolation. The phantom of the cross-shaped array was specially designed to encapsulate

Figure 13.1 The cross-shaped array (Reproduced from Gagnon, J.-C. et al. Dosimetric performance and array assessment of plastic scintillation detectors for stereotactic radiosurgery quality assurance. *Med. Phys.* 39: 429–436; 2012. With permission)

the PSDs in a very compact manner. The structure of the phantom consisted of two horizontal plastic water slabs of an area of 75×60 mm^2, joined together by a vertical plate and spaced vertically by a distance of 20 mm. This structure in two planes allowed bending of the optical fibers below the detection plane without causing excessive mechanical stress to the fibers. The length of optical fiber directly exposed to the radiation beam was therefore strongly reduced during measurements, as was the intensity of the Čerenkov radiation. The 2D PSDs array was designed to be fixed to the support of a motorized water tank and be immersed in water during measurements. Doing so enabled to move the cross-shaped array with a submillimetric precision. This capability was used to increase the number of points of measurements in the dose profiles and to position the array with high precision relative to the radiation beam axis.

The 2D array developed by Guillot et al., shown in Figure 13.2, was primarily designed for IMRT dosimetry [12,15]. It was composed of 781 PSDs inserted vertically into a 3 cm thick plastic water slab, which was fixed into a water tank made of acrylic walls. The PSDs consisted of cylindrical polystyrene scintillating fibers of 1 mm diameter by 3 mm length, emitting in the blue spectral region, which were coupled to Eska Premier optical fibers (Mitsubishi Rayon Co., Ltd., Tokyo, Japan). These optical fibers have a core of PMMA and were specifically chosen because their low spectral attenuation enables highly accurate corrections of the Čerenkov effect when the chromatic removal technique is used [16]. The PSDs were distributed on a regular square grid of 26×26 cm^2, with a spacing of 10 mm, except for two perpendicular lines of detectors centered on the detection plane where the spacing was 5 mm. The PSDs were isolated from external light sources by a thin layer of black acrylic paint. The detector was placed on the treatment couch during measurements and could be irradiated from any direction over a 360° arc. A computed tomography scan showed that the IMRT array was highly homogeneous and water equivalent despite the presence of hundreds of PSDs and their associated collecting optical fibers.

13.2.1.2 Photodetection systems

The photodetection systems of the three arrays of PSDs discussed in Section 13.2.1.1 were all based on the use of charge-coupled devices (CCDs) or complementary metal–oxide–semiconductor (CMOS) cameras as photodetectors and of photographic lenses for collecting the light fluxes released by the PSDs. The performances of these systems are mainly determined by two elements. The first is to optimize the collection efficiency of the light fluxes to maximize the signal-to-noise ratio of the doses measured. The light released by a PSD can be modeled as a divergent cone whose angle is determined by the numerical aperture of the optical fiber and by a Gaussian intensity distribution in the transverse plane. The light

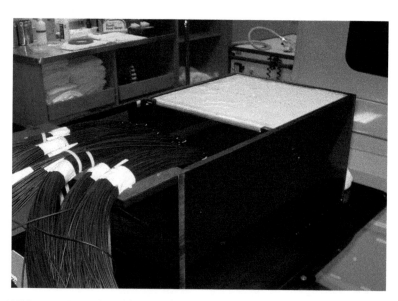

Figure 13.2 The IMRT array. (Reproduced from Guillot et al. A new water-equivalent 2D plastic scintillation detectors array for the dosimetry of megavoltage energy photon beams in radiation therapy. *Med. Phys.* 38: 6763–6774; 2011. With permission.)

collection efficiency can therefore be maximized by minimizing the distance between the proximal end of the optical fiber and the photographic lens, and by orienting the optical fiber so as to point toward the center of the lens. The capability to reduce the fibers-lens distance is however limited by the increase of the magnification of the image produced, which constitutes a challenge when many PSDs are imaged. Therefore, a compromise has to be established between the number of PSDs that can be imaged and the maximum magnification that can be obtained for a given set of parameters of the photodetection system: diameter and focal length of the lens, area of the camera sensor, and so on. The second important point is to have a photodetection system capable of real-time readout to allow efficient and fast analysis of the doses measured. This entails using strategies such as pixel binning and using sensors with interline architectures to reduce the readout time of the image frames and maximize the acquisition frequency.

The photodetection system of the T-shaped array was composed of a double convex lens placed at 10.5 cm from the ends of the optical fibers and a telephoto lens mounted on the CMOS camera and placed at a distance of 100.5 cm from the optical fibers. The diameter of the double convex lens was 50 mm and its focal length was 250 mm. The purpose of the double convex lens was to modify the trajectories of the divergent light cones to focus them toward the telephoto lens. The telephoto lens had a diameter of 59 mm and its function was to form an image, on the sensor of the camera, of the light fluxes released by the PSDs. The photodetection system was placed inside the treatment vault during measurements and the CMOS camera was shielded from ambient radiation inside the room with lead bricks. The parameters of the photodetection system of the T-shaped array were set to read in real time, the doses measured by about 30 of the 50 PSDs, which was sufficient to obtain the depth doses and the half profiles with a single acquisition.

The photodetection system used with the cross-shaped array was originally developed by Lacroix et al. for a 1D array of PSDs [17]. The system consisted of a 16-bit polychrome CCD camera on which was mounted a photographic lens of 6 mm diameter. The distance between the proximal ends of the optical fibers and the lens was 15 cm. The system had the capacity to image simultaneously up to about 2500 PSDs while the coupling efficiency of the photographic lens was between 0.05% and 4.5% depending on the number of PSDs. The light fluxes were separated into blue and green components for correcting the Čerenkov effect with the chromatic removal technique [18–20]. The color separation was done by analyzing separately the grey levels of the color pixels. A dose of about 200 cGy was delivered per acquisition and an image processing algorithm was capable to provide the measured doses few seconds after each acquisition.

The photodetection system of the IMRT detector array was composed of a holder for the optical fibers, a dichroic beam splitter, and two interline CCD cameras. The proximal ends of the 781 optical fibers were fixed compactly into a support of 5×10 cm^2 to be imaged. The light fluxes released by the PSDs were divided in a reflected part and a transmitted part using the dichroic beam splitter and both parts were then directed toward a monochrome CCD camera. A photographic lens used with an iris opening of 36 mm diameter was fixed to each camera. A blue color filter and a green color filter were, respectively, placed in front of the photographic lens associated with the reflected and transmitted light for optimizing the optical spectrum measured. The optical path length between the optical fibers and the lenses was 32 cm. To maximize the light collection efficiency, the holes of the optical fibers holder were drilled at different angles, so that the axes of the fibers were pointing toward the center of the photographic lenses. It was calculated that about 12% of each light flux was collected by the photographic lenses. Measurements showed that the standard deviation of the doses measured was smaller than 1% for the doses greater than 6 cGy on average, over all the PSDs. The CCD cameras were operated at a rate of 1 acquisition per second, each producing images of 781 light spots.

13.2.2 METHODOLOGY

13.2.2.1 Image processing

The images acquired with the cameras consist of multiple light spots separated by dark pixels. Because the cameras are placed inside the treatment room and close to the phantom during measurements, the images acquired are biased by the interactions of ambient radiation with the sensor of the cameras, which produces localized and non-permanent hot pixels. The cameras are usually shielded to reduce the intensity and frequency of those events but it is nonetheless necessary to use a correction algorithm to restore the integrity of the images. Different reconstruction algorithms have been proposed in literature [12,21,22].

One accurate method is the temporal median filter, which consists of repeating at least three times an acquisition for the same situation and calculating the median image. This algorithm was used with the cross-shaped array and also with the IMRT array during the calibration process. Another effective method is the mask-filtering algorithm that was used with the IMRT array for the real-time applications. The mask-filtering algorithm is based on the assumption that the spatial distribution of a given light spot can be described by a reference spatial function unique to that spot, multiplied by an intensity factor. The reference function describing a spot is determined using the temporal filtering algorithm during the calibration process. The mask-filtering algorithm is well suited when the spatial resolution of images is low. For example, this may be the case when the pixels of the sensor are binned during digitalization to reduce the size of the images and consequently the readout time of frames.

A second important element of image processing is the characterization and correction of the background component. Background images consist of images acquired with the same exposure time than during measurements, but without exposition to light. The gray levels obtained in these images are the result of the dark current and the dark noise of the pixels, and the readout noise associated to the digitization of the charges. The dark current and the dark noise are thermal effects that are minimized by controlling the temperature of the camera sensor using a cooling system. Usually, multiple background images are acquired at the beginning of the measurement session and the average or median background image is subtracted of the subsequent images in order to obtain accurate measurements of the intensity of the light spots. The analysis of the background images can also reveal the presence of dead or defective pixels, which can be recognized by either being systematically saturated, insensitive to light, or exhibiting strong fluctuations. One effective approach for dealing with defective pixels and improving the accuracy of measurements consists in ignoring those pixels during the quantitative analysis of the images.

13.2.2.2 Correction of the Čerenkov effect

The accuracy of the doses measured with PSDs is mainly determined by the accuracy of the method used to correct the Čerenkov effect. Correcting the Čerenkov effect is particularly challenging for arrays of PSDs because, unlike systems with a single PSD, the optical fibers are fixed inside the phantom. The user has, therefore, no control on how the optical fibers are irradiated and thus extreme situations occur frequently where the Čerenkov light is the dominant component of the luminous intensity. One example of such situation is when a section of an optical fiber is exposed to the primary field while the scintillator is in the out-of-field region. Another example is when the array is irradiated from different directions because the intensity of Čerenkov light collected in optical fibers greatly depends on the angle between the fibers' axis and the beam axis.

The chromatic removal technique [18,19] was used with the cross-shaped array and the IMRT array. The chromatic method enables discrimination between the scintillation intensity and the Čerenkov intensity, by measuring the light intensity in two different spectral regions, usually the blue and green regions. The decomposition of the light fluxes into color components can be performed directly on the sensor of a polychrome CCD camera, as is the case with the cross-shaped array, or by using a dichroic beam splitter as is the case with the IMRT array. The chromatic removal technique requires the determination of two calibration factors for each PSD of the array: a gain factor and a Čerenkov correction factor. The novel procedure described by Guillot et al. was used for determining the calibration factors of the cross-shaped array and the IMRT array [16]. The first calibration measurement consists of directly irradiating the optical fibers in such a way to maximize the production of Čerenkov light and produce nearly no scintillation. The second calibration irradiation is the inverse. It consists in focusing the radiation on the scintillators to maximize the production of scintillation and minimize the production of Čerenkov light. In the case of the cross-shaped array, the gain factors of the PSDs were determined by irradiating the array at its center with a 6 MV photons field of size 10×10 cm^2. The doses at the PSDs positions were measured with an ion chamber. In the case of the IMRT array, the gain factors of the PSDs were determined by irradiating simultaneously five consecutive rows of detectors with a rectangular field of size 15×40 cm^2. The array was then shifted longitudinally and five other rows of PSDs were irradiated as the previous ones. In total, five irradiations were necessary to cover all the PSDs. The doses deposited at the detectors positions were measured with an ion chamber in the same phantom as the array.

13.2.3 APPLICATIONS AND RESULTS

13.2.3.1 Small field dosimetry

One criterion for classifying a field as small in dosimetry is when the lateral charged particle equilibrium (CPE) is lost on the central axis of the beam due to a lack of lateral scatter caused by the narrowness of the beam. Two properties are necessary for obtaining accurate dose measurements in this situation. First, because the penumbra extends over the entire or a large portion of the field, the sensitive volume of the detector should be small enough to avoid volume averaging effects. Second, the detector should ideally be water equivalent because densities smaller or greater than water will artificially decrease or increase the lateral CPE and thus modify the dose deposited [23,24]. Arrays of PSDs are therefore well adapted to small field dosimetry because of their water equivalence and high spatial resolution.

Yoo et al. have characterized the response of the T-shaped array by measuring the percentage depth doses and the cross-line profiles of a 6 MV photon beam of sizes 1×1 and 3×3 cm^2 [14]. The profiles were measured at the depths of 15, 30, and 50 mm. The results showed that the T-shaped array can measure the changes of the beam flatness, the field size, and the width of the penumbra as a function of depth. For both field sizes, the depth of maximum dose found was close to 15 mm, in agreement with the reference value. This indicates that the depth dose of the beam was not perturbed by the presence of the 10 PSDs on the vertical detection line. It was also determined that the PSDs overestimate the surface dose by about 40% relative to radiochromic films because the volumes of the scintillating fibers were too large. Yoo et al. concluded that systems similar to the T-shaped array could be used for measuring beam profiles and percentage depth doses of small radiation fields prior to carrying out stereotactic radiosurgery treatments.

The response of the cross-shaped array was characterized by Gagnon et al. by measuring the dose profiles of a 6 MV photon beam of circular shapes of 4, 10, and 40 mm diameters used for radiosurgery [13]. Measurements taken with radiochromic films and an unshielded silicon diode were used as references because of their submillimetric spatial resolution. The gamma tests between the cross-shaped array and the radiochromic films, performed with the criteria (2%, 0.3 mm), showed that all the PSDs agreed with the films for the circular collimators of 4 and 10 mm and only two PSDs failed to meet this criteria for the 40 mm collimator. The gamma passing rates for the comparison between the cross-shaped array and the unshielded silicon diode was 100% for all the field sizes studied. Figure 13.3 shows the in-line dose profiles measured with the three detectors for the 4 mm circular field shape. No major volume averaging effect was observed for the cross-shaped array.

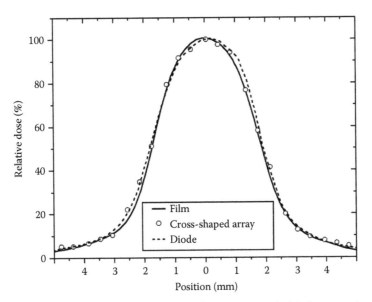

Figure 13.3 Dose profiles of a 4 mm diameter radiosurgery beam measured with the cross-shaped array, a radiochromic film, and a silicon diode.

13.2.3.2 Irradiations from different beam directions

Two elements can contribute to the angular dependence of detector arrays. The first is the intrinsic angular response of the individual detectors. In the case of PSDs, no evidences exist concerning an angular dependence in plastic scintillators because the production of scintillation is a two-step process that involves first, the energy transfer from the radiation to the solvent, and then, from the solvent to the scintillating molecules. However, the intensity of Čerenkov light collected in an optical fiber strongly depends on the angle between the beam axis and the fiber axis. Inaccuracies in the Čerenkov correction can therefore cause angular dependence. The second element is the perturbation of the fluence of particles by the nonwater-equivalent materials present inside the phantom, which depends on the beam direction. The use of water-equivalent materials for the conception of detector arrays is therefore critical.

Guillot et al. have characterized the effect of changes of beam direction on the response of the IMRT array by measuring the doses deposed by a 6 MV photon beam of size 10×10 cm^2 delivered at different gantry angles, from 0° to 180° by step of 10° [12]. The IMRT array was placed on the treatment table and its detection plane was centered relative to the isocentre of the treatment machine. The measured dose distributions were compared to the doses calculated with a treatment-planning system (TPS) by performing gamma tests with tolerances of 3% and 3 mm. The results showed gamma passing rates between 98.2% and 99.6% for the incidences between 0° and 40°, an average passing rate of 99.7% for the incidences between 50° and 120°, and passing rates between 97.8% and 98.9% for the incidences between 160° and 180°. The passing rates were calculated in the high-dose region (doses $\geq$ 10% of maximum dose). A slight decrease in the gamma passing rates was observed when the radiation beam was partly parallel to the fibers axes. This effect was however considered as weak because it corresponded to a decrease of 2–3 PSDs passing the test over a total of about 120 PSDs located in the high-dose regions. Based on the results found, the authors concluded that the IMRT array could be used without angular correction factors when the standard tolerances used in IMRT of 3% and 3 mm were applied [12]. Besides, because the dose distributions calculated with the TPS were for a water medium, it was also concluded that the dose distributions were not perturbed by the presence of the hundreds of PSDs inside the phantom.

13.2.3.3 IMRT dosimetry

The capability of the IMRT array to accurately measure dose distributions produced by intensity modulated beams was assessed by delivering ten step-and-shoot IMRT plans and one volumetric modulated arc therapy plan [15]. Eight of the step-and-shoot plans were for treatments of head and neck cancers and the three other plans were for prostate cancers. The plans were all delivered at the planned gantry angles with the detector array on the treatment table. Each step-and-shoot plan had seven incidences. The doses measured with the IMRT array were compared to the doses calculated with TPSs and to radiochromic films using gamma analysis and dose profiles comparisons. For the 70 step-and-shoot IMRT beams, the results of the gamma analysis between the IMRT array and the TPS showed passing rates (average and standard deviation) of 95.8% ± 2.1% for the criteria 3% and 2 mm, 97.8% ± 1.2% for the criteria 4% and 2 mm and 98.1% ± 1.3% for the criteria 3% and 3 mm. The passing rates were calculated in the high-dose regions. On average, the absolute doses measured were in agreement within 0.6% with the planned doses in the regions of high doses and dose gradients smaller than 0.5%/mm. For the arc therapy plan, the gamma passing rates 3% and 3 mm were 97.2% for the comparison with the TPS and 98.1% for the comparison with the radiochromic film. Figure 13.4 shows the 2D dose distributions and the in-line and cross-line dose profiles obtained with the IMRT array and the TPS for one of the step-and-shoot IMRT plans. The dose maps were interpolated to a resolution of 1 mm from the coordinates of the PSDs for a better visualization. Overall, the results showed excellent qualitative and quantitative agreements between the doses measured and calculated. Guillot et al. concluded that the accuracy of the IMRT array was sufficient to respect the tolerance criteria recommended by the AAPM TG-119 report for IMRT [25].

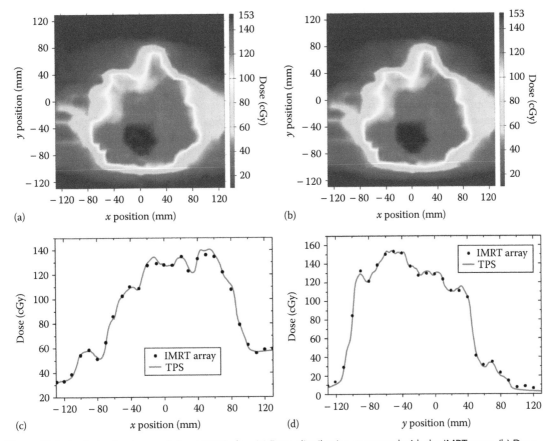

Figure 13.4 Analysis of a step-and-shoot IMRT plan: (a) Dose distribution measured with the IMRT array. (b) Dose distribution calculated with the TPS. (c) Dose profiles along the line $y = -20$ mm. (d) Dose profiles along the line $x = -20$ mm.

13.3 ARRAYS OF LONG SCINTILLATING FIBERS

Scintillating fibers of a few tens of centimeters in length can be used as one-dimensional dose integrators because the amount of scintillation released is a function of the total energy absorbed along the fiber. Two prototypes of 2D arrays of long scintillating fibers have been developed by Goulet et al. The first, in 2010, for monitoring the fluence of beams during IMRT treatments, and the second, in 2012, for IMRT dosimetry applications [26,27]. Compared to arrays of PSDs, the advantages of using long scintillating fibers are that detector arrays can be constructed with fewer detectors, while the dose measurements have higher spatial resolution because the scintillating fibers are spaced from each other in only one direction. This section will discuss the conception and applications of these two types of detectors.

13.3.1 FLUENCE DETECTOR

13.3.1.1 Acquisition setup

The fluence detector is a transmission detector placed between the treatment beam and the patient [26]. Its purpose is to monitor, in real time during patients' treatments, the incident fluence produced by the gap between each pair of opposite MLC leaves. The fluence detector was composed of 60 plastic scintillating fibers of 1 mm diameter and 27 cm length inserted parallel to each other into a grooved acrylic plate of a thickness of 3.2 mm. The lateral spacing between the fibers axes was 3.2 mm for the 40 central fibers and 6.4 mm for the 20 external fibers. The fibers were aligned to match the position and direction of motion of the 60 leaf pairs of a MLC. Both ends of each scintillating fibers were coupled to plastic optical fibers to guide the scintillation produced toward a monochrome CCD camera. The ends of the optical fibers were imaged using a 50 mm diameter photographic lens fixed to the camera. Both the acrylic plate and the camera were fixed to an aluminum frame, which was inserted into the accessory mount of the treatment machine. The source-to-detector distance was 66 cm. Similar fluence detectors based on ion chambers have been proposed in the literature [28–30].

13.3.1.2 Methodology

Because the optical attenuation of scintillating fibers is high, the same dose absorbed in a fiber will generate different light outputs depending on the position of production. Spatial information on the incident fluence can therefore be obtained by analyzing the ratio of the light intensities measured at the two ends of each scintillating fiber. Goulet et al. have developed a mathematical model to describe the production and transport of scintillation in long scintillating fiber detectors. In this model, the light collected at each side (+/−) of a long scintillating fiber detector can be expressed as

$$I_{\pm} = C_{\pm} \cdot \int_{-L/2}^{L/2} \Phi(x) \cdot e^{\pm \mu \cdot x} \cdot dx \qquad (13.1)$$

where:

C is a constant that account for the transmission efficiency of the optical chain and the scintillation efficiency

L is the length of the scintillating fiber

$\Phi(x)$ is the radiation fluence incident at the position x on the fiber

μ is the optical attenuation coefficient of the scintillating fiber

Equation 13.1 assumes a homogeneous scintillation efficiency and a homogeneous optical attenuation with the fiber. From Equation 13.1, Goulet et al. have shown that the one-dimensional integral fluence, Φ, of a measured field, and the central position of the incident fluence, X_c, can be decoupled and expressed as

$$X_c = \frac{1}{2\mu} \cdot \ln\left(\frac{I_+}{I_{N+}} \cdot \frac{I_{N-}}{I_-} \right) \qquad (13.2)$$

$$\Phi = \Phi_N \cdot \sqrt{\frac{I_+}{I_{N+}} \cdot \frac{I_-}{I_{N-}}} \qquad (13.3)$$

where:

N refers to a normalization field centered at $x = 0$ and of known fluence Φ_N

The fluence detector provides therefore two independents parameters that can be compared to the expected values for detecting fluence errors.

Equation 13.1 requires the knowledge of the mean optical attenuation coefficient of the scintillating fibers. The procedure used to measure this factor consists in irradiating all the scintillating fibers simultaneously at different longitudinal positions with a narrow rectangular field of 0.5 or 1.0 cm wide.

One advantage of long scintillating fiber detectors is that the impact of the Čerenkov effect on measurements is sufficiently small to be neglected. This can be explained by the fact that the optical fibers used to guide the scintillation are never exposed to the radiation beam. In addition, the scintillating fibers are always irradiated at normal incidence. The Čerenkov light produced in the scintillating fibers is therefore always proportional to the scintillation and no angular effect related to the capture of the Čerenkov emission cones in the scintillating fibers is observed.

13.3.1.3 Applications and results

Goulet et al. have conducted a series of measurements to determine the sensitivity of the fluence detector to different types of leaf position errors and the effect of the detector on the treatment beam. To determine the sensitivity of the detector to leaf position errors, the intrinsic variation of the detector response was first quantified as a function of the light output collected. These variations are primarily caused by the uncertainties associated to the random process of light detection of the CCD camera (dark, readout, and shot noises). The fluence detector was irradiated with leaf openings between 0.5 and 120 mm and with one to five monitor units (MUs). The results showed that for a leaf opening of 0.5 mm and five MUs, the standard deviation (SD) of the X_c parameter was inferior to 1 mm and the SD of the Φ parameter was inferior to 0.2%. Smaller SDs were obtained for larger leaf gap openings and MUs. The values obtained were used to quantify the smallest fluence errors that could be detected in a statistically significant manner. The sensitivity threshold for errors detection was set to five SDs from the measured values (X_c, Φ) of the error free fields. Leaf position errors of different amplitudes were then voluntary introduced to the segments of step-and-shoot IMRT fields. The results showed that single leaf position errors and single leaf bank position errors of ≥ 1 mm could be detected by analyzing the Φ parameter while leaf pair translations of ≥ 2 mm and translations of both leaf banks of ≥ 1 mm could be detected by analyzing the X_c parameter.

The effect of the fluence detector on the treatment beam was also studied by comparing the doses measured with ion chambers in water, with and without the fluence detector placed in the accessory mount of the medical linear accelerator. The results showed a transmission factor of 98% and no changes greater than 0.5% in the percentage depth dose of a 6 MV photons beam for the depths greater than the depth of maximum dose. An increase of the surface doses of 4%, 11%, and 21%, caused by electrons ejected from the acrylic plate, was observed respectively for the field sizes 5×5, 10×10, and 20×20 cm^2. Overall, the results showed that the fluence detector was among the systems perturbing less the treatment beam compared to other fluence detectors or conventional accessory trays. Goulet et al. concluded that the fluence detector could play an important role in adaptive radiation therapy when the QA of patients' treatment plans cannot be done prior to treatments.

13.3.2 TOMODOSIMETER

Goulet et al. have developed a system that produces tomographic reconstructions of 2D dose distributions from multiple dose projections acquired with an array of long scintillating fibers [27]. Unlike the arrays of PSDs, the spatial resolution of the tomodosimeter is not limited by the inter-detector spacing but rather depends on multiple factors, including the number of projections and the reconstruction algorithm. The main advantages of the tomodosimeter are its water equivalence and high spatial resolution, making it a suitable detector for characterizing dose distributions of irregular shapes and with high dose gradients, as encountered in IMRT.

13.3.2.1 Acquisition setup

The conception of the tomodosimeter was in part similar to the fluence detector. It was composed of 50 polystyrene scintillating fibers of 1 mm diameter inserted parallel to each other in the grooves of an opaque acrylic slab of circular shape. The scintillating fibers had a green emission spectrum and their length varied between 6 and 20 cm in order to form a circular detection zone of 20 cm diameter. The lateral spacing between the scintillating fibers was 3.2 mm for the 40 central fibers and 6.4 mm for the 10 external fibers. Both ends of each scintillating fiber were coupled to a plastic optical fiber to guide the scintillation to a photodetection system composed of a monochrome CCD camera and a photographic lens used with an iris opening of 25 mm in diameter. During the measurements, the detection plane was embedded at a depth of 5 cm in a water-equivalent cylindrical phantom of 30 cm diameter. The scintillating fibers were oriented horizontally and irradiated perpendicularly to their axes.

13.3.2.2 Methodology

The principle of tomodosimetry is similar to the one of computed tomography and positron emission tomography. It consists in the acquisition of multiple dose projections for which the angles between the scintillating fibers axes and the dose distribution orientation change. For example, in the case of the 2D tomodosimeter, the treatment couch was used to rotate the tomodosimeter around the beam axis. The dose distribution deposited in the detection plane was reconstructed by solving the following equation used for discrete tomography:

$$\vec{P} = A \cdot \vec{D} \tag{13.4}$$

where:
$\vec{P}$ is a vector containing the 1D dose integrals measured
$\vec{D}$ is the unknown vector containing the dose of each pixel of the 2D dose distribution
A is the projection matrix that describes the physical process of the dose projections

The length of the $\vec{P}$ vector is equal to the number of detectors, multiplied by the number of projections, and multiplied by two for the two sides of each detector. The dose integrals depend on the light intensities measured on either side of the detectors and can be expressed in terms of a reference field (ref) as

$$p_j = \frac{I_j}{I_{ref}} A_{j*} \cdot \vec{D}_{ref} \tag{13.5}$$

where:
$\vec{D}_{ref}$ is the vector containing the dose of each pixel of the reference field
I_{ref} is the light intensity measured for the reference field
A_{j*} represents the jth row of the projection matrix

The size of the reference field used by Goulet et al. was 14×30 cm^2 and the dose distribution was calculated by a treatment-planning system and validated with film and ion chamber measurements. The projection matrix describes the physics of the dose projections and depends on the properties of each detector. It can be expressed as

$$A_{j,i} = \sqrt{1 - \left(\frac{d_{j,i}}{r_f}\right)^2} \cdot \kappa(u_{j,i}) \cdot e^{\lambda_{\pm}(u_{j,i})} \tag{13.6}$$

where:
$d_{j,i}$ and $u_{j,i}$ are the radial and longitudinal distances between ith dose pixel and the scintillating fiber for the jth dose integral, respectively
r_f is the radius of the scintillating fiber
κ is this scintillation efficiency of the scintillating fiber
λ is the optical attenuation coefficient (– or + side) of the fiber

The square root term in Equation 13.6 is a geometric factor that is null when the dose pixel is outside the volume of the scintillating fiber. Further details about the projection matrix can be found elsewhere [27]. The scintillation efficiency and the optical attenuation of each fiber were determined by a calibration procedure similar to the one used with the fluence detector. Equation 13.4 was then solved using an iterative reconstruction algorithm.

13.3.2.3 Applications and results

Goulet et al. have characterized the response of the tomodosimeter by measuring the doses deposited by two square fields (5×5 and 10×10 cm^2) and seven segments of a 6 MV step-and-shoot IMRT beam. The detector plane was positioned at 100 cm from the source and 5 cm of water-equivalent buildup materials was used. Each tomography consisted in the acquisition of 18 dose projections with steps of $10°$. Eight monitors units were delivered per acquisition. The dose distributions were reconstructed to a resolution of 1×1 mm^2 and about 2 min of calculation were necessary per reconstruction on a single core 2.0 GHz processor. The doses measured were compared to the doses calculated with the treatment-planning system by performing gamma analysis, dose differences, and comparisons of dose profiles. For all the fields, except one, the gamma passing rates were greater or equal to 98.5%, 93.5%, and 87% for the tolerances of 3% and 3 mm, 2% and 2 mm, and 3% and 1 mm respectively. The gamma passing rates were calculated in the high-dose regions. Furthermore, the absolute dose differences were analyzed for the regions with doses greater than 90% of the maximum dose and with dose gradients inferior to 0.3%/mm and the results showed a mean dose difference smaller than 1.2%. Figure 13.5 shows a comparison between the dose profiles measured with the tomodosimeter and calculated with a TPS for a segment of a step-and-shoot IMRT beam. Excellent agreements can be observed in the plateaus and dose gradient regions, illustrating the high spatial resolution of the detector. One application targeted for the tomodosimeter would be to allow easy and quick acquisition of batches of 2D dose distributions for the commissioning of TPSs.

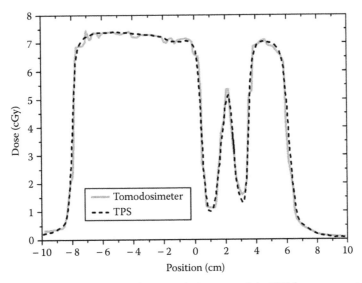

Figure 13.5 Dose profile comparison between the tomodosimeter and the TPS for a segment of a step-and-shoot IMRT beam.

13.4 PLASTIC SCINTILLATOR SHEETS

Other systems are based on plastic scintillator sheets (PSSs) permitting real-time 2D dose measurements of high spatial resolution without the granularity of PSD arrays. One of the first systems was proposed by Boon et al. in 1998 for proton beam control [31] and the use of this kind of device was quickly extended to electron and photon beams [32,33]. The dosimetry concerns are very different between proton and photon irradiations: quenching can occur in the first case [34,35] (Chapter 2), whereas Čerenkov radiation is produced in the second case. Only the latter case corresponding to the most spread studies will be treated here.

13.4.1 ACQUISITION SETUP

All the plastic scintillator systems are based on the same principal elements represented in Figure 13.6. The main component is a PSS of 1–5 mm thickness whose size generally ranges from 15 × 15 to 25 × 25 cm² to match common radiotherapy irradiation fields. The choice of the scintillator depends on various parameters, including its scintillation yield, the spectral sensitivity of the photodetector, the attenuation spectrum of the phantom, or the Čerenkov removing method. The use of EJ-200 or UPS 974R has been reported but the study of more scintillators for dosimetric applications (scintillation yield, emission spectrum, linearity, etc.) have been reported [36]. The scintillator is embedded in a tissue-equivalent and transparent phantom. Polystyrene or water in PMMA tanks has been tested and is perfectly suited to this application. It generally consists of two parts: a front part placed between the scintillator and the photodetector and blocks of different thickness situated on top of the scintillator to perform measurements at depth. Collomb-Patton et al. also used additional side blocks for particular experiments needing electron equilibrium at the edges of the phantom [37]. The light signal produced by the system is then transmitted to the photodetector by a mirror, keeping the photodetector outside of the irradiation field. The mirror can be placed under the system or can be integrated to the phantom for a better compactness. All the proposed 2D PSS systems use a cooled monochrome CCD camera to measure the light distribution. Their low dark current allows long integration times and they generally present very good spatial resolutions. On the bad side, they are sensitive to scattered radiation and in most of the studies, the camera is shielded with lead or a high-density glass lens. The camera lens is also of major importance to limit spatial distortions of the signal map with good light collection efficiency. The study of Collomb-Patton et al. has shown a huge improvement allowed by a high angular aperture objective compared to the standard camera lens used in the former prototype [33,37]. This objective contributed (with image processing) to increase the sensitivity

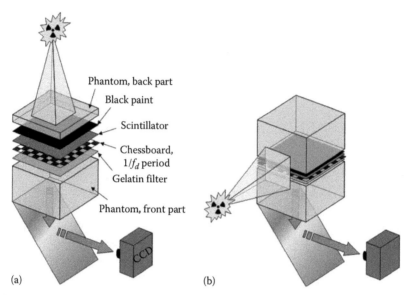

Figure 13.6 Schematic representation of 2D PSS detectors including the different elements (mask, gelatin filter, etc.): (a) the case of vertical irradiation and (b) the case of horizontal irradiation.

of the system by two orders of magnitude. Other elements such as a gelatin filter or a modulation mask are needed to process the signal and deconvoluate the different stem effects occurring in scintillation dosimetry. They will be described in Section 13.4.2. All the components must be optically coupled (with optical grease for instance) and the sides of the phantom must be black coated to reduce the multiple reflections at the interfaces and shelter the device from ambient light.

This setup can be used for various irradiation configurations: the scintillator plane perpendicular to the beam axis to measure dose profiles, or parallel to the beam axis for depth dose measurements. The system can also be vertically translated to provide 3D dose distributions.

13.4.2 METHODOLOGY

Like PSD arrays, 2D PSS dosimeters produce parasitic light besides scintillation, but in much larger quantities. Indeed, in vertical irradiation beams (Figure 13.6a), the volume of irradiated phantom is about 300 times larger than the volume of irradiated scintillator (~30 cm width vs. 1 mm width). Consequently, even if the Čerenkov yield is about 66 times smaller than the scintillation yield (30 photons/path mm vs. 2000 photons/path mm), Čerenkov intensities can be as large as three times the scintillation intensities. Moreover, the Čerenkov radiation intensity is so important that its lowest wavelengths generate additional scintillation not anymore proportional to the dose deposited by the irradiation beam in the scintillator. This parasitic phenomenon, which can reach 60% of the dose-related scintillation, is not observed in PSDs where the Čerenkov amount is moderate. At last, like in PSD arrays, the camera is close to the phantom, leading to numerous hot pixels due to scattered radiations reaching the sensor. In 2D PSS systems, each pixel of the CCD detector represents a position of the dose distribution map, so hot pixels must be accurately filtered. The processing of all these parasitic contributions will be described hereafter.

13.4.2.1 Hot pixels

The limitation of the fluency of scattered radiation reaching the CCD sensor is very important to preserve its lifetime. In the study of Collomb-Patton et al., a high-density glass lens permitted to decrease the proportion of hot pixels due to scattered radiation to less than 3% for 100 ms acquisition with a 2 Gy/min and 30 × 30 cm² irradiation field. It corresponds to a proportion of hot pixel of 3% for 100 ms acquisitions. The remaining hot pixels are an important issue in the light signal map acquisition, especially when an analysis in the frequency domain is needed (see Section 13.4.2.2.3), so an effective filtering is crucial. Most of the studies apply a filtering based on the multiplicity of the acquisition: five acquisitions of 40 MU in the case of Frelin et al. and continuous acquisitions of 100 ms in the case of Collomb-Patton et al. The hot pixels are then suppressed by comparing the successive acquisitions.

13.4.2.2 Čerenkov radiation

Once hot pixels have been removed from raw images, the Čerenkov issue can be treated. The Čerenkov light produced in 2D PSS systems presents various intensities and spatial distributions depending on the irradiation configurations. So, its discrimination must be robust and effective irrespective of the orientation and the beam size. The first step in suppressing the Čerenkov contribution from light signal consists in black painting the back of the scintillator. The effects of the Čerenkov radiation produced in the back part of the phantom are then cancelled in a very simple and effective way. The Čerenkov radiation produced in the front part of the phantom is, of course, of bigger issue and has been managed up to now by three different methods more or less effective.

13.4.2.2.1 Debluring method

In the study presented by Petric et al., the Čerenkov issue is not specifically addressed [38]. Čerenkov radiation is assimilated to scattering light resulting from the multiple reflections between the scintillator and the 45° mirror and handled with methods developed for portal imaging [39]. An antiscatter screen (microlouvre, light control screen, 3M Corporation, Minnesota, USA), consisting of thin parallel optical channels is placed below the scintillator to only transmit the scintillation signal emitted perpendicularly to scintillator plane. In addition, the image is deblurred with an asymmetric kernel fitting the point-spread function with multiple elliptical Gaussians. The method, not addressing the right physical process, is not totally effective in high gradient zone and does not correctly reproduce the scattered doses outside the irradiation field.

13.4.2.2.2 Chromatic method

The chromatic removal technique used in 2D PSS, first developed by Fontbonne et al., relies on the same principle as PSD arrays: the specific emission spectra of Čerenkov radiation and scintillation [18]. In 2D PSS detectors, the decomposition of the light signal into color components is performed, thanks to interferential filters placed in front of the camera lens, and the calibration factors (a gain factor and a Čerenkov correction factor) must be determined for each pixel of the signal map. The calibration procedure requires two measurements: one containing a significant proportion of scintillation compared to Čerenkov radiation and the other containing a small proportion of scintillation. The first configuration is obtained with a 30×30 cm^2 irradiation field delivered by a vertical beam. The second one is performed with a 5×30 cm^2 irradiation field delivered by a horizontal beam below the scintillator. For both configurations, the measurement is performed at a source-to-detector distance of 100 cm and a depth of 2.5 cm. The light signal is decomposed into two wavelength domains (blue and red or green and red) and the reference dose map is given by a dosimetric film placed above the scintillator. This procedure provides two independent relations between light intensities and dose, and the calibration coefficients can be determined for each pixel of the map. This discrimination method has been extensively tested by Frelin et al. [33] and has shown specific limitations for 2D PSS detectors due to the absorption spectrum of the polystyrene in the lower wavelengths leading to important and variable changes of the measured Čerenkov spectrum (see Section 13.4.3.2).

13.4.2.2.3 Subtractive method

Up to now, this is the most accurate method implemented in 2D PSS systems to suppress the Čerenkov contribution from the total light intensity map. It has been first implemented by Frelin et al, and then improved by Collomb-Patton et al. [33,37,40]. This method, comparable to the two-fibers method [41,42], consists in collecting two signals at close range: one containing the whole light signal (scintillation + Čerenkov radiation) and the other containing only Čerenkov radiation. The proximity of the two measurements guaranties consistent Čerenkov intensities between the two points. The application of this method to 2D PSS requires the implementation of a periodic modulating mask alternating transparent and black diagonal lines (or a chessboard in first version) between the scintillator and the front part of the phantom (see Figure 13.6). The transparent zones transmit the scintillation as well as the Čerenkov light, whereas black lines prevent the scintillation to be transmitted and measured. The modulation mask proposed by Collomb-Patton et al. presents a single frequency peak in the frequency space allowing the extraction of the scintillation map from the modulated signal thanks to Hilbert transform. This process has the advantage to be fast and efficient. A calibration factor is then needed to convert the scintillation map into dose map. The procedure providing this factor requires only one irradiation configuration and has been risen by Collomb-Patton et al. to a high level of accuracy. The configuration is roughly the same as the first chromatic calibration measurement but has been improved by adding three polystyrene side blocks to the phantom to ensure electronic equilibrium at the edges of the device and keep the dose variation within the irradiation field below 5%. A reference relative dose distribution $I_{film}^{30\times30}(x,y)$ is measured by a dosimetric film placed above the scintillator and a reference absolute dose D_{ref} is measured at the center of the field by an ionization chamber in the same conditions. Any dose distribution $D(x, y)$ can then be deduced from the scintillation distribution $I_{cam}(x, y)$ with the help of the following relation:

$$D(x,y) = I_{cam}(x,y) \times \frac{I_{film}^{30\times30}(x,y)}{I_{cam}^{30\times30}(x,y)} \times \frac{D_{ref}}{I_{film}^{30\times30}(0,0)} \tag{13.7}$$

Collomb-Patton et al. have shown that this absolute calibration presents uncertainty comparable to the one of ionization chamber (1%–2%). This discrimination method has been shown to be robust and independent on the irradiation configuration despite the large variations of Čerenkov production in terms of intensity and spatial distribution. Moreover, it has been shown that photon as well as electron measurements of various energies can be performed without additional calibration.

13.4.2.3 Parasitic scintillation filtering

The different methods of Čerenkov radiation discrimination, described in the previous paragraph, remove the Čerenkov contribution from the light signal. But as stated earlier, in 2D PSS detectors, a significant proportion of scintillation is produced by the lowest wavelengths of the Čerenkov light. As this parasitic signal is exactly similar to the dose-related scintillation (location, spectrum, etc.) it cannot be simply removed from the measured signal. Its generation must be prevented by a gelatin filter placed between the scintillator and the front part of the phantom. Its transmission spectrum accurately matches the scintillator emission spectrum transmitting a large part of scintillation while absorbing most of Čerenkov light. In the case of the UPS 974-R scintillator, the best compromise has been found with the filter *Fire* manufactured by the Rosco Co. that transmits 82% of the scintillation and absorbs 97% of the Čerenkov light [33].

13.4.2.4 Diffusions and reflections

The main diffusions and reflections occur at the interface between the different parts of the system (plastic/air/plastic). They can be drastically reduced by coupling the different parts with optical grease or glue and by painting in black the back of the scintillator and the edges of the phantom. It has been shown that when the Čerenkov radiation and the parasitic scintillation are correctly suppressed, the remaining signal intensity errors due to diffusion were around 1% [33].

13.4.3 APPLICATIONS AND RESULTS

Because of their promising characteristics (spatial resolution, water equivalence, and real-time measurement), 2D PSS devices have mostly been developed to perform IMRT dose map verifications. Their performances have been evaluated with homogeneous, wedge-filtered, and dynamic MLC IMRT irradiation fields.

13.4.3.1 Homogeneous irradiation fields

Performances of the different 2D scintillation dosimeters have been evaluated with flat field irradiations and have shown the strong impact of the device components (CCD camera, lens, etc.) as well as the processing applied to the various stem effects on the results.

13.4.3.1.1 Linearity

Linearity is a key parameter in dosimetry and was first evaluated for the 2D PSS by Petric et al. Irradiation fields ranging from 3×3 to 12×12 cm² were delivered by a vertical beam and the light signal was measured at a depth of 3 cm and a SAD of 100 cm, for integrated dose ranging from 5 to 200 MU.

The signal intensity at the center of the field was shown to increase linearly with the dose with a correlation factor better than 0.99 and deviations with the average intensity below 5%. In the study made by Collomb-Patton et al., dose linearity was tested with their 2D PSS device and ionization chamber measurements were performed in the same conditions. Irradiation fields of 5 × 5 cm² size were delivered by a vertical 15 MV beam for doses ranging from 20 to 100 MU. The comparison between both detectors results has shown discrepancies smaller than ±1% for doses higher than 50 cGy. Collomb-Patton et al. have also shown that dose linearity can be tested with only two measurements of an irradiation field showing large spatial variations. They used a 10 × 10 cm² field delivered by a 15 MV vertical beam for 100 and 200 MU. The variations in the dose distribution permit to cover different orders of magnitude of dose and the comparison of measurements, pixel by pixel, confirmed that the linearity was within ±1% for about 2 orders of magnitude, with a linear correlation factor of 0.999.

13.4.3.1.2 Precision and reproducibility

The stability of measurements over time is also very important because it contributes to define the precision of the system and the frequency of the calibration. Short- and long-term reproducibility of the signal were characterized by Petric et al. over a 6-month period. The short-term reproducibility was evaluated at the center of 10 × 10 cm² fields, at a depth of 3 cm and a SAD of 100 cm, for 50 MU. Variations of the results were found to have a maximal standard deviation of 1.5%. The long-term reproducibility was evaluated with 10 measurements performed over 6 months with 25 × 25 cm² irradiation fields at a depth of 10 cm, a SAD of 100 cm and for 200 MU. The mean signal measured at the field center was shown to be stable within 1.7%.

The precision and the reproducibility have also been accurately evaluated by Patton et al. Precision has been characterized by calculating local standard deviation for 10 × 10 cm² fields, at a depth of 3 cm and for doses ranging from a few cGy to 183 cGy. The standard deviation was shown to decrease from about 4.5% (at a few cGy) to well below 1% above 40 cGy. Reproducibility between measurements has been shown to be better than ±1% for cumulated doses higher than 50 cGy and better than ±2% above 20 cGy. This stability contributes to set the need of recalibration at a very low frequency: probably a few times over the lifetime of the device.

13.4.3.1.3 Beam orientation independence

Besides its large intensity, Čerenkov radiation spatial distribution strongly depends on the irradiation configuration and on the beam orientation. Hence, it is necessary to validate the angular independence of 2D scintillation dosimetry. Collomb-Patton et al. have shown that discrepancies between two identical 10 × 10 cm² irradiation configurations performed at orthogonal angles (so leading to very different Čerenkov generations) were within ±2%. The exact similarity between both irradiations, especially in terms of electronic equilibrium, was obtained thanks to additional polystyrene blocks placed on the scintillator and on the side of the phantom.

13.4.3.1.4 Linac quality assurance

Above performance characterization, the suitability of 2D PSS detectors to linac QA has been shown by Collomb-Patton et al. Their prototype has been confronted to the MatriXX ionization chamber array from IBA for standard beam profile verifications routinely performed in clinical QA. The flatness, the symmetry, and the penumbra of a flat irradiation field were calculated by both detectors at a depth of 1 cm and a DSP of 99 cm. If agreement between both detectors was good for symmetry, clear differences occurred for penumbra and flatness: penumbra measured with the 2D PSS was smaller than the ones measured with the MatriXX, and the flatness was less favorable with the 2D PSS than with the MatriXX. This can be explained by the limited spatial resolution of the ionization chambers averaging the dose measurements. Consequently, field characteristics measured by 2D PSS are probably more realistic.

13.4.3.2 Wedge-filtered irradiations

The evaluation of wedge-filtered irradiations allowed to highlight the specific limitations of the different systems and signal processing. All 2D PSS measurements of this paragraph were performed with wedge-filtered fields delivered by a vertical beam and compared to dose film measurement acquired in the same conditions by gamma analysis with tolerances of 3% and 3 mm.

Applications for 1D, 2D, and 3D dosimetry

Petric et al. measured the dose distribution delivered by an 8 × 8 cm² wedged field at a SAD of 100 cm, a depth of 3 cm, and for 200 MU. This configuration showed the weakness of the debluring discrimination method leading to an inaccurate reproduction of the shape of the profiles. Gamma tests were made for a zone encompassing the field edges by ±20% but did not meet the typical acceptance criterion of <3%. The Frelin et al. analysis of a 10 × 10 cm² wedged irradiation field provided the first evaluation of chromatic and subtractive Čerenkov discrimination in the particular case of 2D PSS. Cross sections extracted from the dose distributions and represented in Figure 13.7. clearly showed the inefficiency of chromatic discrimination, yet considered as the most effective in PSD arrays [20]. Important biases are observed at the edges of the irradiation field and these discrepancies are larger with the blue and red components than with the green and red components. This can be explained by the absorption spectrum of the polystyrene in the lower wavelengths leading to important and variable changes of the measured Čerenkov spectrum [33]. Thus, the constancy of the emission spectra on which relies the chromatic discrimination is not reliable anymore for 2D PSS detectors.

Meanwhile, the subtractive method shows accurate but noisy results. Despite the underlined biases, for doses higher than 10% of the maximal dose, 83% of the pixels passed the gamma test with subtractive discrimination, 88% with the blue and red chromatic discrimination, and 79% with the green and red chromatic discrimination. Collomb-Patton et al. tested their highly improved system (sensitivity and subtractive processing) in similar irradiation configuration and obtained a far less noisy dose distribution conducting to less than 3% of pixels failing the gamma analysis.

13.4.3.3 IMRT fields

2D PSS was also tested in more relevant IMRT configurations to evaluate its ability to measure high-resolution distributions. As previously, 2D PSS measurements were compared to measurements performed by a film placed on the scintillator by gamma analysis with 3% and 3 mm criteria. The system of Frelin et al. was tested with a nine-segment irradiation field using 185 MU delivered by a 6 MV vertical beam. The dose distribution was measured at a source-to-detector distance of 100 cm and a depth of 5.2 cm, and the irradiation was reproduced five times to remove the hot pixels produced by the scattered radiations. The results confirmed the ones of the Section 13.4.3.2: if the subtractive discrimination suffer from a quite high noise level, the chromatic discrimination lead to biases in the gradient zones because of the variable Čerenkov emission spectrum. It also appeared that the spatial resolution was limited by the period of the demodulation mask and that the smallest patterns of the dose distribution were not accurately reproduced. For doses higher than 5% of the maximal dose, 87% of the pixels passed the gamma test with the subtractive discrimination, 77% with the blue and red chromatic discrimination, and 80% with the green

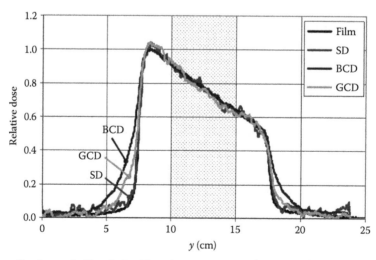

Figure 13.7 Dose profiles for vertical irradiation (Cerenkov maximum) with a wedge filter. The gradients are not correctly reproduced using the blue colorimetric discrimination (BCD) method: blue and red color channels, and the green colorimetric discrimination (GCD) method: green and red color channels. The effect is more important for BCD than for GCD. The substractive discrimination (SD) method gives excellent agreements with the film measurement.

and red discrimination. Collomb-Patton et al. tested two IMRT irradiation fields delivered by a 15 MV vertical irradiation beam: a clinical one composed of 75 segment fields and an experimental one composed of nine segment fields including squares of different sizes (1×1 and 2×2 cm²). In both configurations, almost the whole image passed the gamma test with the exception of a few limited zones of low dose level. In the second configuration, a harsher gamma analysis was performed with (1%, 1 mm) criteria leading to 11.5% of failing pixels, mainly explained by the different spatial resolutions. A precise analysis of the square pattern confirmed that the spatial resolution is determined by the demodulation mask period and here is better than 2×2 mm² on the entire field-of-view of the dosimeter. The dose location accuracy was also shown to be better than 0.25 mm on the whole field-of-view. At last, the continuous acquisition mode of this system (10 frames/s) permitted an accurate dose verification of each segment field.

As shown in this part, the study of 2D PSS and of different Čerenkov discrimination methods led to very satisfying IMRT dose measurements. In the landscape of existing dosimeters, 2D PSS systems have very good properties with water equivalence, making them suitable for photon as well as electron beams, high spatial resolution and real-time measurement allowing an independent control of each irradiation segment.

13.5 CONCLUSION

This chapter covered the principal prototypes of 2D plastic scintillation detectors presented in the literature and discussed their use in varied contexts such as stereotactic radiosurgery, homogeneous fields, static and dynamic IMRT, and beam fluence control. The works done with these prototypes demonstrated the feasibility of developing 2D detectors that are water equivalent, capable of real-time readout, and possess high-spatial resolution and negligible angular dependence. Therefore, those works suggest that the use of plastic scintillators could potentially contribute to increase the quality of dosimetric verifications in megavoltage photon radiation therapy because of their precision and the unique set of properties they possess.

REFERENCES

1. Gao, S., Balter, P. A., Rose, M. and Simon, W. E. Measurement of changes in linear accelerator photon energy through flatness variation using an ion chamber array. *Med. Phys.* **40**: 042101; 2013.
2. Rowshanfarzad, P., Sabet, M., Barnes, M. P., O'Connor, D. J. and Greer, P. B. EPID-based verification of the MLC performance for dynamic IMRT and VMAT. *Med. Phys.* **39**: 6192–6207; 2012.
3. Low, D. A., Moran, J. M., Dempsey, J. F., Dong, L. and Oldham, M. Dosimetry tools and techniques for IMRT. *Med. Phys.* **38**: 1313–1338; 2011.
4. Chandraraj, V., Stathakis, S., Manickam, R., Esquivel, C., Supe, S. S. and Papanikolaou, N. Comparison of four commercial devices for RapidArc and sliding window IMRT QA. *J. Appl. Clin. Med. Phys.* **12**: 338–349; 2011.
5. Poppe, B., Blechschmidt, A., Djouguela, A., Kollhoff, R., Rubach, A., Willborn, K. C. and Harder, D. Two-dimensional ionization chamber arrays for IMRT plan verification. *Med. Phys.* **33**: 1005–1015; 2006.
6. Poppe, B., Stelljes, T. S., Looe, H. K., Chofor, N., Harder, D. and Willborn, K. Performance parameters of a liquid filled ionization chamber array. *Med. Phys.* **40**: 082106; 2013.
7. Jursinic, P. A. and Nelms, B. E. A 2-D diode array and analysis software for verification of intensity modulated radiation therapy delivery. *Med. Phys.* **30**: 870–879; 2003.
8. Elmpt, W., McDermott, L., Nijsten, S., Wendling, M., Lambin, P. and Mijnheer, B. A literature review of electronic portal imaging for radiotherapy dosimetry. *Radiother. Oncol.* **88**: 289–309; 2008.
9. Antonuk, L. E. Electronic portal imaging devices: A review and historical perspective of contemporary technologies and research. *Phys. Med. Biol.* **47**: R31–R65; 2002.
10. Niroomand-Rad, A., Blackwell, C. R., Coursey, B. M., Gall, K. P., Galvin, J. M., McLaughlin, W. L., Meigooni, A. S., Nath, R., Rodgers, J. E. and Soares, C. G. Radiochromic film dosimetry: Recommendations of AAPM Radiation Therapy Committee Task Group 55. *Med. Phys.* **25**: 2093–2115; 1998.
11. Bouchard, H., Lacroix, F., Beaudoin, G., Carrier, J.-F. and Kawrakow, I. On the characterization and uncertainty analysis of radiochromic film dosimetry. *Med. Phys.* **36**: 1931–1946; 2009.
12. Guillot, M., Beaulieu, L., Archambault, L., Beddar, S. and Gingras, L. A new water-equivalent 2D plastic scintillation detectors array for the dosimetry of megavoltage energy photon beams in radiation therapy. *Med. Phys.* **38**: 6763–6774; 2011.
13. Gagnon, J.-C., Thériault, D., Guillot, M., Archambault, L., Beddar, S., Gingras, L. and Beaulieu, L. Dosimetric performance and array assessment of plastic scintillation detectors for stereotactic radiosurgery quality assurance. *Med. Phys.* **39**: 429–436; 2012.

14. Yoo, W. J., Moon, J., Jang, K. W., Han, K. T., Shin, S. H., Jeon, D., Park, J. Y., Park, B. G. and Lee, B. Integral T-shaped phantom-dosimeter system to measure transverse and longitudinal dose distributions simultaneously for stereotactic radiosurgery dosimetry. *Sensors* **12**: 6404–6414; 2012.

15. Guillot, M., Gingras, L., Archambault, L., Beddar, S. and Beaulieu, L. Performance assessment of a 2D array of plastic scintillation detectors for IMRT quality assurance. *Phys. Med. Biol.* **58**: 4439–4454; 2013.

16. Guillot, M., Gingras, L., Archambault, L., Beddar, S. and Beaulieu, L. Spectral method for the correction of the Cerenkov light effect in plastic scintillation detectors: A comparison study of calibration procedures and validation in Cerenkov light-dominated situations. *Med. Phys.* **38**: 2140–2150; 2011.

17. Lacroix, F., Archambault, L., Gingras, L., Guillot, M., Beddar, A. S. and Beaulieu, L. Clinical prototype of a plastic water-equivalent scintillating fiber dosimeter array for QA applications. *Med. Phys.* **35**: 3682–3690; 2008.

18. Fontbonne, J.-M., Iltis, G., Ban, G., Batalla, A., Vernhes, J.-C., Tillier, J., Bellaize, N. et al. Scintillating fiber dosimeter for radiation therapy accelerator. *IEEE Trans. Nucl. Sci.* **49**: 2223–2227; 2002.

19. Frelin, A.-M., Fontbonne, J.-M., Ban, G., Colin, J., Labalme, M., Batalla, A., Isambert, A., Vela, A. and Leroux, T. Spectral discrimination of Čerenkov radiation in scintillating dosimeters. *Med. Phys.* **32**: 3000–3006; 2005.

20. Archambault, L., Beddar, S., Gingras, L., Roy, R. and Beaulieu, L. Measurement accuracy and Cerenkov removal for high performance, high spatial resolution scintillation dosimetry. *Med. Phys.* **33**: 128–135; 2006.

21. Archambault, L., Briere, T. M. and Beddar, S. Transient noise characterization and filtration in CCD cameras exposed to stray radiation from a medical linear accelerator. *Med. Phys.* **35**: 4342–4351; 2008.

22. Klein, D. M., Therriault-Proulx, F., Archambault, L., Briere, T. M., Beaulieu, L. and Beddar, A. S. Technical note: Determining regions of interest for CCD camera-based fiber optic luminescence dosimetry by examining signal-to-noise ratio. *Med. Phys.* **38**: 1374–1377; 2011.

23. Das, I. J., Ding, G. X. and Ahnesjö, A. Small fields: Nonequilibrium radiation dosimetry. *Med. Phys.* **35**: 206–215; 2008.

24. Alfonso, R., Andreo, P., Capote, R., Saiful Huq, M., Kilby, W., Kjäll, P., Mackie, T. R. et al. A new formalism for reference dosimetry of small and nonstandard fields. *Med. Phys.* **35**: 5179–5186; 2008.

25. Ezzell, G. A., Burmeister, J. W., Dogan, N., LoSasso, T. J., Mechalakos, J. G., Mihailidis, D., Molineu, A. et al. IMRT commissioning: Multiple institution planning and dosimetry comparisons, a report from AAPM Task Group 119. *Med. Phys.* **36**: 5359–5373; 2009.

26. Goulet, M., Gingras, L. and Beaulieu, L. Real-time verification of multileaf collimator-driven radiotherapy using a novel optical attenuation-based fluence monitor. *Med. Phys.* **38**: 1459–1467; 2011.

27. Goulet, M., Archambault, L., Beaulieu, L. and Gingras, L. High resolution 2D dose measurement device based on a few long scintillating fibers and tomographic reconstruction. *Med. Phys.* **39**: 4840–4849; 2012.

28. Poppe, B., Thieke, C., Beyer, D., Kollhoff, R., Djouguela, A., Rühmann, A., Willborn, K. C. and Harder, D. DAVID—A translucent multi-wire transmission ionization chamber for in vivo verification of IMRT and conformal irradiation techniques. *Phys. Med. Biol.* **51**: 1237–1248; 2006.

29. Islam, M. K., Norrlinger, B. D., Smale, J. R., Heaton, R. K., Galbraith, D., Fan, C. and Jaffray, D. A. An integral quality monitoring system for real-time verification of intensity modulated radiation therapy. *Med. Phys.* **36**: 5420–5428; 2009.

30. Looe, H. K., Harder, D., Rühmann, A., Willborn, K. C. and Poppe, B. Enhanced accuracy of the permanent surveillance of IMRT deliveries by iterative deconvolution of DAVID chamber signal profiles. *Phys. Med. Biol.* **55**: 3981–3992; 2010.

31. Boon, S. N., Luijk, P. V., Schippers, J. M., Meertens, H., Denis, J. M., Vynckier, S., Medin, J. and Grusell, E. Fast 2D phantom dosimetry for scanning proton beams. *Med. Phys.* **25**: 464–475; 1998.

32. Petric, M. P., Robar, J. L. and Clark, B. G. Verification of IMRT dose distributions using a tissue equivalent plastic scintillator based dosimetry system. *Med. Phys.* **32**: 1891; 2005.

33. Frelin, A.-M., Fontbonne, J.-M., Ban, G., Colin, J., Labalme, M., Batalla, A., Vela, A., Boher, P., Braud, M. and Leroux, T. The DosiMap, a new 2D scintillating dosimeter for IMRT quality assurance: Characterization of two Čerenkov discrimination methods. *Med. Phys.* **35**: 1651–1662; 2008.

34. Beddar, S., Archambault, L., Sahoo, N., Poenisch, F., Chen, G. T., Gillin, M. T. and Mohan, R. Exploration of the potential of liquid scintillators for real-time 3D dosimetry of intensity modulated proton beams. *Med. Phys.* **36**: 1736–1743; 2009.

35. Robertson, D., Mirkovic, D, Sahoo, N. and Beddar, S. Quenching correction for volumetric scintillation dosimetry of proton beams. *Phys. Med. Biol.* **58**: 261; 2013.

36. Frelin, A.-M., Fontbonne, J.-M., Ban, G., Batalla, A., Colin, J., Isambert, A., Labalme, M., Leroux, T. and Vela, A. A new scintillating fiber dosimeter using a single optical fiber and a CCD camera. *IEEE Trans. Nucl. Sci.* **53**: 1113–1117; 2006.

37. Collomb-Patton, V., Boher, P., Leroux, T., Fontbonne, J.-M., Vela, A. and Batalla, A. The DOSIMAP, a high spatial resolution tissue equivalent 2D dosimeter for LINAC QA and IMRT verification. *Med. Phys.* **36**: 317–328; 2009.

38. Petric, M. P., Robar, J. L. and Clark, B. G. Development and characterization of a tissue equivalent plastic scintillator based dosimetry system. *Med. Phys.* **33**: 96–105; 2006.
39. Partridge, M., Evans, P. M. and Symonds-Tayler, J. R. N. Optical scattering in camera-based electronic portal imaging. *Phys. Med. Biol.* **44**: 2381–2396; 1999.
40. Boher, P. and Leroux, T. Device and method for discriminating Cerenkov and scintillation radiation. Patent EP1857836, October 21, 2009.
41. Beddar, A. S., Mackie, T. R. and Attix, F. H. Water-equivalent plastic scintillation detectors for high-energy beam dosimetry: I. Physical characteristics and theoretical considerations. *Phys. Med. Biol.* **37**: 1883–1900; 1992.
42. Beddar, A. S., Mackie, T. R. and Attix, F. H. Water-equivalent plastic scintillation detectors for high-energy beam dosimetry: II. Properties and measurements. *Phys. Med. Biol.* **37**: 1901–1913; 1992.

2D and 3D scintillation dosimetry for brachytherapy

Dirk Flühs, Marion Eichmann, and Assen Kirov

Contents

14.1 INTRODUCTION

14.1.1 BACKGROUND OF BETA AND LOW-ENERGY PHOTON RADIATION AND SOURCES, AND APPLICATION OF THESE SOURCES IN SEEDS AND EYE PLAQUES

Brachytherapy is a method for treating tumors with radiation by placing a radioactive source inside or adjacent to them. Since the irradiation of tumors on or within the eye with radioactive applicators was introduced in the early 1960s, this eye-salvaging brachytherapy modality has become a clinical standard. These applicators, often called ophthalmic plaques, typically consist of a metal calotte as a carrier adapted to the curvature of the eye surface, with varying shapes and sizes (cf. Figure 14.1, for example). The sealed radioactive sources are attached at the concave side of the calotte. The applicator emits radiation mainly into the target volume adjacent to the applicator. The metal carrier absorbs most of the radiation emitted in other directions, thus reducing side effects and problems with radiation protection. The methods described in this chapter were developed and tested specifically for ophthalmic plaques, but with some modifications they can be applied also for other brachytherapy sources.

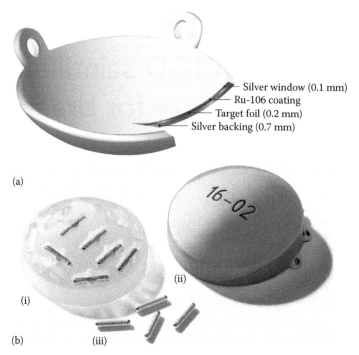

Silver window (0.1 mm)
Ru-106 coating
Target foil (0.2 mm)
Silver backing (0.7 mm)

(a)

16-02

(ii)

(i)

(b)   (iii)

Figure 14.1 (a) ¹⁰⁶Ru eye applicator and (b) Collaborative Ocular Melanoma Study (COMS) eye applicator: (i) COMS insert; (ii) COMS plaque shell; (iii) Iso Seed I-125. (Courtesy of Eckert & Ziegler BEBIG, Berlin, Germany.)

The tumor size treatable with ophthalmic plaques—and the target volume, therefore—is limited to a typical lateral spread of about 20 mm of the tumor basis and an apical height of tumor of about 12 mm. Hence, only emitters of low energy photons, such as ^{125}I and ^{103}Pd, or beta emitters, such as ^{106}Ru/^{106}Rh and ^{90}Sr, are used for radioactive eye plaques today. A comparison between the dose distributions of a ^{106}Ru plaque and an ^{125}I eye plaque is shown in Figure 14.2.

These isotopes provide small radiation fields with large dose gradients. For β-applicators, for instance, the dose rate within the field ranges over 4 orders of magnitude, allowing for a quite effective sparing of neighboring tissue, especially structures at risk. The steep dose fall-off, together with other physical properties of the radiation rather uncommon to the conventional clinical dosimetry, makes eye plaque dosimetry a sophisticated dosimetric application with unique requirements (cf. Figure 14.3).

Only a few types of detectors fit the requirements in this field. Due to their physical characteristics, scintillators have proven to be very useful for the dosimetry of brachytherapy sources. Small-sized plastic scintillation dosimeters in a water phantom, combined with a reliable scanner with high spatial resolution, allow for a fast and direct reading dosimetry of ophthalmic plaques (Flühs et al. 1996, 1997, Bambynek et al. 2000, Eichmann et al. 2009, 2012). They have become established dosimeters, since they are suitable for the basic dosimetry of a plaque and the quality assurance in the hospital as well. These detectors are described in Section 14.2.

A completely different measuring setup based on large plastic scintillator (PS) sheets and a microchannel plate was used to quickly obtain 2D dose distributions around gamma emitting brachytherapy sources (Perera et al. 1992). Due to light scattering processes, the accurate determination of the origin of the scintillation light emission in the sheets is one of the main challenges for such measurement design (Kirov et al. 1999). An alternative approach for the eye plaque dosimetry is the usage of a liquid scintillator (LS). The feasibility and the potential pitfalls of this method are presented in Section 14.3.

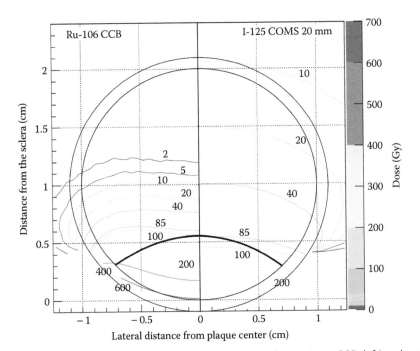

Figure 14.2 Comparison of the dose distributions of a ruthenium applicator (type CCB, left) and an iodine eye-plaque (type COMS 20 mm, right). The dose distribution in a plane, the central axis of the eye plaque (solid vertical line), the eye (two circles), and the tumor (solid bent line) are depicted. Dose distributions are rotationally symmetrical with respect to this axis. Dose distributions derived by Monte Carlo simulations using the Monte Carlo code EGSnrc. (Data from R. M. Thomson et al., *Med. Phys.*, 35, 5530–5543, 2008.)

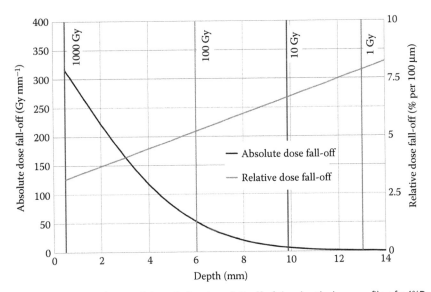

Figure 14.3 Example of the absolute and the relative dose fall-off of the depth dose profile of a ^{106}Ru/^{106}Rh eye plaque, type CCX. Here, the dose fall-off is defined as the negative gradient of the profile, and the relative fall-off is given by the ratio of this value and the dose. The absolute dose fall-off profile corresponds to a dose of 100 Gy at a depth of 6 mm, the typical dosage for a tumor height of 4.8 mm and a sclera thickness of 1.2 mm. The vertical lines indicate the depths at which the depth dose profile shows the specified dose values.

Applications for 1D, 2D, and 3D dosimetry

14.2 APPLICATION—PLASTIC SCINTILLATOR

14.2.1 PHYSICAL PROPERTIES OF SCINTILLATOR MATERIALS WITH REGARD TO THESE RADIATION QUALITIES

PS for eye plaque dosimetry are commonly used within water phantoms and moved by scanning systems. Thus, the signal at a desired position can be read out directly. This allows for a measurement of the absorbed dose to water within water itself as reference medium, which replaces all kinds of eye tissue under measurement conditions. As long as the detector itself, the light guide and the detector housing are dosimetrically water equivalent, the physical interactions within the detector system will be the same as in water, and the particle fluxes will not be disturbed. A measuring signal equivalent to the absorbed dose within the detector will then immediately give a value for the absorbed dose to water, after a careful calibration process.

As typical detectors used for eye plaque dosimetry, scintillators based on polyvinyltoluene (PVT) are used. Their density of 1.04 g/cm^3 is very close to water. With regard to the interactions with electrons, the total mass stopping power has to be compared to the values of water. The energy interesting for eye plaque dosimetry ranges from about 10 keV (energy of secondary electrons of low energy photon emitters) to 3.5 MeV (maximal beta energy of ^{106}Ru/^{106}Rh). In the whole range, the ratio of the mass stopping powers of scintillator and water stays within 2.5% (cf. Figure 14.4), due to a fine agreement of the collision mass stopping power.

The difference of the radiation mass stopping powers between water and PVT is negligible because radiative energy losses contribute about 3% to the total energy loss only at energies higher than 1 MeV, and differences in the produced bremsstrahlung flux do not have to be taken into account since most of the produced photons are absorbed outside the small volume of the scintillator detectors used for this application, which have dimensions of 1 mm or less.

For photon dosimetry, the situation is more complicated, as an analysis of the ratio of the mass absorption coefficients of the PS (based on polyvinyltoluene) and water shows (cf. Figure 14.4). In the range from 150 keV to 3 MeV, a fine agreement of 4% between water and a PS can be observed, since Compton scattering is the dominant interaction which merely depends on the electron density. In the energy range below about 30 keV, however, the photoelectric effect is the dominant interaction. Due to the high dependence of its cross section on the atomic number ($\sim Z^{4.6}$) (Reich 1990), the different atomic compositions

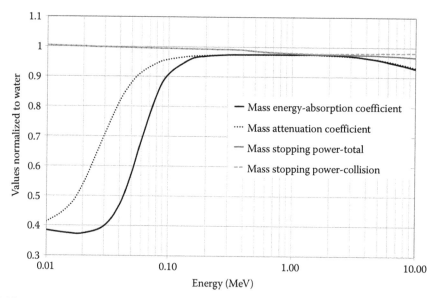

Figure 14.4 Mass-stopping power, mass-absorption, and mass-attenuation coefficient of a plastic scintillator based on polyvinyltoluene, normalized to water. (Data from http://www.nist.gov/pml/data/.)

of scintillator and water result in a significantly smaller value for the ratio of the mass absorption coefficients of scintillator and water. Advantageously, the ratio remains quite constant (~0.37) for energies below 30 keV, allowing a window for the dosimetry with small PS in this energy range, of course, with a calibration value different to the one at higher energies. This energy range covers the photon spectra of the radionuclides ^{125}I and [103]Pd commonly used for ophthalmic plaque therapy, with mean photon energies of 28.4 and 20.7 keV, respectively (Venselaar et al. 2013). For photon emitters with a significant amount of their spectra in the energy range from 35 to 150 keV, an accurate scintillator dosimetry becomes a difficult task. Because of the rapid change of the ratio of the mass absorption coefficients, calculating an exact calibration value for the dose absorbed to water requires an exact knowledge of the spectrum at every measuring point. A solution of this problem can be found by modifying the scintillator composition, so that it matches that of water, while at the same time preserving high scintillator efficiency (Kirov et al. 1999).

Even with a correction for the mass-absorption coefficient, the ratio between the light output of a PVT-based PS and the absorbed dose rate to water is smaller for low photon energies than for energies in the range around 1 MeV, due to quenching effects in the scintillator (Williamson et al. 1999). For BC-400, for instance, a sensitivity decrease of 0.75 is observed, resulting in a total factor of about 3.3 between the light yields at 1 MeV and at 25 keV, compared to a factor of about 2.5 that takes into account the change of the mass absorption coefficients only (Flühs et al. 2004). For practical purposes, these deviations have an influence of about 5% on the measurement of the absorbed dose of eye plaques with photon emitters, as long as the calibration of the PS dosimeter is performed in the corresponding energy range and no significant changes of the photon spectra occur within the measured region (Lessard et al. 2012). Monte Carlo calculations of ^{125}I radiation fields show that in water, the photon energy spectra vary only slightly with increasing distance to the source. For most purposes of eye plaque dosimetry, therefore, the influence of the variation of nonlinearities with energy is small and can be taken into account by an adequate mean value for the calibration factor.

A considerable amount of the dose within the scintillator volume is deposited by electrons, or secondary electrons, respectively, transmitting through or produced by the surrounding media. Thus, the housing of the scintillator, the light guide, the glues connecting scintillator and light guide, and the light tight coating of the scintillator have an important impact on the water equivalence of the detector. These components have to be also water equivalent. This is fulfilled by choosing adequate materials for the housing of the scintillator and light guides, such as RW3 (Goettingen White Water), a water-equivalent plastic (Harder et al. 1988), and PMMA-based light guides. For the light tightness, a paint with pigments based on carbon black is recommended, while oxides of metals with a higher atomic number that disturb the particle flux should be avoided, especially in the case of low-energy photon dosimetry. An example of a suitable setup is described in Section 14.2.2 and Figure 14.5.

To obtain a sufficient spatial resolution and to minimize the influence on the particle fluxes, scintillators for eye plaque dosimetry have small volumes. For a cylindrical detector shape, the value for the diameter is typically 1 mm, while the height varies between 0.3 and 1 mm. Thus, the amount of light originating in the light guide is usually not negligible compared to the scintillation light. It has to be subtracted from the total signal (cf. Figure 14.6). There are two sources of light within the light guide: Cerenkov light and the light due to fluorescence. The latter is quite small—it usually adds an amount of less than 2% of the scintillator signal in

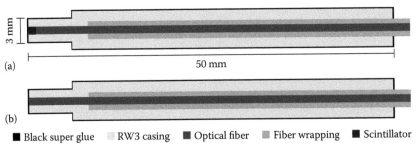

Figure 14.5 Plastic scintillator detector system: (a) Scintillator channel and (b) Cerenkov channel.

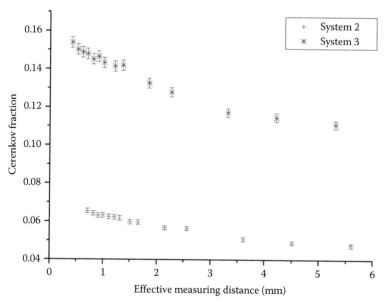

Figure 14.6 Cerenkov fraction I^{CC}/I^{SC} measured for typical CCB plaque on its central axis (system 2: 1 mm detector; system 3: 0.5 mm detector). (Data from M. Eichmann, Entwicklung eines hochpräzisen Dosimetriesystems zur Messung der Oberflächendosisverteilung von Augenapplikatoren, dissertation, Technische Universität Dortmund, 2010.)

a measurement of the radiation field of an ^{125}I seed. Thus, it can be neglected in most cases, depending on the measuring geometry. Cerenkov light is generated when electrons with a kinetic energy above 178 keV move through a typical PMMA light guide. The intensity increases with higher electron energies. Since the amount of Cerenkov light can reach several tens of percent of the scintillator signal, in most measurement situations it has to be measured separately and then be subtracted from the total signal. It is helpful to place the light guide in a way that the generation of Cerenkov light is minimized, for instance along a line with a steep dose fall-off. The details of this procedure and a typical setup are described in Section 14.2.2.

Light tightness of the whole setup consisting of scintillator, light guide, and photomultiplier tube is a basic requirement unless the whole system is placed in an absolutely dark room during the measurement. Light from outside the system can result in another substantial disturbance of the measuring process and even destroy the photomultiplier, in an extreme case. For certain tasks of eye plaque dosimetry, however, it is essential to keep the light-proof housing of the detector as thin as possible. For instance, the measurement of the homogeneity of the surface dose distribution of the radioactive material requires a distance between the scintillator and the surface of the applicator as small as possible. Thus, only thin layers of black paint have to guarantee the light tightness. Since the detector is moved along the metal surface of an applicator, these layers are susceptible to mechanical damage, so they have to be sufficiently resistant.

The lower limit for the dose rate measurable with an optimally constructed PS dosimeter is given by the ratio between the measuring current, depending on the signal from the scintillator and the photomultiplier noise, that is the dark current. To build a sensitive system, it is necessary to purchase a photomultiplier tube with high amplification and low dark current. Selected photomultiplier specimens with a superior performance, with regard to these characteristics, are more advantageous.

14.2.2 DESCRIPTION OF A PS SYSTEM, LIGHT GUIDE BASED, WITH TWO CHANNELS, ONE FOR THE COMPENSATION OF THE CERENKOV SIGNAL

In this section, an example of a typical PS detector system for eye plaque dosimetry is described. It was originally designed at the Technische Universität Dortmund (Flühs et al. 1996, Bambynek et al. 2000, Eichmann et al. 2009), Germany, and has been refined and built at PTB (Physikalisch Technische Bundesanstalt, Braunschweig, Germany).

This system consists of two channels (cf. Figure 14.5): The sensitive part of the detector system is a cylindrical PS (type BC-400 produced by Saint-Gobain Crystals, France) with a diameter of 1 mm and a height of 0.5 mm or 1 mm coupled to an optical fiber (type CWKF-1001 E22 produced by CUNZ GmbH & Co.KG, Germany) transmitting the scintillation light to a photomultiplier tube (type R647-01, Hamamatsu Photonics Deutschland GmbH, Germany). These components of the PS detector system will in the following be called the *scintillator channel* (SC). The Cerenkov light produced by the electrons entering the optical fiber is measured separately by a second optical fiber without a PS called the *Cerenkov channel* (CC) and a second photomultiplier tube. Each measuring head of the two channels is placed in a RW3 casing (water-equivalent plastic) made opaque by a black super glue (LoTyp 480 produced by Loctite/Henkel AG & Co. KGaA, Germany) with a measured thickness varying between 0.1 and 0.2 mm, depending on the system used. The entire measuring system shows a stable performance with 1% variation during a typical measuring session. All channels are corrected for their dark current. For the scintillator detector system containing a plastic detector with a height of 0.5 mm, the dark current is less than 0.4%, for a height of 1 mm it is less than 0.2%, and for the Cerenkov channel it is less than 3% of the signal current (for an applicator activity of 1 MBq).

The effective measuring point of the described PS has been assumed to be in its center (cf. Section 14.2.5). The effective measuring distance, that is, the distance between the applicator surface and the effective measuring point, is determined by the sum of half of the height of the scintillator and the thickness of the black super glue.

For all measurements with the PS detector system, the scintillator and Cerenkov channel need to be calibrated relatively to each other in order to account for the differences in the transmission rates of their optical fibers, their coupling to the photomultiplier, and the amplifications of the photomultiplier tubes. Therefore, the same signal has to be produced in both channels. To achieve this, the Cerenkov signal produced by the β-radiation from a ^{90}Sr/^{90}Y planar source (radioactive check device T48010 provided by PTW Freiburg, Germany) irradiating the light guides of the scintillator and Cerenkov channel is measured. These measurements of currents show a maximal overall uncertainty of 1.8%, including statistical uncertainties and variations in performance and positioning. The relative calibration factor K^{rel} is the ratio of the induced Cerenkov currents in the scintillator $I^{rel,SC}$ and Cerenkov channel $I^{rel,CC}$.

$$K^{rel} = \frac{I^{rel,SC}}{I^{rel,CC}} \tag{14.1}$$

It has a maximal combined uncertainty of 2.5%. The signal current I^{sig} is then determined by subtracting the Cerenkov current I^{CC}, multiplied by the relative calibration factor of the scintillator current I^{SC}.

$$I^{sig} = I^{SC} - K^{rel} I^{CC} \tag{14.2}$$

The signal current has a combined uncertainty of less than 1.8%.

The Cerenkov fraction of the scintillator channel measured on the central axis of a typical CCB plaque is shown in Figure 14.6. For a 1 mm detector, it varies between 4.5% and 6.5% and for a 0.5 mm detector between 11% and 15%.

14.2.3 DESCRIPTION OF A PS SYSTEM, LIGHT GUIDE BASED, WITH ONE CHANNEL

For quality assurance in the hospital, a measuring system is required that allows for a quick dosimetry of any type of new applicator, either delivered by a manufacturer or constructed by the clinical staff members themselves. This measurement has to be sufficiently accurate, within a margin of typically ±5%. The measuring program usually consists of two major tasks: a measurement along the main axis of the applicator (the line perpendicular to the surface, in the center of the applicator), in order to determine the profile of the depth dose as a basis for the clinical dosage, and a check of the homogeneity of the surface distribution, in order to prevent possible over- or under-dosage due to an inhomogeneity (cf. Figures 14.10 and 14.11).

For both tasks, a single-channel scintillation dosimeter is an ideal tool. When the amount of scintillation light and Cerenkov light are compared to each other, a similar dependence with increasing distance to the applicator surface can be found, with the Cerenkov profile slightly steeper. With a cylindrical detector of a diameter and a height of 1 mm each, the intensity of scintillation light is high enough that the difference of the relative depth profiles between the combined signal and the scintillator signal does not exceed ±3%, which is within the typical clinically relevant range, for example, from 0.5 to 8 mm in the case of a ^{106}Ru/^{106}Rh applicator (cf. Figure 14.6). The influence of the Cerenkov light is then negligible within the accepted limits, and can be taken into account by a calibration factor. The calibration itself, in terms of absorbed dose to water, can be reduced to a single measurement at an adequate calibration source, for instance a ^{90}Sr standard. In addition, a scintillation detector of this size provides a sufficient spatial resolution for a qualitative check of the applicator homogeneity. This can be performed, for instance, by a set of 2D measurements in two planes perpendicular to the applicator surface or by a set of measurements at a defined distance from the plaque covering its surface.

14.2.4 CALIBRATION OF SUCH SYSTEMS IN TERMS OF ABSORBED DOSE TO WATER

Since a PS detector system provides only relative readings, an absolute calibration is needed. It requires the proportionality between measured current produced by the light signal of the scintillator and the absorbed dose rate to water. This is shown for a wide range of dose rates (0–80 mGy/s typical applicator contact dose rate 2 mGy/s) in Figure 14.7.

The PS detector system at TU Dortmund and Universitätsklinikum Essen is calibrated with a secondary standard (^{90}Sr/^{90}Y planar source [S/N EY 845]). The secondary standard is certified by PTB by using a specially designed extrapolation chamber (primary standard for beta radiation [Bambynek 2000]). The certificate includes dose rates for 7 depths on the central axis between 1 and 5 mm in RW3 plastic with an uncertainty of 4.1% (1σ).

To obtain the absolute calibration factor K^{abs} of the system, the given dose rate $\dot{D}$ at a certain distance has to be related to a signal current I^{sig} measured at the same distance away from the secondary standard.

To obtain the signal current, the RW3 casing of both channels (cf. Figure 14.5) can be fixed in a support structure, which allows for a precise and reproducible positioning in front of the secondary standard. The overall uncertainty, including variations in performance, positioning, and statistical uncertainties,

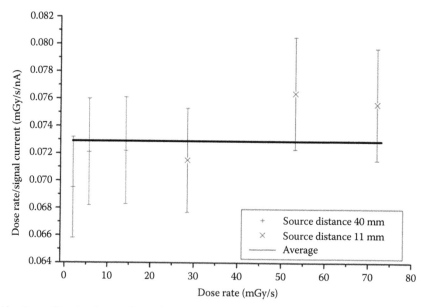

Figure 14.7 Absolute calibration factor of the plastic scintillator detector system in relation to dose rate, measured at PTB secondary standard. (Data from M. Eichmann, Entwicklung eines hochpräzisen Dosimetriesystems zur Messung der Oberflächendosisverteilung von Augenapplikatoren, dissertation, Technische Universität Dortmund, 2010.)

should not exceed 2%. The absolute calibration factor is given by the ratio of the PTB dose rate to water and the measured signal current.

$$K^{\text{abs}} = \frac{\dot{D}}{I^{\text{sig}}} \tag{14.3}$$

Its combined uncertainty is 4.6%. To ensure a precise transfer of the absolute and relative calibration factor measured once for each PS detector system to the dose measurements of the brachytherapy sources, an additional transfer factor is determined. The radioactive check device mentioned above—with a support structure similar to that of the secondary standard—is used to perform this determination. The transfer current $I^{\text{trans,SC,CC}}$ of both channels is measured precisely and reproducibly in front of the radioactive check device during the absolute calibration of the system and during each measurement of a brachytherapy source. The transfer factor $K^{\text{trans,SC,CC}}$ is determined by (separately for scintillator channel and Cerenkov channel)

$$K^{\text{trans,SC,CC}} = \frac{I_{\text{abs}}^{\text{trans,SC,CC}}}{I_{\text{meas}}^{\text{trans,SC,CC}}} \tag{14.4}$$

and the dose rate of a brachytherapy source by

$$\dot{D}_{\text{BS}} = K^{\text{abs}}(K^{\text{trans,SC}}I_{\text{BS}}^{\text{SC}} - K^{\text{rel}}K^{\text{trans,CC}}I_{\text{BC}}^{\text{CC}}) \tag{14.5}$$

The uncertainty of the transfer factor is 1.43% again including variations in performance (1%), positioning (1%), and statistical uncertainties (0.2%) for the transfer current.

14.2.5 UNCERTAINTIES OF MEASUREMENT

Due to the physical characteristics of the radiation fields and the complex detector setup, an accurate analysis of the measuring data obtained from a scintillator system for eye plaque dosimetry is necessary, with respect to the effective measuring point, the positioning accuracy, and the spatial resolution.

The steep fall-off of the dose distributions of ophthalmic plaques requires careful consideration of the position a measuring signal is related to, since a detector always covers a volume in which dose variations cannot simply be neglected. Within a typical detector dimension of 0.5 mm, for instance, the dose change reaches up to 45% in the case of a ^{90}Sr depth dose profile. Advantageously, such applicators are not point sources but usually quite homogeneous area sources. Thus, they typically produce isodose levels that roughly follow the curvature of the plaque, with minimal dose variations within typical lateral detector dimensions of about 1 mm. In regions that are not too close to the edge of the plaque, the direction of the overall dose gradient vector of the radiation field is perpendicular to the plaque surface (cf. Section 14.2.6) and follows an almost exponential law along this direction. Thus, the signal of a cylindrically shaped detector, positioned with its axis perpendicular to the plaque surface, is the result of the integration of an exponential function. With the z-axis chosen in the direction of the detector axis, the position of the effective measuring point on the z-axis is given by the geometric mean of the position values for the lower and upper detector edge along z, therefore. For a small detector of 0.5 mm height, however, the difference between the geometric and arithmetic mean is less than 50 μm. Hence, the arithmetic mean is a sufficiently accurate choice in most practical cases. Because of the steep dose fall-off, however, it has to be taken into account that even slight positioning errors of only 0.2 mm may result in large measuring errors of more than 5%, in case of ^{106}Ru/^{106}Rh plaques, and even more for a ^{90}Sr applicator.

At the edges of an applicator, larger dose gradients in more than one dimension occur. In such cases, the interpretation of the detector signal and the definition of the effective measuring point require a more sophisticated analysis of the data. In general, it can be useful to measure a sufficient number of data at different, closely neighboring points allowing a deconvolution procedure in order to get a better resolution of the dose distribution. Radiation fields with several changes of the direction of the dose gradient vector within the detector volume, however, impose an effective limitation upon the usage of PSs. In such cases, additional measurements using radiochromic film (RCF) can give more detailed information about the dose distribution.

Thus, the desired spatial resolution and the dose gradients impose a limit on the maximal size of a scintillator detector. Typically, a scintillator designed for eye plaque dosimetry should not exceed 1 mm in any dimension. The lower limit is defined by several parameters. First of all, the strength of the light signal of the scintillator must be large enough to be detected and clearly distinguished from the background by the photomultiplier system used, as mentioned before. This signal strength depends on the characteristic light yield of the scintillator, its size, the quality of the light guide tube, and the dose rate of the source in the measuring regions. For a typical dosimeter setup, a minimal scintillator size of 0.3–0.5 mm, at least in one dimension, gives reliable results. Reducing the scintillator size down to 0.1 mm, in order to improve the spatial resolution, yields considerably smaller readings than expected, in several cases. This could be caused by an insufficient concentration of wavelength shifters within the scintillator material after the manufacturing process. For such special measuring tasks novel scintillation materials without wavelength shifters, such as polyethylene naphthalate, can be tested (Flühs et al. 2016). They offer a similar yield of blue light as polyvinyltoluene-based scintillator types that require wavelength shifters.

14.2.6 3D SCANNER FOR PS DOSIMETRY: XYZ MEASURING TABLE AND APPLICATOR SURFACE MEASURING TABLE

The following requirements, which are mainly driven by the high-dose gradients of brachytherapy sources and time and budget constraints, especially important for clinical procedures, have to be considered for a 3D scanner for PS dosimetry of eye plaques. The measurements of dose rate distributions have to be performed in a water phantom in order to be as close as possible to the radiation conditions within human eyes. All materials surrounding the eye plaque and detector must be water equivalent. All mechanical elements of the setup must have minimal impact on the radiation field and at the same time guarantee the mechanical stability of the setup. Because of small structures and steep dose gradients a high spatial resolution is needed.

The following materials have been chosen for the 3D scanner developed at TU Dortmund (Eichmann et al. 2009, 2012). RW3/RW1, which is a water-equivalent plastic for electrons/photons, has been used for all elements directly surrounding or contacting the applicator. All mechanical parts in the area surrounding the eye plaque (distance to plaque surface is greater than 5 mm) are made of anodized aluminum, because it is easier to machine than RW3 and ensures the stability of the guide components. Furthermore, aluminum, a low Z material, is chosen in order to minimize backscattering effects ($\sim\log_{10}(Z + 1)$ [Lee and Reece 2004]). Monte Carlo studies indicate no significant dose perturbation due to aluminum at distances of about 5 mm from the eye plaque surface. At distances of more than 2 cm from the eye plaque surface, the mechanical and stabilizing elements are made of stainless steel since the dose rate does not exceed 1% of the dose rate at the eye plaque surface (for ^{106}Ru). Thus, backscattering effects are negligible.

The measurements have successfully been automated, and only the assemblies of radioactive sources, caliper (distance measuring device), and detector require human intervention.

The surface measuring table (Eichmann et al. 2009, Eichmann 2010) (cf. Figures 14.8a and 14.9a) is a specialized system enabling the measurement of the surface dose profile of spherical shell-shaped applicators such as ^{106}Ru ophthalmic plaques. It moves the PS detector along spherical coordinates at a small, constant, and adjustable [in contrast to (Eichmann et al. 2009) by a miniature translation stage distance between the detector and the plaque surface. The distance can be varied by the translation stage in 1 μm steps (see the xyz-measuring table for more details [Eichmann et al. 2012]). This kind of movement ensures a perpendicular setting between central axis of detector and plaque surface. Such design allows measurements of the surface dose distribution of individual plaques with a high spatial resolution and precision. The resulting dose profiles reflect inhomogeneities in the distribution of ^{106}Ru in the active layer of the plaque distinguishing small-sized spots. Maintaining such small distance is important because of the smearing of the dose profile with increasing distance.

With the xyz-measuring table (Eichmann et al. 2012), the individual 3D dose rate distributions of all kinds of sources of similar dimensions to plaques can be measured with a positioning accuracy of 1 μm (cf. Figure 14.9).

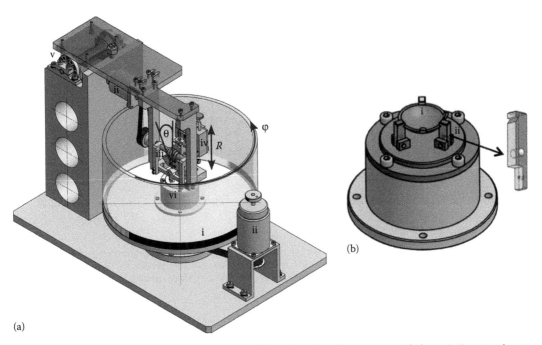

Figure 14.8 (a) Design drawing of surface measuring table, with indicated movements (dark gray): (i) water phantom, (ii) stepper motor, (iii) cogwheel segment and screw, (iv) miniature translation stage, (v) rotary head with ball bearing, and (vi) pedestal with tripod. (b) Mount type with (i) ophthalmic plaque and (ii) pin of tripod with groove.

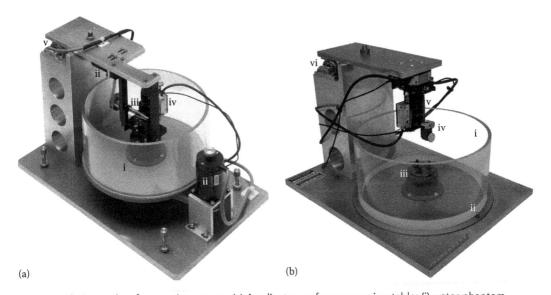

Figure 14.9 Photographs of measuring setups. (a) Applicator surface measuring table: (i) water phantom, (ii) stepper motor, (iii) cogwheel segment and screw, (iv) miniature translation stage, and (v) rotary head with ball bearing. (b) xyz-measuring table: (i) water phantom, (ii) fixation pin, (iii) source mount for an ophthalmic plaque, (iv) detector mount, (v) miniature translation stage, and (vi) rotary head with ball bearing.

To achieve such high resolution, a micropositioning system with piezo inertial drives from mechOnics AG (Munich, Germany) is used. This system consists of three orthogonally located miniature translation stages with a range of 30 mm each and a single-step resolution of 450 nm. In combination with an exposed linear encoder, which controls the positioning, an accuracy of 1 μm is achieved. Accidental shifts due to vibrations, for example, are immediately corrected. The xyz-measuring table setup can be used for rapidly and accurately categorizing small brachytherapy sources, for instance for quality assurance.

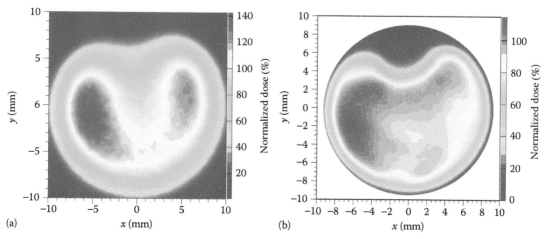

Figure 14.10 (a) Dose profile (normalized to the dose rate at $x = y = 0$) of an ophthalmic plaque type COB measured in the xy-plane, that is, orthogonal to central axis of the plaque, at a distance in z on the central axis to the plaque surface of 7 mm. (b) Normalized surface dose profile with an effective measuring distance of 1 mm. (Data from M. Eichmann et al., *Phys. Med. Biol.*, 57, N421–N429, 2012.)

14.2.7 2D/3D DOSE DISTRIBUTIONS OF EYE PLAQUES, AS EXAMPLES

A 2D dose profile of an ophthalmic plaque (type COB, i.e., plaque with a cutout for the optic nerve) with an activity of 7 MBq has been measured with the xyz-measuring table (cf. Figure 14.10a), and compared to the distance corrected surface dose profile (cf. Figure 14.10a) obtained with the surface positioning system.

For the 2D dose profiles in the *xy*-plane, a distance in z of 7 mm (Figure 14.10a) and a step width of 500 μm were chosen. The 2D dose profile in the *xy*-plane at a distance of 7 mm consists of 1980 points of measurement. At each point, a measuring period of maximum 30 s was chosen, which led to a statistical uncertainty between 1% and 6%. The whole measurement took 16 h.

For the measurement with the surface positioning system (Figure 14.10b), an effective measuring distance (i.e., distance between the plaque surface and the center of the scintillator) of 1 mm was chosen. To receive a nearly constant step width in the *x* and *y* directions of 0.5 mm, for the motion in the polar direction a step width of 2.3° was chosen. In the azimuthal direction, variable step widths depending on the polar angle between 3.6° and 45° had to be used. The combined relative uncertainty of the dose is smaller than 7.5%. To improve comparison between both measurements, the spherical coordinates are transferred to Cartesian coordinates as shown in Figure 14.10b.

In Figure 14.11, surface dose profiles of three different eye applicators (type CCB, standard plaque without cutout) are shown, illustrating typical inhomogeneities in the surface dose distributions of eye applicators. These variations are caused by the manufacturing process of the applicators.

14.3 APPLICATION—LIQUID SCINTILLATOR

The design and the feasibility tests for performing dosimetric measurements around a brachytherapy applicator in 3D using a volume filled with LS are described in the following subsections.

14.3.1 PHYSICAL PROPERTIES OF SCINTILLATOR MATERIALS WITH REGARD TO THESE RADIATION QUALITIES

In contrast to a dosimetry system with a small piece of scintillator moved in a water phantom by a scanner, the usage of large scintillator sheets or LS filled volumes measurement volume would allow faster measurement of complete 2D- and 3D-dose distributions. The scintillation materials for such 2D (Perera et al. 1992) and 3D dosimetry need to be highly efficient, to be dosimetrically tissue (water) equivalent, and to exhibit high localization of the scintillator light emission. If a material is to be used as a 3D dosimeter for radiation therapy it will need to serve simultaneously as a tissue-equivalent phantom material and a

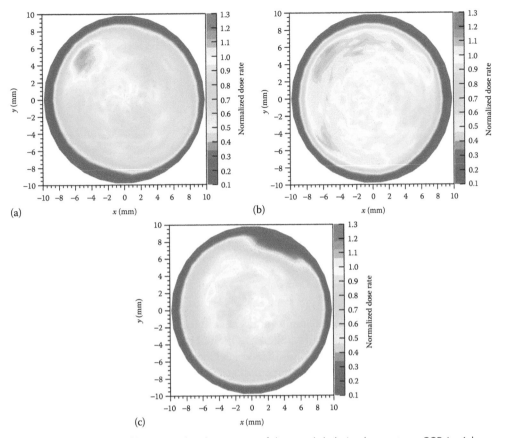

Figure 14.11 Surface dose profiles, normalized to center, of three ophthalmic plaques type CCB (serial number A: 1256 B: 1290 C: 1299): Measured with system 3 in an effective distance of 0.45mm. (Data from M. Eichmann, Entwicklung eines hochpräzisen Dosimetriesystems zur Messung der Oberflächendosisverteilung von Augenapplikatoren, Dissertation, Technische Universität Dortmund, 2010.)

dose sensitive medium (Kirov et al. 2000b). Therefore, tissue equivalence will need to be achieved for both the energy absorption and also for the radiation propagation properties, which are represented by the energy absorption (μ_{en}/ρ) and the linear attenuation (μ/ρ) coefficients. At low energies both coefficients are strongly dependent on the effective atomic number of the LS solutions due to the dominant role of the photoelectric effect cross section.

Achieving dosimetric water equivalence and the related restrictions have been discussed in detail by Kirov et al. (1999, 2000b). It was shown that adding a compound containing a medium atomic number element can bring both the mass energy absorption and linear attenuation coefficients of the scintillating materials closer to water (Figure 14.12). However, the compound needs to be selected in such a way that it does not prevent PS materials to polymerize, does not quench the scintillation light output, and does not affect the localization of scintillation light emission (the distance from the primary radiation interaction point to the point of secondary light photon emission) (Kirov et al. 2000a).

In addition, the following needs to be considered in the additive selection process. First, the atomic number of the additive, Z_{add}, has to be as close as possible to oxygen to match the properties of water. Second, the K-edge of the added element has to be below about 20 keV to avoid abrupt changes in the cross section in the important energy range for brachytherapy dosimetry above this threshold. On the other hand, Z_{add} has to be high enough so that acceptable similarity to water can be achieved with sufficiently low concentration of the additive, so that the other properties of the scintillator are not affected. Several LS solutions and potential loading elements (e.g., Cl, P, and Si) have been considered, which have improved water equivalence and scintillation localization while preserving high efficiency (Kirov et al. 1999, 2000b).

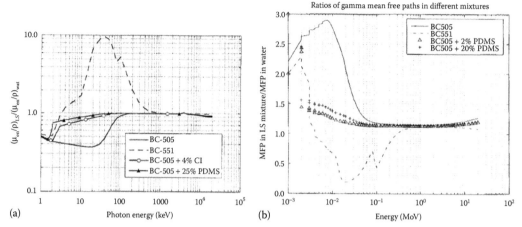

Figure 14.12 Ratios of mass energy absorption coefficients (a) and photon mean free paths (MFP) (b) of different LS mixtures to water. PDMS and PMPS denote two Si containing optical grease compounds described in the text. (Reproduced with permission from A. S. Kirov et al. New water equivalent liquid scintillator solutions for 3D dosimetry. *Med. Phys.* 2000b, 27: 1156–1164. Copyright 2000, American Institute of Physics.)

For loading LS solutions with medium atomic number elements, two Si containing compounds have been tested: poly(dimethylsiloxane) $CH_3[(Si(CH_3)_2O]_2Si(CH_3)_3$ also known as optical grease, and poly(methylphenylsiloxane) $(–Si(CH_3)(C_6H_5)O–)n$ (Aldrich Chemical Company, Inc., Milwaukee, WI) (Kirov et al. 2000b). What makes both compounds attractive is that they both have high optical transparency and in addition to silicon contain also oxygen. On the other hand, however, since collisions between the molecules is one of the two processes dominating non-radiative energy transfer (Hanagodimath et al. 1990, Malimath et al. 1997), increasing the viscosity of the scintillator solution is expected to decrease the energy transfer and the scintillating efficiency of the solution. Therefore, the effect of the viscosity and/or concentration of the additive also need to be considered.

14.3.2 DESCRIPTION OF A LS SYSTEM

A LS solution with tissue or water-equivalent energy absorption and radiation transport properties as described in 14.3.1 provides the possibility for measuring the dose simultaneously in all voxels of the measurement volume (i.e., in 3D). In order to perform this, a tomographic measurement geometry, that is obtaining views of the scintillation volume under multiple angles needs to be designed. This can be realized in several ways, for example, by using many detectors arranged around the scintillating volume similarly to positron emission tomography (PET) or by moving one 2D detector around this volume similarly to single photon emission tomography (SPECT). In the case of brachytherapy such a measurement can be simplified by using a stationary 2D light detector and by rotating the dose distribution inside the measurement volume by rotating the brachytherapy source.

The latter arrangement was suggested and used by Kirov et al for measuring the 3D dose distribution in the vicinity of a Ru-106 eye plaque (Figure 14.13) (Kirov et al. 2005). It is important to note that full tomographic measurement (and therefore the dose distribution measurement), can be performed only for these parts of the volume which can be seen by the detector under all rotation positions of the source. For a source attached from the top (as above [Kirov et al. 2005]) this restricts the measurements only to the part of the LS volume under the source, since above the lower rim of the source the projections will be obstructed by either the source or its holder.

In the above arrangement (Kirov et al. 2005), the measurement volume (under the source rim) was a cube with 1 inch side filled with a LS solution selected in a previous investigation (Kirov et al. 2000b) (see Section 14.3.1). The LS was poured in a rectangular cell of sandblasted black Delrin to minimize light scatter from the cell walls and viewed through a window with high surface accuracy. After pouring into the cell, the liquid scintillator was bubbled with nitrogen to reduce quenching. The scintillation light emitted from the cell was collimated by a 2 mm thick collimating grid with 20 μm diameter hexagonal

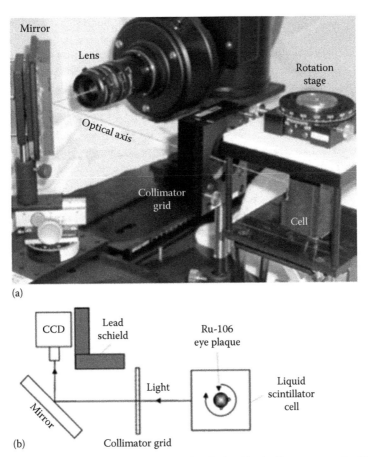

Figure 14.13 A photograph (a) and a top view schematic (b) of a liquid scintillator system for 3D dose measurement in the vicinity of a brachytherapy source attached to the rotation stage. The lead shield protecting the CCD from the source is removed for the photograph. (Reproduced with permission from A. S. Kirov et al. The three-dimensional scintillation dosimetry method: Test for a [106] Ru eye plaque applicator. *Phys. Med. Biol.* 2005, 50: 3063–3081. Copyright Institute of Physics and Engineering in Medicine. All rights reserved.)

holes. The light was detected by a very low noise (liquid nitrogen cooled) charge-coupled device (CCD) camera equipped with low distortion macro lens focused on the front of the collimating grid. To reduce the spikes in the detector produced by gamma rays and scattered bremsstrahlung radiation, the CCD detector was shielded from the source with lead bricks and a front surface mirror is used to ensure a view of the collimator grid. Five exposures with a length of 1 min were performed for 64 equidistant rotation positions of the source positioned on a rotation stage inside the cell just above the measurement volume. After dark image subtraction, the 2D CCD images were corrected for spikes generated by cosmic rays and scattered photons (Devic et al. 1999) and were smoothed to remove *chicken wire* type artifacts from the collimator grid. A modified SPECT maximum likelihood expectation maximization algorithm (MLEM) algorithm (Wallis and Miller 1993) was used to reconstruct the scintillation emission density in voxels with 0.4242 mm side. Post reconstruction were performed corrections for: background due to light scatter inside the cell, edge of cell artifacts, and the point spread function (PSF) of the LS response by performing regularized iterative deconvolution.

14.3.3 UNCERTAINTIES OF MEASUREMENT

For the case of 3D dosimetry performed by optical tomography of a volume filled LS, the system PSF is a major source of dose measurement uncertainty. The following list of possible factors that may affect the PSF and the measurement accuracy was given by Kirov et al. (2005):

1. Light scatter within the cell can produce a slowly varying background signal and therefore can contribute to the tails of the PSF. The amount of scatter can be reduced by special treatment of the measurement cell walls and by estimating and subtracting this background light.
2. Approximations of the collimator aperture function used by the reconstruction algorithm.
3. Approximations of the system PSF and accounting for contributions from all optical components in addition to the collimator grid (cell window, camera lens, and mirror) and their imperfections.
4. Using a flat window of the measurement cell to produce well-defined projections of the measurement volume imposes to use a rectangular cross section cell. For such arrangement, the source needs to be rotated inside the cell and an error will be introduced, since (a) there will be missing projections for part of the dose distribution in the regions of the volume between the cell walls and an inscribed cylinder and (b) there will be difference in handling the projections between the experiment on one hand and the reconstruction algorithm. In the experiment, the detected light rays are always perpendicular to one side of the cubic measurement volume. The MLEM algorithm, however, which uses as an initial approximation, a cube with emission density of 1 for all voxels, *see* the entire nonzero emission density volume under all angles, that is, the forward projections of the volume are not always perpendicular to one side of the cube.
5. Scintillation quenching effects (Birks 1964, Williamson et al. 1999) due to increase of the ionization density at low electron energies, less than about 100 keV (Horrocks 1964, Peron and Cassette 1996), may introduce energy dependence and therefore dependence of the PSF on the distance from the source.
6. Cerenkov light may contribute to the detected light for higher energy beta sources and can also be produced by electrons set in motion by high energy photons (see Chapter 5).
7. The selection of the reconstruction and PSF deconvolution parameters also affect the obtained dose distribution. Careful selection and validation of these parameters is needed.

The magnitude of the errors listed above may be significant and may vary widely between different measurement realizations. Clearly further work in quantifying these errors is needed, prior to attempting 3D dose measurements for clinical brachytherapy measurements.

14.3.4 2D/3D DOSE DISTRIBUTIONS OF EYE PLAQUES, AS EXAMPLES

The 3D liquid scintillation dosimetry (3D SD) method described above was used to measure the relative 3D dose distributions for a CCX Ru-106 eye plaque applicator, which were evaluated by comparison with RCF and diode measurements (Kirov et al. 2005). The evaluation was performed separately along the central axis of the plaque and in planes perpendicular to this axis. Along the plaque axis agreement within the combined measurement uncertainty of the three methods was obtained for the relative dose slope between 5.4 and 7.4 mm from the surface of the plaque. The experimental arrangement of holding the plaque facing down allowed measurements only at distances along the central axis more than 3.3 mm away from its surface due to blocked projection by the plaque and its suture tabs (Figure 14.1). The method was not accurate in the regions between 3.3 and 5.4 mm and beyond 10 mm from the surface of the plaque due to edge effects, plaque positioning accuracy and the other sources of uncertainty (e.g., optical system resolution and light scatter) described above. When the relative 3D SD dose distribution was normalized to RCF measurements at the axis point at 5.4 mm, the agreement with RCF for a volume with 12 mm diameter between 5.4 and 10 mm from the plaque surface was within 25% for most points and ranges up to 45% or 2 mm distance-to-agreement in parts of the volume under the suture tabs of the plaque. Comparisons of the dose distributions measured by RCF and the 3D SD method in two planes perpendicular to the plaque axis are given in Figure 14.14. Comparisons along profiles within these planes also to diode measurements are given in Figure 14.15.

In conclusion, using a LS as both a phantom material and dose measurement media for brachytherapy sources is feasible. Such measurements however require careful characterization and corrections for a large number of physical and numerical phenomena as described above and in more

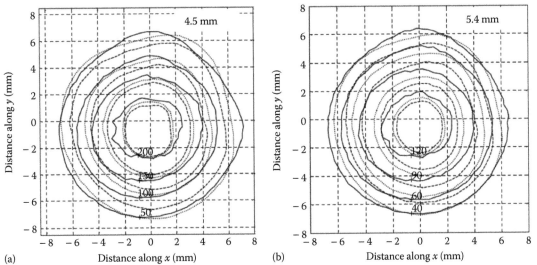

(a) Distance along x (mm)

(b) Distance along x (mm)

Figure 14.14 A comparison of 3D SD measured relative dose-rate contours in slices at (a) 4.5 mm and (b) 5.4 mm from the inner surface of the plaque, to the result from a RCF measurement (thick solid lines). The 3D SD data (dotted and dashed lines) were normalized to the absolute RCF measurements at the peak and were shifted by −0.7 mm (a) and −0.5 mm (b) along y for visual registration of the contours. The dotted lines correspond to the relative emission density, and the dashed lines correspond to the relative dose deduced from the emission density after corrections and liquid scintillator point spread function deconvolution. The contour levels are in cGy/h and correspond to the start of the RCF measurements. (Reproduced with permission from A. S. Kirov et al. The three-dimensional scintillation dosimetry method: Test for a [106]Ru eye plaque applicator. *Phys. Med. Biol.* 2005, 50: 3063–3081. Copyright Institute of Physics and Engineering in Medicine. All rights reserved.)

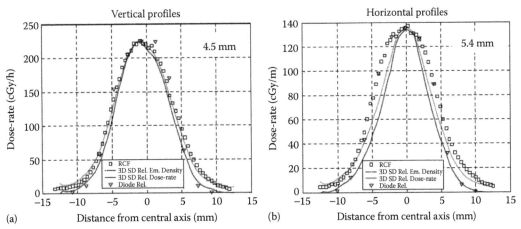

(a) Distance from central axis (mm)

(b) Distance from central axis (mm)

Figure 14.15 A relative comparison of dose-rate profiles obtained with the 3D SD method to RCF and diode profiles through the center of the slices shown in Figure 14.14 at 4.5 mm (a) and 5.4 mm (b) form the plaque surface. The 3D SD and diode profiles are normalized to the absolute RCF measurements at the highest point in each slice. (Reproduced with permission from A. S. Kirov et al. The three-dimensional scintillation dosimetry method: Test for a [106]Ru eye plaque applicator. *Phys. Med. Biol.* 2005, 50: 3063–3081. Copyright Institute of Physics and Engineering in Medicine. All rights reserved.)

detail in the original article (Kirov et al. 2005). As pointed in that article, performing measurements in these parts of the volume, which may have projections that are blocked by the applicator itself, may be approached by providing to the reconstruction algorithm the exact geometry and location of the brachytherapy applicator and by taking steps to eliminate scattering of light from its surface. Since such future work would belong to the research domain, the small PS tip approach described in the beginning of this chapter is closer to using the scintillation process for clinical brachytherapy dosimetry measurements.

REFERENCES

M. Bambynek. Entwicklung einer Multielektroden-Extrapolationskammer als Prototyp einer Primärnormal-Meßeinrichtung zur Darstellung und Weitergabe der Meßgröße Wasser-Energiedosis von Beta-Brachytherapiequellen, dissertation, Dortmund, Germany, Technische Universität Dortmund, 2000.

M. Bambynek, D. Flühs, U. Quast, D. Wegener, C. G. Soares. A high-precision, high-resolution and fast dosimetry system for beta sources applied in cardiovascular brachytherapy. *Med. Phys.* 2000, 27(4): 662–667.

J. B. Birks. *The Theory and Practice of Scintillation Counting.* New York, Pergamon, 1964.

S. Devic, A. Kirov, J. Williamson, Z. Piao, J. Dempsey. A robust algorithm for spikes elimination from 2D CCD low-level light output measurements in the presence of ionizing radiation field (abstract). *Med. Phys.* 1999, 26: 1161.

M. Eichmann. Entwicklung eines hochpräzisen Dosimetriesystems zur Messung der Oberflächendosisverteilung von Augenapplikatoren, dissertation, Dortmund, Germany, Technische Universität Dortmund, 2010.

M. Eichmann, D. Flühs, B. Spaan. Development of a high precision dosimetry system for the measurement of surface dose rate distribution for eye applicators. *Med. Phys.* 2009, 36(10): 4634–4643.

M. Eichmann, T. Krause, D. Flühs, B. Spaan. Development of a high precision xyz-measuring table for the determination of 3D dose rate distributions of brachytherapy sources. *Phys. Med. Biol.* 2012, 57: N421–N429.

D. Flühs, G. Anastassiou, J. Wening, W. Sauerwein, N. Bornfeld. The design and the dosimetry of bi-nuclide radioactive ophthalmic applicators. *Med. Phys.* 2004, 31(6): 1481–1488.

D. Flühs, M. Bambynek, M. Heintz, F. Indenkämpen, H. Kolanoski, D. Wegener, W. Sauerwein, U. Quast. Dosimetry and design of radioactive eye plaques. *Front. Radiat. Ther. Oncol.* 1997, 30: 26–38.

D. Flühs, A. Flühs, M. Ebenau, M. Eichmann. PEN scintillator–A novel detector for the dosimetry of radioactive ophthalmic applicators. *Ocul. Oncol. Pathol.* 2016, 2:5–12.

D. Flühs, M. Heintz, F. Indenkämpen, C. Wieczorek, H. Kolanoski, U. Quast. Direct reading measurement of absorbed dose with plastic scintillators—The general concept and applications to ophthalmic plaque dosimetry. *Med. Phys.* 1996, 23(3): 427–434.

S. M. Hanagodimath, G. S. Gadaginmath, G. C. Chikkur. Role of brownian diffusion and interaction distance on energy transfer and quenching in an organic liquid scintillator. *Appl. Radiat. Isot.* 1990, 41: 817–821.

D. Harder, A. Rubach, K.-P. Hermann, A. Überschär. Wasser- und gewebeäquivalente Festkörperphantome für hochenergetische Photonen und Elektronen. *Med. Phys.* 1988, 325–330.

D. L. Horrocks. Pulse height energy relationship of a liquid scintillator for electrons of energy less than 100 keV. *Nucl. Instr. Meth.* 1964, 30: 157–160.

A. S. Kirov, W. R. Binns, J. F. Dempsey, J. W. Epstein, P. F. Dowkontt, S. Shrinivas, C. Hurlbut, J. F. Williamson. Towards two dimensional brachytherapy dosimetry using plastic scintillator: Localization of the scintillation process. *Nucl. Instr. Meth.* 2000a, A439: 178–188.

A. S. Kirov, C. Hurlbut, J. F. Dempsey, S. B. Shrinivas, J. W. Epstein, W. R. Binns, P. F. Dowkontt, J. F. Williamson. Towards two dimensional brachytherapy dosimetry using plastic scintillator: New highly efficient water equivalent plastic scintillator materials. *Med. Phys.* 1999, 26: 1515–1523.

A. S. Kirov, J. Z. Piao, N. K. Mathur, T. R. Miller, S. Devic, S. Trichter, M. Zaider, C. G. Soares, T. Losasso. The three-dimensional scintillation dosimetry method: Test for a (106)Ru eye plaque applicator. *Phys. Med. Biol.* 2005, 50: 3063–3081.

A. S. Kirov, S. Shrinivas, C. Hurlbut, J. F. Dempsey, W. R. Binns, J. L. Poblete. New water equivalent liquid scintillator solutions for 3D dosimetry. *Med. Phys.* 2000b, 27: 1156–1164.

S. W. Lee and W. D. Reece. Dose backscatter factors for selected beta sources as a function of source, calcified plaque and contrast agent using Monte Carlo calculations. *Phys. Med. Biol.* 2004, 49: 583–599.

F. Lessard, L. Archambault, M. Plamodon, Ph. Després, F. Therriault-Proulx, S. Beddar, L. Beaulieu. Validating plastic scintillation detectors for photon dosimetry in the radiologic energy range. *Med. Phys.* 2012, 39(9): 5308–5316.

G. H. Malimath, G. C. Chikkur, H. Pal, T. Mukherjee. Role of internal mechanisms in energy transfer processes in organic liquid scintillators. *Appl. Radiat. Isot.* 1997, 48: 359–364.

H. Perera, J. F. Williamson, S. P. Monthofer, W. R. Binns, J. Klarmann, G. L. Fuller, J. W. Wong. Rapid two-dimensional dose measurement in brachytherapy using plastic scintillator sheet: Linearity, signal to noise ratio, and energy response characteristics. *Int. J. Rad. Onc. Biol. Phys.* 1992, 23: 1059–1069.

M. N. Peron and P. Cassette. A Compton coincidence study of liquid scintillator response in the 1–20 keV energy range. *Nucl. Instr. & Meth. A* 1996, 369: 344–347.

H. Reich (Pub.). *Dosimetrie ionisierender Strahlung.* Teubner, Stuttgart, Germany, 1990.

R. M. Thomson, R. E. P. Taylor, D. W. O. Rogers. Monte Carlo dosimetry for ^{125}I and ^{103}Pd eye plaque brachytherapy. *Med. Phys.* 2008, 35(12): 5530–5543.

J. Venselaar, A. S. Meigooni, D. Baltas, P. J. Hoskin (Ed.). *Comprehensive Brachytherapy: Physical and Clinical Aspects.* Boca Raton, FL, CRC Press/Taylor & Francis Group, 2013.

J. W. Wallis and T. R. Miller. Rapidly converging iterative reconstruction algorithms in single-photon emission computed tomography. *J. Nucl. Med.* 1993, 34: 1793–1800.

J. F. Williamson, J. F. Dempsey, A. S. Kirov, J. I. Monroe, W. R. Binns, H. Hedtjärn. Plastic scintillator response to low-energy photons. *Phys. Med. Biol.* 1999, 44: 857–871.

3D liquid scintillation dosimetry for photons and protons

Daniel Robertson and Sam Beddar

Contents

15.1 INTRODUCTION

The goal of conformal radiation therapy is to deliver a curative dose to the tumor while keeping the dose to nearby tissues to a minimum. This is accomplished by judicious selection of beam angles and careful shaping of the radiation fields. In recent years, advancements in radiation therapy delivery technology have enabled more detailed sculpting of therapeutic radiation beams in space and time, making it possible to escalate tumor doses and decrease treatment-related side effects. These technological advances include intensity-modulated radiation therapy (IMRT), tomotherapy, volumetric modulated arch therapy (VMAT), stereotactic radiotherapy, passive-scattering proton therapy, and intensity-modulated proton therapy.

While the details of these delivery modalities vary, they all produce complex dose distributions featuring steep dose gradients, irregularly shaped fields, and spatial and temporal heterogeneity. The complexity of these fields makes it difficult to perform secondary monitor unit calculations, and as a result, it has become standard practice to perform patient-specific verification measurements prior to treatment.

The dosimetric complexities of modern radiotherapy modalities also pose a challenge to standard radiation measurement devices such as ionization chambers, ionization chamber and diode arrays, and radiographic and radiochromic films. Ionization chambers are subject to volume averaging effects, which can decrease their accuracy in regions of steep dose gradients. Ionization chambers also perturb the radiation field in the process of measuring it, which places limits on the number and orientation of detectors that can be used simultaneously in a given field. Ionization chamber arrays and diode arrays have limited spatial resolution and they provide measurements only at a single depth. Radiographic and radiochromic films have high spatial resolution, but they also measure only at a single depth, and they are incapable of resolving temporal aspects of radiation delivery. Because today's standard detectors only measure a small subset of locations in a radiation field, they may fail to detect differences between the planned and delivered dose distributions.

A large-volume, high-resolution, three-dimensional (3D) dosimeter would be a valuable tool in radiation dosimetry. Such a detector could be used to measure the entire dose distribution in a radiation field at once, providing much more information than current dosimeters. It could be used for full 3D dosimetric measurements of patient treatment fields as well as a variety of machine-related quality assurance (QA) measurements. Temporal resolution would provide increased utility to such a detector, making it capable of measuring the dynamic aspects of beam delivery.

Currently available 3D dosimeters include Fricke (Kelly et al. 1998) and polymer gel dosimeters (Maryański et al. 1996, Oldham et al. 2001) and solid radiochromic dosimeters (Guo et al. 2006). Despite many years of development, gel dosimeters have yet to be used widely. This is most likely due to the difficulty of preparing and working with the gels, their sensitivity to temperature and preparation conditions (Maryanski et al. 1997), and (in the case of Fricke dosimeters) diffusion of the activated material (Pedersen et al. 1997). Radiochromic dosimeters are a more recent development, and they are not subject to the preparation and handling-related difficulties associated with polymer gel dosimeters. However, they are only available in limited sizes (up to 2000 cm^3) and tend to be expensive.

Gel and radiochromic dosimeters require readout on a CT, MRI, or optical CT scanner (Oldham et al. 2001, 2003, Hilts and Duzenli 2004). While they have historically been single-use dosimeters, reusable formulations of gel and radiochromic dosimeters have recently been developed (Pierquet et al. 2010). Such dosimeters fade to background with a time scale between a day and 2 weeks, allowing the dosimeter to be reused a number of times. Another limitation of polymer and radiochromic dosimeters is that they are integrating dosimeters and are not capable of measuring any time-dependence in the dose distribution.

Volumetric scintillation dosimetry is a relatively new area of study with the potential to provide new capabilities to the field of 3D dosimetry (Beddar 2015). The basic concept of volumetric scintillation dosimetry is the use of a large volume of scintillator to convert the delivered dose distribution into a visible light distribution, which can then be measured using cameras or other light-sensitive detectors (Figure 15.1). When an organic liquid or plastic scintillator material is used, the detection volume doubles

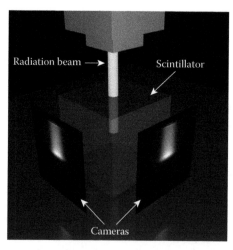

Figure 15.1 Volumetric scintillation dosimetry employs a large volume of scintillator, which serves as a water-equivalent phantom as well as a radiation-sensitive medium. A radiation beam is delivered to the scintillator volume, and the emitted light distribution is imaged by cameras.

as a near-water-equivalent phantom. This enables the radiation field to be measured without the need to account for the perturbation of a non-water-equivalent detector.

Another significant advantage of volumetric scintillation dosimetry is that the scintillation light is emitted immediately (typically within a few nanoseconds) following irradiation. This fast response makes it possible to measure the temporal details of treatment delivery, even in the case of rapidly changing radiation fields such as spot-scanning proton beams.

The objectives of this chapter are to introduce the concept of volumetric scintillation dosimetry, to discuss various aspects of practical systems including instrumentation, image processing, and 3D reconstruction, and to review prior work and potential applications.

15.2 INSTRUMENTATION

The two key components of a volumetric scintillation dosimetry system are the scintillator itself and the detectors used to collect the scintillation light. In designing such a system, it is important to carefully select a scintillator material, and a light detection system and configuration that are appropriate for the intended application. Once this has been established, the detection system must be calibrated to establish its spatial scale, dose response, and dynamic range. If the detector is to be used to measure the temporal dynamics of radiation delivery, additional steps must be taken to synchronize the measurement electronics with the timing of beam delivery.

15.2.1 SELECTION OF THE SCINTILLATOR MATERIAL

The scintillator in a volumetric scintillation dosimetry system fills the role of the detector as well as that of the measurement phantom. Therefore, it is important to select a scintillator with favorable properties for both of these roles. Some important properties for the radiation detection role include the scintillation efficiency, emission spectrum, temporal response, and optical clarity of the scintillator. Important properties for the measurement phantom role include the physical density, the elemental composition, and radiological properties such as stopping powers and attenuation coefficients for the particles and energy range of interest. Other practical considerations include physical and chemical properties such as the phase of the scintillator (solid or liquid) and its chemical safety and compatibility with other materials.

15.2.1.1 Scintillator response properties

The scintillation efficiency (ε) is the ratio of the energy emitted as visible light (E_{em}) to the energy deposited within the scintillator (E_{dep}):

$$\varepsilon = \frac{E_{em}}{E_{dep}} \tag{15.1}$$

Historically, anthracene crystals were one of the most efficient and most commonly used scintillating materials, so scintillation efficiency is often given as a percentage relative to the scintillation efficiency of anthracene. A considerable amount of work has been devoted into developing scintillators with high efficiency, with the goal of increasing the signal-to-noise ratio in situations where the signal is very weak. Fortunately, in volumetric scintillation dosimetry, we usually operate far from the low-signal limits of the scintillator or detector system. We are generally concerned about measuring irradiations on the order of cGy, rather than individual particle interactions. Most modern scintillator materials provide plenty of light output in the dose range of interest for external beam radiation therapy. Indeed, it is probably more common to struggle with detector saturation than with insufficient signal.

That being said, there are a few cases in which sufficient signal may be a concern. One such case is the measurement of low-dose-rate brachytherapy sources. Another situation where high scintillation efficiency may be important is in high-frame-rate measurements of dynamic radiation fields, where the dose delivered during a given image frame may be low. There are also systems such as synchrotron-based proton and carbon ion accelerators that may deliver radiation in a large number of very small bursts or *spots*. Accurate measurements of individual low-dose beam spots may also require a scintillator with high scintillation efficiency.

Table 15.1 **Physical properties of some common organic scintillators**

	BC-531	BC-400	BCF-12	BCF-60
Form	Liquid	Plastic	Plastic fiber	Plastic fiber
Solvent	Linear alkylbenzene	Polyvinyltoluene	Polystyrene	Polystyrene
ε (% Anthracene)	59	65	50	44
λ_p (nm)	425	423	435	530
l (m)	3.5	1.6	2.7	3.5
ρ (g/cm³)	0.87	1.03	1.05	1.05
H:C ratio	1.63:1	1.104	1	1

The scintillation efficiencies of several scintillators used in previous volumetric scintillation detectors are listed in Table 15.1. Of course, scintillation efficiency is only one aspect of the detector efficiency. Equally important is the geometry and efficiency of the light collection and measurement equipment.

The spectrum of the scintillation light is another important property to consider. The primary concern with the spectrum is that it overlaps well with the spectral sensitivity of the light detection device. The peak emission wavelength, λ_p, of most organic scintillators is in the near-UV or blue spectral region. Some scintillators incorporate wavelength shifters, which may provide better spectral matching with various detectors. Another potential spectral concern is related to Cerenkov light. If the detector is to be used in high-energy photon beams or other settings with non-negligible Cerenkov light emission, it may be desirable to select a scintillator with an emission spectrum that will aid in discrimination between the scintillation and Cerenkov light (see Chapter 5 for details). Table 15.1 lists the peak emission wavelengths of several organic scintillators.

The temporal response of a scintillator is relatively unimportant in volumetric scintillation dosimetry. While some rare applications may require frame rates as high as hundreds of frames per second, nearly all organic scintillators have decay times of a few nanoseconds. While it is good to be aware of the decay time of the selected scintillator, it is unlikely to impact the suitability of a given scintillator for this application.

One final property of note is the degree of self-attenuation and scatter in the scintillator. An ideal scintillator will emit light at a longer wavelength than it absorbs, allowing it to be effectively transparent to its own emissions. Organic scintillators generally perform very well in this regard, but none is perfectly transparent. Because volumetric scintillation dosimetry employs fairly large volumes of scintillator, it is important to select a material in which the emitted light will not experience significant attenuation or scatter as it propagates from one end of the detector to the other. The self-attenuation of a scintillator can be defined by its attenuation length, l, which is the distance the scintillation light can travel within the scintillator before it is attenuated to $1/e$ of its initial intensity. Table 15.1 includes the attenuation lengths of several scintillators.

15.2.1.2 Radiological properties

The standard measurement medium in radiotherapy physics is water. Therefore, while performing measurements in a medium other than water, it is important to consider the differences in the radiation interactions in the medium relative to water—in other words, its *water-equivalence*. While the only truly water-equivalent material is H_2O, there are a number of other materials (typically organic polymers or liquids) that interact in a manner quite similar to water within the energy range of interest in radiation therapy. The specific interactions and cross-sections of interest for evaluating radiological water-equivalence depend upon the particle species and energy range. However, a good first approximation of the water-equivalence of a material can be made by considering its physical density, ρ, and elemental composition (often given as a ratio). Because the majority of the scintillator material is the solvent, the properties of a given scintillator are essentially the same as its solvent. Table 15.1 includes physical densities, elemental compositions, and solvent materials of some commonly used liquid and plastic scintillators.

In the 1–20 MeV energy range of interest for external-beam photon radiotherapy, Compton scattering is the dominant interaction mechanism. Therefore, the most important aspects of radiological water-equivalence for photon therapy detectors are the mass energy absorption coefficient and the electron mass collisional stopping power. In addition to these parameters, when measuring high-energy electron beams, it is important to consider the mass angular scattering power of the medium. As shown in Figures 15.2–15.4, organic scintillators exhibit a high degree of water-equivalence by these metrics (Attix 1986, Beddar et al. 1992).

In the 1–300 MeV energy range of interest for proton and heavy ion therapy, the most relevant radiological property for a phantom material is its scintillator-to-water linear stopping power ratio. The linear stopping power ratio typically does not change substantially in the energy range of interest, making it a concise metric to compare the water-equivalence of different materials. In addition, the scintillator-to-water linear stopping power ratio is related to the water-equivalent thickness of a material, so it can be used to easily determine the scintillator-to-water depth scaling of depth-dose measurements. The proton linear stopping power ratios for polystyrene- and polyvinyltoluene-based scintillators are shown in Figure 15.5.

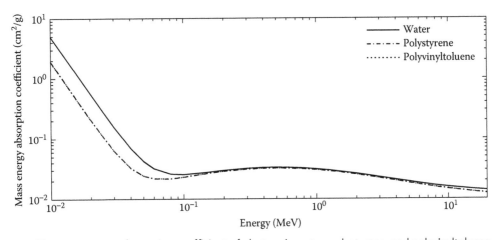

Figure 15.2 The mass energy absorption coefficient of photons in water, polystyrene, and polyvinyltoluene-based plastic scintillator. (Data from Berger, M.J. et al. *Stopping-Power and Range Tables for Electrons, Protons, and Helium Ions.* National Institute of Standards and Technology, 1998.)

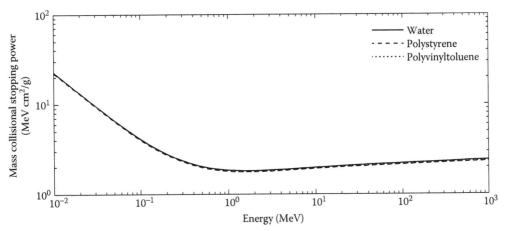

Figure 15.3 The mass collisional stopping power of electrons in water, polystyrene, and polyvinyltoluene-based plastic scintillator. (Data from Berger, M.J. et al., *Stopping-Power and Range Tables for Electrons, Protons, and Helium Ions.* National Institute of Standards and Technology, 1998.)

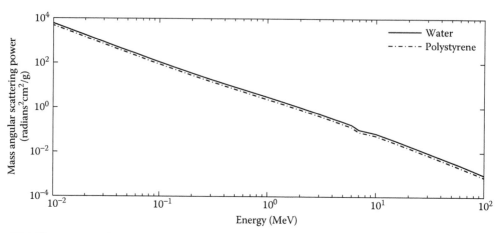

Figure 15.4 The mass angular scattering power of electrons in water and polystyrene. (Data from ICRU, *Radiation Dosimetry: Electron Beams with Energies between 1 and 50 MeV*. Bethesda, MD: ICRU, 1984.)

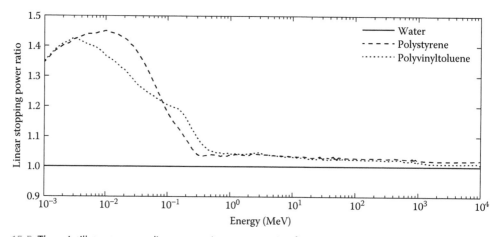

Figure 15.5 The scintillator-to-water linear stopping power ratio of protons in water, polystyrene, and polyvinyltoluene-based plastic scintillator. (Data from Berger, M.J. et al., *Stopping-Power and Range Tables for Electrons, Protons, and Helium Ions*. National Institute of Standards and Technology, 1998.)

Stopping powers and ranges of ions have been measured for some of the bulk solvent materials used in organic scintillators (Berger et al. 1998). In the cases where measured data is not available, stopping powers and ranges can be calculated using a linear combination of the stopping powers of the atomic constituents, with correction factors to account for the molecular bonds of the material (Ziegler et al. 2010). One difficulty of this approach is that scintillator manufacturers may not provide sufficient details about a scintillator's chemical composition to provide the required input for these calculations. In this situation, some assumptions about the chemical composition must be made, which increases the uncertainty of the calculation. The stopping power ratio can also be determined experimentally based on measurements of the water-equivalent thickness of the scintillator (Zhang et al. 2010, Sánchez-Parcerisa et al. 2012). While stopping power calculations are easier to perform, measurements tend to be more accurate and are therefore preferred for material characterization in volumetric scintillation dosimetry (Ingram et al. 2015).

15.2.1.3 Physical and chemical properties

The choice between plastic and liquid scintillator materials has an impact on many of the details of detector design, use, and performance. The majority of volumetric scintillation dosimetry research has been done using liquid scintillators (Kirov et al. 2005, Beddar et al. 2009, Ponisch et al. 2009, Robertson et al. 2013, 2014, Hui et al. 2014), but solid plastic scintillators have also been used (Fukushima et al. 2006, Goulet et al. 2014). Liquid scintillators offer many benefits for this application. They are inexpensive, their size and shape are limited only by their container, and they are not subject to manufacturing defects and physical wear-and-tear in the way that large blocks of plastic scintillator are. Because the liquid continually mixes, liquid scintillators are more resistant to radiation damage than plastic scintillators, and any damage that does occur will be distributed homogeneously throughout the scintillator. Plastic scintillators eventually discolor in regions receiving high cumulative doses, which leads to spatial inhomogeneity of the light propagation within the scintillator.

While liquid scintillators have many advantages for volumetric dosimetry, they also have drawbacks. The largest component of most organic liquid scintillators is an aromatic hydrocarbon solvent. From a health hazard point of view, these solvents range from irritants to carcinogens, and they also tend to be flammable. Newer scintillator formulations are less toxic than older varieties, but precautions are still warranted while handling them. Liquid scintillators are also volatile, and they should be tightly contained to avoid exposure to fumes and spillage. It is also important to consider the chemical compatibility of all materials that will come into contact with the scintillator.

15.2.2 DETECTOR CONFIGURATION

The detector configuration should meet the requirements of the application. This involves decisions regarding the size and shape of the scintillator, the type and number of light sensors, and the orientation of the sensors with respect to the scintillator volume.

15.2.2.1 Scintillator volume

Ideally, the scintillation volume should be larger than the largest radiation field that will be delivered. However, in the case of most external beam radiotherapy machines, this would lead to a prohibitively large and heavy scintillator volume. In the water tanks used for dosimetry measurements, this problem is solved by filling the tank from a faucet or an external reservoir with a pump. Solid phantoms are usually composed of multiple small slabs that are easy to work with. Volumetric scintillation dosimetry is not amenable to either of these approaches. As a consequence, a compromise is typically made so as to provide a large enough detection volume for most common radiation fields, while not exceeding a manageable weight and size.

Most studies to date have employed cubic or rectangular scintillator geometries (Kirov et al. 2005, Beddar et al. 2009, Archambault et al. 2012), which are easy to construct and work with. These shapes also provide a flat detector-air interface, which simplifies the effects of light refraction at the surface of the scintillator volume. Figure 15.6 gives an example of a system developed by the authors. One limitation of a cubic detector geometry is that the number of viewing angles is effectively limited by the number of

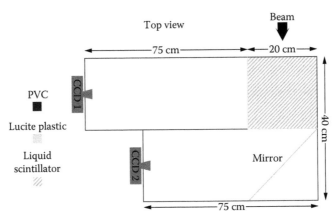

Figure 15.6 A schematic of a volumetric scintillation dosimetry system consisting of a large tank of liquid scintillator in a light-tight housing. Two adjacent cameras view orthogonal faces of the tank. A mirror re-directs the light from one face to the corresponding camera.

faces. One study employed a scintillator in the shape of a hexagonal prism to provide a larger number of unique viewing angles (Kroll et al. 2013). A drawback to increasing the number of sides is the decrease in size of each side, which decreases the portion of the volume interior that can be viewed through a given side.

15.2.2.2 Light sensors

The high resolution required for volumetric scintillation dosimetry dictates the use of cameras for detecting the scintillation light. Charge-coupled device (CCD) and complementary metal–oxide–semiconductor (CMOS) cameras have been the sensors of choice for most investigators. Many important properties of the detector, including its sensitivity and its spatial and temporal resolution (not to mention its cost), are directly determined by the selection of cameras.

The number, orientation, and specifications of the cameras depend on the application. More viewing angles will lead to improved 3D reconstruction, but the bulk, complexity, and cost of the detector also increase with the addition of more cameras, placing practical limits on the number of sensors in the detector configuration. It is also advisable to keep the cameras and other electronics out of the direct line of the radiation beam (Kroll et al. 2013). Mirrors can be used to redirect the light from the beam direction, allowing the measurement of a beam's-eye-view without exposing the camera directly to the radiation beam, as shown in Figure 15.6.

15.2.3 MEASUREMENT OF DYNAMIC RADIATION BEAMS

The high temporal resolution of the scintillator-camera system introduces the possibility of what might be termed *4D scintillation dosimetry*, or measurement of a 3D dose distribution as it changes over time. This capability is most useful in radiation beams that change significantly over time, such as VMAT and scanning proton beams. In these delivery systems, the overall dose distribution is comprised of many small segments or beamlets with different shapes and spatial locations, which are delivered sequentially to the target. Integrating dosimeters can only measure the overall dose distribution. However, in order to fully characterize the performance of these systems and precisely detect the source of delivery errors, it would be beneficial to independently measure the contribution of each beamlet as it is delivered in real time.

Volumetric scintillation dosimetry is uniquely qualified for dynamic 3D radiation measurements. However, the measurement of such beams places additional requirements on the detector system. The frame rate of the cameras must be sufficient for the desired temporal resolution, and the lens aperture and image sensor sensitivity must be large enough to measure the light emitted in each temporal step with a good signal-to-noise ratio.

Other important considerations are the triggering of the camera acquisitions and appropriate management of the cameras' dead time. If the goal of the detector is to measure the light emitted from the entire beam delivery, divided into smaller time steps, then the camera-on periods must match the appropriate beam-on periods, and the camera-off periods must occur while the beam is off. All digital video cameras have some dead time between images during which the image readout takes place. Technologies such as frame transfer and interline transfer CCDs have been developed to minimize this camera dead time, but it has not yet been completely eliminated.

In order to correctly time the camera acquisitions and synchronize them with one another, some form of triggering is required. The most effective triggering method is to activate the cameras with signals from the accelerator that indicate the beginning and end of each beamlet. Cameras with a very short readout period may be used without a triggering signal from the accelerator, but this will increase the uncertainty in the measurements and may lead to overlap between sequential beamlets in a given image.

15.3 CALIBRATION AND IMAGE PROCESSING

A central problem in volumetric scintillation dosimetry is the translation of pixel locations and intensities from the camera images into scintillation light emission values at the correct spatial coordinates within the detector. This spatial and intensity information can then be used as the input for a 3D reconstruction algorithm. It can also be used to provide much valuable information about the radiation beam without the need for a complete 3D reconstruction. However, before any useful information can be obtained from the images, the detector must be calibrated and various sources of optical artifacts must be accounted for and corrected. These artifacts include photon scattering, refraction at the scintillator-air interface, camera perspective, lens distortion, vignetting, and artifacts related to the camera sensor itself.

15.3.1 DETECTOR CALIBRATION

The detector calibration defines the relationship between the light intensity measured by a given camera pixel and the light intensity emitted at a given location in the bulk of the scintillator. Important aspects of the calibration include the pixel size, location of the optical axis, ratio of the delivered dose to the measured scintillation intensity, and the degree of attenuation of the scintillation light as it passes through the scintillator and other detector materials.

The spatial aspects of the calibration can be encompassed by the pixel size and by the offset between an image feature (such as the image center or edge) and a physical landmark in the scintillation volume (such as its center or edge). A spatial calibration can be performed by imaging a calibration test object of known size at various locations within the scintillator volume (Robertson et al. 2014). An advantage to using a liquid scintillator material is the ability to place a ruler or test pattern inside the scintillator volume. When using a solid plastic scintillator, the spatial calibration can be performed using a laser mounted on an indexing stage (Kroll et al. 2013). Unless a telecentric lens is used, the effective pixel size will increase with distance from the camera, so it is important to characterize the variation of the pixel size with depth in the scintillator.

The relationship between the delivered radiation dose and the measured light intensity is determined by many factors, including the scintillation efficiency, the quality of the radiation (see Chapter 2), the sensitivity and spectral response of the camera chip, the size of the lens aperture, and the geometry of the detector assembly. As volumetric scintillation dosimetry is typically concerned with relative measurements, the absolute dose-to-light intensity ratio is of secondary importance. However, this value is important for setting parameters such as the monitor units of radiation to deliver and the camera acquisition time, in order to keep the light signal within the dynamic range of the camera chip. For a relative dosimeter, it is also important to establish the linearity of the scintillator light response with dose.

Another important aspect related to the measured light intensity is the self-attenuation of the scintillator. A good scintillator will exhibit very little attenuation of its own light emissions. However, in a large-volume detector the self-attenuation may be non-negligible, and a correction factor can be applied to the scintillation signal based on its depth of origin within the scintillator.

15.3.2 BLURRING

While scintillators are generally transparent to their own light emissions, some scattering and absorption/re-emission events occur as the light travels through the scintillator. This leads to a depth-dependent blurring of the scintillator signal. In addition to scattering within the scintillator, the optical system also adds some blurring to the image. However, contrary to photon scattering, this lens blurring does not vary with depth in the scintillator. Both blurring components can be measured and corrected in order to improve the sharpness of the measured scintillation light distributions.

Measuring the blurring due to the lens and optical system is quite straightforward. A simple and accurate method involves measuring the edge spread function by imaging the border between black and white regions on a test pattern in air.

Measuring the blurring due to scatter in the scintillator is more challenging. In liquid scintillators, the edge spread function technique may also be used, but the test pattern must be immersed in the scintillator and imaged at various depths in order to characterize the relationship between scatter and distance traveled in the scintillator. In order to obtain the most accurate results, the light used to image the edge pattern should match the emission spectrum of the scintillator. An additional practical hurdle is to design a test pattern that is resilient to the organic solvent base of the liquid scintillator.

Once the blurring has been determined, it can be corrected by deconvolving the measured point spread function from the given image. For lens-based blurring, this is straightforward, as the point spread function is approximately independent of the depth of the light emission in the scintillator. In the case of photon scattering in the scintillator, the point spread function increases with depth, complicating the correction process. Fortunately, for many scintillators the scattering of light is negligible, and the blurring in the system is primarily due to the camera optics (Robertson et al. 2014).

15.3.3 PERSPECTIVE AND REFRACTION

Most camera-lens systems exhibit perspective, where objects at a greater distance appear smaller. Perspective is a result of the camera lens design that collects and focuses light rays that converge toward the lens aperture. Because volumetric scintillation detectors consist of a camera viewing a large scintillator volume, the apparent size of an object (or a pixel) near the front of the scintillator volume is larger than the same object near the back of the volume. Perspective must be appropriately addressed in order to achieve spatial accuracy in volumetric scintillation dosimetry.

It is possible to eliminate perspective by using telecentric lenses, which effectively exclude diverging rays from the image. However, a fundamental requirement of telecentric lenses is that the field of view can be no larger than the primary lens. Consequently, telecentric lenses that are large enough to image the large scintillators used in volumetric scintillation dosimetry are prohibitively bulky, heavy, and expensive. A more practical option is to account for perspective by calibrating the variation in pixel size with depth in the scintillator volume.

The issue of perspective is complicated by refraction of the scintillation light at material boundaries. The index of refraction of liquid and plastic scintillators is typically about 1.5, resulting in a significant amount of refraction at the boundary between the air and the scintillator or its container. The degree of refraction increases from zero on the camera's central axis to a maximum value at the edge of the field of view. Refraction changes the apparent location of the light origin, including its depth, resulting in a corresponding change in pixel size and intensity (Figure 15.7). Perspective and refraction are inter-related in their effects, and they can both be described analytically and corrected using a pin-hole approximation of the camera (Robertson et al. 2014).

15.3.4 LENS ARTIFACTS

In addition to perspective and image blurring, the lens introduces distortions and vignetting to the image. Lens distortion is a nonlinear mapping between object space and image space. It is primarily radial because of the radial symmetry of the lens system. However, it can include tangential components. The most common forms of lens distortion are barrel distortion, in which the image magnification decreases with distance from the image center (Figure 15.8a), and pincushion distortion, in which the magnification increases with distance from the image center (Figure 15.8b).

Lens distortion can be corrected through the process of lens calibration, which is the subject of a substantial body of literature. One particularly convenient method uses vanishing-line information from a planar checkerboard pattern viewed at multiple unknown orientations (Caprile and Torre 1990, Wang and Tsai 1990, Tsai 1992, Bouget and Perona 1998). Depending on the application, the use of modern low-distortion lenses can decrease lens distortion to the point that calibration and correction are unnecessary.

Vignetting is a decrease in brightness with increased distance from the image center, as illustrated in Figure 15.9. Vignetting is sometimes introduced to photographs for artistic effect by mechanically blocking the light at the periphery of the image. However, even without this mechanical blocking, all cameras experience natural vignetting, which is caused by the divergence of light as it travels through the camera.

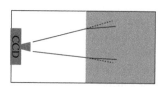

Figure 15.7 Refraction changes the apparent position of the scintillation light origin as viewed by the camera.

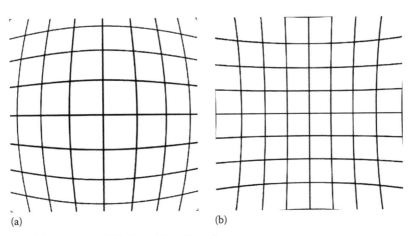

(a)   (b)

Figure 15.8 (a) Barrel distortion and (b) pincushion distortion.

Figure 15.9 Vignetting is a decrease in image brightness with distance from the center.

Vignetting can be modeled by a $\cos^4(\theta)$ function (Ray 1994), where θ is the angle between the optical axis and the ray from the exit pupil to the image sensor (see Figure 15.10). Vignetting for a given location in the image can be calculated by

$$V_{i,j} = \cos^4(\theta_{i,j}) = \frac{a^4}{\left(a^2 + d_{i,j}^2\right)^2}$$

where:
    $V_{i,j}$ is the fractional brightness decrease at pixel i,j
    a is the distance from the exit pupil to the image sensor
    $d_{i,j}$ is the distance from the principal point ($\times$) to pixel i,j

The $\cos^4(\theta)$ model works well for many lenses and may be used to correct images for vignetting. For lenses that are not well-described by the $\cos^4(\theta)$ model, it may be more appropriate to measure the natural vignetting by imaging a flat, uniform light field and then using the measured vignetting profile for image correction.

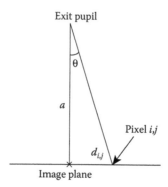

Figure 15.10 Schematic of the cos⁴(θ) rule of vignetting, where θ is the angle between the optical axis and the ray from the exit pupil to pixel *i,j*, *a* is the distance from the exit pupil to the image center, and $d_{i,j}$ is the distance from the principal point (×) to pixel *i,j*. (Reproduced with permission from Robertson, D. et al. 2014. Optical artefact characterization and correction in volumetric scintillation dosimetry. *Physics in Medicine and Biology* 59: 23, Copyright Institute of Physics and Engineering in Medicine. All rights reserved.)

15.3.5 STRAY RADIATION

Volumetric scintillation dosimetry requires the cameras to be in the treatment room adjacent to the radiation beam. CCD and CMOS chips are sensitive to stray radiation in the form of photons, neutrons, and charged particles. When stray radiation strikes the camera chip, it may deposit its energy, leading to *hot pixels* and *blooming* artifacts, where the charge deposited in a single pixel overflows into adjacent pixels (see Figure 15.11). This can lead to corruption of individual pixels or groups of pixels in a given image frame. Stray radiation can also cause permanent damage to individual pixels or other components in the digital camera electronics. This is a greater problem in radiation beams with a large amount of neutron scatter, such as passive scattering proton therapy.

The imaging artifacts caused by stray radiation may be decreased by increasing the distance of the cameras from the radiation source or by adding shielding between the camera and radiation source—for example, by viewing the scintillator indirectly using a mirror (Kirov et al. 2005, Goulet et al. 2014). It is also important to keep the cameras out of the direct path of the beam, as the direct radiation beam would fill the detector with blooming artifacts and render that camera image useless (Kroll et al. 2013).

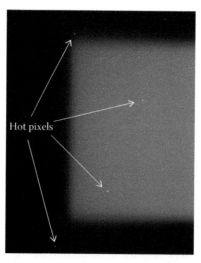

Figure 15.11 Hot pixels and streaks appear in CCD images when stray radiation impinges on the image sensor.

It is also possible to remove hot pixels and streaks through image processing. For example, it has been found that a simple 3×3 spatial median filter was sufficient to remove the majority of these artifacts without significantly affecting the underlying scintillation light distribution (Robertson et al. 2014).

15.4 3D RECONSTRUCTION

The greatest potential benefit from volumetric scintillation dosimetry is the possibility of performing rapid 3D measurements of radiation dose distributions. The primary challenge for 3D reconstruction in this setting is the limited number of viewing angles. As discussed earlier, the number of unique viewing angles perpendicular to the scintillator surface is limited by the number of non-parallel faces of the scintillator volume. A cubic volume provides three such unique viewing angles, and a hexagonal prism provides four. The addition of more faces or more cameras per face tends to decrease the fraction of the scintillator volume in which 3D reconstruction can be accomplished. A cylindrical scintillator volume could be imaged by multiple cameras at different viewing angles, but the curved surface would make refraction correction very difficult. An additional difficulty of using multiple cameras from different directions is the added bulk and complexity of the setup. With all of these considerations, it is most practical to limit the number of cameras to three or four.

Three or four projections is not typically considered a sufficient number to produce an accurate tomographic 3D reconstruction. However, there are some considerations in the case of volumetric scintillation dosimetry that can bring an accurate reconstruction from three or four viewpoints to within the realm of possibility.

The first consideration is the fact that the scintillator is a uniform medium with low attenuation. There is no need to perform attenuation correction to compensate for non-uniform attenuation in different materials, as is required for accurate reconstruction in positron emission tomography.

The second helpful consideration is the fact that the projections are not limited to axial directions, as is the case in most 3D medical imaging modalities. It is possible to acquire a *beam's-eye view* in addition to axial views of the radiation beam. This *beam's-eye-view* adds a great deal of independent information that is complementary to the axial views.

Three studies have developed 3D reconstruction methods for volumetric scintillation dosimetry of external radiotherapy beams. Kroll et al. (2013) constructed a small-scale prototype detector using a plastic scintillator in the shape of a hexagonal prism, with a height of 10 cm and a side length of 5 cm. This was imaged by three cameras facing the rectangular sides and one camera facing the hexagonal end (Figure 15.12). Beams of X-rays, electrons, and protons were directed toward the hexagonal-facing camera, and the scintillation light distribution was reconstructed using a maximum-likelihood estimation maximization algorithm. Although their results were subject to geometric errors introduced by uncorrected optical artifacts, they were able to reconstruct the 3D scintillation light distribution with sub-millimeter resolution within a few minutes.

The second study employed a novel acquisition and reconstruction scheme using a plenoptic camera and an EPID detector in conjunction to perform 3D reconstruction of the light emission from external beam radiation fields (Goulet et al. 2014). This study, its method, and results are discussed in Section 6.5 in Chapter 6.

The third study was a simulation study of a detector using three cameras that faced the orthogonal sides of a large cubic scintillator volume, as illustrated in Figure 15.13 (Hui et al. 2014). Monte Carlo-calculated 3D dose distributions of proton pencil beams were used to produce the initial camera projections. A maximum a-posteriori (MAP) iterative algorithm was used for reconstruction, and a novel profile-based technique was used to obtain the initial estimate for the first step of reconstruction. This method was able to accurately reconstruct 3D scintillation light distributions from single proton pencil beams as well as a single energy layer from an intensity modulated proton therapy (IMPT) treatment plan. Figure 15.14 shows a slice of the initial simulated light distribution (Figure 15.14a), the reconstructed light signal (Figure 15.14b), the difference between the two (Figure 15.14c), and the results of a gamma analysis comparison with gamma criteria of 3% dose difference and 3 mm to agreement (Figure 15.14d). The scintillator volume in this study was much larger than in the Kroll study, resulting in a larger number of voxels and consequently longer reconstruction times.

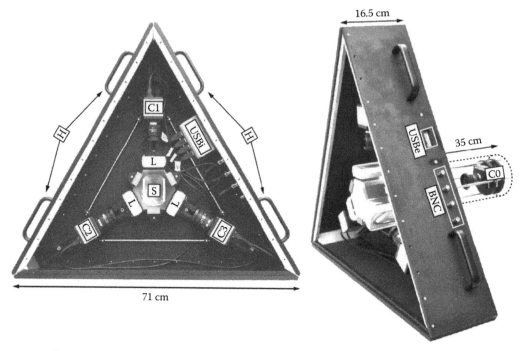

Figure 15.12 Dimensions and composition of a prototype 3D scintillation dosimetry system: Scintillator (S), CCD-Cameras (C0–C3), Lenses (L), Handles (H), USB hub (USBi), external connectors of the hub (USBe), BNC feedthrough (BNC); housing of C0 depicted with dashed lines; front cover not shown. (From Kroll, F. et al., *Medical Physics*, 40, 082104, 2013.)

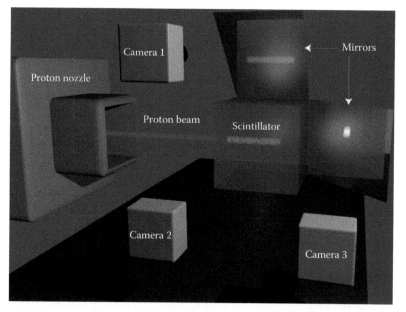

Figure 15.13 A volumetric scintillation dosimetry system design using three cameras and two mirrors to simultaneously measure the projected scintillation light from three orthogonal directions. (Reproduced with permission from Hui, C. et al. 2014. 3D reconstruction of scintillation light emission from proton pencil beams using limited viewing angles—A simulation study. *Physics in Medicine and Biology* 59: 4477, Copyright Institute of Physics and Engineering in Medicine. All rights reserved.)

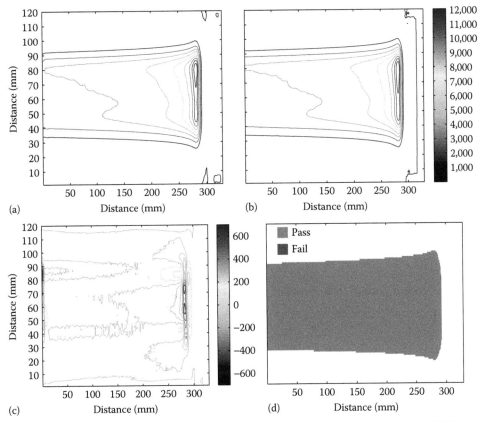

Figure 15.14 (a) Sagittal profile of the scintillation light from a single energy layer of a prostate IMPT plan. (b) The reconstructed axial profile based on simulated projections of the data. (c) The difference between the distributions in (a) and (b). (d) A corresponding slice of a 3D gamma map with gamma criteria of 3% dose difference and 3 mm to agreement.

15.5 APPLICATIONS

The unique capabilities of volumetric scintillation dosimetry make it well suited for certain radiation measurement tasks. These include machine QA and patient-specific QA measurements of photon and proton radiation beams. In this section, we will discuss prior work in these areas and mention other applications where volumetric scintillation dosimetry could be advantageous.

15.5.1 PHOTON BEAM MEASUREMENTS

Volumetric scintillation dosimetry offers several advantages for measurements of therapeutic photon beams. Modern highly conformal photon radiation treatments such as IMRT, VMAT, and tomotherapy deliver highly heterogeneous radiation distributions with temporal variations. Because of the complexity of their 3D radiation dose distributions, these modalities require detailed QA measurements (Low et al. 2011) and would benefit greatly from a real-time, high-resolution 3D detector.

Radiation therapy using small fields, including stereotactic body radiotherapy, CyberKnife and Gamma Knife, is another area of potential value for volumetric scintillation dosimetry. Ionization chambers are not ideal for measuring small radiation fields, because the chamber perturbs the radiation beam, and ionization chambers with the requisite resolution have a poor signal-to-noise ratio (Alfonso et al. 2008, Das et al. 2008). Volumetric scintillation dosimetry systems are capable of sub-millimeter resolution without perturbing the radiation beam, making them ideal for dosimetry of small fields.

Two studies have applied volumetric scintillation dosimetry to external photon beam sources. The first study employed a large liquid scintillator detector to measure 6 MV photon beams (Ponisch et al. 2009).

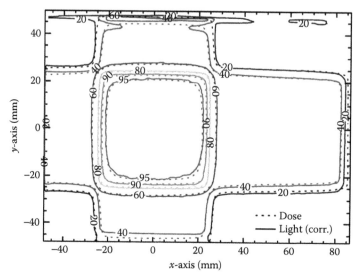

Figure 15.15 Comparison of the corrected light signal to the calculated 2D dose distribution of a four-field-box irradiation. (From Ponisch, F. et al., *Medical Physics*, 36, 1478–1485, 2009.)

The authors characterized their system, showing a linear light-to-dose response and no dose-rate dependence. They then measured beam parameters including lateral profiles, depth-dose profiles, and the projected light signal from a four-field box treatment plan. The measured lateral and depth-dose profiles matched well with the profiles from the treatment planning system, and the measurement of the four-field box achieved a 96% pass rate when compared with the treatment plan using gamma analysis with gamma criteria of 3% dose difference and 3 mm distance to agreement (Figure 15.15).

The second study, as described above, employed a plenoptic camera and an EPID detector to perform 3D reconstruction of the light emission from a large plastic scintillator (Goulet et al. 2014). This method was used to reconstruct the light distributions from clinical brain tumor treatment fields using IMRT and VMAT techniques.

15.5.2 SCANNED PROTON BEAM MEASUREMENTS

Volumetric scintillation dosimetry is particularly well suited to scanned proton beam measurements because many of the strengths of this detector type address challenges that are specific to scanned proton beams. The high resolution of volumetric scintillation detectors is valuable for measuring the steep dose gradients of proton Bragg peaks and IMPT treatment fields. The fast temporal response of these detectors makes it possible to measure energy layers or even individual proton pencil beams independently during treatment delivery, increasing the ability of the system to detect delivery errors and pinpoint their source.

Figure 15.16 illustrates one measurement approach for spot scanning proton fields. The camera acquires images at a high frame rate during beam delivery (Figure 15.16a). Ideally, the camera would be synchronized with the beam delivery so that each spot is measured independently. All images from a given energy layer are summed (Figure 15.16b). The integrated images from each of the energy layers (Figure 15.16c) are combined to form a single image of the total light emission by the treatment field (Figure 15.16d). With this approach, the treatment field can be analyzed spot-by-spot, one energy layer at a time, or in its entirety.

A major advantage of volumetric scintillation dosimetry for proton beam measurements is the ability to obtain 3D dose data in a single measurement. The dose distribution of a single photon beam can be characterized effectively using only a 2D detector, because the lateral shape of the dose distribution changes relatively little with depth. However, because of their finite range that varies with beam energy, proton

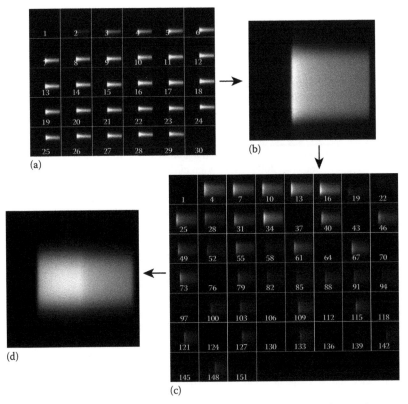

Figure 15.16 A scanning proton beam treatment field can be measured using a high-speed camera to capture images of each pencil beam in a given energy layer (a). All beams in a given energy layer can be summed (b), and the distribution from each energy layer (c) can be combined to view the distribution from the entire field (d).

beams require measurements at multiple depths in order to effectively characterize their dose distribution. This makes treatment verification measurements difficult and very time-consuming with conventional detectors. For example, it is common practice to measure a few 2D dose profiles of IMPT treatment fields as a part of patient-specific QA. This process may take several hours per patient while only measuring the dose at a small subset of points in highly inhomogeneous IMPT dose distributions. As a result, patient-specific QA can become a bottleneck in the treatment workflow, limiting the number of patients who can be treated in a given time frame. Volumetric scintillation dosimetry has the potential to make proton therapy QA measurements faster and more comprehensive.

A significant challenge for scintillation dosimetry of proton beams is the linear energy transfer-dependence of the scintillator, which leads to an under-response or *quenching* of the proton Bragg peak. While the beam range, Bragg peak location, and lateral profiles of proton beams are unaffected by quenching, the depth-dose distribution is adversely affected, making it difficult to measure 3D proton dose distributions with scintillator detectors. Chapter 2 describes quenching in greater detail and discusses approaches that have been taken to correct for quenching in scintillation dosimetry of proton beams.

Volumetric liquid scintillation detectors have been used to measure the range, lateral position, lateral profile, and relative intensity of individual proton pencil beams from a clinical proton therapy accelerator (Beddar et al. 2009, Archambault et al. 2012, Robertson 2014). These studies have shown that important proton beam parameters can be measured for dozens of beam energies in a matter of seconds using volumetric scintillation dosimetry. This capability makes it possible to dramatically increase the quantity of beam data that can be measured in a clinical proton therapy system while simultaneously decreasing the required measurement time.

REFERENCES

Alfonso, R., P. Andreo, R. Capote et al. 2008. A new formalism for reference dosimetry of small and nonstandard fields. *Medical Physics* 35 (11): 5179–5186.

Archambault, L., F. Poenisch, N. Sahoo et al. 2012. Verification of proton range, position, and intensity in IMPT with a 3D liquid scintillator detector system. *Medical Physics* 39 (3): 1239–1246.

Attix, F. H. 1986. *Introduction to Radiological Physics and Radiation Dosimetry*. Weinheim, Germany: Wiley-VCH.

Beddar, A. S., T. R. Mackie, and F. H. Attix. 1992. Water-equivalent plastic scintillation detectors for high-energy beam dosimetry.1. Physical characteristics and theoretical considerations. *Physics in Medicine and Biology* 37 (10): 1883–1900.

Beddar, S. 2015. Real-time volumetric scintillation dosimetry. *Journal of Physics: Conference Series* 573 (1): 012005.

Beddar, S., L. Archambault, N. Sahoo et al. 2009. Exploration of the potential of liquid scintillators for real-time 3D dosimetry of intensity modulated proton beams. *Medical Physics* 36 (5): 1736–1743.

Berger, M.J., J. S. Coursey, M.A. Zucker, and J. Chang. 1998. *Stopping-Power and Range Tables for Electrons, Protons, and Helium Ions*. Gaithersburg, MD: National Institute of Standards and Technology.

Bouget, J., and P. Perona. 1998. *Closed-Form Camera Calibration in Dual-Space Geometry*. Pasadena, CA: California Institute of Technology.

Caprile, B., and V. Torre. 1990. Using vanishing points for camera calibration. *International Journal of Computer Vision* 4 (2): 127–139.

Das, I. J., G. X. Ding, and A. Ahnesjo. 2008. Small fields: Nonequilibrium radiation dosimetry. *Medical Physics* 35 (1): 206–215.

Fukushima, Y., M. Hamada, T. Nishio, and K. Maruyama. 2006. Development of an easy-to-handle range measurement tool using a plastic scintillator for proton beam therapy. *Physics in Medicine and Biology* 51 (22): 5927–5936.

Goulet, M., M. Rilling, L. Gingras et al. 2014. Novel, full 3D scintillation dosimetry using a static plenoptic camera. *Medical Physics* 41 (8): 449–461.

Guo, P. Y., J. A. Adamovics, and M. Oldham. 2006. Characterization of a new radiochromic three-dimensional dosimeter. *Medical Physics* 33 (5): 1338–1345.

Hilts, M., and C. Duzenli. 2004. Image filtering for improved dose resolution in CT polymer gel dosimetry. *Medical Physics* 31 (1): 39–49.

Hui, C., D. Robertson, and S. Beddar. 2014. 3D reconstruction of scintillation light emission from proton pencil beams using limited viewing angles—A simulation study. *Physics in Medicine and Biology* 59 (16): 4477.

ICRU (International Commission on Radiation Units and Measurements). 1984. *Radiation Dosimetry: Electron Beams with Energies between 1 and 50 MeV*. Bethesda, MD: ICRU.

Ingram, W. S., D. Robertson, and S. Beddar. 2015. Calculations and measurements of the scintillator-to-water stopping power ratio of liquid scintillators for use in proton radiotherapy. *Nuclear Instruments and Methods in Physics Research Section A: Accelerators, Spectrometers, Detectors and Associated Equipment* 776: 15–20.

Kelly, R. G., K. J. Jordan, and J. J. Battista. 1998. Optical CT reconstruction of 3D dose distributions using the ferrous–benzoic–xylenol (FBX) gel dosimeter. *Medical Physics* 25 (9): 1741–1750.

Kirov, A. S., J. Z. Piao, N. K. Mathur et al. 2005. The three-dimensional scintillation dosimetry method: Test for a Ru-106 eye plaque applicator. *Physics in Medicine and Biology* 50 (13): 3063–3081.

Kroll, F., J. Pawelke, and L. Karsch. 2013. Preliminary investigations on the determination of three-dimensional dose distributions using scintillator blocks and optical tomography. *Medical Physics* 40 (8): 082104.

Low, D. A., J. M. Moran, J. F. Dempsey, L. Dong, and M. Oldham. 2011. Dosimetry tools and techniques for IMRT. *Medical Physics* 38 (3): 1313–1338.

Maryanski, M. J., C. Audet, and J. C. Gore. 1997. Effects of crosslinking and temperature on the dose response of a BANG polymer gel dosimeter. *Physics in Medicine and Biology* 42 (2): 303.

Maryański, M. J., Y. Z. Zastavker, and J. C. Gore. 1996. Radiation dose distributions in three dimensions from tomographic optical density scanning of polymer gels: II. Optical properties of the BANG polymer gel. *Physics in Medicine and Biology* 41 (12): 2705.

Oldham, M., J. H. Siewerdsen, S. Kumar, J. Wong, and D. A. Jaffray. 2003. Optical-CT gel-dosimetry I: Basic investigations. *Medical Physics* 30 (4): 623–634.

Oldham, M., J. H. Siewerdsen, A. Shetty, and D. A. Jaffray. 2001. High resolution gel-dosimetry by optical-CT and MR scanning. *Medical Physics* 28 (7): 1436–1445.

Pedersen, T. V., D. R. Olsen, and A. Skretting. 1997. Measurement of the ferric diffusion coefficient in agarose and gelatine gels by utilization of the evolution of a radiation induced edge as reflected in relaxation rate images. *Physics in Medicine and Biology* 42 (8): 1575.

Pierquet, M., A. Thomas, J. Adamovics, and M. Oldham. 2010. An investigation into a new re-useable 3D radiochromic dosimetry material, Presage (REU). *Journal of Physics. Conference Series* 250 (1): 1–4.

Ponisch, F., L. Archambault, T. M. Briere et al. 2009. Liquid scintillator for 2D dosimetry for high-energy photon beams. *Medical Physics* 36 (5): 1478–1485.

Ray, S.F. 1994. *Applied Photographic Optics: Lenses and Optical Systems for Photography, Film, Video, and Electronic Imaging.* Woburn, MA: Focal Press.

Robertson, D. 2014. *Volumetric Scintillation Dosimetry for Scanned Proton Beams.* Houston, TX: The University of Texas Health Science Center at Houston.

Robertson, D., C. Hui, L. Archambault, R. Mohan, and S. Beddar. 2014. Optical artefact characterization and correction in volumetric scintillation dosimetry. *Physics in Medicine and Biology* 59 (1): 23.

Robertson, D., D. Mirkovic, N. Sahoo, and S. Beddar. 2013. Quenching correction for volumetric scintillation dosimetry of proton beams. *Physics in Medicine and Biology* 58 (2): 261–273.

Sánchez-Parcerisa, D., A. Gemmel, O. Jäkel, K. Parodi, and E. Rietzel. 2012. Experimental study of the water-to-air stopping power ratio of monoenergetic carbon ion beams for particle therapy. *Physics in Medicine and Biology* 57 (11): 3629.

Tsai, R. Y. 1992. A versatile camera calibration technique for high-accuracy 3D machine vision metrology using off-the-shelf TV cameras and lenses. In *Radiometry*, edited by L. B. Wolff, S. A. Shafer, and G. Healey, 221–244. Burlington, MA: Jones & Bartlett Publishers.

Wang, L. L., and W. H. Tsai. 1990. Computing camera parameters using vanishing-line information from a rectangular parallelepiped. *Machine Vision and Applications* 3 (3): 129–141.

Zhang, R., P. J. Taddei, M. M. Fitzek, and W. D. Newhauser. 2010. Water equivalent thickness values of materials used in beams of protons, helium, carbon and iron ions. *Physics in Medicine and Biology* 55 (9): 2481.

Ziegler, J. F., M. D. Ziegler, and J. P. Biersack. 2010. SRIM—The stopping and range of ions in matter (2010). *Nuclear Instruments and Methods in Physics Research Section B—Beam Interactions with Materials and Atoms* 268 (11–12): 1818–1823.

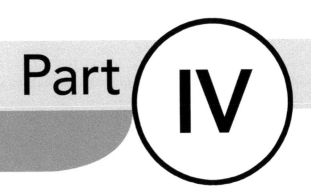

Part **IV**

Other luminescence-based applications

Fiber optic-based radiochromic dosimetry

Alexandra Rink and David A. Jaffray

Contents

16.1 RATIONALE FOR OPTICS-BASED DOSIMETRY

Dosimeters play an important part in radiotherapy. They help establish and monitor linear accelerator (LINAC) performance by measuring calibration factors, profiles, and percent depth curves, which provide information for beam modeling used for treatment planning. Some dosimeters verify and commission brachytherapy sources by measuring air kerma strength and 2D/3D dose distribution. And some dosimeters can be used for *in vivo* measurements (Mijnheer 2013, Tanderup 2013), either as part of standard patient quality assurance (QA) for certain procedures (such as total body irradiation) or to measure absorbed dose in a newly implemented technique. Despite a wide availability of dosimeters and dosimetry systems that can be utilized for *in vivo* measurements, there is still a lack of a commercially available system that can be used in all *in vivo* scenarios and across all radiotherapeutic energies in external beam radiation therapy (EBRT) and brachytherapy without performing energy-specific calibration. While 2D/3D dosimetry using an electronic portal imaging device (EPID) is a promising candidate for use in external beam radiotherapy, it cannot be practically implemented for brachytherapy procedures, especially those using low-energy sources. This means that *in vivo* dose verification in brachytherapy is largely limited to point-based (or array) dosimetry, provided that the location of dosimeter can be accurately known and tied back to the planned dose. To help diagnose sources of error, dose or dose-rate information as a function of treatment time is of great use, instead of a single measurement of total absorbed dose. Real-time data can provide information per individual intensity modulated radiation therapy (IMRT) segment, or volumetric modulated arc therapy arc, or a particular source dwell position. In addition, real-time dosimetry allows for quick intervention and interruption of treatment in case a gross error has been detected. This is of particular use in brachytherapy, where higher doses and fewer fractions are used compared to standard external beam fractionation.

To limit the invasiveness of point-based *in vivo* measurement, dosimetry for external beam is often performed on the skin, either in the beam (entrance or exit dose), or outside of the beam (to assess dose to organs at risk or structure of interest, such as a pacemaker). Although large errors in delivery may be detected in this manner, the actual points of interest are typically elsewhere. We can theoretically deduce the dose at the point of interest from skin measurements, but must make certain assumptions to do so (e.g., no breathing or internal motion present and distance between measurement and point of interest is same at every fraction). Each assumption will increase the uncertainty in our deduced dose at the point of interest, possibly adding up to be greater than the error we may be trying to detect. Furthermore, skin measurements in brachytherapy are typically too far away from the source and are too error-prone to provide any meaningful measurements, unless it is the skin dose itself that is of interest (as is the case with breast treatments). It stands to reason that if we are interested in a dose at a particular point within the patient, the best way to know that dose is to actually accurately measure it. Whether such measurements are required for every patient and every fraction, as part of QA and record of actual dose delivered, remains to be seen. But point-based real-time *in vivo* dosimeter would definitely be useful in investigations of new techniques, and prudent in procedures with large doses and few fractions. Some argue that sufficient QA takes place before the patient is treated, and that provided all such checks are performed, one can be confident that the dose delivered during treatment is correct. Philosophical debate on this topic, although interesting, is outside of the scope of this chapter. We will therefore accept that a point-based real-time *in vivo* dosimeter is desirable, in at least some of the instances, and that having one such dosimeter would advance the field of radiotherapy.

Measurements performed *ex vivo*, or ones that do not involve patients, (e.g., calibration of the LINAC in a water tank, or IMRT QA using a phantom), can have dose-measuring devices that perturb the field, provided that appropriate correction factors are available to then improve the accuracy of the measurement. However, in an *in vivo* measurement, perturbation of dose due to a measuring device may have greater consequences. Even if an accurate measurement is performed by using appropriate correction factors, the dose to patient would have been modified by performing the measurement in the first place. In a hypofractionated EBRT or brachytherapy treatment, significant modification of even a single fraction may have consequences to the overall outcome of treatment, and would be considered unacceptable. Just like any quality measurement device, *in vivo* dosimeters should meet certain criteria to be a useful tool. The criteria, built in part on basic engineering principles as well as on the requirements in specific applications, are shown in Table 16.1.

Dosimeters using optical fibers are particularly desirable given that they can easily meet the size criterion, and are more likely to meet the atomic composition criterion than an electronic device. The dosimetric material chosen at the tip must be sensitive enough so as to not make the volume of interest too long. Otherwise, volume averaging will occur, which can be problematic in high-gradient areas, making measurements difficult to interpret. Because of excellent sensitivity to dose of radiochromic materials commercially used in GafChromic® films, they have been considered for the sensitive material in a fiber-optic dosimeter probe.

Table 16.1 **List of desired criteria for an *in vivo* point-based real-time dosimeter**

CRITERION	COMMENTS
Small size (Dutreix and Bridier 1985)	Minimize dose perturbations from dosimeter presence; minimize trauma if used interstitially (either alone or via catheter/needle)
Appropriate atomic composition (Dutreix and Bridier 1985)	Minimize dose perturbations from dosimeter presence; ideally sensitive material of detector and wall is same as surrounding medium; in *in vivo* dosimetry, water is a decent approximation to most tissue (NIST)
Fast kinetics and stable response	Requirement for real-time read-out
Signal proportional to dose	Ideally linear to eliminate need for total dose tracking and for ease of conversion from signal to dose; simple function is acceptable
Dose resolution	1 cGy resolution for radiotherapy applications
Dose rate independence	Dose accuracy maintained regardless of rate of dose delivery; can be used for permanent seed applications as well as high dose rate brachytherapy and EBRT
Insensitive to environmental conditions	Insensitivity to temperature, humidity, and light variations allows for easier implementation in the clinic; position on skin as well as *in situ* provides accurate read-out
User-friendly	Robust and easy to handle, disposable or reusable (application specific), sterile or sterilizable (application specific), low cost

16.2 RADIOCHROMIC PROCESSES AND IMPLICATIONS FOR USE IN DOSIMETRY

Radiochromic process refers to a change in color of the medium due to absorption of energy, without the need for any other instigator. For the purposes of radiation dosimetry, materials that undergo a radiochromic process when exposed to directly or indirectly ionizing radiation is of primary interest. The radiochromic process within GafChromic films is due to a chemical progression from a diacetylene monomer in a crystalline state to an acetylene polymer (Figure 16.1), and falls into a category of *topochemical polymerization*, where the crystal structure of the polymer is controlled by the packing of the surrounding monomer molecules within the crystal lattice. These types of reactions for a variety of monomers are described in detail elsewhere (Bloor 1984, Guillet 1985, Sixl 1985). In short, the monomer molecules of diacetylenes are packed in a ladder-like manner, with ends of one triple-bond system within ~4 Å of the next nearest triple-bond system. Polymerization can be randomly initiated throughout the lattice chemically, by thermal radiation (provided that temperature is below melting point), and by ionizing radiation. For polymerization to proceed, the lattice must be sufficiently mobile that the individual monomer molecules can be within 3 Å of each other at some time; else the reaction will cease if the separation becomes much greater than this (Bloor 1984). The polymer chains grow independently within each ladder of monomers, without crosslinking with other reactions in adjacent rows, thus following a *homogeneous solid-state* reaction. The radical reaction from b → d (Figure 16.1) happens quickly once there are more than six units in a polymer (Huber and Sigmund 1985, Sixl 1985), and results in the observed deep color due to the π-electron generated absorbance spectrum from the conjugated backbone of the acetylene polymer. Because of the need for a particular geometry and separation between monomers in order for the reaction to proceed, and because the crystalline packing of such monomers is controlled by

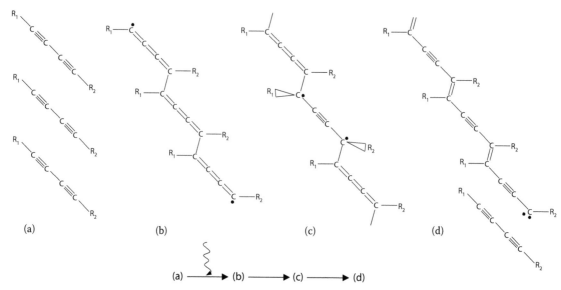

Figure 16.1 (a) Diacetylene monomers, upon exposure to ionizing radiation, polymerizes into (b) butatriene structure polymer; as the polymer chain grows, it rearranges via (c) an intermediate between butatriene structure and acetylene structure, into (d) acetylene structure polymer. (From Rink, A. et al., *Medical Physics*, 32, 1140–1155, 2005b. With permission.)

the R1 and R2 groups (Bloor 1984), only certain crystal forms of certain monomers can actually undergo topochemical polymerization. Pentacosa-10,12-diynoic acid (PCDA) and lithium pentacosa-10,12-diynoate (LiPCDA) are two such monomers.

The rate of reaction is a function of distance between the monomers, controlled by the R1 and R2 groups. As polymerization in a crystal made of PCDA monomers progresses, the polymer chain deforms enough to shorten in the overall length. This causes the distance between the last unit in a polymer chain and the next available monomer to increase, and the polymerization reaction slows down. The separation between polymer chains and the monomer matrix around them causes mechanical stresses and further change in the polymer structure, evidenced by a shift in absorbance band to shorter wavelengths as dose continues to increase (McLaughlin 1996a, Saylor et al. 1988).

In aqueous solutions, the relationship between absorbance at a particular wavelength and concentration of absorbing molecules is related through Beer's law (Skoog et al. 1998):

$$A = \epsilon bc$$

where:

ϵ is the molar absorptivity in L mol^{-1} cm^{-1}
b is the optical path length in centimeters
c is molar concentration

Interpretation of absorbance for a solid-state monomer and polymer mixture is a little bit more complicated than that. Some authors showed that the spectrum of LiPCDA film can be explained by as many as eight individual peaks overlapping (Devic et al. 2007, 2010). Perhaps because of this difficulty in linking the absorbance at any given wavelength to number or concentration of polymers, these radiochromic substances typically require calibration.

16.3 2D RADIOCHROMIC FILM DOSIMETRY

Comprehensive review of PCDA films (referred to as GafChromic DM-1260, HD-810 and MD-55) was published by AAPM Radiation Therapy Committee Task Group 55 (Niroomand-Rad et al. 1998). Subsequently, various authors addressed performance of LiPCDA films, such as GafChromic EBT and

EBT-2 (Devic et al. 2005, 2007, 2010, Todorovic et al. 2006, van Battum et al. 2008, Zeidan et al. 2006). This section aims to give the reader a brief summary of the typical results. The chapter will focus primarily on dosimetry for photon beams and emitters, since this comprises majority of radiotherapy. Readers interested in electron and proton dosimetry are encouraged to seek original publications (Arjomandy et al. 2010, Piermattei et al. 2000, Richter 2009).

A typical film dosimetry system is composed of the film, a scanner or densitometer, and some kind of processing software to allow quantification of results. For the purposes of considering the radiochromic materials in fiber-optic dosimetry, only the results dependent on film properties will be discussed. Performance relying solely on the choice of processing software or scanner that has no bearing on real-time point-based dosimetry will be neglected. This section will also cover some results that have been obtained with real-time measurements if the results are not unique to this type of read-out (e.g., energy dependence).

16.3.1 SMALL SIZE

Although the films are typically cut to at least 1×1 cm^2 pieces, McLaughlin et al. (1991) illustrated that an absorbance or optical density signal can be measured from submicron width of film, and only 6 µm thickness of PCDA-based sensitive material, by showing resolution capability of greater than 1000 lines per mm. It stands to reason that a very small volume of this material can be used as a point dosimeter on a tip of an optical fiber, provided the dose is high enough.

16.3.2 APPROPRIATE ATOMIC COMPOSITION

The GafChromic film has evolved over several decades from DM-1260 to EBT-3, with many versions (some modified for a particular application) in between. Along the way, the thickness and construction of the films have been changed (Devic et al. 2005, 2006, 2012, Niroomand-Rad et al. 1998), as well the chemical composition of monomers and additives within the suspension (Devic et al. 2012, Lindsay et al. 2010, Niroomand-Rad et al. 1998). PCDA was used in films up to and including versions of MD-55-2, and LiPCDA in EBT and subsequent versions. The films had a variety of energy dependencies, ranging from near independent over 75 kVp to 18 MV range (Lindsay et al. 2010, Rink et al. 2007a) as shown in Figure 16.2, to films that are geared toward increased sensitivity to dose at a particular energy range (Rampado et al. 2006). The sensitive layer in the original version of the film, DM-1260, was calculated to have mass energy-absorption coefficients within ~3% of water and muscle, while the polyethylene terephthalate (Mylar) film base was 6%–7% lower than water over 0.15–1.25 MeV energy range (McLaughlin et al. 1991). Electron mass collision stopping powers of the sensitive layer was calculated to be within ~2% of water and muscle over a similar (0.1–1 MeV) range (McLaughlin et al. 1991). It was also illustrated that the increase in absorbance of the film irradiated with 10 MeV electrons is within uncertainty of the increase in absorbance of the film irradiated with ^{60}Co γ-rays to the same dose. Experimentally, the sensitivity (netOD per unit of dose, typically Gy) has been shown to decrease with decreasing energy for MD-55 films, dropping by 30%–40% at ~30 keV compared to photon energies above 100 keV (Chiu-Tsao et al. 1994, Muench et al. 1991).

Additives to sensitive layer suspension (such as chlorine, bromine, sodium or potassium) play a crucial role in establishing energy independence (Lindsay et al. 2010). It may be feasible, although not illustrated in literature, to achieve energy independence of PCDA-based radiochromic medium with the right additives. This has not been pursued commercially likely due to LiPCDA's faster kinetics and greater sensitivity to dose, which has obvious benefits for both real-time and non-real-time dosimetry.

16.3.3 KINETICS AND STABILITY OF RESPONSE

As discussed in Section 16.2, the likelihood and speed of polymerization is a function of the monomers' arrangement within the crystal structure. The three-dimensional packing of monomers is affected by the diacetylene side groups (R1 and R2), and in the case of LiPCDA, also by the level of hydration (Rink et al. 2008). Stability of absorbance spectra of PCDA and LiPCDA was investigated and reported on by various authors. It has been found that absorbance continues to increase even after irradiation of films has been completed. Traditional 2D measurements showed that PCDA will continue to increase

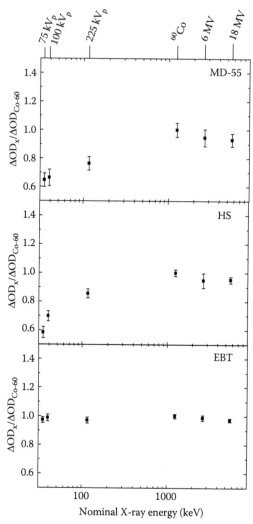

Figure 16.2 Sensitivity of PCDA (MD-55 and HS) and LiPCDA (EBT) films across a range of photon energies. (From Rink, A. et al., *Physics in Medicine and Biology*, 52, N523–N529, 2007a. With permission.)

as much as 16% on the first day, and ~4% thereafter for approximately two weeks (Klassen et al. 1997 McLaughlin et al. 1991). The exact magnitude of this change will vary with wavelength, dose level, and storage temperature (Saylor et al. 1988), as well as read-out method and time of initial ($t = 0$ s) measurement with respect to end of radiation. As an example, real-time measurements over a period of 18 h immediately after irradiation showed an increase in net optical density (denoted as netOD typically, and is final optical density minus initial optical density) of MD-55 by more than 25% (Rink et al. 2005a), rather than the previously reported 16%. This is because the polymerization speed immediately at the end of irradiation is highest, contributing a significant amount to final ($t = \infty$) optical density. But this immediate post-irradiation contribution is not captured if the $t = 0$ read-out occurs even 15 min after irradiation has been completed.

In contrast to PCDA films, the change in netOD of EBT increased by ~12.5% over the same 18 h after irradiation (Rink et al. 2005a). The rapid polymerization in LiPCDA-based film is favorable for real-time measurements, as this directly affects the extent of dose-rate dependence observed (discussed in Section 16.4.3). Furthermore, a shift of the main absorbance band to lower wavelengths is observed with increased dose for PCDA (McLaughlin et al. 1991, Saylor et al. 1988), but not for LiPCDA (Devic et al. 2007).

This spectral shift is due to restructuring of the monomer/polymer system, as discussed in Section 16.2. This restructuring also manifests itself in the fact that kinetics of film darkening was shown to be faster at the lower doses (Ali et al. 2003, Dempsey et al. 2000), where the polymer to monomer fraction is still low, and thus the structure of monomer crystal still dominates.

Upon review of literature, Task Group 55 of American Association of Medical Physicists recommended a delay of at least one day following irradiation to allow for accurate dosimetry with PCDA films (Niroomand-Rad et al. 1998). LiPCDA films, on the other hand, tend to stabilize much faster and measurements after 3 h are considered to be accurate (Todorovic et al. 2006). In real-time dosimetry, waiting after irradiation is not feasible and is contradictory to the goal. It is therefore important to use the material that minimizes the uncertainty in real-time measurements due to dose-rate variation.

16.3.4 SIGNAL LINEARITY AND SENSITIVITY TO DOSE

Net optical density as a function of dose depends not only on the sensitive material (PCDA or LiPCDA), but also on the choice of wavelengths (McLaughlin et al. 1991, 1996b, Muench et al. 1991, Saylor et al. 1988). This is true for both film and real-time dosimetry, although the specifics of the read-out device and method can affect the exact values for both. In general, sensitivity (netOD/Gy) has been shown to be greatest when a narrow band around the main absorbance peak is used (~675 nm for PCDA and ~633 nm for LiPCDA) (McLaughlin et al. 1996b, Rink et al. 2005b, Saylor et al. 1988). The sensitivity decreases and dynamic range increases if the wavelength range is picked in the green or blue part of the spectrum (Devic et al. 2007, 2009, McLaughlin et al. 1996b), or the entire visible band is used (Todorovic et al. 2006). For the same sensitive material and choice of wavelength, netOD measured for a given dose can be easily increased by increasing the thickness of the sensitive material (Rink et al. 2005b). This is seen by an increase in sensitivity when the films have transitioned from HD-810 to MD-55-2 format (increase from 7 to 30 μm total).

PCDA, although showing a lower netOD for a given absorbed dose (Devic et al. 2005, Rink et al. 2005a, Todorovic et al. 2006), has a linear response from 0 to 300 Gy (McLaughlin et al. 1994b, 1996b, Muench et al. 1991) at 633 nm and up to 6 Gy at 676 nm, corresponding to the main absorbance peak (Klassen et al. 1997). The loss of linearity after this point is easily explained by the shift in the absorbance peak to shorter wavelengths with increased dose, thus showing a 12% decrease in sensitivity from 6 to 15 Gy. On the other hand, netOD of LiPCDA with dose is not linear, even if the measurement band is increased to 600–670 nm around the 636 nm peak (Rink et al. 2005a). It is typically characterized by other functions (Borca et al. 2013, Devic et al. 2005), with the coefficients dependent on read-out technique.

16.3.5 DOSE RATE INDEPENDENCE

Dose-rate independence of PCDA was investigated (Saylor et al. 1988) over a wide range of dose rates (0.02–198 Gy/min). With the measurements performed once the film response has stabilized, no significant difference within experimental uncertainty of 5% was reported.

16.3.6 INSENSITIVITY TO ENVIRONMENTAL CONDITIONS

The three environmental factors of concern when performing film dosimetry are ambient light, humidity, and temperature. Sensitivity to ambient light has been dealt with by the manufacturer, with the modern version of radiochromic films having a protective Mylar on both sides of the sensitive layer. The users are further recommended to keep film in black opaque envelopes to prevent any inadvertent exposure to light (McLaughlin et al. 1991, Niroomand-Rad et al. 1998), which can cause polymerization and therefore cause inaccuracies in dose reading. Variations in humidity and temperature, however, are typically out of user's direct control in a standard clinical environment. Effect on dose readout due to an increase in temperature, both during irradiation and scanning, was investigated by several authors (Klassen et al. 1997, McLaughlin et al. 1991). When temperature was varied between 20°C and

60°C for irradiations, but films were subsequently stored and read-out at 20°C, the relative response (compared to that at 20°C) generally first increased with temperature, but then gradually decreased, followed by a sharp drop in sensitivity at 60°C (likely due to destruction of the crystalline structure). The exact temperature at which the decrease in sensitivity started to occur was both dose and wavelength dependent. McLaughlin et al. (1991) also demonstrated that when the films were irradiated at normal room temperature (20°C), but the spectrophotometry performed at temperatures varying between 23°C and 40°C, a shift of the absorbance peaks to lower wavelengths was observed (Figure 16.3). Thus, depending on the wavelength of readout, the absorbance will either decrease or increase with temperature, as was also illustrated by Klassen et al. (1997). The magnitude of temperature dependence is a function of the dose received by the film, with greater effect observed at higher doses. This shift in absorbance peak due to increase in read-out temperature is reversible. However, increase in absorbance due to polymerization induced by increase in temperature (either during irradiation or measurement) is not. It is therefore important to take some precautions in not initiating heat-induced polymerization to maintain dose prediction accuracy.

16.3.7 OPTICAL ISSUES

Due to the presence of fast and slow axes in Mylar layers used in the commercial film, and because the analyzing light in most detectors used for film dosimetry is at least partially linearly polarized, the optical density measured varies with film orientation (Butson et al. 2009, Klassen et al. 1997, Lynch 2006). While this is an issue requiring careful and reproducible alignment in film dosimetry using digitizers or scanners, this does not affect optical fiber-based measurements. First, light does not remain polarized after traveling through several meters of regular silica optical fiber, the kind used between light source and dosimeter probe, or between dosimeter probe and detector. Second, in optical fiber-based dosimetry, use of Mylar is not necessary as other means are used to isolate the radiosensitive gel from the surroundings. Further, even if LiPCDA crystals themselves are optically active when aligned, the monomers do not align with each other when using other layering techniques appropriate for optical fiber probes.

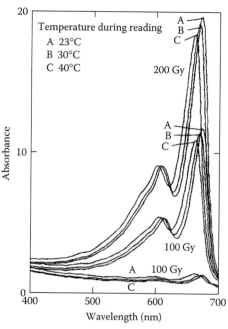

Figure 16.3 Absorption spectra of PCDA, for unirradiated and irradiated to indicated doses films, at specified temperatures during reading. (From McLaughlin, W.L. et al., *Nuclear Instruments and Methods in Physics Research*, A302, 165–176, 1991. With permission.)

16.4 REAL-TIME PERFORMANCE OF RADIOCHROMIC MATERIALS

Both PCDA and LiPCDA were investigated for real-time *in vivo* applications, and a summary of the results is shown in Table 16.2. The real-time data for GafChromic films were obtained with transmission measurements, with a single 50 μm-core fiber illuminating a piece of film with an approximate spot size of 650 μm in diameter, and the transmitted light collected by a separate 1.5 mm-core fiber (Rink et al. 2005b).

16.4.1 FEASIBILITY OF SMALL SIZE DETECTOR

By using a single piece of commercially available film, it was shown that PCDA can measure >25 cGy with 32 μm (two layers, 16 μm each) of sensitive material (Rink et al. 2005b), and LiPCDA can measure >5 cGy with 34 μm (two layers, 17 μm each) (Rink et al. 2007b). None of the investigations focused specifically on finding the lower limit of detection, and sensitivity can be easily increased with increasing the thickness of material. Doubling the thickness of material and also changing from simple transmission measurements to a transmission to reflector, and then transmission back (thus also doubling the photon path length) would increase the signal fourfold.

16.4.2 DETECTION LIMIT AND DOSE RESOLUTION

One generally accepted definition of *detection limit* in analytical chemistry is as minimum concentration or mass of analyte that can be detected at a known confidence level, and is typically taken as signal from blank plus three times standard deviation (1σ) of a blank (Skoog et al. 1998). If the signal is greater than this value, then one can say with >95% confidence that the analyte is present. If the signal is below this value, then analyte is considered undetected. As discussed in Section 16.2, the increase in netOD of radiochromic material is due to increase in the number of conjugated polymer backbones, which can be considered similar to a chemical analyte that has specific absorbance properties. Intensity of transmitted photons through an unirradiated film can therefore be used as the blank, and ideally this value is as close to the

Table 16.2 **Summary of real-time results for PCDA and LiPCDA films**

CRITERION	PCDA	LiPCDA
Small size	650 μm diameter and ~30 μm thick sensitive material to measure >25 cGy	650 μm spot size and ~30 μm thick sensitive material to measure >5 cGy
Appropriate atomic composition	Energy dependence	Energy independence can be achieved over 75 kVp–18 MV range tested within 3% through appropriate additives to the sensitive layer
Fast kinetics and stable response	Increase in OD real time; subsequent increase in OD of 25% over 18 h	Increase in OD real time; subsequent increase in OD of 12.5% over 18 h
Signal proportional to dose	Linear with dose up to ~300 Gy (upper limit depends on thickness of material and wavelength of read-out)	Non-linear
Dose resolution and limit of detection	5.4×10^{-4} cGy^{-1}; 8 cGy detected for 32 μm photon path	6.1×10^{-3} cGy^{-1} initial resolution; 1 cGy detected for 34 μm photon path
Dose rate independence ($\alpha = 0.05$)	Dose rate dependent for dose ≥ 190 cGy over 95–571 cGy/min dose rate	Dose rate dependent if varies by >8-fold (16–520 cGy/min range) for all doses in 5–1000 cGy range
Insensitive to environmental conditions	Main absorbance peak shift with temperature	Main absorbance peak shift with temperature; correction possible with peak tracking and correction factor

maximum detectable signal as possible to use most of the detector range. The standard deviation of the blank (σ_{bl}) is just the noise due to light and spectrometer fluctuations. This will be dependent on the stability of the light source, stability of the spectrometer, and other parameters such as integration time and averaging. The noise can be decreased by choosing a fan-cooled light source and a cooled spectrometer if necessary, adding to the overall cost of the back end system. Regardless of which components are chosen, however, LiPCDA will have a lower detection limit compared to PCDA, since it has been shown to be approximately eight times more sensitive at low doses when using the main absorbance peaks to perform measurements (Rink et al. 2005a). The measurements typically start with the blank—unirradiated film or probe—having a high transmitted photon count (maximum of 2^{16} for the spectrometer currently used with the system) compared to that obtained for irradiated dosimeter. With the current system, once the light source has stabilized after approximately 20–40 min, the signal-to-noise ratio of the spectrometer at full signal (2^{16} counts) is 300:1, and represents the largest source of noise. This makes the detectable signal (I_m) = 64882, which is the minimum signal required to distinguish it from the blank with >95% confidence. The corresponding minimum detectable optical density change, $netOD_m$, is then 0.0044 as

$$\sigma_{bl} = \frac{2^{16}}{300} = 218$$

$$I_m = 2^{16} - 3 \cdot 218 = 64882$$

$$netOD_m = -\log\left(\frac{I_m}{2^{16}}\right) = 0.0044$$

For a single 16 μm coating of PCDA on the tip of the optical fiber, and using the main absorbance peak for detection, this works out to be approximately 8 cGy [32 μm long photon path through PCDA has netOD of 5.4×10^{-4} cGy^{-1} (Rink et al. 2005b)]. Experimentally, it was confirmed that 25 cGy can be measured using main absorbance peak and PCDA film in transmission mode (Rink et al. 2005b), and doses > 67 cGy were measured with 5% accuracy (1σ). For 17 μm coating of LiPCDA on the tip of the optical fiber, the minimum detectable dose is ~0.7 cGy (6.1×10^{-3} cGy^{-1} initial sensitivity for unirradiated EBT film, Rink et al. 2007a), and experimentally as low as 5 cGy was measured. Doses <25 and <5 cGy were not tested for PCDA- and LiPCDA-based films, respectively. It is worth noting that reducing noise in signal detection and correcting for other uncertainties, as well as increasing the thickness of sensitive material used, will decrease the lowest detectable dose for both PCDA and LiPCDA.

Dose resolution (netOD/cGy or netOD/Gy), often referred to as sensitivity, for both PCDA and LiPCDA varies with dose. At the onset of radiation, dose resolution is greatest, especially if the main absorbance peak is used for netOD measurements. However, as dose increases, a deviation from linearity and a decrease in dose resolution is seen for both types of radiochromic systems. This occurs at a much higher dose for PCDA and thus, for most clinical applications and dose ranges, PCDA response is considered linear. Some strategies to improve dose resolution at high doses involve switching to other parts of the spectrum for netOD measurements (Andrés et al. 2010, Borca et al. 2013, Devic et al. 2009). The exact dosimetric point where the main absorbance peak ceases to provide the highest dose resolution depends on the detection method, including the type of detector, the spectral range used for netOD calculation, light source, and so on.

16.4.3 DOSE RATE INDEPENDENCE

Rate of polymerization is particularly important when the radiochromic materials are used for real-time measurements. Increase in absorbed dose by the dosimeter can be thought of as individual pulses of dose deposition, each characterized by a sharp increase in netOD within 2 ms, followed by logarithmic increase (McLaughlin et al. 1994a). This is not typically seen in real-time dosimetry because the time between pulses is significantly smaller than the temporal resolution of our measurements. However, it can be easily understood that changing the dose rate affects how much of this logarithmic development has occurred after the pulses at the beginning of the irradiation. Using a low dose rate allows for more of *post-irradiation* coloration to occur than if the dose rate was high. It is therefore highly preferable that the radiochromic dosimeters used for real-time measurements have as fast of polymerization as possible. If this post-irradiation logarithmic increase stabilizes nearly instantly, then technically there would be no dose rate dependence for the radiochromic dosimeters at all. The experimental results show that this is not the case.

Doses measured immediately at the end of irradiation of PCDA films show a trend of increasing netOD with decreasing dose rate (Figure 16.4). The films were irradiated at 95, 286, and 571 cGy/min to doses up to 381 cGy. Using type I error of 1% ($\alpha = 0.01$) and the F-test (Milton and Arnold 1990), the netOD measurements were indistinguishable for the entire dose range tested (Rink et al. 2005b). Using $\alpha = 0.05$ (type I error of 5%), doses ≥ 190 cGy showed dose-rate dependence.

For LiPCDA films, netOD measured immediately at the end of irradiation showed a statistically significant difference ($\alpha = 0.05$) only when the dose-rate variability increased or decreased by eightfold or more (such as between 16 and 130 cGy/min) (Rink et al. 2007b). As with PCDA films, decreasing the dose rate increased the real-time value of netOD for a given delivered dose. The extra uncertainty due to this intra-irradiation polymerization was in the order of 1% for all the doses tested (5–1000 cGy).

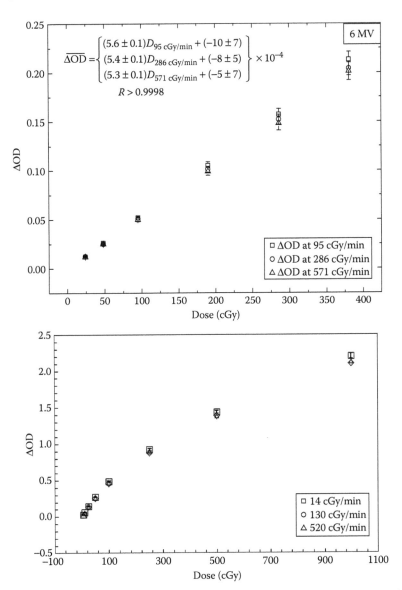

Figure 16.4 (Top) Dose-rate effect on real-time measurements with PCDA. (From Rink 2005b. With permission.) (Bottom) Dose-rate effect on real-time measurements with LiPCDA. (Data from Rink, A. et al., *Physics in Medicine and Biology*, 52, N523–N529, 2007b.)

16.4.4 SENSITIVITY TO ENVIRONMENTAL CONDITIONS

Changes in the surrounding environment can affect the rate of reactions occurring within a dosimeter, as well as other properties. In the case of radiochromic materials PCDA and LiPCDA, it is the rate of polymerization and the resulting shape of spectrum that are of interest. Unlike in standard film measurements, where irradiation temperature and read-out temperature are separate because the film is read-out some time later, in real-time measurements they are one and the same. There is no way to separate out the reversible read-out temperature effects from the irreversible irradiation temperature effects (see Section 16.3.6). Understanding how temperature affects real-time *in vivo* measurements is important since there is a substantial difference between ambient room temperature (~22°C) and temperature within the patient (~37°C). Correction for this is not simple, as even *in vivo* measurements can have variable temperatures (e.g., on skin vs. inside the patient).

To understand the overall impact of temperature variation on PCDA and LiPCDA signal in real-time dosimetry, one must consider several simultaneous effects. Since PCDA and LiPCDA are three-dimensional crystalline structures, increase in temperature may affect the packing due to the deformation of each individual monomer. Packing on its own can determine whether polymerization proceeds or not (see Section 16.2), and increasing the temperature past the crystals' melting points can completely destroy the structure such that it ceases to be useful as a dosimeter. At the same time, increasing temperature may also increase the rate of polymerization by allowing the monomers within a crystal to be more mobile and approach the next monomer in the chain in order for polymerization proceed. Heat can induce brand new polymerization reactions as well, causing increase in netOD that has nothing to do with radiation dose. These effects would be taking place at the same time, and whether increase in temperature increases or decreases the rate of polymerization would be dependent not only on the side groups and initial structure but also on the temperature range in question.

The next thing to consider is the signal that is being measured. It has been observed that varying the read-out temperature of the film (Section 16.3.6) can either decrease or increase the netOD, depending on which wavelength or range of wavelengths is used for analysis. This is in part due to the fact that with increasing temperature, the absorbance peaks shift to lower wavelength (higher energies) as illustrated in Figure 16.3 (McLaughlin et al. 1991). Therefore, performing a narrow-band measurement centered on the main absorbance peak would underestimate the dose.

All of the effects discussed in the last two paragraphs would be present in a real-time measurement, making the problem of temperature dependence a challenging one to understand and resolve. The main absorbance peak shifts to lower wavelength with increased temperature for both PCDA and LiPCDA, as shown in Figure 16.5a and b (Rink et al. 2005b, 2008). The decrease is linear with increasing temperature

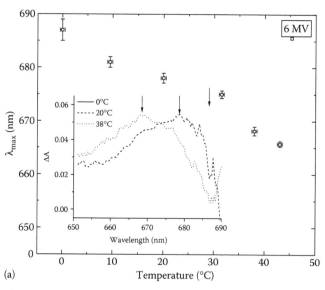

(a)

Figure 16.5 (a) Position of main absorbance peak for PCDA as a function of irradiation/measurement temperature. (From Rink, A. et al., *Medical Physics*, 32, 1140–1155, 2005b. With permission.) *(Continued)*

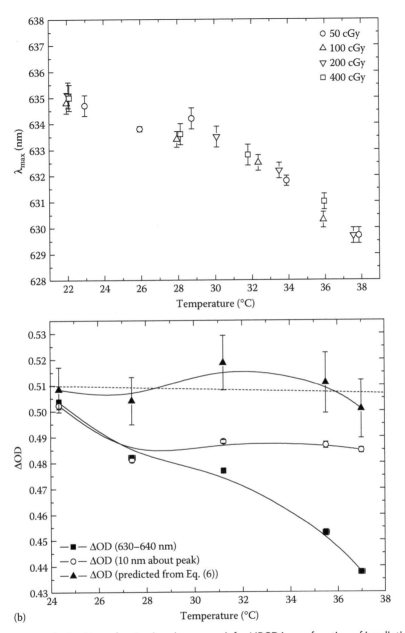

Figure 16.5 (Continued) (b) Position of main absorbance peak for LiPCDA as a function of irradiation/measurement temperature (top); application of two temperature correction algorithms (bottom). (From Rink, A. et al., *Medical Physics*, 35, 4545–4555, 2008. With permission.)

over the 22°C–38°C range expected in a variety of *in vivo* measurements. Since the wavelength of the main absorbance peak is a decent predictor of temperature, this information along with a temperature-dependent correction factor can be used to compensate for the overall effect. As shown in Figure 16.5b with LiPCDA, this correction method works well, although some residual decrease in netOD is still observed. Simply shifting the spectral range over which measurement is performed to coincide with the shifting absorbance peak does not provide sufficient correction, further proving that the net result is due to more than just a shift in the main absorbance peak.

Small variations in humidity, typical of normal operations, are unlikely to affect the radiochromic sensitive layer within the sealed system of the fiber-optic probe, although this has not been explicitly tested. What has a higher probability of affecting the quality of the dosimeter is the final hydration of the sensitive layer achieved during manufacturing. It has been illustrated that water plays an important role in how the monomers arrange themselves within the three-dimensional crystal of LiPCDA (Rink et al. 2008). Desiccating the sensitive layer resulted in both a shift in main spectral peak to higher wavelengths and a decrease in netOD of nearly 70% (for a 3 Gy dose). Reintroducing water into the sensitive layer by exposing it to moisture resulted in another shift in spectrum, this time with main absorbance peak moving to lower wavelengths, but not quite to the original ~635 nm. Although desiccation has drastic effects both on spectral properties and on sensitivity of LiPCDA, it requires effort to achieve this. The results described above were typical for desiccation over calcium chloride in a 50°C oven for 24 h or more. Therefore, while an important issue to consider for manufacturing purposes, humidity fluctuations are not likely to be an issue for the end user.

16.4.5 CONSIDERATION OF CERENKOV RADIATION

One of the difficulties in performing optical real-time measurements is the potential interference of Cerenkov radiation. Cerenkov radiation occurs when a charged particle travels through a medium faster than photon phase velocity in the same medium, and is typically observed in high-energy beams with optical fibers of refractive index greater than one ($n > 1$). The resulting light, occurring mainly in visible and near-visible part of the spectrum and appearing as a pale blue *glow*, is due to constructive interference from electromagnetic pulses set up along the electron path (Jelley 1958). Some of this light is captured by the optical fiber and is relayed to the detector, potentially interfering with optical measurements. The more fiber is present in high-energy beam, the larger is the effect.

The Cerenkov spectrum decreases in intensity (total number of photons) with increasing wavelength, therefore being strongest in the near UV and blue region, and much weaker in the red and near-infrared regions. Since radiochromic materials discussed in this chapter absorb in the red region of the spectrum, Cerenkov interference is negligible compared to the number of photons used for read-out (which is controlled by the user to have a high signal-to-noise ratio). For more discussion on Cerenkov radiation and significance in fiber-optic dosimetry, consult Chapter 5.

16.5 FIBER-OPTICS-BASED RADIOCHROMIC DOSIMETRY PROTOTYPES

Two different sets of prototypes were made using radiochromic materials (Figure 16.6). Both prototypes operated in a transmission-reflection mode, whereby the light from a light source enters the plastic optical fiber of the probe, gets transmitted through the radiochromic thin-film at the tip, is reflected from a dielectric mirror, is transmitted back through the radiochromic thin-film and back along the same optical fiber toward a detector (Croteau 2011a,b). For the construction of prototypes, LiPCDA was deemed a more appropriate choice due to (1) faster polymerization kinetics, resulting in decreased dose-rate dependence; (2) higher sensitivity per given thickness of material; and (3) lack of peak shift with dose, allowing for temperature-based correction.

In short, the entire real-time dosimetry system consists of the radiochromic fiber-optic probe (total length of 50 cm from connector to tip) connected to a custom-made fiber-optic splitter, and two optical cables between the splitter and back end, one connected to a light source and the other to the light detector. All fiber-to-fiber connections are SMA to SMA. A white light source with a 500 nm high pass filter and a PC-operated shutter to prevent unnecessary exposure of probes to light is used. The optical cables between the back end and the splitter are silica fibers of 600 μm core diameter. Silica fibers were chosen to have better resistance to radiation damage, and therefore a longer lifetime. Since these cables are not expected to be within radiation field (for external beam therapy) or near a source (for brachytherapy), their lack of energy independence should not compromise the accuracy of the measurement.

The spectrometer is an OceanOptics CCD (USB4000, Dunedin, FL), with 350–1040 nm range and ~0.2 nm step width. The splitter is a seven-fiber assembly, with each fiber being 100 μm in diameter (Caron et al. 2012). The seven-fiber port is connected to the probe, whereas the four- and the three-fiber ports are connected to the optical cables leading to the back end. This configuration showed a significant

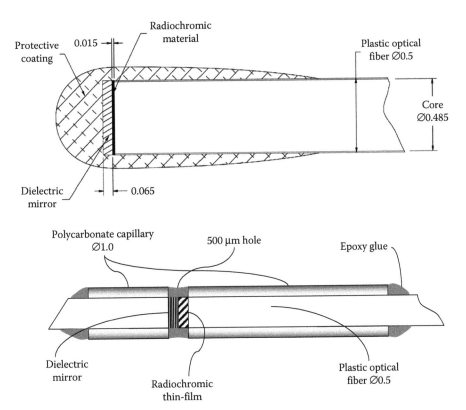

Figure 16.6 Schematics of radiochromic fiber-optic prototypes. Prototype A (top) and prototype B (bottom). (Modified from Croteau, A. et al., Real-time optical fiber dosimeter probe. In *Optical Fibers, Sensors, and Devices for Biomedical Diagnostics and Treatment XI, Proceedings of SPIE*, vol. 7894, 2011b.)

reduction in the amount of light reflected toward the detector, as well as better reproducibility in the reflected amount. In a transmission-reflection prototype, controlling the amount of light transferred to the detector from other sources other than the radiation-sensitive medium itself is crucial to the accuracy of the measurement.

16.5.1 SIZE

To meet the small size requirement, a fiber of 500 μm diameter was used in both prototypes. The second prototype (prototype B, Figure 16.7) is thicker in the region of the polycarbonate capillary, totaling 1000 μm in diameter. Using the capillary increased construction reproducibility and allowed for a more robust tip, at the cost of the overall probe thickness. The submillimeter size of either of the prototypes allows the probe to be positioned in a catheter (such as 5F or 6F catheters used in interstitial brachytherapy), potentially to be used within the patient and not just on the surface as is most common with larger dosimeters. The thickness of radiochromic material used in the prototypes can be coated to be anywhere between 2 and 20 μm, with the netOD at a given dose proportional to the thickness of material (Croteau 2011b).

16.5.2 COMPOSITION

All the materials for the probe were carefully chosen. As already discussed in Section 16.3.2, the radiochromic suspensions can be energy independent over a large range, provided that the appropriate elements are added to the suspension to achieve this. However, in engineering an *in vivo* probe, it is not just the sensitive material that is of interest but also the entire probe. Ideally, its presence should not significantly disturb the deposition of dose in the surrounding region. Additionally, due to an increase in image-guided radiotherapy in recent years, the probes should not interfere with image quality during

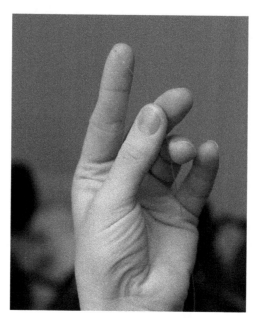

Figure 16.7 Photograph of prototype B.

CT or MR imaging. For this reason, low atomic number non-ferromagnetic materials were selected. For optical fibers, plastic was selected over silica. Although plastic optical fibers get damaged with prolonged exposure to ionizing radiation, to observe a drop in photon transmission a much higher dose is required than needed for the radiochromic material (Sporea et al. 2012). After a certain dose (<30 Gy), long before coloration of the fibers becomes an issue, the probes are intended to be disposed off, and therefore fiber damage due to radiation is not considered to be an issue. For the reflective material at the tip of the probe, a non-metallic film composed of multi-layer polymers (dielectric mirror by Vikuiti) is used. The high reflectivity of this polymer across the entire visible spectrum is appealing since the rest of the spectrum may be needed (e.g., for the dynamic range extension of the probes). In the first prototype (prototype A), the radiochromic suspension was spin-coated onto dielectic mirror, dried, and the reflector with radiochromic thin-film was then adhered to a 500 µm plastic optical fiber. A protective coating, made of a biocompatible epoxy with a well-established usage in medical applications, was applied to the tip. In the second prototype (prototype B), a 1 mm outer diameter and 500 µm inner diameter custom-made polycarbonate capillary (Paradigm Optics, Vancouver, WA) is used to align the dielectric mirror with radiochromic thin film to the optical fiber. Although the materials for the prototypes were chosen carefully to allow their use across a wide range of energies, the exact range over which these prototypes are energy independent remains to be verified experimentally at the time of writing.

16.5.3 REAL-TIME RESPONSE

As expected, the thin-film on the tip of an optical fiber was shown to have the same response to absorbed radiation dose as the commercially manufactured films. That is, an increase in absorbance is seen for the main and secondary peak at 636 and 585 nm, respectively, (Croteau 2011a). The absorbance and optical density increased in real-time, and sensitivity of probe was controlled by the thickness of material.

For a dosimeter to be useful for real-time measurements, the signal must be measured and processed sufficiently fast as to provide the user with dose estimate without any perceived delay. Although an important aspect of real-time dosimetry, review of various light detectors and processing algorithms is outside the scope of this chapter. In short, the spectrometer is directly and continuously accessed by custom-written software, installed on a computer or a laptop in the treatment console area. The software accesses the spectrometer signal (intensity at each wavelength and time stamp) from the moment measurement is initiated. Using the measured light intensity, it calculates the netOD and then using calibration factors, shows measured dose and dose rate to the end user on the computer screen. There is a trade-off between read-out frequency and noise and dose accuracy. The measurement frequency depends on several settings on spectrometer, such as integration time and averaging. A high integration time (range of milliseconds to seconds) allows for measurements with very low light intensity. Although useful in some scenarios, high integration time is not required in the current setup as the photon count from the light source, even after many meters of optical cable is sufficiently high. Averaging signal from the spectrometer increases its signal-to-noise ratio, decreasing the uncertainty in predicted dose. On the other hand, averaging for too long and over a substantial dose increase produces an error in dose estimate by the time it is displayed to the user. It is therefore important to select the appropriate detector and establish parameters (e.g., integration time, averaging, and spectral smoothing) to achieve the most accurate dose prediction with as high as possible read-out frequency. With the current configuration of light, spectrometer, and software, the system updates the dose on the display at 2–5 Hz, but can be easily increased to 10 Hz by decreasing spectral averaging.

16.5.4 REPRODUCIBILITY

The variability in material thickness between probes appears to be one of the biggest contributors to the overall uncertainty in the measured dose. The reproducibility in signal for the same probe, assessed by disconnecting and connecting it back up multiple times, is shown to be in the order of 2% (1σ). This includes light and spectrometer noise, as well as variation in alignment between probe and splitter. In comparison, experiments demonstrated the 1σ in netOD for probes of the same batch to be in the order of 6%. This includes the 1σ of ~3.5% (8.5 ± 0.3 μm) in thickness of sensitive material as measured during manufacturing, on top of the other uncertainties mentioned above.

The 1σ of 6% will translate into >10% overall uncertainty in dose (at 95% confidence level) unless some kind of a correction for film thickness is made. One common technique in radiation dosimetry is to use a double irradiation (or double-exposure) method, which has been shown by Zhu to be particularly useful in improving precision of GafChromic MD-55 film (Zhu 1997). Using this method, instead of relying on a factory calibration factor or batch calibration factor for a particular dosimeter, it is calibrated with a small known dose by the user and then that calibration factor is used for subsequent measurements with the probe. If the signal is linear with dose, then this process is a simple method to account for any minor variation in sensitivity between dosimeters. Although the process is a bit more complicated for nonlinear response of LiPCDA with dose, performing pre-irradiation calibration decreased the 1σ in netOD to <2%. This was illustrated by measuring the 1σ of netOD for probes irradiated twice, five minutes apart, without disconnecting and reconnecting at the splitter-probe interface. The 1σ of netOD was 5.1% and 5.3% for first and second irradiation, respectively. However, when the netOD for second irradiation was normalized by that obtained during first irradiation, thus eliminating any variation in material thickness between probes, the 1σ dropped to 1.3%. If the probe is disconnected after calibration and reconnected prior to measurement, as would be typical of normal use within the clinic, the 1σ will increase back up to ~2% due to extra uncertainty at probe-splitter connection.

These types of calibrations are time consuming for the user and can also be error prone if the calibration factors and probes are mixed up. To limit the likelihood of error, an auto-recognition and a means for encoding calibration factors within the software for each dosimeter should be built into the system. Other methods for eliminating or decreasing the effect of thickness variation exist, such as factory-based calibrations, binning, and optical measurements of material thickness immediately prior to use.

16.5.5 ANGLE DEPENDENCY

Often dosimeters suffer from angle dependency, where the measurement obtained varies with the direction of irradiation beam. Dosimeters that have cylindrical symmetry, as is typical of those made with optical fibers, are not expected to be affected by a change in azimuth angle. However, there may be a significant difference between measurements with beam oriented perpendicular to the cylindrical axis, and beam that is parallel to the cylindrical axis. This difference may be even more pronounced at low X-ray energies, where the dependence of photoelectric mass attenuation coefficient on the atomic number can play a large role in overall photon interaction probability and amount of energy absorbed. In the case of the radiochromic dosimeter prototypes, this may occur due to the presence of the dielectric mirror or the polycarbonate capillary at the tip or the optical fiber.

To assess angle dependency of prototype B, netOD for probes irradiated with LINAC beam parallel and perpendicular to probe axis ($0°$ and $90°$, respectively) were compared. To eliminate uncertainty due to variable thickness of radiochromic thin-film, each probe's response was normalized by its netOD for to 2 Gy calibration irradiation delivered at $90°$. The second set of irradiations was performed five minutes after the end of the first. The normalized netOD for 2 Gy dose delivered with beam at $0°$ (going through the reflector and part of capillary) was 2.9% higher than the normalized netOD for irradiation at $90°$, but was found to be statistically insignificant ($\alpha = 0.05$). It was therefore concluded that there is no significant angle dependency at 6 MV for the prototype B dosimeter. At the time of writing, angle dependency at kilovoltage energies remains to be tested.

16.5.6 MRI COMPATIBILITY

Prototype A dosimeter probes were imaged (Table 16.3) within agarose gel in a GE Signa Infinity 1.5 T TwinSpeed and a Siemens Magnetom Verio 3T MRI scanner, using clinically relevant acquisition sequences and resolution (Rink et al. 2011), and on a Bruker Bio-Spec 7 T research magnet at high resolution. Two probes were in direct contact with the gel (*in situ*), and the other two were within plastic brachytherapy catheters 2.0 mm in outer diameter (6F Pro-Guide sharp needles, Elekta Brachytherapy). Both the *in situ* probes, and the probes in needles corresponded to a drop in MR signal consistent with signal void, typical of air or plastic (Figure 16.8). Plastic catheters containing the probes were easily identifiable. On the other hand, *in situ* results demonstrated difficulty in localizing the dosimeters (500 μm in diameter) using clinically achievable resolutions on 1.5 and 3.0 T scanners, except at the very tip, where the prototypes had protective epoxy and therefore larger in size. Given the larger outer diameter of the prototype B probes, we anticipate being able to easily identify the capillary portion on MRI *in situ* if the dosimeters ever become used in that fashion (without catheter). Because of the choice of MR-compatible materials for the construction of the probes, no other signal artifacts were observed. Identifying *in situ* probes was possible at a higher resolution on a 7.0 T scanner, however, the radiation sensitive volume (500 μm in diameter, 15 μm thick) was still impossible to locate due to a lack of any distinctive feature on MR imaging. Despite the fact that the radiochromic thin film contains water, it was insufficient to provide any kind of discernible signal to localize the sensitive volume within the 3D scan.

Table 16.3 **MRI acquisition parameters used in assessing MRI compatibility**

SEQUENCE	VOXEL RESOLUTION (AP/RL/SI mm)	SCAN DETAILS
GE Signa Infinity 1.5 T TwinSpeed (3D)		
FSPGR	0.8^3	TE = 5.14 ms, TR = 12.056 ms, BW = 31.3 kHz
FR-FSE	0.8^3	TE = 112 ms, TR = 3000 ms, BW = 62.5 kHz
FIESTA	$0.8 \times 0.8 \times 0.4$	TE = 2.172 ms, TR = 4.468 ms, BW = 125 kHz
Siemens Magnetom Verio 3T (3D)		
VIBE	0.5^3	TE = 2.72 ms, TR = 7.09 ms, BW = 56.1 kHz
SPACE	0.6^3	TE = 138 ms, TR = 1000 ms, BW = 92.7 kHz
MPRAGE	$0.6 \times 0.6 \times 0.7$	TE = 3.19 ms, TR = 1400 ms, BW = 43.5 kHz
Bruker Bio-Spec 7 T (coronal)		
2D-RARE	$0.5 \times 0.15 \times 0.15$	TE = 16 ms, TR = 4000 ms, BW = 81.5 kHz
2D-FLASH	$0.5 \times 0.15 \times 0.15$	TE = 6 ms, TR = 250 ms, BW = 81.5 kHz

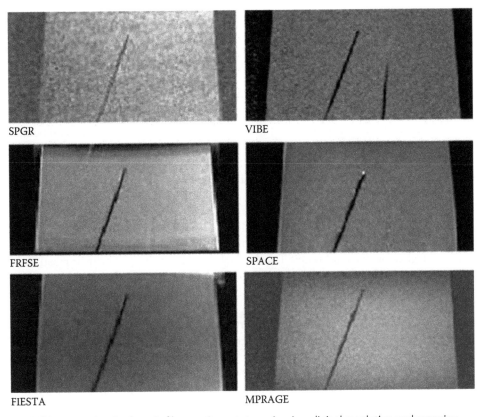

SPGR

VIBE

FRFSE

SPACE

FIESTA

MPRAGE

Figure 16.8 MR images of radiochromic fiber-optic prototype A using clinical resolution and scanning parameters.

It was therefore concluded that the radiochromic fiber probes are MR compatible, but identifying the dosimeter volume within the image requires another solution since it could not be identified visually.

16.6 CLINICAL APPLICATIONS

The development of a robust fiber-optic radiochromic dosimeter has the promise to provide real-time near tissue-equivalent dosimetry in sub-mm^3 volumes under a wide range of settings, including *in vivo* conditions and in the presence of high magnetic fields. This combination is unique, and there are many clinical applications that can be pursued.

The small format and limited perturbation of the photon and electron fluence makes the fiber optic radiochromic dosimeter adaptable to a variety of geometries, while still maintaining remote, real-time readout. One area of application is in the direct integration of the sensor into radiotherapy devices. This approach creates *dosimetry enabled devices*. This is of particular interest in the field of conventional brachytherapy and/or electronic brachytherapy. Through the creation of a dedicated *dosimetry channel*, the expected dose to the dosimeter's location in the applicator could be tracked and compared to the predicted dose corresponding to the patient's treatment plan. The real-time nature of the readout scheme allows *control point-by control-point* monitoring. The value of this approach is heightened in the electronic brachytherapy setting, which cannot rely on the stability of the source decay physics to provide the same confidence in delivered dose. This approach has the advantage of not being significantly influenced by the relative placement of the dosimeter and source—a challenge in brachytherapy where the dose gradients are very steep. The same concept could also be applied to other radiotherapy applications. For example, the direct integration of low-cost, dose integration technology into patient immobilization systems (e.g., head and neck mask). Such an approach would provide both real-time readout and long-term dose accumulation records for a given patient.

The continued development of volumetric dosimetry systems that can be read by either 3D CT-based optical readout (Ibbott et al. 1997, Oldham et al. 2005) or MR-based relaxation measurements (Gore 1984) will require validation and linkage to absolute dosimetry. Fiber-optic readouts using the well-understood radiochromic approach may prove useful. The co-localization of the fiber sensor and the optical CT or MR readout voxel is relatively straightforward.

The fiber-optic device would have application in the measurement of dose at surfaces and boundaries that induce electronic disequilibrium. The remarkably small measurement volume (~500 μm diameter, <20 μm thick) could be optimized to make high resolution measurements at surfaces or boundaries such as adjacent to a dense or high-Z metal, while minimizing the effect on photon fluence. The ability to make measurements at internal surfaces (i.e., mucosa) in the head and neck has been raised as a potential application. These measurements may become of greater interest as the field wrestles with the challenges of the *electron return effect* (ERE) associated with MR-guided radiation therapy systems (Raaymakers 2004).

The emerging field of MR-guided radiotherapy will place significant challenges on current *in vivo* dosimetry technologies. Diodes and MOSFET technologies will be challenged by their metal components which could disturb the MR-field and the potential for RF-induced heating of the conductive components during MR imaging. Optical readout isolates the dosimeter from the MR imaging process and should allow real-time readout without interference. Applications in external beam (regardless of ^{60}Co or linear accelerator approaches) are quite straightforward as a replacement of current *in vivo* dosimetry techniques. The growth in MR-guided brachytherapy creates another application for the technology.

The past 10 years has seen a growing interest in the development of radiosensitizing nanoparticle technology (Chithrani et al. 2010, Hainfeld et al. 2004). The flexibility of the fiber-optic readout schemes could provide unique configurations for radiobiological research and validation of radiochemical effects in the pre-clinical setting. This is an interesting area of development that remains unproven, but requires greater quantification of the applied dosimetry.

Finally, there are logistical and financial elements of the technology that could drive broader uptake. Of specific interest is the demand of radiotherapy in low and middle income countries. It is estimated that less than 10% of the population in Africa has access to radiotherapy (Zubizarreta et al. 2015). A major barrier to access is the lack of expertise, including expertise in medical physics. This is compounded by the limited financial support for test and evaluation equipment to support safety and quality. The creation of a low-cost, easily applied and broadly useful dosimeter could provide a very reliable safety net for the patient, as well as, the vulnerable and emerging radiotherapy programs being established. The radiochromic technology allows operation without the need for power and holds its recorded dose in an analog form. Taken together, the development of a robust, fiber-optic dosimeter could be highly beneficial provided cost could be kept to a minimum.

REFERENCES

Ali, I., Costescu, C., Vicic, M., Dempsey, J.F., Williamson, J.F. 2003. Dependence of radiochromic film optical density post-exposure kinetics on dose and dose fractionation. *Medical Physics* 30: 1958–1967.

Andrés, C., del Castilo, A., Tortosa, R., Alonso, D., Barquero, R. 2010. A Comprehensive study of the Gafchromic EBT2 radiochromic film. A comparison with EBT. *Medical Physics* 37: 6271–6278.

Arjomandy, B., Tailor, R., Anand, A. 2010. Energy dependence and dose response of Gafchromic EBT2 film over a wide range of photon, electron, and proton beam energies. *Medical Physics* 37: 1942–1947.

Bloor, D. 1984. The preparation and properties of ordered and disordered diacetylene polymers. In *Quantum Chemistry of Polymers—Solid State Aspects*, eds. J. Ladik, and J.M. André, 191–220. Boston, MA: D. Reidel Publishing Company.

Borca, V.C., Pasquino, M., Russo, G. et al. 2013. Dosimetric characterization and use of GafChromic EBT3 film for IMRT dose verification. *Journal of Applied Clinical Medical Physics* 14: 158–171.

Butson, M.J., Cheung, T., Yu, P.K. 2009. Evaluation of the magnitude of EBT Gafchromic film polarization effects. *Australasian Physical & Engineering Sciences in Medicine* 32: 21–25.

Butson, M.J., Yu, P.K.N., Cheung, T., Metcalfe, P. 2003. Radiochromic film for medical radiation dosimetry. *Materials Science and Engineering* R41: 61–120.

Caron, S., Croteau, A., Rink, A., Jaffray, D., Mermut, O. 2012. Selecting the appropriate splitter for a reflective optical fiber dosimeter probe. In *Photonics North 2012, Proceedings of SPIE*, ed. J. Kieffer, vol. 8412, Bellingham, WA: SPIE.

Chithrani, D.B., Jelveh, S., Jalali, F. et al. 2010. Gold nanoparticles as radiation sensitizers in cancer therapy. *Radiation Research* 173: 719–728.

Chiu-Tsao, S., de la Zerda, A., Lin, J., Kim, J.H. 1994. High-sensitivity GafChromic film dosimetry for ^{125}I seed. *Medical Physics* 21: 651–657.

Croteau, A., Caron, S., Rink, A., Jaffray, D., Mermut, O. 2011a. Fabrication and characterization of a real-time optical fiber dosimeter probe. In *Novel Biophotonic Techniques and Applications, Proceedings of SPIE*, eds. H. Sterenborg and I.A. Vitkin, vol. 8090, Bellingham, WA: SPIE.

Croteau, A., Caron, S., Rink, A., Jaffray, D., Mermut, O. 2011b. Real-time optical fiber dosimeter probe. In *Optical Fibers, Sensors, and Devices for Biomedical Diagnostics and Treatment XI, Proceedings of SPIE*, ed. I. Gannot, vol. 7894, Bellingham, WA: SPIE.

Dempsey, J.F., Low, D.A., Mutic, S. et al. 2000. Validation of a precision radiochromic film dosimetry system for quantitative two-dimensional imaging of acute exposure dose distributions. *Medical Physics* 27: 2462–2475.

Devic, S., Aldelaijan, S., Mohammed, H. et al. 2010. Absorption spectra time evolution of EBT-2 model GARCHROMIC™ film. *Medical Physics* 37: 2207–2214.

Devic, S., Seuntjens, J., Abdel-Rahman, W. et al. 2006. Accurate skin dose measurements using radiochromic film in clinical applications. *Medical Physics* 33: 1116–1124.

Devic, S., Seuntjens, J., Sham E. et al. 2005. Precise radiochromic film dosimetry using a flat-bed document scanner. *Medical Physics* 32: 2245–2253.

Devic, S., Tomic, N., Aldelaijan, S. et al. 2012. Linearization of dose-response curve of the radiochromic film dosimetry system. *Medical Physics* 39: 4850–4857.

Devic, S., Tomic, N., Pang, Z., Seuntjens, J., Podgorsak, E.B., Soares, C.G. 2007. Absorption spectroscopy of EBT model GAFCHROMIC™ film. *Medical Physics* 34: 112–118.

Devic, S., Tomic, N., Soares, C.G., Podgorsak, E.B. 2009. Optimizing the dynamic range extension of a radiochromic film dosimetry system. *Medical Physics* 36: 429–437.

Dutreix, A. and Bridier, A. 1985. Dosimetry for external beams of photon and electron radiation. In *The Dosimetry of Ionizing Radiation*, vol 1, ed. K.R. Kase, B.E. Bjarngard, and F.H. Attix, 163–228. Toronto, Ontario, Canada: Academic Press.

Gore, J.C., Kang, Y.S., Schulz, R.J. 1984. Measurement of radiation dose distributions by nuclear magnetic resonance (NMR) imaging. *Physics in Medicine and Biology* 29: 1189–1197.

Guillet, J. 1985. Photopolymerization. In *Polymer Photophysics and Photochemistry: An Introduction to the Study of Photoprocesses in Macromolecules*, 295–313. Cambridge: Cambridge University Press.

Hainfeld, J.F., Slatkin, D.N., Smilowitz, H.M. 2004. The use of gold nanoparticles to enhance radiotherapy in mice. *Physics in Medicine and Biology* 21: N309–N315.

Huber, R. and Sigmund, E. 1985. Excitons in short-chain polydiacetylene molecules. In *Electronic Properties of Polymers and Related Compounds*, ed. H. Kuzmany, M. Mehring, and S. Roth, 249–252. Berlin, Germany: Springer-Verlag.

Ibbott, G.S., Maryanski, M.J., Eastman, P. et al. 1997. Three-dimensional visualization and measurement of conformal dose distributions using magnetic resonance imaging of BANG polymer gel dosimeters. *International Journal of Radiation Oncology, Biology, Physics* 38: 1097–1103.

Jelley, J.V. 1958. *Cerenkov Radiation and Its Applications*. New York: Pergamon Press.

Klassen, N.V., van der Zwan, L., Cygler, J. 1997. GafChromic MD-55: Investigated as a precision dosimeter. *Medical Physics* 24: 1924–1934.

Lindsay, P., Rink, A., Ruschin, M., Jaffray, D. 2010. Investigation of energy dependence of EBT and EBT-2 Gafchromic film. *Medical Physics* 37: 571–576.

Lynch, B.D., Kozelka, J., Ranade, M.K., Li, J.G., Simon, W.E., Dempsey, J.F. 2006. Important considerations for radiochromic film dosimetry with flatbed CCD scanners and EBT GafChromic® film. *Medical Physics* 32: 4551–4556.

McLaughlin, W.L., Al-Sheikhly, M., Lewis, D.F., Kovacs, A., Wojnarovits, L. 1994a. A radiochromic solid-state polymerization reaction. *Polymer Preprints* 35: 920–921.

McLaughlin, W.L., Al-Sheikhly, M., Lewis, D.F., Kovacs, A., Wojnarovits, L. 1996a. Radiochromic solid-state polymerization reaction. In *Irradiation of Polymers: Fundamentals and Technological Applications*, eds. R.L. Clough and S.W. Shalaby, 152–166. Washington, DC: American Chemical Society.

McLaughlin, W.L., Puhl, J.M., Al-Sheikhly, M. et al. 1996b. Novel radiochromic films for clinical dosimetry. *Nuclear Technology Publishing* 66: 263–268.

McLaughlin, W.L., Soares, C.G., Sayeg, J.A. et al. 1994b. The use of a radiochromic detector for the determination of stereotactic radiosurgery dose characteristics. *Medical Physics* 21: 379–388.

McLaughlin, W.L., Yun-Dong, C., Soares, C.G., Miller, A., Van Dyk, G., Lewis, D.F. 1991. Sensitometry of the response of a new radiochromic film dosimeter to gamma radiation and electron beams. *Nuclear Instruments and Methods in Physics Research* A302: 165–176.

Mijnheer, B., Beddar, S., Izewska, J., Reft, C. 2013. *In vivo* dosimetry in external beam radiotherapy. *Medical Physics* 40: 070903.

Milton, J.S. and Arnold, J.C. 1990. *Introduction to Probability and Statistics: Principles and Applications for Engineering and the Computing Sciences.* 2nd ed. Toronto, Ontario, Canada: McGraw-Hill.

Muench, P.J., Meigooni, A.S., Nath, R., McLaughlin, W.L. 1991. Photon energy dependence of the sensitivity of radiochromic film and comparison with silver halide film and LiF TLDs used for brachytherapy dosimetry. *Medical Physics* 18: 769–775.

Niroomand-Rad, A., Blackwell, C.R., Coursey, B.M. et al. 1998. Radiochromic film dosimetry: Recommendations of AAPM Radiation Therapy Committee Task Group 55. *Medical Physics* 25: 2093–2115.

Oldham, M., Kim, L., Hugo, G. 2005. Optical-CT imaging of complex 3D dose distributions. *Journal of Physics* 5745: 138–146.

Piermattei, A., Miceli, R., Azario, L. et al. 2000. Radiochromic film dosimetry of a low energy proton beam. *Medical Physics* 27: 1655–1660.

Raaymakers, B.W., Raaijmakers, A.J., Kotte, A.N., Jette, D., Lagendijk, J.J. 2004. Integrating a MRI scanner with a 6 MV radiotherapy accelerator: Dose deposition in a transverse magnetic field. *Physics in Medicine and Biology* 49: 4109–4118.

Rampado, O., Garelli, E., Deagostini, S., Ropolo, R. 2006. Dose and energy dependence of response of Gafchromic® XR-QA film for kilovoltage X-ray beams. *Physics in Medicine and Biology* 51: 2871–2881.

Richter, C., Pawelke, J., Karsch, L., Woithe, J. 2009. Energy dependence of EBT-1 radiochromic film response for photon (10 kVp–15 MVp) and electron beams (6–18 MeV) readout by a flatbed scanner. *Medical Physics* 36: 5506–5514.

Rink, A., Croteau, A., Simeonov, A., Ménard, C., Jaffray, D.A. 2011. Preliminary investigations of the MR-compatible real-time optical fiber dosimetry probe for MR-guided brachytherapy procedures. *Brachytherapy* 10: S68.

Rink, A., Lewis, D.F., Varma, S., Vitkin, I.A., Jaffray, D.A. 2008. Temperature and hydration effects on absorbance spectra and radiation sensitivity of a radiochromic medium. *Medical Physics* 35: 4545–4555.

Rink, A., Vitkin, I.A., Jaffray, D.A. 2005a. Characterization and real-time optical measurements of the ionizing radiation dose response for a new radiochromic medium. *Medical Physics* 32: 2510–2516.

Rink, A., Vitkin, I.A., Jaffray, D.A. 2005b. Suitability of radiochromic medium for real-time optical measurements of ionizing radiation dose. *Medical Physics* 32: 1140–1155.

Rink, A., Vitkin, I.A., Jaffray, D.A. 2007a. Energy dependence (75 kVp to 18 MV) of radiochromic films assessed using a real-time optical dosimeter. *Medical Physics* 34: 458–463.

Rink, A., Vitkin, I.A., Jaffray, D.A. 2007b. Intra-irradiation changes in the signal of polymer-based dosimeter (GafChromic EBT) due to dose rate variations. *Physics in Medicine and Biology* 52: N523–N529.

Saylor, M.C., Tamargo, T.T., McLaughlin, W.L., Khan, H.M., Lewis, D.F., Schenfele, R.D. 1988. A thin film recording medium for use in food irradiation. *Radiation Physics and Chemistry* 31: 529–536.

Sixl, H. 1985. Electronic structures of conjugated polydiacetylene oligomer molecules. In *Electronic Properties of Polymers and Related Compounds*, ed. H. Kuzmany, M. Mehring, and S. Roth, 240–245. Berlin, Germany: Springer-Verlag.

Skoog, D.A., F.J. Holler, T.A. Nieman. 1998. *Principles of Instrumental Analysis.* 5th ed. Toronto, Ontario, Canada: Harcourt Brace College Publishers.

Sporea, D., Sporea, A., O'Keeffe, S., McCarthy, D., Lewis, E. 2012. Optical fibers and optical fiber sensors used in radiation monitoring. In *Selected Topics on Optical Fiber Technology*, ed. M. Yasin, 607–652. Rijeka, Croatia: InTech.

Tanderup, K., Beddar, S., Andersen, C.E., Kertzscher, G., Cygler, J.E. 2013. *In vivo* dosimetry in brachytherapy. *Medical Physics* 40: 070902.

Todorovic, M., Fischer, M., Cremers, F., Thom, E., Schmidt, R. 2006. Evaluation of GafChromic EBT prototype B for external beam dose verification. *Medical Physics* 33: 1321–1328.

van Battum, L.J., Hoffmans, D., Piersma, H., Heukelom, S. 2008. Accurate dosimetry with GafChromic™ EBT film of a 6 MV photon beam in water: What level is achievable? *Medical Physics* 35: 704–716.

Zeidan, O.A., Stephenson, S.A.L., Meeks, S.L. et al. 2006. Characterization and use of EBT radiochromic film for IMRT dose verification. *Medical Physics* 33: 4064–4072.

Zhu, Y., Kirov, A.S., Mishra, V., Meigooni, A.S., Williamson, J.F. 1997. Quantitative evaluation of radiochromic film response for two-dimensional dosimetry. *Medical Physics* 24: 223–231.

Zubizarreta, E.H., Fidarova, E., Healy, B., Rosenblatt, E. 2015. Need for radiotherapy in low and middle income countries—The silent crisis continues. *Clinical Oncology* 27: 107–114.

17 Fiber-coupled luminescence dosimetry with inorganic crystals

Claus E. Andersen

Contents

17.1 INTRODUCTION

The purpose of this chapter is to outline the basics of how inorganic crystals can be used for fiber-coupled luminescence dosimetry, for example, during brachytherapy or external beam radiotherapy with megavoltage (MV) photons. Inorganic crystals generally offer two luminescence signals that can be used for dosimetry: prompt radioluminescence (RL) and optically stimulated luminescence (OSL). The RL signal is present only during the irradiation, whereas the OSL signal can be generated any time after the irradiation by stimulation with an external light source. Under ideal conditions, the RL signal provides an online signal that is proportional to the absorbed dose rate at the position of the crystal, whereas the OSL signal is proportional to the passively integrated absorbed dose. Fiber-coupled luminescence dosimetry with inorganic crystals is similar to the techniques based on organic plastic scintillators described elsewhere in this book.

In both cases, the instrumentation can be placed several meters away from the radiation sensing element, and this element can be made relatively small (submillimeter size). Inorganic crystals such as carbon-doped aluminum oxide or copper-doped quartz are less water equivalent than organic plastic scintillators (in terms of both effective atomic number and mass density), but the inorganic crystals offer improved methods for stem signal suppression, and they also seem to provide a higher sensitivity. This chapter first discusses the basics of luminescence physics using carbon-doped aluminum oxide as the key example. Other materials and their applications in medical dosimetry are briefly discussed in Section 17.4.

17.2 LUMINESCENCE PHYSICS

17.2.1 TERMINOLOGY

Broadly speaking, the luminescence from inorganic crystals that arises in response to ionizing radiation can be classified as being either prompted or stimulated RL [1,2].

The first type of phenomenon, the prompted RL, is often simply called radioluminescence or RL because it occurs simultaneously with the irradiation, and in all of the following, we will follow this convention. RL is similar to the scintillation light from organic plastic scintillators although the fluorescence in organic scintillators originates from transitions in single molecules, whereas the luminescence phenomena in inorganic crystals are best understood using solid-state physics and band-gap models. The luminescence processes in inorganic crystals require a regular crystal lattice (i.e., a host material), and these processes cannot occur if the detector material is not crystalline.

The second type of phenomenon, the stimulated RL, is light that is emitted after the irradiation when the crystal is thermally or optically stimulated. No signal is produced before the stimulation is switched on. In principle, the OSL signal can therefore be read out at any later time after the irradiation. However, sometimes the potential OSL signal will slowly vanish in time in the absence of any stimulation. This is called fading. In the context of fiber-coupled dosimetry, it is almost exclusively the OSL that is of interest, and thermoluminescence (TL) will not be discussed here. The resetting of the OSL signal with a sustained source of light is called bleaching. This is the optical equivalent to annealing where a crystal is heated to a high temperature (e.g., 900°C).

17.2.2 PROCESSES

RL and OSL can be understood as a three-step process [1,2]: (1) electron–hole pair creation; (2) trapping, stimulation, and escape of electrons and holes; and (3) electron–hole recombination and relaxation of the excited recombination center under emission of luminescence. These processes will be discussed in Sections 17.2.2.1 through 17.2.2.3. Figure 17.1 shows a simple sketch that represents a band-gap model of a crystal.

17.2.2.1 Step 1: Electron–hole creation

The energy absorbed in the crystal upon exposure to ionizing radiation can be used to lift an electron from the valence band to the conduction band. The empty electron state in the valence band is called a hole. The electron–hole pair creation in luminescence dosimetry is equivalent to the ionization process in gas-filled ionization chambers. The crystals of interest have a wide band gap (typically 5–9 eV), and these crystals are electrical insulators at room temperature. The creation of electron–hole pairs therefore can lead to a significant increase in the electrical conductivity of the exposed crystal [3]. If an electron is lifted from a state deep in the valence band to some free state high in the conduction band, then the electron–hole pair will subsequently dissipate energy through phonon creations in the crystal lattice (i.e., the energy heats up the crystal). During this thermalization process, the electron and the hole will therefore move toward the band-gap edges, and eventually, electrons will end up in the bottom of the conduction band, whereas holes will end up at the top of the valence band.

17.2.2.2 Step 2: Trapping and stimulation

Electrons in the conduction band and holes in the valence band can become trapped in the crystal within the forbidden band gap. These traps are best viewed as localized centers associated with imperfections in an otherwise perfect host crystal structure. Real crystals always contain traps, and synthetic crystals can be

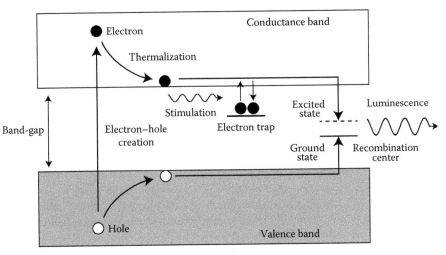

Figure 17.1 Simple band-gap model illustrating some of the processes that take place in inorganic crystals (see text).

designed, for example, using different dopants or crystal growth techniques, to have an abundance of traps as needed for particular dosimetry applications [4]. Trapped charge carriers may escape again at a later stage with the supply of sufficient stimulation energy. Electrons return to the conduction band, and holes return to the valence band.

Besides the distinction of traps as being either electron traps and hole traps, traps are further characterized as being thermally or optically active, depending on whether there is a significant route for thermal or optical energy to make the charge carriers escape from the trap in question. Optically and thermally active traps can be used for OSL and TL dosimetry, respectively.

Depending on how much energy is required for escape, traps may finally be broadly classified as being shallow traps, dosimetry traps, or deep traps. Shallow traps require an energy for charge carriers to escape that is comparable to the thermal energy associated with crystal vibrations at room temperature (0.025 eV). Hence, the trapped charge in shallow traps will decay with time, and this decay is affected by the temperature of the crystal. Such traps cannot therefore generally be used for dosimetry. Shallow traps are, for example, responsible for the phosphorescence (afterglow) often observed immediately after irradiation of inorganic crystals. At the other extreme, deep traps generally require a very high energy before escape is possible, and deep traps can therefore only be emptied if the crystal is annealed at a high temperature (e.g., 900 °C). Hence, these traps are not directly useful for fiber-coupled dosimetry. However, they may influence dosimetry applications as they may compete for the charge carriers during the trapping process. Dosimetry traps require an energy for escape that is in between the energy required for escape from shallow and deep traps. This energy is sufficiently large to provide stability at room temperature, and the energy is sufficiently low to allow a high degree of escape probability during stimulation with low-power light sources used in fiber-coupled medical dosimetry (e.g., blue or green milliwatt lasers or light-emitting diodes).

17.2.2.3 Step 3: Recombination and luminescence production

In a perfect crystal, the direct return of an electron from the valence band to the conduction band with the emission of a photon is an inefficient process, and the band-gap energy is generally too large for the resulting photon to lie in the visible range [5] where the crystal may be transparent. However, real crystals used in luminescence dosimetry have recombination centers where electrons and holes can recombine with a high probability. Some of these centers (called color centers) produce luminescence upon recombination. As indicated in Figure 17.1, this is a two-step process: First, the electron and the hole recombine, and a certain energy is released. It is useful to think about the luminescence center as a local entity that has been excited from its ground state to some higher level by the recombination energy. Second, the relaxation of the luminescence center from this excited state back to the ground state produces the luminescence (RL or OSL). Therefore, the resultant emission is characteristic for the luminescence centers present in the crystal.

The transition normally results in a broad spectrum of light much like the broad spectral distribution for organic scintillators. The relaxation is associated with a certain probability, and the different luminescence centers therefore have certain characteristic lifetimes (typically in the microsecond-to-millisecond range).

17.2.2.4 Competition and quenching

For an ideal dosimeter, the route from energy absorption to signal generation should be constant within a wide range of conditions. One of the reasons why gas-filled ionization chambers are so useful in medical dosimetry is that the energy required to ionize an air molecule is almost independent of the energy of the electron that produces the ionization. RL/OSL dosimetry with inorganic crystals is less perfect, and the signal yield is generally not independent of the microdosimetric nature of dose delivery [6–8]. Therefore, if one knows the ratio between the luminescence signal and the absorbed dose to the crystal for one beam quality (e.g., during calibration in a cobalt-60 beam), then it is not given that this signal-to-dose ratio will be accurate also at other beam qualities having different ionization densities. Increasing the ionization density generally leads to less luminescence, and this phenomenon is therefore often referred to as ionization density quenching. Studies of these problems can be performed, for example, with a mono-energetic gamma source and a so-called Compton spectrometer [9] or with a conventional beam of kilovoltage X-rays. The latter method requires accurate control of the dosimetry as discussed by Tedgren et al. [10]. Studies can also be conducted in proton and heavy-charged particle beams [11,12], where the ability to resolve Bragg peaks can be used as a benchmark for detector efficiency and volume averaging (size).

In track structure modeling, one attempts to model the reduced luminescence signal in, for example, heavy-charged particle beams on the basis of the detector gamma response and the knowledge about the detailed dose deposition around individual tracks [13]. Temperature can also influence the generation of luminescence signals. If the temperature of the crystal is increased, the excited luminescence centers may relax back to the ground state along alternative routes without emitting any luminescence, and this is generally referred to as thermal quenching [14,15]. An additional nonideal complication for the inorganic crystals is that the concentrations of traps and recombination centers may be dynamically influenced and changed by the irradiation [16].

17.2.3 READ-OUT MODES

Figure 17.2 shows a prototypical instrument for fiber-coupled RL/OSL measurements. The inorganic crystal is attached directly to a long optical fiber cable [17]. Typical dimensions are $1 \times 0.5 \times 0.5$ mm^3 for the crystal, and the optical fiber cable is typically 15 m long and has a 500 μm diameter core. The whole dosimeter probe has to be lighttight. OSL-based systems will need both a stimulation source (e.g., a focused laser) and a light detection system with enough sensitivity to record luminescence. A beam splitter is normally required to couple both the light source and the detection system to the same fiber. Strong filtration is required to protect the light detection system from any direct or scattered light from the stimulation light source. Specific RL/OSL systems have been described by several authors

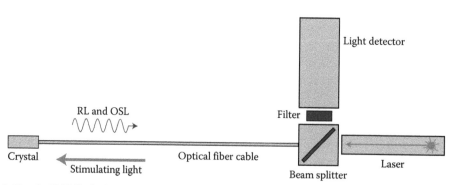

Figure 17.2 Simple RL/OSL dosimetry system.

[12,17–25]. For crystals with combined RL/OSL capabilities, several readout modes can be used: (1) pure OSL, (2) pure RL, or (3) combined RL/OSL. These modes are described in Sections 17.2.3.1 through 17.2.3.3.

17.2.3.1 Pure OSL

To better understand some of the virtues of OSL dosimetry, it may be useful first to highlight that optical stimulation can be delivered over short periods of time and with a well-defined energy and intensity (e.g., a short laser pulse). This is in somewhat contrast to the properties of thermal stimulation where the stimulation is always associated with a broad Boltzmann distribution of energies characteristic for the given temperature.

OSL readout with a constant stimulation intensity (i.e., continuous-wave OSL [CW-OSL]) gives rise to OSL decay curves of the type shown in Figure 17.3. The OSL signal peaks during the initial part of the curve where the concentration of trapped charge is at a maximum. During the readout, the OSL signal becomes weaker, and eventually after some time of stimulation, the recorded signal may vanish or it may reach a near-constant level, indicating that this part of the curve is not related to the detrapping of the dosimetry trap system but rather to deep traps or non-OSL phenomena excited by the stimulation. Therefore, although such OSL decay curves may look simple, their shape will reflect the kinetics of the detrapping of charge and the production of light, and they can therefore be quite complex. The complexity increases with the number of trapping systems and recombination centers. The shape of the OSL decay curve may change with, for example, dose and ionization density [7]. It is therefore not always straightforward to identify an optimal metric that best represents the OSL signal. Typical candidates are (1) the peak value, (2) the full integral of the curve, and (3) an integral over a certain part of the curve. The peak value is of particular interest because this value is available after a short time of stimulation (e.g., less than 1 s). However, the peak value may not have the best reproducibility because it is sensitive to minute changes in stimulation intensity.

For fiber-coupled OSL dosimetry, the idea is that the dosimeter probe can be used many times without annealing the crystal. It is therefore paramount that a protocol is designed for how long to bleach the crystal between consecutive irradiations. Because of the inherent memory associated with inorganic crystals, it is not enough to demonstrate that the OSL signal for identical irradiations has a high reproducibility; it must also be demonstrated what happens if doses vary widely from one irradiation to the next. Randomized tests are therefore generally needed to get a realistic assessment of the performance of fiber-coupled OSL systems [26,27]. An ideal system, of course, provides an OSL signal that is independent of dose history.

For some applications, it is of interest to modulate the stimulation intensity during the OSL readout. In pulsed OSL [46], the stimulation intensity is switched on and off, and often the on-time is much shorter

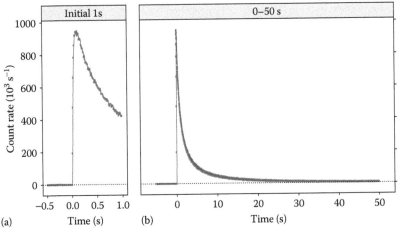

(a)　Time (s)　(b)　Time (s)

Figure 17.3 OSL decay curve for Al$_2$O$_3$:C during stimulation with green light (532 nm) and detection in a narrow band about 420 nm. (a) Shows the initial first second and (b) shows the initial 50 s.

than the off-time. This has two implications. First, the sensitivity of the system increases because the signal is focused to short periods of time where the background can safely be assumed to be constant. Second, the method provides improved separation between stimulation and detection if the latter is carried out only during the off-period of the stimulation.

Normally, the OSL readout is performed after the irradiation, and in this case, we obtain a measure of the integrated dose because the crystal was last reset. Essentially, the crystal acts as a passively integrating dosimeter, and the integration will take place at all time, even if the system is without electrical power. Some systems have also been proposed to allow for OSL readout during the irradiation [20,28–30], and these designs allow for real-time dose rate measurement capability. An important challenge for such systems is to separate the OSL signal from the RL and the stem signal. In this context, the pulsed OSL is of particular interest. If the dose rate is not changing significantly within the on-periods of the optical stimulation, then it can be assumed that the light immediately before and after an on-period reflects RL and stem signals, and that the net increase during the on-period is a clean OSL signal useful for dosimetry. Certain materials such as KBr:Eu have a very fast OSL decay time, and Klein et al. [20] were able to acquire the full OSL signal at a rate of 10 Hz in a study of computed tomography dosimetry.

17.2.3.2 Pure RL

The RL signal provides a measure of the real-time dose rate. However, the signal is less straightforward to use for dosimetry in comparison with the signal from organic scintillators. First of all, the luminescence lifetime for inorganic crystals can be relatively long. For example, the ~35 ms lifetime for Al_2O_3:C means that the dose rate variations faster than about 100 ms cannot be resolved with good accuracy. However, the main problem for RL dosimetry (besides the stem signal problem) is the potential influence of traps. This problem originates from the competition between dosimetry traps and shallow and deep traps, and the dynamic creation of recombination centers during irradiations. This means that repeated identical irradiations do not necessarily give the same RL output. This is, for example, seen for a pristine, or annealed Al_2O_3:C crystal, where the RL signal per dose rate unit systematically increases with dose [22,31,32]. However, after a large dose (typically about 1 kGy) when all relevant traps have been filled and when the concentration of recombination centers has stabilized, a stable RL signal can be obtained. This saturated RL signal is useful for real-time dosimetry [32], and the saturated RL protocol for Al_2O_3:C does not have the memory effect discussed below for the combined RL/OSL readout protocol.

Shallow traps directly influence the time resolution of RL dosimetry [26] as shown in Figure 17.4. If we use RL to monitor the constant dose rate from a radioactive source, we will note that the apparent RL signal persists after the source is closed. This afterglow reflects the emptying of shallow traps after the irradiation has stopped. Likewise, when the irradiation starts again, we will notice that the RL takes a little while to reach a stable level. This is because the shallow traps initially consume a significant part of the change carrier production. In other words, a stable RL signal is not reached before the shallow traps have been filled (i.e., a certain degree of equilibrium has been established). The influence of shallow traps is generally temperature dependent and for Al_2O_3:C, for example, there is no apparent influence at about 35°C, which means that these effects cannot be seen during *in vivo* dosimetry.

17.2.3.3 Combined RL/OSL

It is possible to combined RL/OSL readout such that the RL signal is used for real-time recording of the dose rate during the irradiation, whereas the OSL is used for recording the integrated dose. However, this can be a relatively complex matter as the OSL readout can disturb the RL sensitivity. For Al_2O_3:C, algorithms have been developed to account for the change in RL sensitivity [19] or to take advance of it [22]. However, Damkjær and Andersen [26] concluded that it was not possible to obtain reproducible results with the combined RL/OSL readout protocol for Al_2O_3:C. First, a test dose was selected randomly among nine predetermined values in the change from 0 to 4 Gy. Then this test dose was administrated 7 times while recording both RL and postirradiation CW-OSL. Then a new test dose was randomly selected among the same nine predetermined values, and the whole procedure was repeated many times. The key result is shown in Figure 17.5. It is seen that the OSL signal had a

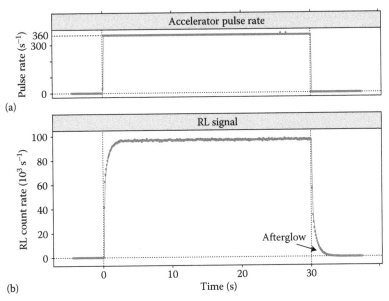

(a)

(b)

Figure 17.4 RL signal from Al$_2$O$_3$:C using the saturated RL protocol during a 2.5 Gy irradiation with 6 MV X-rays. (a) The linear accelerator pulse rate and (b) the RL signal. The measurements were recorded at room temperature (22°C), and the effect of shallow traps is visible when the irradiation starts and stops. (From Andersen CE et al., *Radiation Measurements*, 46, 1090–1098, 2011.)

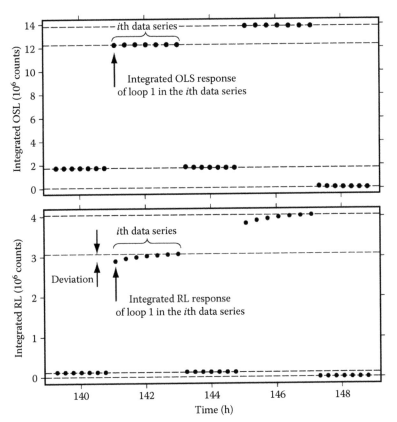

Figure 17.5 Memory effect for the RL signal from Al$_2$O$_3$:C in the RL/OSL readout protocol. In the plot, the first seven irradiations were at the 0.4 Gy dose level. The subsequent sets of seven irradiations were at 3.4, 0.4, 4, and 0 Gy, respectively. Notice how the RL signal systematically changes the value in the step from 0.4 to 3.4 Gy, and again in the step from 0.4 to 4 Gy. (Damkjær SMS, Andersen CE. *Radiation Measurements*, 45, 671–673, 2010.)

high degree of reproducibility, whereas the RL signal changed systematically when the dose level was changed. This problem is the so-called memory effect of RL. It should be pointed out that several previous tests without the sevenfold repeats of identical doses had failed to demonstrate the systematic changes in the RL response. It therefore seems that it is best to use a readout protocol based purely on either OSL or RL. The combined RL/OSL protocol can be problematic, at least for a system as complex as Al_2O_3:C.

17.3 INSTRUMENTATION

17.3.1 SIGNAL DETECTION IN PULSED BEAMS

A range of detector technologies are available for detection of luminescence as described elsewhere in this book. The long lifetime of certain inorganic crystals can have implications for what light detection technique to use. It can, for example, be problematic to apply photomultiplier tubes operating in an event-counting mode in pulsed linear accelerator beams if the material in question has a very fast lifetime (ns). For example, if the photomultiplier tube has a maximum counting frequency of 10 MHz, and if the linear accelerator has a pulse duration of 5 μs and a pulse repetition frequency of 360 Hz, then the system will reach the nonlinear domain for a luminescence center with a short lifetime already for an average counting frequency of

$$\frac{10\,\mathrm{events}}{1\,\mu s} \times 5\,\mu s \times 360\,\mathrm{Hz} = 18\,\mathrm{kHz} \tag{17.1}$$

However, if the lifetime of the luminescence center is long, like for Al_2O_3:C, then the RL pulses will be evenly distributed in time, and the system will be in the linear domain up to an average counting frequency of 10 MHz. Notice that a counting system that has a linear response to the RL signal from an inorganic crystal with long luminescence lifetime may not have a linear response for the stem signal. To test for linearity, it is useful to see if there is proportionality between the recorded stem signal and the length of fiber cable in the primary beam.

17.3.2 SEPARATION OF STIMULATION LIGHT AND LUMINESCENCE

For OSL systems, it is imperative with a strong separation between the stimulation light and the luminescence. This separation can be in time as in pulsed OSL, where the luminescence detection system can be deactivated during the short stimulation pulse. However, the main technique is to use a detection filter (e.g., a narrow bandpass filter), which does not overlap with the spectrum of the stimulation light source.

17.3.3 STEM SIGNAL SUPPRESSION TECHNIQUES

Fiber-coupled dosimetry with inorganic crystals is also subject to errors from the stem signal, that is, Cerenkov light and fluorescence generated in the optical fiber cable in response to irradiation with ionizing radiation. This problem has already been discussed in the context of organic plastic scintillators elsewhere in this book and material-specific details will be given in Section 17.4. Here, we will here just highlight stem effect reduction techniques that are unique or of particular relevance for systems based on inorganic crystals.

First of all, postirradiation OSL techniques are completely insensitive to the stem effect because the stem signal is not present during the OSL readout. The OSL signal can therefore be used to quantify the significance of the stem signal in the RL/OSL protocol [33].

For pulsed accelerator beams, it is possible to separate the RL from the stem signal if the luminescence lifetime is long compared with the accelerator pulse width. Medical accelerators typically have a pulse duration of 5 μs and a typical delay between pulses of 2–3 ms. As the stem signal consists of Cerenkov light and fluorescence, the stem signal is essentially only present during the actual accelerator pulses. This technique has been demonstrated for materials such as Al_2O_3:Cr [34,35], SiO_2:Cu [18,36], and Al_2O_3:C [19,32].

It is useful to point out that the chromatic removal technique [37] originally developed for organic scintillators can also be applied for materials such as Al_2O_3:C [38], and this technique facilitates stem signal removal in, for example, Ir-192-based brachytherapy. As discussed by Veronese et al. [39] (see Section 17.4), Eu-doped SiO_2 offers interesting features for improved chromatic stem signal removal because the emission is in the red region (i.e., well separated from the bluish Cerenkov light) and because the emission is relatively narrow. Furthermore, Veronese et al. [56] have demonstrated that infrared RL from Yb^{3+}-doped silica displays a sharp emission line at about 975 nm, which is essentially completely separated from any stem signal.

17.4 MATERIALS AND APPLICATIONS

Several inorganic crystals have been used for fiber-coupled luminescence dosimetry, and Table 17.1 provides an overview of the key characteristics for some of the main materials: emission wavelength, density, effective atomic number, and luminescence lifetime. In medical dosimetry, we are often interested in the absorbed dose to water, and detector materials with a high degree of water equivalence are therefore particularly useful. Note that this is not just a question of having an effective atomic number close to that of water (7.5); it is sometimes more important that the density is close to unity [40,41]. The host material determines the dosimetric properties of the crystal (e.g., density and effective atomic number). The dopants are present only at trace level concentrations, but they play a key role for optical properties of the crystal (e.g., traps and luminescence emission wavelength, intensity, and lifetimes).

17.4.1 SiO_2-BASED PHOSPHORS

17.4.1.1 Cu-doped silica fiber

Justus et al. (2004) [18] developed a fiber-coupled dosimetry system based on the detection of RL from Cu^{+1}-doped silica fibers. The primary intended use of the system was dosimetry during radiotherapy with megavoltage linear accelerators. A short Cu^{+1}-doped silica fiber was attached to a 1 m multimode optical fiber (400 µm core diameter) using a plasma fusion splicer. This technique provided an attachment without any evidence of interface between the doped fiber piece and the optical fiber cable. Although the signal is described to arise from phosphorescence, Tanyi et al. [36] was unable to register any temperature dependence of the signal in the range from 22°C to 50°C during irradiations in a 15 MV photon beam. The decay of the luminescence is well described by a biexponential decay function with lifetimes of 51 and 104 µs. It is therefore possible to separate the RL from the stem signal in beams of pulsed linear accelerators and to obtain an estimate of the dose associated with each

Table 17.1 **Key characteristics of selected inorganic crystals used for fiber-coupled medical dosimetry**

MATERIAL	MODE	Z_{EFF}	DENSITY (g/cm³)	EMISSION PEAK (nm)	LUMINESCENCE LIFETIME (ms)	REFERENCES
BeO	RL/OSL	7.2	3.0	280 (RL) 370 (OSL)	0.027 (OSL)	[23,15]
Al_2O_3:C	RL/OSL	11.3	4.0	420	35	[2]
Al_2O_3:Cr_2O_3	RL	11.3	4.0	694	~3	[34,35]
SiO_2:Ce	RL	11.8	2.3	450		[42]
SiO_2:Cu	RL/OSL	11.8	2.3	500	~0.1	[18]
SiO_2:Eu	RL	11.8	2.3	620		[39]
GaN	RL	29.1	6.1	370	~0	[43]
KBr:Eu	OSL	31.5	2.8	420	<1	[20]

An extended list of phosphors and references can be found elsewhere (Andersen CE, *AIP Conference Proceedings*, 1345, 100–119, 2011.)

individual accelerator pulse. The time separation of the two signals may be carried out using a gated counter controlled by (1) the electrical synchronization signal from the linear accelerator [18] or (2) a fast scintillator that triggers on scattered radiation in the accelerator room [36]. To avoid any influence of variations in the time structure of the accelerator pulsing on the response of the dosimetry system, it is important that the gating window is made significantly wider than the radiation pulses. Justus et al. [18] used a gating window of 8 μs.

17.4.1.2 Eu-doped silica fiber

Veronese et al. [39] developed a dosimetry system based on RL from Eu-doped (Eu^{+3}) silica optical fibers. This material has an emission spectrum that is dominated by a relatively narrow peak at 620 nm. An important feature of this emission is that it can be separated from the Cerenkov light which is weak in the red region. Hence, an improved separation of scintillation light and stem signal can be achieved by just considering the net scintillator signal in a narrow window around 620 nm (see Figure 17.6). This technique therefore does not rely on the critical assumption: the stem signal has a fixed spectral distribution [45] within the irradiation conditions studied (e.g., from one field size, irradiation angle, depth, or beam energy to another). This is a significant advance compared with the chromatic removal technique developed by Fontbonne et al. [37]. Veronese et al. [39] demonstrated how the net signal in the 620 nm window remained constant regardless of irradiation angle when irradiating the scintillator crystal at a constant dose rate from several angles with 6 MeV electrons (i.e., under conditions where the stem signal will change in intensity and spectral distribution).

17.4.1.3 Ce-doped silica fiber

Carrara et al. [42] developed a dosimetry system for brachytherapy using a 10 mm-long Ce-doped silica fiber (emission spectrum from 380 to 750 nm with a peak at about 500 nm). The detector signal was found to increase by 0.2% per kelvin, and the system had a reproducibility of about 0.6% (one standard deviation). The authors were not able to detect any significant energy dependence for the detector in a conventional Ir-192 depth dose curve experiment.

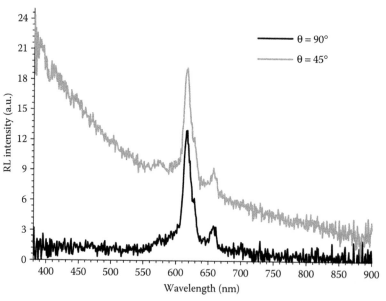

Figure 17.6 Example of a combined stem signal and RL spectrum from a 10 mm Eu-doped silica fiber when the fiber is exposed to 6 MeV electrons at different angles. At θ = 90°, the stem signal from Cerenkov light is much less than for θ = 45°. The peak at 620 nm is the Eu emission peak. (Data from Veronese I et al., *Journal of Physics D: Applied Physics*, 46, 015101, 2013.)

17.4.2 Al_2O_3:C

Al_2O_3:C is one of the most widely used luminescence materials for OSL dosimetry, and several systems and services are currently available on a commercial basis from Landauer Inc. (Glenwood, IL) [46]. The material was originally developed for TL dosimetery by the Ural Polytechnic Institute (Ekaterinburg, Russia). However, it was found that the TL signal was unstable when exposed to light. This problem, however, could be circumvented and the refinement of OSL dosimetry based on Al_2O_3:C was initiated.

Al_2O_3:C is used both for routine dosimetry in radiation protection and for medical dosimetry. For example, the Radiological Physics Center in the United States uses OSL dosimetry for remote auditing of medical linear accelerators [47]. It has also been investigated if Al_2O_3:C would be suitable for fiber-coupled dosimetry [19,22,29,30,48–51]. It can be used as a single (bulk) crystal with typical dimensions ($1 \times 1 \times 2$ mm^3) that can be attached to an optical fiber cable. Alternatively, a 100 μm layer of Al_2O_3:C grains can be embedded in a host matrix of ultraviolet polymethyl methacrylate (PMMA) polymer [11] attached onto a PMMA fiber cable. The latter technique provides a thin detector with good spatial resolution [11,12]. As has already been discussed, Al_2O_3:C can be used in at least three readout modes: (1) RL only, (2) OSL only, or (3) combined RL/OSL. The original idea with the combined RL/OSL mode was to use the RL for time-resolved dose rate measurements during the treatment, whereas the OSL would provide an integral of the absorbed dose. However, it has been found [32] that the OSL readout was too time consuming for clinical work, that the OSL disturbed the RL signal [26], and that the RL signal required large corrections [19,22], far outside the range of corrections normally found acceptable in radiotherapy dosimetry. The most recent studies [12,52] with fiber-coupled Al_2O_3:C therefore seem to focus on the saturated RL protocol.

Systems based on Al_2O_3:C have been designed for *in vivo* dosimetry during pulsed dose rate (PDR) and high dose rate (HDR) brachytherapy [38,53,54], and their use has been demonstrated in smaller patient studies [33,52]. Because of the high sensitivity of Al_2O_3:C, it was feasible to design dosimeter probes that could fit in standard applicators such as brachy needles with a 1 mm inner diameter. Hence, the main feature of the technique was that dosimetry could be carried out directly within the tumor region. In the initial study [33], a combined RL/OSL protocol was used, whereas in the later study [52], the saturated RL protocol was used in combination with the chromatic filtration technique for stem signal reduction. Systems for measurements in diagnostic kilovoltage X-ray beams [50], and MV photon and electron beams [21,22] have also been designed and tested. Recently, fiber-coupled Al_2O_3:C has been applied in a study organized by the *International Atomic Energy Agency* (IAEA) [55] of detector correction factors related to small field dosimetry. The results for Al_2O_3:C for field sizes down to 0.6 cm^2 were within 2% of the reference alanine dosimeters.

17.4.3 BeO

BeO is a very interesting material for medical dosimetry because its effective number (7.2) is close to that of water (7.5). It can be used for both RL and OSL fiber-coupled dosimetry [23,24]. Systems for point measurements based on passive BeO chips (i.e., not fiber-coupled system) are commercially available from the Helmholtz Zentrum München in Germany.

The RL has a broad emission wavelength 200–400 nm with a peak at about 280 nm [15], which is difficult to transport in plastic fibers and which also makes it difficult to reduce the influence of the stem signal using the chromatic removal technique [37]. Santos et al. [23] therefore reduced the stem signal (which amounted to about 40% of the RL signal) using separate measurements with a blank fiber. Interestingly, the OSL emission spectrum (peak at about 370 nm) is considerably different from the RL emission spectrum, indicating that different luminescence processes are taking place in BeO ceramics [15].

17.4.4 GaN

Si-doped gallium nitride (GaN) emits RL with a high-light yield, and because it also has a high density and a high atomic number, it is possible to design detectors of a very small volume (<0.1 mm^3) [43]. GaN coupled to silica fiber has been studied in the context of external beam radiotherapy, brachytherapy, and interventional radiology [25]. The RL emission from GaN is centered in a relatively narrow band around 370 nm, which offers possibilities for chromatic stem removal similar to the technique developed by Veronese et al. [39] for Eu-doped silica fibers.

17.5 SUMMARY AND OUTLOOK

The physical origin of luminescence from inorganic crystals is different from the luminescence phenomena in organic scintillators. As has been discussed in this chapter, an important implication of this difference is that inorganic crystals can store charge carriers (i.e., holes and electrons) in traps in the crystal, whereas organic scintillators essentially have no memory of previous irradiations. This feature of traps in inorganic crystals is directly exploited in OSL dosimetry where the stored charge in the ideal case is directly proportional to the absorbed dose. For the RL signal, however, such memory can be problematic as it can lead to a light yield that changes with dose history of the crystal. Another important difference between organic scintillators and inorganic crystals is that inorganic crystals tend to have luminescence lifetimes that are much longer than the molecular transitions in organic scintillators. Long luminescence lifetimes are of practical interest because they enable time separation of the RL signal from the stem signal in pulsed linear accelerator beams.

The main application of fiber-coupled inorganic crystals in medical dosimetry is *in vivo* systems, rather than reference dosimetry or machine-specific quality assurance systems. The main reason seems to be the high sensitivity per volume of material and the improved capabilities for stem signal removal compared with organic plastic scintillators. Although fiber-coupled luminescence dosimetry with inorganic crystals has a range of potential applications in medical dosimetry, these applications have been demonstrated in the literature based on prototypes or systems designed for research rather than on commercially available systems.

REFERENCES

1. Bøtter-Jensen L, McKeever SWS, Wintle AG. *Optically Stimulated Luminescence Dosimetry.* Elsevier, Amsterdam, The Netherlands, 2003.
2. Yukihara EG, McKeever SWS. *Optically Stimulated Luminescence: Fundamentals and Applications.* Wiley, West Sussex, UK, 2011.
3. Larsen NA, Btter-Jensen L, McKeever SWS. Thermally stimulated conductivity and thermoluminescence from Al_2O_3:C. *Radiation Protection Dosimetry* 84(1–4), 87–90; 1999.
4. Akselrod MS, Bruni FJ. Modern trends in crystal growth and new applications of sapphire. *Journal of Crystal Growth* 360, 134–145; 2012.
5. Knoll GF. Radiation detection and measurement. Third edition. Wiley, Hoboken, NJ, 2000.
6. Olko P. Microdosimetry, track structure and the response of thermoluminescence detectors. *Radiation Measurements* 41, S57–S70; 2007.
7. Yukihara EG, McKeever SWS. Ionisation density dependence of the optically and thermally stimulated luminescence from Al_2O_3:C. *Radiation Protection Dosimetry* 119(1–4), 206–217; 2006.
8. Jain M, Bøtter-Jensen L, Thomsen KJ. High local ionization density effects in x-ray excitations deduced from optical stimulation of trapped charge in Al_2O_3:C. *Journal of Physics Condensed Matter* 19(11), 116201; 2007.
9. Frelin A-M, Fontbonne J-M, Ban G, Colin J, Labalme M. Comparative study of plastic scintillators for dosimetric applications. *IEEE Transactions on Nuclear Science* 55(5), 2749–2756; 2008.
10. Tedgren ÅC, Hedman A, Grindborg J-E, Carlsson GA. Response of LiF:Mg,Ti thermoluminescent dosimeters at photon energies relevant to the dosimetry of brachytherapy. *Medical Physics* 38, 5539; 2011.
11. Klein FA, Greilich S, Andersen CE, Lindvold LR, Jäkel O: A thin layer fiber-coupled luminescence dosimeter based on Al_2O_3:C. *Radiation Measurements* 46(12), 1607–1609; 2011.
12. Nascimento LF. On-line dosimetry for radiotherapy using non-invasive optical fibre sensors with Al_2O_3:C RL/OSL detector. PhD thesis, Ghent University, Ghent, Belgium, 2015.
13. Greilich S, Hahn U, Kiderlen M., Andersen CE, Bassler N. Efficient calculation of local dose distributions for response modeling in proton and heavier ion beams. *The European Physical Journal D* 68, 327; 2014.
14. Akselrod MS, Agersnap Larsen N, Whitley V, McKeever SWS. Thermal quenching of F-center luminescence in Al_2O_3:C. *Journal of Applied Physics* 84(6), 3364; 1998.
15. Yukihara EG. Luminescence properties of BeO optically stimulated luminescence (OSL) detectors. *Radiation Measurements* 46(6–7), 580–587; 2011.
16. Pagonis V, Lawless J, Chen R, Andersen C. Radioluminescence in Al_2O_3:C—Analytical and numerical simulation results. *Journal of Physics D: Applied Physics* 42, 175107; 2009.
17. Marckmann CJ, Andersen CE, Aznar MC, Bøtter-Jensen L. Optical fibre dosemeter systems for clinical applications based on radioluminescence and optically stimulated luminescence from Al_2O_3:C. *Radiation Protection Dosimetry* 120(1–4), 28–32; 2006.

18. Justus BL, Falkenstein P, Huston AL, Plazas MC, Ning H, Miller RW. Gated fiber-optic-coupled detector for *in vivo* real-time radiation dosimetry. *Applied Optics* 43(8), 1663–1668; 2004.
19. Andersen CE, Marckmann CJ, Aznar MC, Bøtter-Jensen L, Kjær-Kristoffersen F, Medin J. An algorithm for real-time dosimetry in intensity-modulated radiation therapy using the radioluminescence signal from Al_2O_3:C. *Radiation Protection Dosimetry* 120(1–4), 7–13; 2006.
20. Klein D, Peakheart DW, McKeever SWS. Performance of a near-real-time KBr:Eu dosimetry system under computed tomography x-rays. *Radiation Measurements* 45, 663–667; 2010.
21. Magne S, Auger L, Bordy JM, de Carlan L, Isambert A, Bridier A, Ferdinand P, Barthe J. Multichannel dosemeter and Al_2O_3:C optically stimulated luminescence fibre sensors for use in radiation therapy: Evaluation with electron beams. *Radiation Protection Dosimetry* 131(1), 93–99; 2008.
22. Magne S, Deloule S, Ostrowsky A, Ferdinand P. Fiber-coupled, time-gated Al_2O_3:C radioluminescence dosimetry technique and algorithm for radiation therapy with LINACs. *IEEE Transactions on Nuclear Science* 60(4), 2998–3007; 2013.
23. Santos AMC, Mohammadi M, Asp J, Monro TM, Shahraam Afshar V. Characterisation of a real-time fibre-coupled beryllium oxide (BeO) luminescence dosimeter in X-ray beams. *Radiation Measurements* 53–54, 1–7; 2013.
24. Santos AMC, Mohammadi M, Shahraam Afshar V. Investigation of a fibre-coupled beryllium oxide (BeO) ceramic luminescence dosimetry system. *Radiation Measurements* 70, 52–58; 2014.
25. Pittet P, Jalade P, Balosso J, Gindraux L, Guiral P, Wang R, Chaikh A et al. Dosimetry systems based on Gallium Nitride probe for radiotherapy, brachytherapy and interventional radiology. *Innovation and Research in BioMedical Engineering* 36(2), 92–100; 2015.
26. Damkjær SMS, Andersen CE. Memory effects and systematic errors in the RL signal from fiber coupled Al_2O_3:C for medical dosimetry. *Radiation Measurements* 45(3–6), 671–673; 2010.
27. Andersen CE, Edmund JM, Damkjr SMS. Precision of RL/OSL medical dosimetry with fiber-coupled Al_2O_3:C: Influence of readout delay and temperature variations. *Radiation Measurements* 45, 653–657; 2010.
28. Polf JC, Yukihara EG, Akselrod MS, McKeever SWS. Real-time luminescence from Al_2O_3 fiber dosimeters. *Radiation Measurements* 38, 227–240; 2004.
29. Gaza R, McKeever SWS, Akselrod MS. Near-real-time radiotherapy dosimetry using optically stimulated luminescence of Al_2O_3:C: Mathematical models and preliminary results. *Medical Physics* 32(4), 1094–1102; 2005.
30. Gaza R, McKeever SWS. A real-time, high-resolution optical fibre dosemeter based on optically stimulated luminescence (OSL) of KBr:Eu, for potential use during the radiotherapy of cancer. *Radiation Protection Dosimetry* 120(1–4), 14–19; 2006.
31. Edmund JM, Andersen CE, Marckmann CJ, Aznar MC, Akselrod MS, Bøtter-Jensen L. CW-OSL measurement protocols using optical fibre Al_2O_3:C dosemeters. *Radiation Protection Dosimetry* 119(1–4), 368–374; 2006.
32. Andersen CE, Damkjær SMS, Kertzscher G, Greilich S, Aznar MC. Fiber-coupled radioluminescence dosimetry with saturated Al_2O_3:C crystals: Characterization in 6 and 18 MV photon beams. *Radiation Measurements* 46(10), 1090–1098; 2011.
33. Andersen CE, Nielsen SK, Lindegaard JC, Tanderup K. Time-resolved in vivo luminescence dosimetry for online error detection in pulsed dose-rate brachytherapy. *Medical Physics* 36(11), 5033–5043; 2009.
34. Jordan KJ. Evaluation of ruby as a fluorescent sensor for optical fiber-based radiation dosimetry. *Proceedings of SPIE* 2705, 170–178; 1996.
35. Teichmann T, Sommer M, Henniger J. Dose rate measurements with a ruby-based fiber optic radioluminescent probe. *Radiation Measurements* 56, 347–350; 2013.
36. Tanyi JA, Krafft SP, Ushino T, Huston AL, Justus BL. Performance characteristics of a gated fiber-optic-coupled dosimeter in high-energy pulsed photon radiation dosimetry. *Applied Radiation and Isotopes* 68, 364–369; 2010.
37. Fontbonne JM, Iltis G, Ban G, Battala A, Vernhes JC, Tillier J, Bellaize N et al. Scintillating fiber dosimeter for radiation therapy accelerator. *IEEE Transactions on Nuclear Science* 49(5), 2223–2227; 2002.
38. Kertzscher G, Andersen CE, Edmund JM, Tanderup K. Stem signal suppression in fiber-coupled Al_2O_3:C dosimetry for Ir-192 brachytherapy. *Radiation Measurements* 46(12), 2020–2024; 2011.
39. Veronese I, Cantone MC, Catalano M, Chiodini N, Fasoli M, Mancosu P, Mones E, Moretti F, Scorsetti M, Vedda A. Study of the radioluminesence spectra of doped silica optical fibre dosimeters for stem effect removal. *Journal of Physics D: Applied Physics* 46, 015101; 2013.
40. Scott AJD, Kumar S, Nahum AE, Fenwick JD. Characterizing the influence of detector density on dosimeter response in non-equilibrium small photon fields. *Physics in Medicine and Biology* 57, 4461–4476; 2012.
41. Underwood TSA, Winter HC, Hill MA, Fenwick JD. Detector density and small field dosimetry: Integral versus point dose measurement schemes. *Medical Physics* 40, 082102; 2013.
42. Carrara M, Cavatorta C, Borroni M, Tenconi C, Cerrotta A, Fallai C, Gambarini G, Vedda A, Pignoli E. Characterization of a Ce^{3+} doped SiO_2 optical dosimeter for dose measurements in HDR brachytherapy. *Radiation Measurements* 56, 312–315; 2013.

43. Pittet P, Lu G-N, Galvan J-M, Loisy J-Y, Ismail A, Giraudd J-Y, Balosso. Implantable real-time dosimetric probe using GaN as scintillation material. *Sensors and Actuators* A 151, 29–34; 2009.

44. Andersen CE. Fiber-coupled luminescence dosimetry in therapeutic and diagnostic radiology. *AIP Conference Proceedings* 1345(1), 100–119; 2011.

45. Therriault-Proulx F, Beaulieu L, Archambault L, Beddar S. On the nature of the light produced within PMMA optical light guides in scintillation fiber-optic dosimetry. *Physics in Medicine and Biology* 58(7), 2073–2084; 2013.

46. Akselrod MS, Lucas AC, Polf JC, McKeever SWS. Optically stimulated luminescence of Al_2O_3. *Radiation Measurements* 29(3–4), 391–399; 1998.

47. International Atomic Energy Agency. Standards, applications and quality assurance in medical radiation dosimetry (IDOS). *Proceedings of International Symposium*, Vol. 2. IAEA, Vienna, Austria, 2010. http://www-pub.iaea.org/MTCD/Publications/PDF/P1514_web/p1514_vol2_web.pdf.

48. Ranchoux G, Magne S, Bouvet JP, Ferdinand F. Fibre remote optoelectronic gamma dosimetry based on optically stimulated luminescence of Al_2O_3:C. *Radiation Protection Dosimetry* 100(1–4), 255–260; 2002.

49. Andersen CE, Aznar MC, Bøtter-Jensen L, Bäck SÅJ, Mattsson S, Medin, J. Development of optical fibre luminescence techniques for real time in vivo dosimetry in radiotherapy. *Proceedings of International Symposium on Standards and Codes of Practice in Medical Radiation Dosimetry*, Vienna, 2002, Vol. 2. IAEA, Vienna, Austria, 353–360; 2003.

50. International Atomic Energy Agency. Standards and codes of practice in medical radiation dosimetry. *Proceedings of International Symposium*, IAEA, Vienna, Austria, 2002. http://www-pub.iaea.org/MTCD/publications/PDF/Pub1153/Start.pdf.

51. Aznar MC, Andersen CE, Bøtter-Jensen L, Bäck SÅJ, Mattsson S, Kjær-Kristoffersen F, Medin J. Real-time optical-fibre luminescence dosimetry for radiotherapy: physical characteristics and applications in photon beams. *Physics in Medicine and Biology* 49(9), 1655–1669; 2004.

52. Kertzscher G, Andersen CE, Tanderup K. Adaptive error detection for HDR/PDR brachytherapy: Guidance for decision making during real-time in vivo point dosimetry. *Medical Physics* 41, 052102; 2014.

53. Andersen CE, Nielsen SK, Greilich S, Helt-Hansen J, Lindegaard JC, Tanderup K. Characterization of a fiber-coupled Al_2O_3:C luminescence dosimetry system for online in vivo dose verification during 192Ir brachytherapy. *Medical Physics* 36(3), 708–718; 2009.

54. Kertzscher G, Andersen CE, Siebert FA, Nielsen SK, Lindegaard JC, Tanderup K. Identifying afterloading PDR and HDR brachytherapy errors using real-time fiber-coupled Al_2O_3:C dosimetry and a novel statistical error decision criterion. *Radiotherapy and Oncology* 100(3), 456–462; 2011.

55. Azangwe G, Grochowska P, Georg D, Izewska J, Hopfgartner J, Lechner W, Andersen CE et al. Detector to detector corrections: A comprehensive experimental study of detector specific correction factors for beam output measurements for small radiotherapy beams. *Medical Physics* 41(7), 072103; 2014.

56. Veronese I, Mattia C, Fasoli M, Chiodini N, Mones E, Cantone MC, Vedda A. Infrared luminescence for real time ionizing radiation detection. *Applied Physics Letters* 105, 061103; 2014.

Other luminescence-based applications

(18) OSL point dosimeters for *in vivo* patient dosimetry

Josephine Chen

Contents

18.1 INTRODUCTION

Personal dosimeters based on optically stimulated luminescence (OSL) have been widely used for many years to monitor the occupational exposure of radiation workers. More recently, their use has been investigated and adopted for *in vivo* dosimetry of patients undergoing medical procedures, such as radiation therapy or diagnostic imaging. This chapter focuses specifically on the characteristics and the clinical uses of small OSL dosimeters used as passive point detectors for *in vivo* patient dosimetry. Currently, there is only one commercial source of these detectors. The nanoDot™, produced by Landauer Inc. (Glenwood, Illinois), utilizes a 5 mm-diameter, 0.2 mm-thick disk containing Al_2O_3:C. As seen in Figure 18.1, the disk is mounted inside a light-tight plastic case, measuring approximately $10 \times 10 \times 2$ mm in dimensions. Landauer also produced a larger OSL Dot, which contained a 7 mm-diameter Al_2O_3:C disk inside a $24 \times 12 \times 2$ mm plastic case. The physics of OSL dosimetry has been described in detail in Chapter 17. In simple terms, radiation can excite electrons out of the conduction band, creating free electrons and holes. Some of these electrons and holes become stably trapped in isolated energy states created by the lattice defects in the crystal. Stimulation of the material with a light source will release trapped electrons,

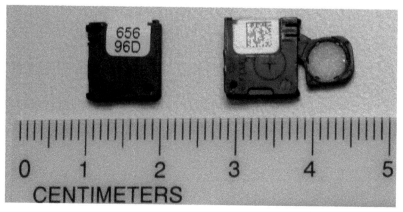

Figure 18.1 Picture of Landauer nanoDots. The OSL disk mounted inside the case has been swung out of the nanoDot on the right. During normal usage, the OSL disk remains in the case during transport and exposure, and is only rotated out of its case while inside the microStar reader during readout. The paper label on the case contains a bar code and serial number that specifies the manufacturer-supplied sensitivity factor for that individual detector.

which can then recombine with the holes creating the luminescence signal. The emitted light is a measure of the radiation dose that was experienced by the OSL detector. Landauer produced readers to perform this measurement on their detectors. The microStar™ reader accepts the nanoDot, swings the mounted Al_2O_3:C disk out of its case, and exposes it to 1 s of continuous stimulation from light with peak energy of 540 nm, provided by green light-emitting diodes with a bandpass filter. The OSL signal is simultaneously measured by a photomultiplier tube, filtered to provide a peak sensitivity at 420 nm.

In the following sections, the basic dosimetric properties of the Landauer nanoDot system will be summarized. It is important to note that some of these properties may depend on the specifics of the Al_2O_3:C material or the readout technique. Other readout protocols have been proposed and used for Al_2O_3:C. Landauer's Luxel personal dosimetry system uses a pulsed stimulation technique in which the OSL signal is only detected while the stimulation is off. The choice of stimulating light energy, stimulation duration/power, and filtration of the detected signal may also vary. These particular choices will affect the electronic transitions that contribute to the detected signal, and this in turn may modify the dosimetric properties. Thus, caution should be used when applying the following results to a different dosimetry system, even one that employs the same OSL material.

18.2 DOSIMETRIC CHARACTERISTICS OF OSL nanoDOTs

18.2.1 TIME DEPENDENCE OF THE OSL SIGNAL

Although the simplified picture presented in Section 18.1 explains the basic OSL process, the electronic landscape for Al_2O_3:C is quite complex, and there are numerous transitions that occur simultaneously and in competition. In particular, there are low-energy traps that are highly unstable. Over several minutes, electrons can escape these traps without stimulation. Thus, the OSL signal detected can change rapidly in the time interval immediately after irradiation. Several researchers have reported on this rapidly decaying signal (Jursinic 2007, Reft 2009, Omotayo et al. 2012), as demonstrated in Figure 18.2. This transient signal was found to depend on the particular dosimeter measured (Jursinic 2007) and on the dose to which the dosimeter was exposed (Omotayo et al. 2012). All measurements indicated that the transient signal decays after 10 min postirradiation, and it is recommended to perform readout after this time period.

Escape from the more stable traps can still occur spontaneously at a lower rate, creating a longer term fading effect. Reported rates for fading have varied. Viamonte (Viamonte et al. 2008) reported a 2% reduction in signal in the first 5 days after exposure to 50 cGy from a Co-60 beam and then a stable signal through day 21. Reft found that the signal decayed approximately 0.3% per day for the first 11 days postirradiation to doses of 0.5–4 Gy using a 6 MV beam (Reft 2009). Al-Senan (Al-Senan and Hatab 2011)

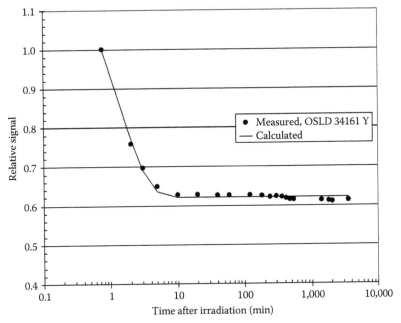

Figure 18.2 Dots represent repeated measurement of a single OSL dosimeter exposed to 100 cGy in a 6 MV beam, normalized to the initial measurement performed 45 s after irradiation. The measurements were corrected for the signal depletion due to the repeated readings. The solid line represents a calculation based on an empirical model. (Reproduced from Jursinic P., *Medical Physics*, 34, 4594–604, 2007, Figure 3. With permission.)

exposed nanoDots to low doses, on the order of 1 cGy, using diagnostic energy beams, and reported a 4% reduction in signal over the first 24 h postirradiation. Kerns (Kerns et al. 2012) exposed nanoDots to doses of 25–500 cGy in a 250 MeV proton beam. Between 3 and 6 h postirradiation, the signal fell 1.5% and then remained stable for several days. Thus, there appears to be some dependence of the fading effect on the type of beam used for irradiation and possibly on the magnitude of dose delivered. For improved accuracy, it is advisable to verify the fading rate for typical doses that will be delivered to the nanoDot and to use that information to determine an appropriate postirradiation time window for readout.

18.2.2 LINEARITY

The OSL signal is approximately linear at low doses, but at greater doses it exhibits supralinearity; the signal is greater than expected from linear extrapolation. Figure 18.3 depicts an example of this behavior. The presumed cause of the supralinearity is the progressive filling of nonradiative electron/hole traps leading to reduced competition from nonradiative transitions. Al-Senan (Al-Senan and Hatab 2011) documented a linear response for nanoDots exposed to low doses, 2–40 mGy, using diagnostic X-ray beams. For radiotherapy photon beams, a linear response at low doses followed by supralinearity above a threshold dose has been documented for exposures up to 10 Gy (Jursinic 2007, Viamonte et al. 2008, Reft 2009, Omotayo et al. 2012). Reported threshold doses for the start of supralinearity have varied from 2 Gy (Reft 2009, Omotayo et al. 2012) to 4 Gy (Viamonte et al. 2008). The reported magnitude of the suparlinearity has also varied. Although all groups reported using nanoDots or the larger OSL Dots with the microStar reader, it may be that differences in the readout effectively modified the energy spectrum that was detected and altered the observed dose–response curves. It is standard practice that a unique calibration curve is obtained for each reader. Supralinearity has also been reported for radiotherapy proton beam exposures up to 10 Gy with a threshold of approximately 2 Gy (Reft 2009, Kerns et al. 2012). At very large doses, above 60 Gy, the OSL sensitivity begins to decrease due to further changes in the available electronic states (Jursinic 2010). In terms of practical measurements, however, the limitations of the reader components become the determining factor before these changes are seen. For the microStar reader, doses of approximately 15 Gy can be measured in the low-intensity

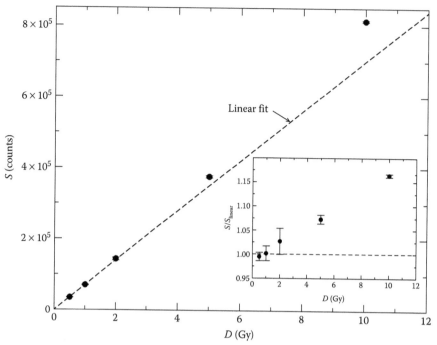

Figure 18.3 Example of a plot showing the supralinearity of the OSL signal (S) as a function of dose (D) from Omotayo et al. The magnitude of supralinearity may not be consistent across different readers, and the determination of a unique calibration curve for each reader is recommended. The dots in the main graph represent the signal detected from nanoDots exposed to various doses in a 6 MV beam, compared to a linear fit through the 1 Gy measurement. The inset plots the deviation of the sensitivity from the 1 Gy reference sensitivity (S_{linear}). (Reproduced from Omotayo A et al., *Medical Physics*, 39, 5457–68, 2012, Figure 4. With permission.)

mode before reader saturation begins to occur. The reading of greater doses can be performed by modifying the commercial hardware (Jursinic 2010).

18.2.3 BEAM ENERGY DEPENDENCE

Al_2O_3:C OSL materials are not tissue equivalent, and thus exhibit a sensitivity dependence on the type and energy of the irradiating beam. However, for typical accelerator-generated radiotherapy photon and electron beams, the energy dependence is minimal, and several studies using the nanoDot and microStar reader have found that the sensitivity is approximately constant for 6–18 MV photon beams and 5–20 MeV electron beams (Jursinic 2007, Viamonte et al. 2008, Reft 2009). A previous study using Al_2O_3:C-coated films showed a 3.6% difference in the sensitivity for electron beams and a 4% difference between 6 and 18 MV photon beams (Schembri and Heijmen 2007). However, the studies performed specifically using nanoDots and the microStar reader all consistently demonstrated sufficient energy independence to justify the use of a single calibration curve for typical radiotherapy accelerator beams.

With an effective atomic number of 11.2, Al_2O_3:C dosimeters demonstrate an increased sensitivity for lower energy beams, consistent with an increasing contribution from the photoelectric effect. Both Reft (2009) and Viamonte (Viamonte et al. 2008) found a 4% increase in sensitivity for Co-60 beams relative to a 6 MV accelerator beam. The sensitivity measured at a 7 cm distance from an Ir-192 source was 6% higher than for a 6 MV beam (Jursinic 2007). With decreasing energy, the energy dependence becomes even more pronounced. The measured sensitivity varied by a factor of 2 between the 250 and 125 peak kilovoltage (kVp) beams from an orthovoltage radiotherapy unit (Reft 2009). The observed sensitivity also varied by a factor of 2 by varying the tube potential from 50 to 120 kVp and using variable filtration to modify the half-value layer from 2.0 to 9.8 mm Al on a radiographic unit (Al-Senan and Hatab 2011). Figure 18.4 combines the energy-dependence data from Reft and Al-Senan, and illustrates the sharp

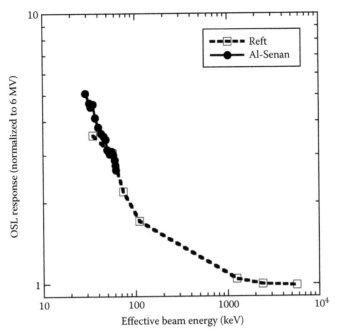

Figure 18.4 OSL response as a function of effective beam energy. Data were compiled from the works of Reft and Al-Senan. Data from Al-Senan were renormalized to match those from Reft at 57 keV. (Data from Reft C., *Medical Physics*, 36, 1690–9, 2009, Table I; Al-Senan R, Hatab M., *Medical Physics*, 38, 4396–405, 2011, Table II.)

increase in sensitivity with decreasing energy below 100 keV. The energy dependence of the detector has been successfully modeled by Scarboro (Scarboro and Kry 2013) using Burlin cavity theory and the mass stopping powers and mass energy absorption coefficients for Al_2O_3 and water. The simulations also underscored the importance of the energy spectrum of the beam. Predictions based on the full spectrum differed from those based on a single mean beam energy and were closer to measured values. The rapidly changing sensitivity of nanoDots at the lower energy ranges complicates their use for *in vivo* dosimetry for patients undergoing diagnostic imaging exams. Use of generic energy-dependent correction factors is likely inadequate. The most accurate results will be obtained by calibrating the detectors in the specific beam used for their exposure. When using correction factors from the literature or other sources, it would be better to match the half-value layer of the beam and not the nominal energy.

The sensitivity of nanoDots in radiotherapy proton beams has also been investigated. Reft found the sensitivity to be independent of proton beam energy for energies from 47 to 250 MeV using measurements from several institutions. The sensitivity in the proton beams was about 6% higher than the sensitivity in a 6 MV photon beam (Reft 2009). As illustrated in Figure 18.5, Kerns (Kerns et al. 2012) also found the sensitivity to be fairly independent of proton beam energy, both when changing the beam energy from 160 to 250 MeV and when modulating the beam to create spread-out Bragg peaks of widths 4–10 cm.

18.2.4 ANGULAR DEPENDENCE

Studies have also been performed to investigate the detector dependence on the angle of incidence of the irradiating beam. Jursinic (2007) found no angular dependence when testing the larger OSL Dot detectors in a 6 MV photon beam. However, Kerns (Kerns et al. 2011) found a small angular dependence for the nanoDot detectors. They reported a 4% lower response in a 6 MV photon beam and a 3% lower response in an 18 MV beam when the nanoDot was irradiated edge-on compared to when the beam was perpendicular to the flat surface of the detector. Intermediate responses were found for the angles in-between. Monte Carlo simulations performed by Kerns suggest that the source of this angular

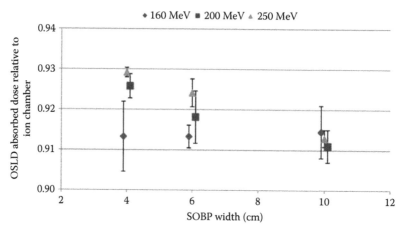

Figure 18.5 Points represent the relative OSL signal as a function of proton beam energy, varied both by changing the nominal accelerated beam energy and by modulating the beam to different spread-out Bragg peak (SOBP) widths. (Reproduced from Kerns J et al., *Medical Physics*, 39, 1854–63, 2012, Figure 5. With permission.)

dependence is a reduction in energy deposited by low-energy electrons when the OSL disk is parallel to the beam. The effect was found to be partially compensated for by the air gap surrounding the disk in the plastic container.

The angular dependence becomes more pronounced at lower beam energies. Al-Senan (Al-Senan and Hatab 2011) investigated the angular dependence of nanoDots exposed using a 25 kVp mammography beam, 80 and 120 kVp beams from a general radiography unit, and 80 and 120 kVp beams from a computed tomography (CT) scanner. The angular dependence was the strongest for the mammography beam with a maximum reduction of 70% for on-edge irradiation; the maximum reduction for the 80 kVp beam from the radiography unit was 40%. For exposure to CT beams, the observed angular dependence was more moderate, with a maximum reduction on the order of 10%. As seen in Figure 18.6, the nanoDot sensitivity was also dependent on which edge was facing the beam, with the lowest response observed when the upper edge around which the label is wrapped (designated 90° in the graphs) was facing the beam source. The strong angular dependence, especially at the lowest diagnostic energies, would need to be considered when positioning the detectors for *in vivo* dosimetry.

18.2.5 DOSE RATE DEPENDENCE

Variations in the dose rate experienced by an *in vivo* detector can occur because the dose rate of the beam produced by the equipment is modified, the distance of the detector from the source is changed, or the amount of attenuating material above the detector is altered. For accelerator-produced radiotherapy beams, the beams are not continuous but are pulsed, and changes to the nominal dose rate for a particular mode can be modified by adjusting the pulse repetition frequency or the dose per pulse. The instantaneous dose rate experienced by the detector is proportional to the dose per pulse. Viamonte (Viamonte et al. 2008) demonstrated that the sensitivity of the nanoDot remained constant in a 6 MV beam when changing the dose rate from 200 to 600 cGy/min by varying the pulse repetition frequency. Some detectors are not sensitive to changes in pulse repetition frequency but are sensitive to changes in the instantaneous dose rate. Viamonte also found the sensitivity to be constant when changing the distance from a Co-60 source from 85 to 105 cm, a 34% change in the instantaneous dose rate. Jursinic (2007) varied the dose per pulse of a 6 MV beam by over 2 orders of magnitude by adjusting the distance from the source and attenuation above the detectors. As illustrated in Figure 18.7, the sensitivity was found to be constant over this large range.

18.2.6 TEMPERATURE DEPENDENCE

In vivo dosimeters are usually placed directly on the patient's body, and thus experience changes in temperature. Jursinic (2007) found that the sensitivity was constant when the nanoDot was exposed at temperatures ranging from 10°C to 40°C. As with all results reported here, the OSL signal was measured

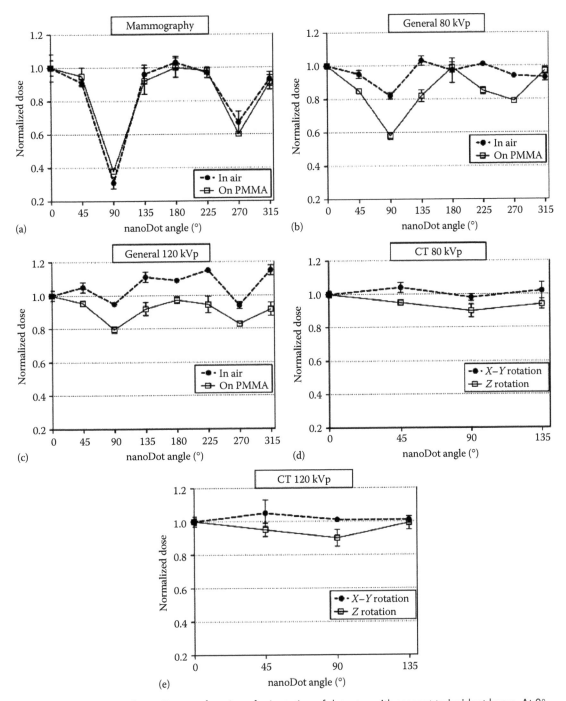

Figure 18.6 Response of nanoDot as a function of orientation of detector with respect to incident beam. At 0°, the beam was incident on the face with the serial number (see Figure 18.1). At 90°, the beam was incident on the edge containing the paper label. Angular dependence was evaluated using (a) a mammography unit, (b) 80 kVp from a general radiography unit, (c) 120 kVp from a general radiography unit, (d) 80 kVp from a CT scanner, and (e) 120 kVp from a CT scanner. PMMA, polymethyl methacrylate. (Reproduced from Al-Senan R, Hatab M., *Medical Physics*, 38, 4396–405, 2011, Figure 8. With permission.)

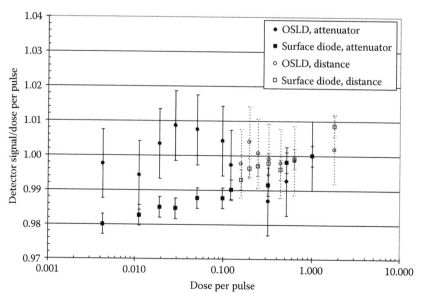

Figure 18.7 Comparision of the dose response of optically stimulated luminescence dosimeters (OSLD) (circles) and diodes (squares) to changes in dose per pulse in a 6 MV beam. The dose per pulse was varied by adding attenuating material to the beam (solid) and varying the distance from the source (open). (Reproduced from Jursinic P., *Medical Physics*, 34, 4594–604, 2007, Figure 10. With permission.)

after an 8 min wait postirradiation. It is possible that changes in temperature could affect the rapidly decaying transient signal caused by the low-energy traps, but does not affect the signal that is used for dosimetric measurements.

18.2.7 SUMMARY OF DOSIMETRIC PROPERTIES

As detailed in this section, Al_2O_3:C nanoDots have many desirable properties as an *in vivo* dosimeter for radiotherapy applications. The small temperature and angular dependence allow them to be placed directly on curved or angled parts of the body. The low beam energy dependence in this energy range and the minimal dose rate dependence allow the use of a single calibration curve for a variety of treatment modalities and measurement scenarios. However, at the lower beam energies used for diagnostic imaging techniques, the angular and energy dependences increase sharply. For *in vivo* applications for these techniques, additional care will be required to calibrate the detector for specific beams and to position the detectors using standard protocols.

18.3 ACCURACY AND PRECISION OF SINGLE AND MULTIPLE READINGS

18.3.1 SINGLE AND REPEATED READINGS OF AN EXPOSED nanoDot

One convenient feature of OSL dosimetry is that the short stimulation employed by the reader only depletes a small portion of the trapped electrons, and thus, the measurement can be performed multiple times. Repeated readings can be used to reduce measurement uncertainty and to estimate the precision of a single reading. Viamonte (Viamonte et al. 2008) found that fluctuations were on the order of 1% (one standard deviation) for repeated readings of the OSL Dots. Fluctuations may be caused by instabilities in the reader stimulation and detection of the light as well as the inherent statistical nature of photon production and detection.

Although the depletion of the trapped electrons is small with each reading, a systematic reduction in signal can be observed with numerous repeated readings. The reduction in signal depends on the strength and the duration of the stimulating light. The microStar reader employs a low-intensity mode to measure detectors exposed to higher doses to prevent saturation of the reader. In the low-intensity mode, depletion rates of 0.05% per reading (Jursinic 2007) and 0.019%–0.035% per reading (Omotayo et al. 2012) have been reported. For the high-intensity mode, a depletion rate of 0.287% per reading has been reported for dosimeters exposed to 10 cGy in a 6 MV beam (Omotayo et al. 2012), and a rate of 0.5% per reading has been reported for dosimeters exposed to 8 mGy using a 120 kVp radiography beam (Al-Senan and Hatab 2011). Thus, for higher doses typical of radiotherapy treatments, the depletion from repeated readings is generally negligible. For low doses, it may be necessary to consider the depletion effect when averaging multiple repeated readings.

18.3.2 ACCURACY OF nanoDot SENSITIVITY

The accuracy of the nanoDot reading is effected by how accurately the user knows the true sensitivity of that particular detector. The sensitivity of each detector is determined by the electronic structure of the OSL material, including the energy levels and the concentration of electron/hole traps, and the amount of the OSL material in each plastic disk. The Landauer nanoDot detectors are labeled with a nominal sensitivity factor. The nanoDot detectors, unscreened or prescreened, may be purchased by the manufacturer. At the time of writing, the manufacturer reports ±10% accuracy for the unscreened detectors and ±5% accuracy for the screened detectors. Exposing a batch of 47 unscreened detectors to a field with 2% uniformity, Al-Senan (Al-Senan and Hatab 2011) reported the variability of the response was 4.3%–4.8% (one standard deviation), which falls within the manufacturer's specifications.

The accuracy of the detectors can be improved by measuring the sensitivity of each individual nanoDot. This can be accomplished by exposing the detectors to a small known dose and reading the response (Reft 2009, Kerns et al. 2011). The nanoDot can then be used for the patient measurement, subtracting the test dose measurement from the total reading. The signal from the test exposure can also be removed prior to the patient measurement by optically bleaching the OSL material, as described in Section 18.3.3. Reft (2009) found that if the sensitivity of a detector was measured several times by repeatedly exposing it to a small test irradiation and then optical bleaching the detector, the measured sensitivity factors varied by less than 1% (one standard deviation) from the average value. This approximately matches the variability found by Viamonte when repeatedly measuring the same exposed detector. Thus, the accuracy of the sensitivity measured using a test exposure is limited by the precision of the detector readout.

18.3.3 SENSITIVITY VARIATIONS AFTER REPEATED EXPOSURES AND OPTICAL BLEACHING

The OSL detectors may be optically bleached after each patient exposure to deplete the trapped charges and reused multiple times. However, depending on the bleaching protocol and the total accumulated dose to the detector, there may be an increasing residual signal as well as changes to the sensitivity of the detector. Jursinic found that using a 1 min illumination with a 150 W tungsten–halogen lamp effectively reduced the OSL signal by greater than 98%. Using this bleaching protocol, he found that the sensitivity of the detector was stable for the first 20 Gy of accumulated dose, but then dropped 4% per 10 Gy of additional accumulated dose (Jursinic 2007). Reft (2009) found a similar change in sensitivity with accumulated dose using a 22 W fluorescent lamp or a 150 W tungsten–halogen lamp for optical bleaching. Omotayo (Omotayo et al. 2012) investigated several different bleaching procedures using 22 W fluorescent bulbs, both unfiltered and filtered. The results of these experiments determined that the residual signal and changes in sensitivity of the detectors varied depending on the spectrum of bleaching light, the bleaching time, and the accumulated dose. Therefore, if the nanoDot is optically bleached and reused, the sensitivity should be either retested or standardized bleaching protocols, and tracking of accumulated dose must be implemented to determine the detector sensitivity.

18.3.4 IMPLEMENTING A CLINICAL DOSIMETRY PROGRAM

Many radiotherapy clinics have implemented clinical programs using nanoDots to perform *in vivo* patient dosimetry. Although the data presented earlier can provide an expectation of how the detectors will perform, it is advisable for any new user to perform measurements verifying the same behavior. A unique calibration curve is created for each reader. Test measurements should be performed for typical doses and beam energies to verify accuracy. Detectors should be read out after a minimum 10 min wait period postirradiation. Existing data suggest that for typical radiotherapy beams and exposures, the reading can be performed within a few days of exposure without significant reduction in signal. At the time of writing, the manufacturer specifies a ±5% and a ±10% accuracy for screened and unscreened detectors, respectively, when using the manufacturer-provided calibration factor. If greater accuracy is needed, individual sensitivity factors can be measured. As discussed earlier, if individual sensitivity factors will be determined by the user and/or if the detectors will be bleached and reused, standard procedures will need to be put in place to ensure the correct tracking of the factors with each detector. As with any dosimetry program, annual checks of the stability of the calibration curves should also be scheduled. With a fairly modest amount of preparation, nanoDots can be an efficient and accurate tool for clinical *in vivo* dosimetry. In Section 18.4, the use of nanoDots for specific clinical applications will be discussed.

18.4 CLINICAL APPLICATIONS

18.4.1 *IN VIVO* DOSE MEASUREMENTS DURING MEDICAL PROCEDURES

As discussed in Chapters 8 and 9, the ability to perform an *in vivo* patient dose measurement is an important tool in radiation therapy. Although not employed routinely at all clinics, it is important to be able to perform a measurement on a patient if a particular concern arises. *In vivo* dosimetry is not commonly employed in diagnostic or interventional radiology. With increasing awareness of patient dose concerns, there is growing interest in this application as well. Sections 18.4.2 through 18.4.5 discuss the use of OSL nanoDots for various clinical *in vivo* dosimetry applications.

18.4.2 SUPERFICIAL DOSE MEASUREMENTS DURING RADIOTHERAPY

For some patients undergoing radiotherapy, the skin and the superficial tissue beneath the skin are targets for high radiation dose. Radiation dose at or near the surface may not be accurately calculated using commercial treatment planning software, and thus, patient *in vivo* measurements can be taken to verify the dose to the skin. Measurement of dose at the patient skin during external photon beam radiotherapy can be difficult because the dose deposited by high-energy photon beams rapidly changes with depth from the surface of incidence. The measured dose from a dosimeter does not reflect the dose on the surface of the detector but the dose at an effective depth. Depending on the dosimeter construction, this effective depth can be made very small, for example, the effective depth is below a tenth of a millimeter for some parallel-plate ion chambers. Effective depth measurements on the larger OSL Dots were performed by Jursinic (2007) and Viamonte (Viamonte et al. 2008), both finding an effective depth of approximately 0.4 mm compared to parallel-plate ion chamber measurements in 15 MV and Co-60 beams, respectively. This effective depth roughly corresponds to the thickness of the plastic case that surrounds the disk containing the OSL material. Figure 18.8 compares nanoDot measurements with a Markus parallel-plate ion chamber measured at several depths of solid water irradiated with a 6 MV beam. The plotted nanoDot measurements are shifted by an effective depth of 0.5 mm and agree within the measurement error to the Markus chamber measurements. As illustrated in Figure 18.8, even with a 0.4–0.5 mm effective depth, the nanoDot measurement will systematically overestimate the true surface dose. For the depicted test measurement, this corresponds to a 40% overestimation. Thus, the effective depth of the nanoDot is not a negligible effect and should be considered when used to measure superficial doses.

When superficial tissues are targeted for high radiation dose, frequently bolus material is placed over the patient's skin during treatment. In this case, the patient's skin is no longer at the surface of incidence but at a depth equal to the bolus thickness. As seen in Figure 18.8, with increasing depth, the depth dose curve becomes shallower and the 0.5 mm effective measurement depth of the nanoDot starts to become negligible. Therefore, nanoDots can be placed on the patient's skin, under the bolus, to estimate the skin dose. An example of this is the use of nanoDots to measure skin dose for postmastectomy breast patients treated with bolus. Detectors can be placed at several locations on the chest. Treatment beams are directed tangentially across the chest. Depending on where the nanoDots are placed, the nanoDots will lie at an angle to the incident beam. The small angular dependence of the detector, within 3%–4% for radiotherapy

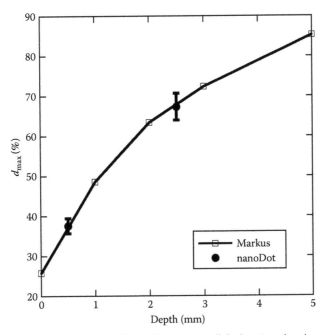

Figure 18.8 Measurements from nanoDots and from a Markus parallel-plate ion chamber as a function of depth in solid water with irradiation by a 6 MV beam. For comparison, the nanoDot measurements are shifted by a 0.5 mm effective depth. Measurements are normalized by the value at d_{max} depth.

beams, makes the nanoDot suitable for this application. The detectors are also not affected by the change in temperature experienced when placed on a patient's skin. However, for skin measurements of breast patients treated *without* bolus, detectors with a smaller effective depth of measurement are advised.

18.4.3 ENTRANCE/EXIT DOSE MEASUREMENTS DURING RADIOTHERAPY

In vivo dosimetry can also be used as part of a quality assurance program to document the accuracy of the delivered radiation dose to the patient. In some countries, this is a regulatory requirement. For many patients, the high dose region is not near the surface but is deep within the body. Measurement of the superficial dose is not the most relevant information in these cases. The dose deep in the patient cannot be measured using detectors outside the body. However, the entrance or exit dose, the dose at a depth of d_{max} from the entrance or exit surface of a beam, can be estimated by using external detectors with appropriate buildup caps. The d_{max} depth, which varies with beam energy, is the depth from the surface at which the dose from a perpendicularly incident beam reaches a maximum before beginning to decrease due to beam attenuation. The measured entrance or exit dose can be compared with the calculated dose to validate the accuracy of the treatment delivery. Jursinic (Jursinic and Yahnke 2011) tested the use of buildup caps made from various materials to measure the entrance/exit dose with nanoDots. The response was not sensitive to the material used to make the buildup cap as long as the effective thickness was equal to d_{max}. Field size correction factors were low, varying by 4% for field sizes ranging from 3 × 3 to 30 × 30 cm. In addition, the independence of the detector response over a large range of dose rates also makes the nanoDot suitable for measurements during low dose rate treatments, such as typically used for total body irradiation.

18.4.4 PERIPHERAL OR SHIELDED DOSE MEASUREMENTS DURING RADIOTHERAPY

Sometimes, it is desirable to document the doses delivered to regions that are far from the treatment field or under shielding. Examples include the dose to the pelvic region for a pregnant patient or the dose to the shielded, healthy testicle for a seminoma patient. The high sensitivity of the OSL detectors is advantageous for these measurements, as the doses received in these areas are much lower than the therapeutic doses. The nanoDots also have demonstrated little dependence with dose rate, which is also much lower in these regions. The nanoDots however do have a dependence on beam energy, which increases with decreasing photon energy. For regions outside the field, a larger portion of the photons will be scattered photons with lower energy. Scarboro (Scarboro and Kry 2013) measured an approximately 35% greater detector response at a point outside a 6 MV field versus inside the field. For most clinical applications, however, a rough estimate of the peripheral or shielded dose is sufficient, and the nanoDot measurement can be useful as an estimated upper limit.

18.4.5 DOSE RECEIVED DURING KILOVOLTAGE CT IMAGING

CT imaging techniques are used extensively in diagnostic radiology and are also increasingly used for pretreatment patient position verification in radiation therapy. The doses are low compared to therapeutic radiation doses, but with their increasing utilization, there is growing awareness and concern over the dose delivered to the patient during these imaging procedures. NanoDots, with their high sensitivity, are of potential interest for measuring these low patient doses. However, the beam energy dependence in the kiloelectronvolt energy range complicates their use. Al-Senan (Al-Senan and Hatab 2011) found that the nanoDot response differed by approximately 14% for a 120 kVp CT scan versus an 80 kVp CT scan. With the appropriate correction factors applied, the nanoDot dose measurements in a CT phantom agreed reasonably well with thermoluminescent dosimeter measurements. Ding (Ding and Malcolm 2013) found that the nanoDot response differed by as much as 28% when using different kVp and filters on a kilovoltage cone-beam CT imager attached to a radiotherapy unit. Corrections factors were dependent not only on the kVp but also on the use or absence of bowtie filters. Thus, for more accurate dose measurements, the nanoDots should be calibrated for the particular beam used during imaging, rather than using generic calibration factors based on kVp. Based on these experiments, it seems possible to use nanoDots for *in vivo* patient dosimetry during cone-beam and convention CT imaging. At the time of writing, this form of dosimetry has not been clinically implemented.

18.5 CONCLUSION

Al_2O_3:C nanoDots have proven themselves to be useful devices for *in vivo* patient dosimetry for radiation therapy applications. There is also interest in expanding their use to monitor the patient dose during imaging procedures performed with kilovoltage beams. Greater beam energy and angular dependence at the lowest beam energies complicate this effort. However, there have been promising results using nanoDots to measure the dose during conventional fan-beam CT and cone-beam CT. With growing interest of the medical community in the monitoring and verification of patient radiation doses, there will likely be increasing use and wider application of Al_2O_3:C nanoDots for *in vivo* dosimetry in the future.

REFERENCES

Al-Senan R, Hatab M. 2011. Characteristics of an OSLD in the diagnostic energy range. *Medical Physics* 38: 4396–405.
Ding G, Malcolm A. 2013. An optically stimulated luminescence dosimeter for measuring patient exposure from imaging guidance procedures. *Physics in Medicine and Biology* 58: 5885–97.
Jursinic P. 2007. Characterization of optically stimulated luminescent dosimeters, OSLDs, for clinical dosimetric measurements. *Medical Physics* 34: 4594–604.
Jursinic P. 2010. Changes in optically stimulated luminescent dosimeter (OSLD) dosimetric characteristics with accumulated dose. *Medical Physics* 37: 132–40.
Jursinic P, Yahnke C. 2011. In vivo dosimetry with optically stimulated luminescent dosimeters, OSLDs, compared to diodes; the effects of buildup cap thickness and fabrication material. *Medical Physics* 38: 5432–40.
Kerns J, Kry S, Sahoo N. 2012. Characteristics of optically stimulated luminescent dosimeters in the spread-out Bragg peak region of clinical proton beams. *Medical Physics* 39: 1854–63.
Kerns J, Kry S, Sahoo N, Followill D, Ibbott G. 2011. Angular dependence of the nanoDot OSL dosimeter. *Medical Physics* 38: 3955–62.
Omotayo A, Cygler J, Sawakuchi G. 2012. The effect of different bleaching wavelengths on the sensitivity of Al2O3:C optically stimulated luminescence detectors (OSLDs) exposed to 6 MV photon beams. *Medical Physics* 39: 5457–68.

Reft C. 2009. The energy dependence and dose response of a commercial optically stimulated luminescent detector for kilovoltage photon, megavoltage photon, and electron, proton, and carbon beams. *Medical Physics* 36: 1690–9.

Scarboro S, Kry S. 2013. Characterisation of energy response of Al2O3:C optically stimulated luminescent dosemeters (OSLDs) using cavity theory. *Radiation Protection Dosimetry* 153: 23–31.

Schembri V, Heijmen B. 2007. Optically stimulated luminescence (OSL) of carbon-doped aluminum oxide (Al2O3:C) for film dosimetry in radiotherapy. *Medical Physics* 34: 2113–18.

Viamonte A, da Rosa L, Buckley L, Cherpak A, Cygler J. 2008. Radiotherapy dosimetry using a commercial OSL system. *Medical Physics* 35: 1261–6.

19 Scintillating quantum dots

Claudine Nì. Allen, Marie-Ève Lecavalier, Sébastien Lamarre, and Dominic Larivière

Contents

Semiconductor quantum dots (QDs) are a fascinating nanomaterial that goes much beyond its proclaimed properties of size-tunable light emission. Not only the size but also the composition of the bulk semiconductor crystal used to make the QDs modify all their optical properties. Indeed, modifying the composition of the bulk semiconductor means that it is no longer the same material and obviously results in different material characteristics affecting the bandgap among others. With technological advances allowing the preparation of crystalline materials at the nanometer scale with a limited number of atoms, typically between 1,000 and 10,000 in QDs, it is now common knowledge that material properties are independent of size only down to a minimum threshold. Below that minimum, the size dependence is illustrated in Figure 19.1 specifically for the optical property of the light emitted by the QDs under ultraviolet (UV) or visible (Vis) excitation: the emission can be tuned to shorter wavelengths with decreasing QD size, as represented by the colored arrows.

This additional tuning is due to the quantum mechanical coherence of the valence electron cloud in a semiconductor. The whole nanocrystal (NC) behaves as a single artificial atom with discrete states as a direct application of the *particle-in-a-box* model. Whereas natural atoms have fixed equilibrium positions of electrons and nuclei, hence fixed sizes and energy states, the cloud of valence electrons within the QD *box* is affected by the size and shape of said box. The associated energy states are thus continuously adjustable. These semiconductor NCs are thus known as quantum *boxes* or QDs, and if they are suspended in a liquid, they are specified as colloidal QDs (cQDs). In the end, combining the size- and composition-dependent properties, it is possible to selectively obtain cQD luminescence from the UV to the near-infrared. This aspect of the cQDs is very attractive for scintillation applications.

Section 19.1 is devoted to further exploration of quantum properties and provides the essential background to understand the prolific literature on semiconductor QDs and other quantum confined systems. It includes phenomenological explanations of the required concepts in condensed matter and advanced materials, first from a physics perspective, then from a chemical perspective. Section 19.2 is

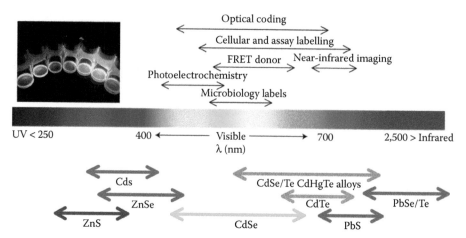

Figure 19.1 List of semiconductor materials typically used to prepare cQD cores and examples of biological applications. Colored arrows (bottom) indicate the spectral tuning range of the emitted light when these semiconductors have small sizes on the nanometer scale, and the inset displays an example based on CdSe. Black arrows (top) represent wavelength ranges often used in the corresponding biological applications. (Reprinted by permission from Macmillan Publishers Ltd. *Nature Materials*, Medint, Igor L. et al., 2005, copyright 2005; picture courtesy of Dany Lachance-Quirion.)

devoted to benefits of cQDs in the context of scintillation counting and dosimetry. Section 19.3 focuses on the recent developments of QD scintillators. Section 19.4 offers insights and perspectives on the emerging challenges and the outlook for research on QD scintillators.

19.1 INTRODUCTION TO THE PHYSICS AND CHEMISTRY OF QDs

19.1.1 QUANTUM CONFINEMENT

To properly understand the nature of quantum confinement of electrons in nanostructures underlying their peculiar properties often known to be size tunable, one has to look back at the origins of electrical conductivity in solids. Indeed, the multitude of oppositely charged electrons and ions should stick together due to the Coulomb attraction, but instead electrons behave as if they were ionized into a free gas within the solid. This explains why the relationship between the current and the applied voltage on a resistive wire is linear, that is, Ohm's law. Circa 1900, Paul Drude borrowed from the kinetic theory of gases to model this microscopic behavior of electrons in metals with a periodic crystal lattice (Drude, 1900). He assumed the valence electrons in solids to be completely free like a simple gas of particles and independent of the chemical nature of the ionized atoms left behind; only the interactions through collisions with the static crystal lattice ions were kept in his model (Ashcroft et al., 1976). That some electrons in matter can be considered free or nearly so is a key concept to explain quantum confinement, namely, the direct application of the particle-in-a-box model where these free particles become trapped in the small QD *box*.

The regular and well-ordered crystal lattice in good conductors creates an approximately homogeneous charge distribution resulting in only a weak net Coulomb force acting on each valence electron. Whereas core electrons are strongly bound to the atomic nuclei and shield its charge, valence electrons are delocalized into a nearly free electron gas and leave a lattice of ions behind. In other words, summing the Coulomb potentials of the lattice ions averages out to a weak and periodic potential seen by the valence electrons. Combined with a quantum treatment of the motion of electrons in a solid, this periodic potential leads to solutions of the Schrödinger equation that are nearly those of free electron wavefunctions. Therefore, this basis of functions provides a model supporting the free and independent electron gas approximation. This result is known as the

Bloch theorem (Bloch, 1929). More precisely, it states that electrons in a periodic potential are represented by spatial wavefunctions $\psi(r)$ taking the product of two functions:

1. A plane wave $e^{i\vec{k}\cdot\vec{r}}$ characteristic of free electrons with crystal momentum $\vec{p}$ proportional to the $\vec{k}$-wavevector $\left(\vec{p}=\hbar\vec{k}\right)$.
2. A function $u(r)$ having the periodicity of the crystal lattice.

Therefore, the electronic probability density, or electron cloud density, is nearly constant in space meaning each independent electron could be found equally anywhere in the crystal on average.

Basically, crystalline metals or semiconductors confine a free electron gas in an attractive potential over the whole solid, which can be approximated by a large square well along any given direction, but with a periodic variation at the bottom. Bulk crystals are so large compared to the atomic scale that their surfaces are typically neglected. Therefore, it suffices for the delocalized electron wavefunctions to satisfy the periodic boundary conditions of the ion lattice modulating the potential well bottom. This yields a quantization in energy and $\vec{k}$-momentum, but the resulting states are so close together in energy that they become quasi-continuums known as *bands*. In this situation, it is not possible to directly probe individual energy levels. We mainly have access to overall averaged material properties depending chiefly on the following:

- The number of states per interval of energy, known as the density of states (DOS)
- The occupation of these states with electrons as given by their probability distribution over energy (Fermi–Dirac distribution around the Fermi level)

Whereas a macroscopic crystal already exhibits confinement with its surfaces to keep its electrons within, it is not considered quantum confinement until the DOS becomes altered by the finite size of the crystal. Indeed, the approximation of infinite bulk crystals without surfaces does not hold as their size is decreased down to NCs, and then, the solutions to the Schrödinger equation are no longer akin to free electron wavefunctions. Instead, their characteristics are closer to bound electron wavefunctions, such as atomic-like orbitals with discrete quantum numbers replacing the $\vec{k}$-wavevector, but still quickly oscillating in space with the periodicity of the crystal lattice. Accordingly, the DOS is no longer a quasi-continuum; it becomes divided into discrete and distinct energy levels. It is then easier to probe these states individually.

In principle, there are no fundamental restrictions preventing a smooth progression of the DOS from quantum confined discrete states in nanometer-sized crystals to the quasi-continuum of bulk macrocrystals. However, the nearly free electron approximation does not always hold as a crystal gets larger because its periodic lattice of ions is rarely perfect on long scales. Such irregularities in real crystals can come from impurities, vacancies, interstitials, dislocations, lattice vibrations (phonons), and so on. These defects can either scatter or trap the nearly free electrons. In the former case, the electrons are still delocalized but may have transitioned between the crystal's energy states (e.g., through phonon interaction), whereas in the latter case, the electrons are trapped into a local state of the defect: they are no longer free. Therefore, experiments probing parameters related to the DOS do not always yield exactly the results predicted by the nearly free electron model. However, with well-controlled experimental conditions, very pure copper, for example, can reach millimeter size before starting to be affected by electron scattering (Ibach and Lüth, 2009). By contrast, if a crystal is so small that it consists of a number of atoms comparable to molecules, the concept of a periodic lattice seems to be limited. Nonetheless, Ohm's law has been shown to hold for silicon wires as small as four atoms wide and one atom high (Weber et al., 2012).

Back to the ideal periodic crystal lattice, the passage from free electron behavior to bound electron cloud states in three-dimensional (3D) quantum confinement potentials prompted the introduction of terms such as *zero-dimensional (0D) systems* to describe semiconductor NCs. It is this parallel with ideal dots that builds up to the well-known *quantum dot (QD)* expression. Similarly, electrons can be quantum confined in one dimension with a very thin semiconductor layer (nanoplatelets, quantum wells) or in two dimensions with a very narrow semiconductor wire (nanowires, quantum wires) as illustrated in Figure 19.2. This results in other interesting changes in the DOS. However, to conclude on 0D quantum confined systems, we came full circle starting with discrete states of atoms turning into a quasi-continuum of states for the macroscopic crystal lattice and then back to discrete states with the bound envelope wavefunctions of QDs. There is, however, a net gain: we can manipulate these states at will with our fabrication methods. Hence, QDs are sometimes portrayed as artificial atoms.

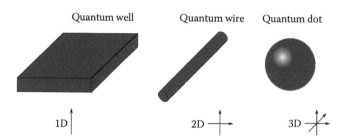

Figure 19.2 Illustrations of possible geometries for the quantum confinement of electrons in one, two, or three dimensions. The arrows specify the confinement directions where the surfaces define the potential well of the particle-in-a-box model.

19.1.2 NANOCRYSTALLINE SEMICONDUCTORS

All QDs are NCs, but not all NCs are QDs. Following the introduction of quantum confinement in Section 19.1.1, the NCs must support nearly free electron wavefunctions delocalized over their whole volume to exhibit optoelectronic properties tunable with their size and shape. As such, NCs whose properties of interest derive from color centers or other impurity dopants are not QDs, because the active local states are provided by an isolated trapping atom or a lattice site. Hence, the wavefunctions involved are not delocalized over the whole NC. In the end, impurity-free nanometals and nanosemiconductors can both meet the requirements for quantum confinement of electrons, but only the latter are labeled QDs and bring the possibility of interband transitions often stimulated by light. However, in this era of functional materials, a large number of hybrids of all the above are developed and a strict categorization might prove surprisingly challenging. In case of doubt, one can always rely on the more general terms: nanoparticles to enclose almost anything at the nanometer scale, NCs if said nanoparticles have a periodic lattice, and/or nanoemitters if they can emit light. To be slightly more specific, nanoemitters could be categorized into photoluminescent fluorophores when the exciting radiation is low-energy infrared or Vis light versus radioluminescent scintillants when the excitation is ionizing radiation.

From here on, we will discuss only nanocrystalline semiconductors. Even in this very specific category, different fabrication methods give rise to three distinct types of QDs:

1. Potential barriers electrostatically defined by metallic gates above a thin semiconductor layer (lateral QDs) (Van der Wiel et al., 2002)
2. Epitaxial growth on a semiconducting substrate of a different lattice constant resulting in surface tension and strain relaxed by forming nanoislands (Fafard, 2000; Stangl et al., 2004), much like droplets of water appearing on a cold flat surface once enough water vapor has condensed on it
3. Synthetic nucleation reaction of semiconductor precursors in solvents producing nanocolloids

The second type of QDs is known as *self-assembled QDs (SAQDs)* or *epitaxial QDs (eQDs)*, whereas the third type is known as *colloidal QDs (cQDs)*. An important distinction between these two types is the nature of their surroundings: the former is entirely embedded in a solid bulk semiconductor matrix with high relative permittivity (dielectric constant), whereas the latter is suspended in a dielectric liquid of lower relative permittivity. The confining potential well of eQDs is thus defined by an interface between two semiconductors which yields a more defect-free and stable environment with limited electromagnetic fluctuations. By contrast, a deeper confining potential well is obtained with the surface of cQDs preventing electrons from escaping the well through thermal activation. This gives the benefit of stronger irradiance, that is, brighter light emission, at room temperature. These solution-based matrices also enable the concentration control of cQDs with flexible and cost-effective processing into various solid matrices, spin-coated layers, other solvents, and so on.

So far, the discussion of nanosemiconductors has taken a top-down perspective based on the assumption of a perfect and infinite crystal in the nearly free electron model. However, local and finite surface-related effects cannot be neglected in real nanostructures because of their high surface/volume ratio: it scales with the inverse radius for approximately spherical cQDs. It is therefore necessary to also consider how bands arise from an atomistic bottom-up perspective. As isolated atoms get closer and their electron clouds start overlapping to form a crystal, the orbitals are mixing and their energy levels split mainly due to the exchange interaction originating from the Pauli exclusion principle. This principle requires the total wavefunction, including spin, of all electrons (fermions) in the system to be antisymmetric under the exchange of particles. In the new basis of states after the atoms have bonded, the higher energy orbitals are antibonding, whereas the lower energy ones are bonding. In the ground state of bulk semiconductors, the lower ones are completely filled, setting the Fermi level, and are separated by a bandgap from the empty antibonding states. This Fermi level for the crystal is therefore related to the concept of valence electrons for atoms. The energy states nearest the Fermi level are usually those most suitable for the nearly free electron model, namely, the valence band below it and the conduction band above it in semiconductors. Considering only the spatial part of the many-electron wavefunctions (orbitals) for the whole crystal, this part still can be either a symmetric or an antisymmetric function under parity inversion because the spin part will adjust accordingly. For typical tetravalent systems undergoing sp^3 hybridization when bonding into a crystal, such as group-IV, III–V, and II–VI semiconductors, wavefunctions for states near the top of the valence band keep a p-like symmetry, whereas those near the bottom of the conduction band preserve an s-character as depicted in Figure 19.3 (Kane, 1982; Streetman and Banerjee, 2005).

For semiconductor scintillation, once the electrons have relaxed nonradiatively from very high energies down to the edge of typical bandgaps of a few electronvolts or less, the emitted light comes from an

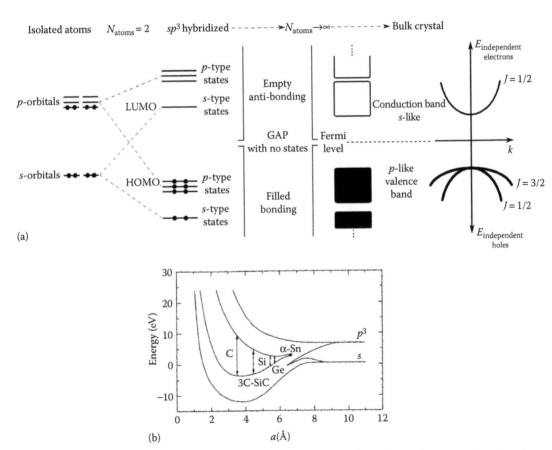

Figure 19.3 Simplified diagrams depicting how bands arise from the bonding of atoms for group-IV semiconductors in their ground state. (a) A different basis set of states emerges from the bonding of atoms to prevent a violation of the Pauli exclusion principle. As the number of bonded atoms grows, so does the number of one-particle states that are getting closer in energy and eventually merging into the quasi-continuum known as a band. These bands are represented for the bulk crystal both as the sum of all k-states for a given energy (black and white rectangles) and as with the E–k dispersion relation. The total angular momentum value J is indicated to illustrate the more complicated nature of the valence band, including heavy holes (J = 3/2) and light holes (J = 1/2). (b) Graph giving a more quantitative description with the isolated atom energy levels on the right side progressively splitting as the lattice constant a decreases. The actual bandgaps of a few group-IV semiconductors are indicated by arrows. (From Adachi, Sadao: Properties of Group-IV, III-VI and II-VI Semiconductors. p. 118. 2005. Copyright John Wiley & Sons. Reproduced with permission.)

allowed transition near $\vec{k} \sim 0$ from the s-type conduction band to the p-type valence band, thus bringing the system back to its ground state. Given this ground state in the valence band consists of a huge number of electrons, it is common to introduce the concept of the positively charged hole as an approximation of the many-body valence electron system minus one electron. Indeed, it is simpler to think of a bubble moving up in your favorite drink than about the cohesive displacement of all the liquid molecules around it, though one should not take this analogy too literally. With these holes and electrons as charge carriers, we readily see that the conduction band-to-valence band radiative transition obeys charge conservation: 1 electron$^-$ + 1 hole$^+$ $\rightarrow$ 1 photon^0. This highlights the Coulomb attraction between the electron–hole (e–h) pair, classically known as an electric dipole often called exciton, underlying the interaction with light. The full treatment of the charge density distribution and correlated e–h pairs is beyond the scope of this introduction, but we note that optical transition selection rules are obtained via the dipole moment, that is, Fermi's golden rule that includes the electric dipole matrix elements. In particular, low-energy (less than a few electronvolts) linearly polarized optical transitions across the bulk bandgap obey $\Delta\vec{k} = 0$ and the Laporte selection rule with $\Delta l = \pm 1$. In the former case, the wavelength of UV–Vis photons is too large compared to the crystal's lattice constant to significantly transfer $\vec{k}$-momentum to the charge carriers. In the latter case regarding parity, the s-like conduction band and the p-like valence band provide the change in symmetry required for allowed optical transitions. For spherical cQDs, the $\vec{k}$-momentum of free charge carriers is replaced by the n, l, m orbital quantum numbers of bound states for the envelope part of the spatial wavefunction. This implies that the transition between the lowest energy levels for electrons and holes in cQDs is not optically forbidden even if $\Delta l = 0$. In fact, the required change in parity is still taken care of by the s-type and p-type atomic part $u(r)$ of the spatial wavefunction for the conduction and valence bands, respectively (Davies, 1997).

From such considerations on Coulomb interaction, we deduce the first of five main characteristics of semiconductor cQDs: quantum confinement decreases the average distance between electrons and holes, thus promoting e–h correlation and increasing their binding energy. Therefore, brighter light should be emitted from the lowest cQD excited state than the relaxation across the bandgap of the equivalent bulk semiconductor. The total amount of light extracted from the sample also depends on the transparency of the host matrix (solvent) and its refractive index contrast relative to the nanoemitters. Contrary to atoms, a second characteristic of cQD ensembles is the inhomogeneous broadening of their emission and absorption lines. The spectral line broadening seen in atomic spectra mainly comes from scattering events affecting every identical atom equally: it is thus homogeneous. In spectrocopic studies of polydisperse cQD ensembles, the line broadening also has an inhomogeneous cause: the size and shape of each cQD is not exactly the same, and thus, their eigenstates and transition energies vary from one to another. The approximately normal cQD size distribution yields a Gaussian spectroscopic line profile. Another spectral characteristic of most cQDs is their significant shift between absorption and emission spectra due to contributions of different and optically forbidden electronic transitions within the fine structure in addition to the Stokes shift (Efros, 1996). The fourth characteristic concerns the cQD absorption spectrum which is much broader than those of any other fluorophore and thus gives greater flexibility to choose a nonionizing excitation source. Combined with the broadening mentioned above, the transition lines overlap into a large spectral range of absorption especially when the energy states get closer, as seen in Figure 19.4, further away from the band edges.

The last, but most well-known, characteristic of cQDs is their size-tunable emission frequency (color) but we emphasize that it does not change the bandgap of the underlying semiconductor. Rather, the tuning essentially comes from the change in zero-point energy E_c and E_v of quantum confined bond states as stated by the particle-in-a-box model as illustrated in Figure 19.4. This tuning of the emission energy, hence wavelength, can be best understood with an analogy to a vibrating guitar string changing pitch with its length. On the one hand, we hear high-frequency treble sounds when a guitar player holds the string against the fretboard near the guitar body to shorten the vibrating length: this corresponds to the blue-emitting cQDs with small electron clouds instead of strings. On the other hand, low-frequency bass sounds are played with longer vibrating lengths by holding the string near the headstock, as mirrored by larger electron clouds in bigger cQDs with red or even infrared light emission.

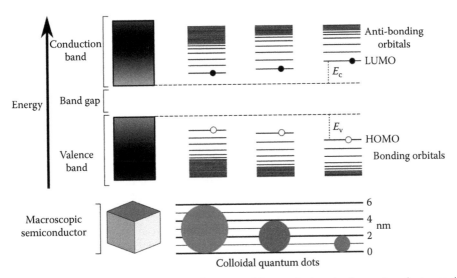

Figure 19.4 Transition from the quasi-continuum of states making up the bands of a semiconductor to the discrete states resulting from 3D quantum confinement of charge carriers in cQDs. The change in zero-point energy E_c, E_v with cQD size is also represented. Charge carriers are shown in the lowest excited state, and the light of energy $E = E_{bandgap} + E_c + E_v$ would be emitted by their relaxation between cQD states equivalent to the lowest unoccupied molecular orbital (LUMO) and the highest occupied molecular orbital (HOMO).

The bottom-up perspective was introduced here to tackle local atom-scale effects that are often detrimental to the charge carrier delocalization and thus to the stability of the emitted light. All nanoemitters are sensitive to their environment because light emission is fundamentally coupled to electric charges, and thus to the generally fluctuating electric fields around. This will affect the electronic energy states and transition dynamics within any nanoemitters. One possible mechanism in cQDs is the trapping of a charge carrier by the local charged environment or by a crystal lattice defect. Because this carrier comes from a photoexcited *e–h* pair, a delocalized charge is left behind in the cQD that will quench subsequent light emission: it is replaced by the exchange of kinetic energy with the new photogenerated charge carriers (Auger processes). Locally trapped charges can also get activated into the excited cQD states, returning it to a neutral and light-emitting status (bright). Given the fluctuating nature of these capture processes, this results in an on/off blinking of the emitted light. In the worst-case scenario, the charge trapping may be irreversible. For example, unprotected cQD surfaces can undergo photoactivated chemical reactions with the environment that irreversibly quench their light emission, a phenomenon known as photobleaching. Thanks to their larger number of constituent atoms, cQDs are often more robust against this type of chemical degradation than most molecular fluorophores, but they still do bleach when continuously photoexcited by nonionizing radiation. In the end, the further away the environment is from the cQD confinement potential, the more bright and photostable it will be. This requires surface passivation strategies, but even then, the charge carrier wavefunctions are never quite null outside the cQD. The relative permittivity of the surrounding matrix or solvent should thus be taken into account to assess the impact of fluctuating environmental electric fields. The topic of surface treatments to improve and use cQDs could probably fill an entire book, but we will only briefly cover the four main themes in Section 19.1.3.

19.1.3 SURFACE PASSIVATION AND FUNCTIONALIZATION

The discussion has focussed so far on the inorganic semiconductor composing the core of the cQDs, but another important intrinsic component is the interface between the core and its environment. Nanoscale objects have a large surface/volume ratio amplifying some surface-related effects, in particular the instabilities discussed in Section 19.1.2. The surface atoms do not have a complete coordination compared to atoms inside the crystal. This leaves partly filled orbitals with an unexploited bonding potential, called dangling bonds, that can act as scattering or trapping defects. Indeed, their high reactivity can cause the capture of delocalized charge carriers in local trap states leading to quenching of the light emission.

However, this reactivity also allows the adsorption of molecules called ligands. By filling the dangling bonds, the ligands reduce their reactivity, thus passivating the cQD surface by inhibiting the capture of charge carriers and reducing interactions with the environment (Medintz et al., 2005).

These ligands fulfill the following four roles:

1. Passivating the cQD surface.
2. Enabling further semiconductor growth on this surface.
3. Keeping the cQDs in colloidal suspension.
4. Allowing cQD functionalization in order to target or avoid binding on specific sites, most often of a biological nature.

Even if the ligands stabilize the cQD surface, they are labile: a ligand is not irreversibly bonded to the surface and is able to diffuse in the solvent as well as bond to another cQD. The degree of stability of a ligand on the surface is dictated by the bonding strength between ligands and cQD surface as well as the stability of the ligands in the solvent. The weaker is the bond on the semiconductor surface, the more labile is the ligand. Such labile ligands are therefore in thermodynamic equilibrium between the bonded state on the surface and the free state in the solvent. As a result, the cQD surface is still in interaction with its environment and can always undergo chemical reactions, especially under photoexcitation.

Because of this lability and incomplete cQD surface coverage by long flexible organic ligands that cannot be closely packed (steric hindrance), the surface passivation needs to be improved. Fortunately, the ligand lability enables controlled semiconductor growth at the surface because inorganic ions can also bind unsaturated surface dangling bonds. Inorganic ions are more permanently bonded to the cQD surface because they are far less reactive than their organic counterpart. They also provide a better passivation barrier that can be grown into a full semiconductor shell often thicker than the cQD core radius. Since Hines and Guyot-Sionnest (1996) successfully passivated CdSe cores with an inorganic ZnS shell several material combinations have been explored over the years to minimize surface trap states and defects, especially for cadmium-based semiconductor colloids (Reiss et al., 2009).

The ideal passivation scenario would require an infinite potential well keeping the excited charge carriers delocalized indefinitely within the core. In practice, two semiconductor parameters set the finite potential well depth of a core–shell cQD heterostructure: the bandgap energy and how the bands are aligned relative to one another. This band alignment is often quantified by the valence band offset between core and shell semiconductors. With the correct band alignment to obtain confining potential wells for both electrons and holes, the shell keeps their wavefunctions more localized in the core and thus decreases the surface charge density. This results in a lower probability of charge carrier trapping in surface defects of cQDs. Usually, the larger semiconductor bandgaps required to increase the shell potential barriers are associated with smaller interatomic distances (a) as seen in Figure 19.3b. The obvious passivation strategy of only maximizing the shell bandgap is thus inadequate, because shells with much larger bandgaps than the core are also likely to have a much smaller crystal lattice constant. Such a significant lattice mismatch between the core and the shell induces strain leading to the generation of point defects and dislocations. This basically renders the passivation ineffective. A small lattice mismatch may also cause enough strain to disrupt the shell growth, thus limiting its thickness. This also reduces the passivation effectiveness because the probability of finding charge carriers at the shell surface decreases with its distance from the core (Fafard, 2000; Pistol et al., 1999). Hence, superior passivation is obtained with thicker shells. Therefore, a compromise between a high bandgap semiconductor shell and its lattice mismatch with the core must be reached to optimize the passivation.

Two general approaches of shell design address this passivation issue. First, discrete successive layering (multishell) using inorganic semiconductors was explored, such as CdSe/CdS/Zn_xCd_{1-x}S/ZnS core/multishell cQDs (Xu et al., 2011). Each shell layer increases incrementally the bandgap and lattice constant to create a stepped confinement potential. Second, the progressive variation of a semiconductor alloy was used to obtain a graded shell bandgap resulting in a smooth confinement potential (Panda et al., 2011). Both shell structures are, respectively, represented by the top and bottom parts of the cQD illustrated in Figure 19.5. Such passivation strategies balance the lattice strain with maintaining an acceptable level of charge carrier confinement. A last semiconductor parameter to consider for effective passivation is the

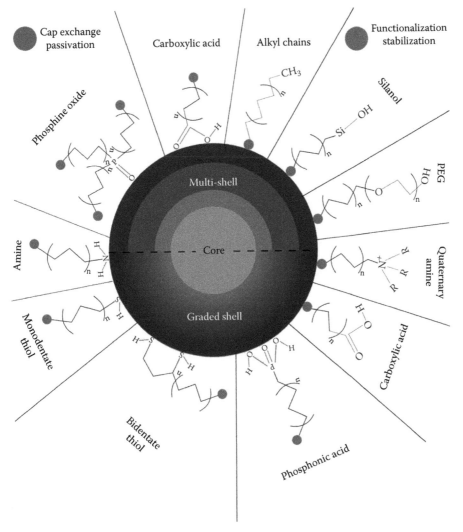

Figure 19.5 Passivation of a cQD with possible ligand anchoring groups (red circles or moieties) and the semiconductor multishell or graded shell. The remainder of the ligand enables colloidal suspension and may also include terminal groups for further functionalization (green circles or moieties). Many other ligands are suitable beyond the subset presented here; typically more than one kind is used at once. PEG, polyethylene glycol.

relative permittivity given its role in Coulomb interactions with other charge carriers as was discussed at the end of Section 19.1.2. However, a detailed discussion is beyond the scope of this introduction. In the end, optimizing cQD passivation in various environments is still a very active research field in terms of semiconductor heterostructure design with elaborate shells (Reiss et al., 2009), many-body dynamic studies (Bonati et al., 2005; García-Santamaría et al., 2011) and surface ligands with more complex chemistry. Figure 19.5 provides but a taste of the vast possibilities (Jorge, 2007).

Beyond the passivation aspects, the ligands also assume the third role of ensuring dispersion of the cQDs in liquids to form colloidal suspensions. As the surface bonding is ensured by an anchoring group illustrated in red in Figure 19.5, the rest of the ligand, often the terminal group in particular (green in Figure 19.5), acts as a stabilizer. Depending on the chemical nature of the matrix in which the cQDs are to be dispersed, different groups must be used to provide colloidal stability. A general rule of the thumb is that solvents will like molecules of similar nature or of similar polarity. A few examples are represented in Figure 19.5, but broadly speaking, two categories of stabilizers emerge according to the relative permittivity of the solvent. For apolar solvents such as toluene or alkylbenzene, alkyl chains (hydrocarbons) are the most used stabilizing groups. For polar solvents such as water or alcohols, polar or charged groups are much more suitable. Also, the specific choice of stabilizing groups in polar solvents will be dictated by the pH. Depending on its value, some chemical moieties will change charge. As an example, carboxylic acid will become negatively charged by losing an ion H^+ (deprotonated) at basic pH, which means that this functional group can only be used at a pH above 6. The situation is reversed for ligands with amine functionality that are positively charged at acidic pH. Therefore, to remain charged and dispersed over a wide range of pH, a ligand must possess both cationic and anionic moieties. Several other ligands of a different chemical nature could be grafted to the cQD surface, in particular, quaternary amines that are always charged or polyethylene glycol which is a neutral hydrophilic polymer. All these principles also apply to solid polymeric matrices because most begin as a liquid monomer mixture with the cQDs. Then, they undergo solidification during the polymerisation process, bringing additional challenges to prevent the issues of cQD segregation and loading inhomogeneity.

Not only terminal groups can enhance the colloidal stability of the cQDs, but they can also act as a functionalizing group, that is, the fourth and final role fulfilled by the ligands. To further functionalize the cQD surface, amine and carboxylic groups are commonly used after an activation step. Therefore, the cQD surface becomes an organic chemistry substrate and allows versatile functionalization for many applications. For example, in bioimaging, functional groups are used to bind cQDs to proteins, DNA, or other biomolecules to better image or track processes occurring in or near cells, using the cQD luminescence. Some functional groups can also bind to metal ions such as radionuclides, thus enabling the detection of radioactivity in the environment. With appropriate groups, affinity for a specific metal ion or biological label can therefore be achieved, thus providing the means to selectively target the desired object of an experiment. The benefits of this functionalization will be further discussed in the context of cQD scintillation in Section 19.2.

19.2 BENEFITS OF cQDs FOR SCINTILLATION COUNTING AND DOSIMETRY

With the current knowledge about the physics and chemistry of cQDs, we foresee new opportunities for these scintillating materials that may compare advantageously to existing organic fluorophores. Whereas organic fluorophores are composed of mainly C, H, O, N, and S, cQDs are made of heavier nuclei with a higher density of electrons, which should favor interactions with ionizing radiation. Their incorporation into a variety of solid and liquid supports is also relatively simple with the proper surface passivation and functionalization. However, most of all, the photoluminescence (PL) emission spectrum of cQDs can be continuously tuned by their chemical composition, size, and shape, thanks to the quantum confinement phenomenon discussed earlier.

The usefulness of tuning the emission wavelength of the cQDs is apparent in situations that require multiplexing, where different structures, information, analytes, and so on need distinct detection channels. This is routinely used in biological imaging where multiple fluorophores, emitting various colors, can produce beautiful micrographs such as Figure 19.6, identifying different formations in a specimen (Liu et al., 2013).

In scintillation-related applications, distinct radionuclides could be labeled with cQDs of different radioluminescence (RL) wavelengths, to track their distribution in a host (such as in cancerous cells). Alternatively, radioisotopes (e.g., ^{35}S) could be included directly within the cQDs during the synthesis. In both instances, two different modalities are combined:

1. Medical treatment with the exciting α or β particles
2. Monitoring of the treatment location with the RL emitted

To track several different locations at once, it is desirable to have as many detection channels as possible, and this becomes feasible with the precise and reproducible control of the central wavelength of the cQD emission spectrum. The emission peak of each cQD ensemble defines a detection channel, and the closer they can be brought together without overlapping significantly (no cross talk), the greater the number

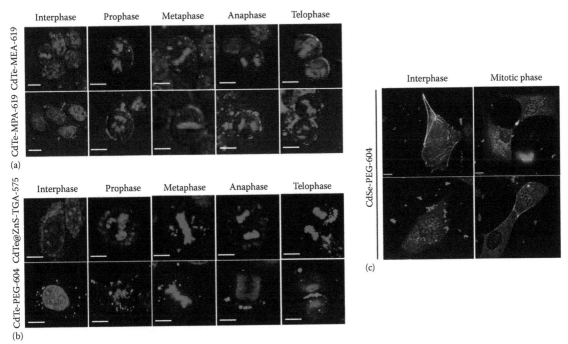

Figure 19.6 HeLa cells stained with CdSe, CdTe, and CdTe/ZnS cQDs as well as dyes. (From Liu, Yuexian et al.: The influence on cell cycle and cell division by various cadmium-containing quantum dots. *Small.* 2013. 9. 2447. Copyright Wiley-VCH Verlag GmbH & Co. KGaA. Reproduced with permission.)

of channels obtained in a given spectral range. With a narrow emission bandwidth, cQDs could be used for multipoint detectors to measure the radiation dose over different points more easily than organic fluorophores. Also, the cQD emission spectrum can be tuned into the spectral range of maximum sensitivity for the detector used in a given application. In the very low fluence conditions typical of scintillator excitation, complex detection schemes, such as coincidence circuitry, are required, thus making it easier to adapt the cQDs to the detectors, instead of the detectors to the cQDs.

The flexible processing of cQDs into various supports is achieved through functionalization of their surface with ligands that ensure suspension and integration into the desired matrix. This is performed using tethering strategies that do not affect the cQDs' internal composition and crystal structure (Figure 19.5), thus preserving their optical properties as much as possible when they are properly passivated. This strategy cannot be achieved easily with organic fluorophores unless the molecule is redesigned or modifiers are added during the preparation of the specific scintillating materials. Combining this versatility of processing with prospects of linear response to dose and resistance to degradation, the cQDs are therefore suitable for all sorts of scintillation applications under various experimental conditions.

The structure and composition of semiconducting cQDs can be optimized for interaction with ionizing radiation. Much of the extensive knowledge gathered on synthesis and characterization of cQDs with UV–Vis excitation is directly applicable in the context of scintillation.

Topics currently studied in the optical regime, but still relevant to RL, include the following:

- Minimization of defects in the crystal structure
- Choice of heterostructure design with passivating shells of different semiconductors
- Concentration and dispersion of cQDs in their host matrix to maximize conversion of an incident radiation into luminescence with minimal reabsorption

New aspects will need to be taken under consideration in the cQD design to increase their interactions with ionizing radiation. The role played by the host matrix should also be carefully studied because it can transfer energy or charges left there by the radiation to the cQDs. All the above topics are scarcely explored in the context of scintillation at this point, and Section 19.3 will cover the current state of the research.

Other luminescence-based applications

19.3 RECENT DEVELOPMENTS OF cQD SCINTILLATORS

Over the past 10 years, several research groups and organizations have begun to experiment with the potential of cQDs as scintillators for radiation detection. Table 19.1 presents a summary of the various cQD systems studied for scintillation applications.

These studies were performed on cQDs of diverse chemical nature incorporated or dispersed in various matrices, with specific ionization radiation sources (α, β, γ, and X-rays). The overview of the current research in the field will be done based on the type of scintillator support to highlight the challenges originating from the dispersion of cQDs in these supports.

Table 19.1 **Summary of cQD systems used for scintillation applications**

TYPE OF SCINTILLATOR SUPPORT	cQD SYSTEM USED	DISPERSION MATRIX	EMISSION WAVELENGTH (NM)	EXPOSED RADIATION TYPE	REFERENCE
Glass	CdSe/ZnS	Porous glass	540	α ($^{243-244}$Cm, 0.2 uCi)	Létant and Wang (2006)
	CdSe/ZnS	Porous glass	510	γ (^{241}Am, 1 uCi) α ($^{243-244}$Cm, 0.2 uCi)	Létant and Wang (2006)
	ZnS	Lithiated gel	380	NA	Dai et al. (2002)
	CdSe/ZnS	Lithiated gel	590	α(^{210}Po, 5,41 MeV)	Dai et al. (2002)
Polymer	CdSe/ZnS	Polystyrene	520	γ (^{137}Cs, 661 KeV) α Proton beam (45 MeV) X-ray, (100 KV)	Park et al. (2014)
	CdSe/ZnS	Polystyrene	472	γ (^{241}Am, 1 uCi) α (^{252}Cf, 3 mCi)	Brown et al. (2007)
	CdTe	PMMA	547	γ (^{241}Am, 59.5 KeV; ^{137}Cs, 661 KeV)	Wagner et al. (2012)
	CdSe-ZnSe	MEH-PPV	550	β (3 keV, CL)	Campbell and Crone (2006)
Liquid	CdSe/CdS/CdZnS/ZnS	Hexane/water	605	(^{226}Ra source) γ (^{60}Co, 1.17 MeV; 1.33 MeV)	Lecavalier et al. (2013)
	CdSe/ZnS	Hexane	524	γ (^{60}Co, 1.17 MeV; 1.33 MeV)	Stodilka et al. (2009)
	CdS + PPO	Toluene	360–420	β (^{90}Sr, 0.5 MeV; ^{90}Y, 2.28 MeV)	Winslow and Simpson (2012)
	CdSe/ZnS	Hexane	579	γ (^{137}Cs, 661 KeV)	Withers et al. (2008)

MEH-PPV, poly[2-methoxy-5-(2′-ethylhexyloxy)-*p*-phenylene vinylene]; NA, not available; PMMA, polymethyl methacrylate; PPO, 2,5-diphenyloxazole.

19.3.1 GLASS SUPPORT

One challenge in the design of cQD-doped glass scintillator is how to homogeneously incorporate cQDs into a glassy matrix. Common glass-based scintillators (e.g., cerium-activated lithium glass scintillators) are manufactured by incorporating the inorganic scintillant into the glass matrix through fusion at high temperatures, typically above 1000°C (Kang et al., 2013). However, this process is incompatible with the production of cQD-doped glass scintillators, because cQDs would be degraded at such temperatures even if the melting temperatures of their bulk material components, (e.g., CdS), were much higher than 1000°C (Haynes, 2013). As an example, Goldstein et al. (1992) have observed a large decrease in the melting temperature of CdS cQDs (from 1200°C to 600°C) with decreased size (from 40 to 10 nm). To overcome this manufacturing disadvantage, nanoporous glass was tried as a scintillation support in which the cQDs could diffuse. Typically, porous glass with nanopores 10–20 nm in diameter was generated by slowly dissolving porous glass with interconnected pores of 4 nm diameter using aqueous solution of hydrofluoric acid and ethanol (1% and 20%, respectively). Using a diffusion approach, Létant and Wang in 2006 estimated the QD density in the porous glass close to 10 mg/cm^3. Porous glass is suitable for scintillation applications due to its inert character, its transparency, and the ability to hold guest molecules while isolating them through a succession of nanometer-sized cavities, which prevents self-quenching effects.

Létant and Wang (2006) used this strategy to incorporate CdSe/ZnS cQD, suspended in toluene at a concentration of 10 mg/mL, as guest dopants in the interconnected pores of their glass support. The performances of this new glass-based nanocomposite were assessed for the detection of α and β exposure through a photomultiplier tube (PMT). The α-source and the nanocomposite must be in close proximity to ensure that enough particles reach the scintillant. The data gathered showed that the energy of the α-ray radiation is converted into Vis photons by the cQDs, but at much lower output than rhodamine B, a laser dye. Through simulation code, they found that 0.4% of the photons generated in the cQD were amplified by the PMT. An energy resolution of 2% was calculated through modeling CdSe/ZnS QD behavior under α-ray irradiation as a result of their Vis bandgap.

In a subsequent experiment, the same cQD system was subjected to α- and γ-ray irradiation (Létant and Wang, 2006). The signal generated by nanocomposite samples was integrated for 10 and 72 h for α- and γ-irradiation, respectively*. The spectra obtained are presented in Figure 19.7. For γ-ray detection, Létant and Wang (2006) reported an energy resolution ($E/\Delta E$) at 59 keV of 15%. Under similar conditions,

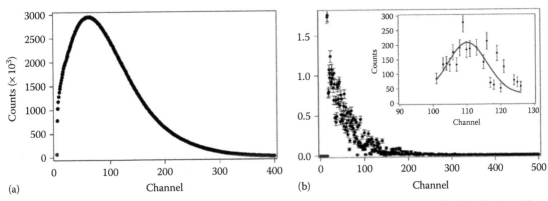

Figure 19.7 Scintillation output of a 25 mm-thick cQD-nanoporous glass composite under (a) α-irradiation with ^{243,244}Cm for 10 h and (b) γ-irradiation with an ^{241}Am source. A Gaussian fit of the 59 keV line of ^{241}Am is shown in inset. (Reprinted with permission from Létant, Sonia E., and Tzu-Fang Wang, *Nanoletters*, 6, 2877, 2006. Copyright 2006 American Chemical Society.)

---

* The difference in integration time is the result of the lower stopping power of γ-rays compared to alpha particles.

the signal generated by a 1″-thick NaI crystal and which was integrated for only 2 h led to an energy resolution of 30%. This would place cQD energy resolution for γ-rays between cooled semiconductor detectors and inorganic scintillant crystals. γ-Ray detector performances are governed by two criteria: detection efficiency and energy resolution. cQD has better energy resolution than NaI but lower detection efficiency than both types of γ-detectors. Létant and Wang (2006) proposed that the comparative lack of detection efficiency of CdSe/ZnS cQD indicative of its lower stopping power is the result of the material's low density, the lower, on average, atomic number Z of its constituents, and its smaller thickness, but other explanations in terms of cQD surface passivation could be considered. They also concluded that, to compete with inorganic scintillant crystals, new nanocomposite materials with higher densities, designed around heavier elements and of increased thickness, would need to be developed. A stronger shift between the emission and absorption spectra of the cQD would also be needed to minimize losses through reabsorption in thicker materials.

In recent years, the sol–gel process has been recognized as an extremely useful strategy to encapsulate scintillant in a glassy matrix, providing an alternative to the porous glass diffusion method used by Létant and Wang (2006). The term *sol–gel* refers to a process in which solid nanoparticles dispersed in a liquid (a sol) agglomerate together to form a continuous 3D network extending throughout the liquid (a gel). The method used to prepare the sol–gel uses a scintillant that cannot typically be introduced into a high-temperature glass melt.

Earlier, preparing micrometer-sized inorganic semiconductors in inorganic matrices such as sol–gels resulted in optically opaque gels. This opacity resulted in a lower RL (quantum yield), and consequently lower analytical performances (Germai, 1964). Here, the transition from the micrometric to the nanometric scale could provide a solution to this issue because the size of scintillants becomes much smaller than the wavelength of Vis light, and thus significantly reduces light scattering issues. CdSe/ZnS cQD sol–gel scintillators were prepared by Dai et al. (2002) to assess their performances under α irradiation using a ^{210}Po source. Although the authors did not report any specific figures of merit other than the pulse height spectrum for their system, they stated that the signals generated by the sol–gel scintillator were comparable in amplitude to those produced in existing inorganic scintillators.

19.3.2 POLYMERIC SUPPORT

Plastic scintillator materials have been widely used as nuclear detectors.
- They can be produced simply at low cost.
- They provide a fast response time.
- They can be relatively easily doped with other materials to enhance their properties.

Based on this premise, Park et al. (2014) used a thermal polymerization approach proposed by Basile (1957) to incorporate cQDs in plastic scintillators doped with 2,5-diphenyloxazole (PPO) and 1,4-bis(5-phenyloxazol-2-yl)benzene (POPOP) in order to modulate the emission wavelength of the doped material to align better with the maximum quantum efficiencies of various photosensors (i.e., PMT, photodiodes, avalanche photodiodes, and charge-coupled devices).

Seven styrene-based plastic scintillators were produced and exposed to X-rays, photons, and 45 MeV proton beams. As little is known on the luminescence mechanisms of the cQD-doped plastic scintillators, they were designed to test the following hypotheses:
1. The emission wavelength shifts from the PPO spectrum to the cQD spectrum.
2. The luminescence is directly transferred from the styrene to the cQDs.

Although the latter hypothesis was not validated in the conditions tested, results showed that Förster resonance energy transfer (FRET) occurred between the PPO and cQDs. The optimal energy transfer occurs when PPO is 4 times more abundant (in mass) than cQDs. The RL decay time could also be modulated through the presence of cQDs in the plastic scintillators (Table 19.2), suggesting the applicability of such polymer scintillant in radiation detection, nuclear, and high-energy physics applications.

In opposition to Park's vision of the transfer between the PPO and the cQDs, Campbell and Crone designed a cQD/organic semiconductor composite that produces excitation upon exposure to ionizing radiation. The excitation is predominantly in the cQDs and is transferred via FRET to the organic semiconductor. CdSe/ZnSe core–shell cQDs and a conjugated polymer of

Table 19.2 **Measured RL decay time after exposure to a Cs-137 γ-ray excitation for several doped plastic scintillators**

PLASTIC SCINTILLANT COMPOSITION (WT.%)				RL DECAY TIME (NS)	
Styrene	PPO	POPOP	CdSe/Zns	Fast component	Slow component
~99	0.4	0.01	–	4.40	16.0
~99	0.4	–	–	3.90	13.3
~99	0.4	–	0.1	2.40	11.4

Source: Park, Jong M. et al., *Journal of Luminescence*, 146, 157–161, 2014.

poly[2-methoxy-5-(2′-ethylhexyloxy)-*p*-phenylenevinylene] (MEH-PPV) were mixed and spin-cast on the sapphire substrate to form a thin composite film. To demonstrate the energy transfer from the cQDs to the organic semiconductor and the optimal fraction of cQDs in the composite, four volume fractions were characterized with absorption spectra, PL spectra, and cathodoluminescence (CL) measurements. MEH-PPV and CdSe/ZnSe cQDs were chosen based on the overlaps between the absorption spectrum of the MEH-PPV and the emission spectrum of the cQDs—the overlap allows the FRET process to occur. The CL measurement shows that the signal of the mixture containing the cQDs was twice as large as that from the pure MEH-PPV and was obtained with a cQD volume fraction of 0.15.

Wagner et al. (2012) used CdTe cQD in a polymethyl methacrylate (PMMA) plastic to perform the detection of γ radiation with an ^{241}Am source. To prevent aggregation and thus maintain material transparency at energies below the cQD emission, they used an engineered polymerizable material to lower the interfacial tensions to bound the surface ligand of the nanoparticles through either electrostatic or covalent interaction. This surfactant played a dual role: it aided the dispersability of the NC in the monomeric solution, and prevented agglomeration and segregation of the cQD during the polymerization process through immobilization. The authors used octadecyl-vinylbenzyl-dimethylammonium chloride as a polymerizable surfactant for thiol-capped cQD. Although transparency was achieved through this strategy, their results show a low counting efficiency for the proposed material because the PMT used was not sensitive to photons near the peak emission of CdTe-PMMA at 547nm.

Finally, Brown et al. (2007) used CdSe/ZnS cQDs with an emission maximum at 472 nm embedded in a polystyrene matrix to achieve the detection of α particle from the ^{241}Am source. The spectra obtained show a very low relative pulse height (0.1 channel number/gain) in comparison with BC-400™ (2.5), a plastic scintillator composed of organic fluors, PPO and POPOP in polyvinyl chloride (not shown). cQDs were also compared to Y_2O_3:Ce@sol–gel. These materials also showed very low relative pulse heights, not significantly different from those of the CdSe/ZnS cQDs. $LaPO_4$:Ce NCs were also prepared and exhibited the highest relative pulse heights and the number of counts of all the materials investigated, with similar resolution to cQDs (Figure 19.8). Based on their experiments, Brown et al. (2007) concluded that the difference in optical quality of the nanocrystalline samples investigated renders the assessment of determining factors for the detector design based on this approach inconclusive.

19.3.3 LIQUID SUPPORT

Materials with cQDs dispersed in glass or polymeric supports may be potential replacements for solid dosimeters. Dispersing cQDs in solvents may work for alternatives or additives for liquid scintillation cocktails. The latter aspect would be a more natural application for cQDs because synthesis methods typically suspend them in toluene, an excellent scintillator. Due to the chemical and physical challenges associated with the dispersion of cQDs in polar solvents, only a few chemical varieties of cQDs in solvents have been investigated so far (Table 19.1).

Withers et al. (2008) evaluated the use of CdSe/ZnS cQDs in hexanes as scintillators for γ irradiation. One of their objectives was to determine the speed of degradation of their cQDs using a PL approach. They observed a strong decrease in light output as the received dose, in kiloroentgens, increased (Figure 19.9). After converting the exposure into the absorbed dose in rads, they calculated that 50% of the light output was lost after approximately 11.5 krad, which is about 20-fold less than the radiation hardness

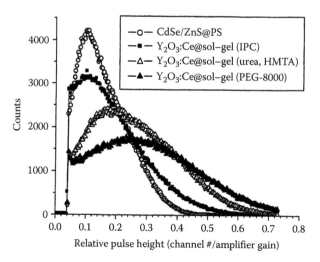

Figure 19.8 Pulse height spectra from α particle detection by CdSe/ZnS core/shell and $Y_{1.90}Ce_{0.1}O_3$ nanocrystals prepared by different methods. (Reprinted with permission from Brown, Suree S. Adam J. Rondinone, and Dai Shengheng. 2007. *Applications of nanoparticles in scintillation detectors*, Washington, DC: Oxford University Press. Copyright 2007 American Chemical Society.)

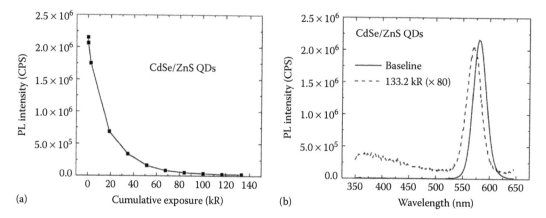

Figure 19.9 (a) Peak PL intensity in count/second for CdSe/ZnS cQDs as a function of cumulative exposure *E* in kiloroentgens. (b) Spectral changes in PL response of CdSe/ZnS cQDs induced by 662 keV γ-ray irradiation. (Reprinted with permission from Withers, Nathan J. et al., *Journal of Applied Physics Letters*, 93, 173101, 2008. Copyright 2008, American Institute of Physics.)

of GaN/InGaN. By comparison, an NaI:Tl scintillating crystal showed a rapid loss of light output after exposure to merely 500 rad originating from a ^{60}Co source (Normand, 2007). In addition, exposed cQDs exhibited a radiation-induced blueshift from 579 to 570 nm and a decrease in PL quantum yield from 23.4% to 0.2% after 120 krad. The authors postulated that this may be related to partial strain relaxation of the cQD crystalline structure and γ-irradiation-induced nonradiative recombination channels, respectively.

Following the efforts of Withers et al. (2008), Stodilka et al. (2009) also reported the optical degradation of CdSe/ZnS cQDs in hexanes upon γ irradiation. They reported quantitative changes in PL characteristics of these cQDs even for radiation exposures as low as 0.1 Gy, suggesting that cQDs could be potentially used as dosimeters that report the cumulative dose as a function of change in emission intensity. The cQDs prepared were not water soluble and often precipitated during the experiment. They concluded that surface modification of cQDs would result in water-soluble NCs and should provide materials with improved performance with respect to stability.

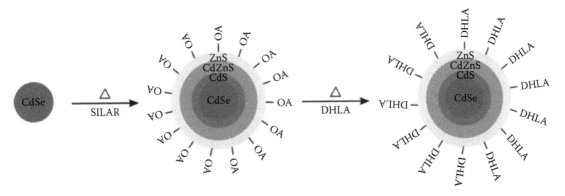

Figure 19.10 The growth schematic of a core/multishell cQD by SILAR followed by the DHLA ligand exchange. (Lecavalier, Marie-Eve et al., *Chemical Communications*, 49, 11629–11631, 2013. Reproduced by permission of The Royal Society of Chemistry.)

Based on these findings, Lecavalier et al. (2013) proposed a multilayer cQD system (CdSe/CdS/CdZnS/ZnS) prepared through the successive ionic layer adsorption and reaction (SILAR) synthesis. This system offers an improved defect-free surface passivation without introducing too much strain and can be converted into a water-soluble system through ligand exchange using dihydrolipoic acid (DHLA) (Figure 19.10). Using this strategy, they were able to report a calibration curve for ^{226}Ra in a borate buffer at pH = 8 (precipitation occurred in acidic conditions). A maximum detection efficiency of 58% was calculated, roughly 10-fold lower than commercial scintillation cocktails (437%–682%). The developed cQDs remained stable for months in these conditions. They also observed a clear linear relationship between the level of light generated and the dose they were exposed (0.05–3 Gy) for their cQDs in hexane under ^{60}Co irradiation. The fact that little degradation in the signal was measured with these cQDs suggests that the supposition of Stodilka et al. was correct in that both the functionalization of the surface (with oleic acid [OA] in the case of cQDs in hexane) and the passivation with multiple semiconductor shells were factors in this improvement. Using multishell cQDs also eliminates the irradiation-induced blueshift noted by Withers et al. (2008). Based on the results presented by Lecavalier et al. (2013), future applications of cQDs may include liquid scintillation counting, especially in biological fluids where the pH is relatively neutral.

Winslow and Simpson (2012) reported the use of a cQD-doped liquid scintillator for the detection of neutrino. Because such experiments require a large scale, the use of a diluted cQD system is extremely attractive. Here, 2,5-diphenyloxazole (PPO) was used as a *conventional* scintillator, and CdS cQDs in toluene were used as a secondary wavelength shifter. As they were concerned about the potential interference of the cQDs on the primary scintillator process, Winslow and Simpson used a ^{90}Sr pin source suspended in the sample to compare the performances of toluene-PPO and toluene-PPO-cQD systems. As expected, due to their opaque nature and considering their elevated concentrations (1.25 g/L) in the mixture, light yield measurements show reduced output when cQDs were present (Figure 19.11). Nonetheless, the authors believe that cQDs hold promise as neutrino detectors, especially for neutrinoless double β decay as quantum efficiencies continue to improve.

19.4 CHALLENGES AND OUTLOOK

As illustrated in Section 19.3, research on scintillation applications for cQDs has barely begun. One reason for this late bloom in comparison with other cQD applications is the insufficient understanding of the interaction between the cQDs and the ionizing radiation, especially particle radiation (α and β). We believe that a better understanding of this interaction should help to position cQDs as an interesting alternative to current scintillators. The cQDs may then be developed for niche applications complementary to those already fulfilled by organic fluorophores and bulk semiconductor crystals.

Among the potential applications of cQDs in nuclear medicine, the possibility to functionalize cQD surfaces to strongly bind pure α- or β-emitters for medical treatment is certainly interesting. The use of

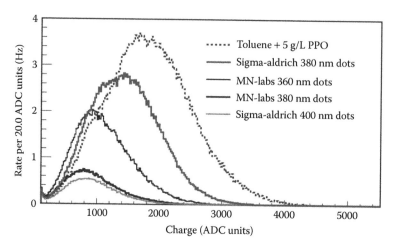

Figure 19.11 The total light yield of different scintillator mixtures obtained with a ^{90}Sr source. The NN-Labs 380 nm dots are the sample from June 2010. (Reprinted with permission from Winslow, Lindley, and Raspberry Simpson, Characterizing quantum-dot-doped liquid scintillator for applications to neutrino detectors, *Journal of Instrumentation* 7, no. 7, P07010, 2012. Copyright SISSA Medialab Srl. Copyright 2012, Institute of Physics.)

radioisotopes (e.g., ^{35}S) during the synthesis of the cQDs could also be an interesting avenue. Although the ionizing radiation could play its role in the treatment of cancer, its interaction with the cQDs could lead to external visualisation, depending on the emission wavelength chosen. Their small size could make cQDs suitable for detection or oncological treatment inside cells. However, cytotoxicity or biocompatibility issues need to be assessed, but alternative heavy metal-free cQDs are currently developed. For nuclear medicine, because cQDs are relatively simple to include in various matrices, they could become an inherent part in the development of scintillating phantoms.

Although the detection of light originating from cQDs due to their exposure to ionizing radiation has been demonstrated by several research groups, reported conversion factors are generally much lower than other fluorophores and inorganic scintillants. As the mechanisms behind this energy transfer are still unclear, the design of cQDs with optimal RL properties and composition is still a hit-or-miss type of research. Yet, two broad conclusions have started to emerge:
1. Passivation and functionalization of the surface reduce the irreversible effects of the irradiation on the cQDs;
2. The response of cQDs is measurable for the most common types of ionizing radiation.

The variability in the cQD heterostructure and in irradiation strategies employed so far makes it challenging to draw more general conclusions. Finally, in addition to the interesting features of cQDs for scintillation purposes mentioned in Section 19.2, it is foreseen that they could overcome volume limitation issues. These limitations on opacity are intrinsic to macroscopic semiconductor and inorganic crystal growth, whereas cQDs can be dispersed in a transparent matrix. This strategy was used by Winslow and Simpson (2012) for the design of a large-scale neutrino detector.

ACKNOWLEDGMENTS

The authors thank Marie-Ève Delage, Jean-Michel Ménard, and Dany Lachance-Quirion for their helpful reviews and discussions.

REFERENCES

Adachi, Sadao. 2005. *Properties of Group-IV, III-VI and II-VI Semiconductors*, 118. England: John Wiley & Sons.
Ashcroft, Neil W., and David N. Mermin. 1976. *Solid State Physics*. New York: Holt, Rinehart and Winston.
Basile, Louis J. 1957. Characteristics of plastic scintillators. *Journal of Chemical Physics* 27: 801–806. doi:10.1063/1.1743832.

Bloch, Felix. 1929. Über die quantenmechanik der elektronen in kristallgittern. *Zeitschrift für physik* 52, no. 7–8: 555–600. doi:10.1007/BF01339455.

Bonati, Camelia, Mona B. Mohamed, Dino Tonti, Goran Zgrablic, Stephan Haacke, Frank Van Mourik, and Majed Chergui. 2005. Spectral and dynamical characterization of multiexcitons in colloidal CdSe semiconductor quantum dots. *Physical Review B* 71, no. 20: 205317. doi:10.1103/PhysRevB.71.205317.

Brown, Suree S., Adam J. Rondinone, and Dai Shengheng. 2007. *Applications of Nanoparticles in Scintillation Detectors*. Washington, DC: Oxford University Press.

Campbell, Ian. H., and Brian K. Crone. 2006. Quantum dot/organic semiconductor composites for radiation detection. *Journal of Advanced Materials* 18: 77–79. doi:10.1002/adma.200501434.

Dai, Sheng, Suree Saengkerdsub, Hee-Jung Im, Andrew C. Stephan, and Shannon M. Mahurin Dai. 2002. Nanocrystal-based scintillators for radiation detection. *AIP Conference Proceedings* 632: 220–224. doi:10.1063/1.1513973.

Davies, John H. 1997. *The Physics of Low-dimensional Semiconductors: An Introduction*. Cambridge: Cambridge University Press.

Drude, Paul. 1900. Zur Elektronentheorie der Metalle. *Annalen der Physik* 306: 566–613. doi:10.1002/andp.19003060312.

Efros, Alexander L., M. Rosen, M. Kuno, Manoj Nirmal, David J. Norris, and Moungi Bawendi. 1996. Band-edge exciton in quantum dots of semiconductors with a degenerate valence band: Dark and bright exciton states. *Physical Review B* 54, no 7: 4843–4856.

Fafard, Simon. 2000. Near-surface InAs/GaAs quantum dots with sharp electronic shells. *Applied Physics Letters* 76: 2707–2709. doi:10.1063/1.126450.

García-Santamaría, Florencio, Sergio Brovelli, Ranjani Viswanatha, Jennifer A. Hollingsworth, Han Htoon, Scott A. Crooker, and Victor I. Klimov. 2011. Breakdown of volume scaling in Auger recombination in CdSe/CdS heteronanocrystals: The role of the core–shell interface. *Nano Letters* 11, no. 2: 687–693. doi:10.1021/nl10380e.

Germai, Georges. 1964. Quelques facteurs influençant l'efficacité de comptage des ^{137}Cs, ^{90}Ssr et ^{131}I sous forme de sels minéraux en suspension dans les gels scintillants. *Bulletin de la société royale des sciences de liège* 10: 563–578.

Goldstein, Avery. N., C. M. Echer, and Paul A. Alivisatos. 1992. Melting in semiconductor nanocrystals. *Science* 256:1425–1427. doi:10.1126/science.256.5062.1425.

Haynes, William. M. 2013. *CRC Handbook of Chemistry and Physics*. Boston, MA: CRC Press.

Hines, Margaret A., and Philippe Guyot-Sionnest. 1996. Synthesis and characterization of strongly luminescing ZnS-capped CdSe nanocrystals. *The Journal of Physical Chemistry* 100: 468–471. doi:10.1021/jp9530562.

Ibach, Harald, and Hans Lüth. 2009. *Solid-State Physics: An Introduction to Principles of Materials Science*, 269–272. Berlin, Germany: Springer.

Jorge, Pedro, Manuel António Martins, Tito Trindade, José Luís Santos, and Faramarz Farahi. 2007. Optical fiber sensing using quantum dots. *Sensors* 7, no. 12: 3489–3534. doi:10.3390/s7123489.

Kane, E. O. 1982. Energy band theory. In *Handbook on Semiconductor*, ed. T. S. Moss, 193–217. Amsterdam, The Netherlands: Elsevier.

Kang, Zhitao, Robert Rosson, M. Brooke Barta, Jason Nadler, Brent Wagner, and Bernd Kahn. 2013. GgdBbr3: Ce in a glass wafer as a nuclear radiation monitor. *Health Physics* 104: 504–510. doi:10.1097/HP.0b013e318286c062.

Lecavalier, Marie-Eve, Mathieu Goulet, Claudine Nì Allen, Luc Beaulieu, and Dominic Larivière. 2013. Water-dispersable colloidal quantum dots for the detection of ionizing radiation. *Chemical Communication* 49: 11629. doi:10.1039/C3CC46209A.

Létant, Sonia E., and Tzu-Fang Wang. 2006. Semiconductor quantum dot scintillation under γ-Ray irradiation. *Nanoletters* 6: 2877–2880. doi:10.1021/nl0620942.

Létant, Sonia E., and Tzu-Fang Wang. 2006. Study of porous glass doped with quantum dots or laser dyes under alpha irradiation. *Applied Physics Letters* 88: 103110. doi:10.1063/1.2182072.

Liu, Yuexian, Peng Wang, Yue Wang, Zhening Zhu, Fang Lao, Xuefeng Liu et al. 2013. The influence on cell cycle and cell division by various cadmium-containing quantum dots. *Small* 9: 2440–2451. doi:10.1002/smll.201300861.

Medintz, Igor L., H. Tetsuo Uyeda, Ellen R. Goldman, and Hedi Mattoussi. 2005. Quantum dot bioconjugates for imaging, labelling and sensing. *Nature Materials* 4: 435–446. doi:10.1038/nmat1390.

Normand, Stephan, Ana Iltis, François Bernard, Teresa Domenech, and Philippe Delacour. 2007. Resistance to γ irradiation of LaBr₃:Ce and LaCl₃:Ce single crystals. *Nuclear Instruments and Methods in Physics Research Section A: Accelerators, Spectrometers, Detectors and Associated Equipment* 572: 754–759. doi:10.1016/j.nima.2006.11.060.

Panda, Subhendu K., Stephen G. Hickey, Christian Waurisch, and Alexander Eychmüller. 2011. Gradated alloyed CdZnSe nanocrystals with high luminescence quantum yields and stability for optoelectronic and biological applications. *Journal of Materials Chemistry* 21: 11550–11555. doi:10.1039/C1JM11375E.

Park, Jong M., Hyo-Joong Kim, H. J,. Yoontae S. Hwang, Dennis H. Y. S. Kim, D. H. Parkand, and Hyun W. Park. 2014. Scintillation properties of quantum-dot doped styrene based plastic scintillators. *Journal of Luminescence* 146: 157–161. doi:10.1016/j.jlumin.2013.09.051.

Pistol, Mats-Erik, Pedro Castillo, Dan Hessman, José A. Prieto, and Lars Samuelson. 1999. Random telegraph noise in photoluminescence from individual self-assembled quantum dots. *Physical Review B* 59: 10725. doi:10.1103/PhysRevB.59.10725.

Regulacio, Michelle D., and Ming-Yong Han. 2010. Composition-tunable alloyed semiconductor nanocrystals. *Accounts of Chemical Research* 43: 621–630. doi:10.1021/ar900242r.

Reiss, Peter, Myriam Protiere, and Liang Li. 2009. Core/shell semiconductor nanocrystals. *Small* 5: 154–168. doi:10.1002/smll.200800841.

Stangl, Julian, Vàclav Holý, and Günther Bauer. 2004. Structural properties of self-organized semiconductor nanostructures. *Reviews of Modern Physics* 76: 725. doi:10.1103/RevModPhys.76.725.

Stodilka, Robert Z., Jeffrey J. L. Carson, Kui Yu, Md Badruz Zaman, Chunsheng Li, and Diana Wilkinson. 2009. Optical degradation of CdSe/ZnS quantum dots upon gamma-ray irradiation. *Journal of Physical Chemistry* 113: 2580–2585. doi:10.1021/jp808836g.

Streetman, Ben, and Sanjay Banerjee, 2005. *Solid State Electronic Devices.* 6th ed. Upper Saddle River, NJ: Prentice Hall.

Van der Wiel, Wilfred G., S. De Franceschi, J. M. Elzerman, Toshimasa Fujisawa, Seigo Tarucha, and Leo P. Kouwenhoven. 2002. Electron transport through double quantum dots. *Reviews of Modern Physics* 75, no. 1: 1–22. doi:10.1103/RevModPhys.75.1.

Wagner, Brent K., Zhitao Kang, Jason Nadler, Robert Rosson, and Bernd Kahn. 2012. Nanocomposites for radiation sensing. *SPIE Defense, Security, and Sensing* 83730K.

Weber, Bent, Suddhasatta Mahapatra, Hoon Ryu, S. Lee, A. Fuhrer, T. C. G. Reusch et al. 2012. Ohm's law survives to the atomic scale. *Science* 335: 64–67. doi:10.1126/science.1214319.

Winslow, Lindley, and Raspberry Simpson. 2012. Characterizing quantum-dot-doped liquid scintillator for applications to neutrino detectors. *Journal of Instrumentation* 7, no. 7: P07010.

Withers, Nathan J., Krishnaprasad Sankar, Brian A. Akins, Tosifa A. Memon, Tingyi Gu, Jiangjiang Gu et al. 2008. Rapid degradation of CdSe/ZnS colloidal quantum dots exposed to gamma irradiation. *Journal of Applied Physics Letters* 93: 173101. doi:10.1063/1.2978073.

Xu, Shasha, Huaibin Shen, Changhua Zhou, Hang Yuan, Changson Wang, Hingzhe Liu et al. 2011. Effect of shell thickness on the optical properties in CdSe/CdS/Zn$_{0.5}$Cd$_{0.5}$S/ZnS and CdSe/CdS/Zn$_x$Cd$_{1-x}$S/ZnS core/multishell nanocrystals. *The Journal of Physical Chemistry C* 115: 20876–20881. doi:10.1021/jp204831y.

20 Čerenkov for portal imaging dosimetry in radiation therapy

Geordi G. Pang

Contents

20.1 INTRODUCTION

Shortly after the discovery of X-rays by Wilhelm Röntgen in 1895 (Nitske 1971), it was found that X-rays could be used not only for imaging but also for treating disease. One of the earliest clinical examples was to use X-rays to treat skin malignancies (Freund 1904). Although the mechanism involved in using X-rays to treat malignancies is very complex (Hall 1994), clinical evidence has shown that the use of X-rays, or more generally, radiation to treat cancer (now known as radiation therapy) is effective for a large number of cancer patients. Radiation therapy has now become one of the main treatment modalities for cancer patients (Holleb et al. 1991).

There are two main steps involved in modern radiotherapy treatment: treatment planning and treatment delivery. The purpose of treatment planning for a given patient is to determine where to treat and how to treat (Bentel 1996). A computed tomography machine is typically used to scan the patient to determine the treatment target(s). In some cases, other imaging modalities such as magnetic resonance imaging and positron emission tomography (PET) are also used. Once the location and extent of the treatment target are determined, a treatment planning computer system is used to design and optimize the arrangement of radiation beams to be delivered to the target. The planning software also calculates the planned radiation dose distribution in the patient. A reference image or image set, which will be used during treatment delivery for treatment verification, is generated during planning.

Once treatment planning is completed, the patient is taken to a treatment room and the treatment delivery begins. The treatment typically lasts for a few weeks with one fraction per day, depending on the oncologist's prescription (Hall 1994). Portal imaging and portal dosimetry, which will be discussed in Sections 20.1.1 and 20.1.2, are used in the treatment room to verify the geometric and dosimetric accuracies of the treatment, respectively.

20.1.1 PORTAL IMAGING

Megavoltage (MV) X-rays generated by linear accelerators (LINACs; see Figure 20.1) are currently most often used for cancer treatments (Van Dyk 1999). In a LINAC, electrons emitted by an electron gun are accelerated (using microwave technology) to millions of volts (MeV) before they collide with a metal target. When the collision happens, MV X-rays are produced at the metal target (i.e., the location of the X-ray source). These X-rays are then shaped using a multileaf collimator to conform to the shape of the patient's tumor as they exit the machine (here MV X-rays are often used in order to treat deep-seated tumors inside the patient). The patient usually lies on a movable treatment couch, and the gantry of the LINAC can be rotated around the patient to deliver MV X-ray beams from different directions to maximize the dose to the tumor and minimize the dose to the surrounding healthy tissues.

To verify that the treatment beam is indeed delivered to the tumor or treatment target (i.e., the breast in the example shown in Figure 20.1) as planned, various imaging systems are used in the treatment room. One of the most often used is called portal imaging. Portal imaging is accomplished by taking X-ray images of the radiation port (i.e., the shape of the treatment beam) with respect to the patient anatomy using an electronic portal imaging device (EPID) attached to the gantry of the LINAC (see Figure 20.1).

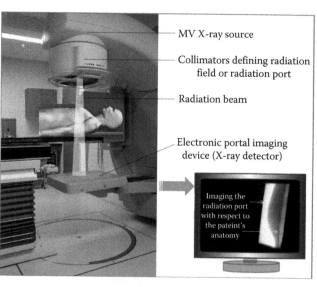

Figure 20.1 Picture of a LINAC and illustration of portal imaging.

Here the MV source is not only used for treatment but also for imaging, and the EPID is used as an X-ray detector (Herman et al. 2001). When MV X-rays are incident on the EPID, some of the incident X-rays interact with the EPID, depositing energy in the detector, which in turn generate image signals. The amount of X-rays per unit area that are attenuated by the detector is given by

$$\Delta I = I_0(1 - e^{-\mu d}) \tag{20.1}$$

where:
  I_0 is the incident X-ray fluence on the detector
  μ is the X-ray attenuation coefficient of the detector
  d is the detector thickness

In the case when the product μd is constant across the detector, and the image signal from the detector for a given pixel is *linearly* proportional to ΔI; the detector records and displays an incident X-ray fluence map or the variation of I_0 across the detector surface. The variation of I_0 is in turn determined by the radiation port and the object (i.e. the patient) in the radiation beam, resulting in an X-ray image (or a gray value map) of the radiation port with respect to the patient anatomy on the computer screen (see Figure 20.1). In reality, both μ and the relationship between the image signal and ΔI depend strongly on the incident X-ray spectrum, which may vary across the detector plane. Thus, what a detector really records is only an *approximate* incident X-ray fluence map at the detector surface.

By turning on the MV X-ray beam for a very short period of time (<1s) just prior to the treatment, a portal image is obtained using an EPID (Herman et al. 2001). This portal image is then compared to the reference image generated during treatment planning to determine if the treatment setup (i.e., the spatial relation between the radiation beam and the patient anatomy) matches exactly what has been planned. If not, the patient's position will be adjusted using the movable couch to match the plan before delivering the treatment. The geometric accuracy of the treatment is thus verified.

20.1.2 PORTAL DOSIMETRY

Other than geometric verification, there is also a need for dosimetric verification, that is, verifying the dose delivered to the patient in the treatment room. Here, the dose is defined as the mean energy absorbed per unit mass of irradiated material. Dosimetric verification can be done by detecting X-rays exiting the patient because what is coming out of the patient is related to what is absorbed (or the dose) inside the patient if the total energy of X-rays incident on the patient is known. When the exiting MV X-rays interact with an EPID, some of the X-ray energy is absorbed in the detector. The absorbed energy is then converted into the image signal S per unit area (or per pixel), that is,

$$S = \eta D_d M \tag{20.2}$$

where:
  η is the energy conversion efficiency
  M is the effective mass of the detector per unit area
  D_d is the absorbed dose to the detector as a result of interaction between X-rays and the detector

In the case when the conversion efficiency η is 1 or a nonzero constant, the image signal S is also a measure of the absorbed dose D_d (apart from a constant ηM). Thus, an EPID can also be used as a dosimeter, in additional to an imaging tool. Similar to other dosimeters, EPIDs are usually not ideal dosimeters. For example, the energy conversion efficiency in EPIDs may not be a nonzero constant (e.g., some X-rays attenuated by an EPID may not generate any image signal at all, and the conversion efficiency may vary with the incident X-ray spectrum). Furthermore, the dose to detector (i.e., D_d) is usually not equal to the dose to water (or dose to tissue), where water is usually used as the standard dose calibration medium for radiation therapy equipment. Thus, EPIDs as dosimeters should be used with caution. However, EPIDs do have some advantages over other conventional dosimeters, including (1) it is readily available in the treatment room and convenient to use, (2) it is a 2D large area (40 cm × 40 cm) dosimeter with a high spatial resolution (with a detector pixel size of typically <1 mm × 1mm), and (3) it is digital and online with a high image acquisition speed (approximately seven frames per second or higher).

There is a large body of literature devoted to the use of EPIDs for dosimetric applications in radiation therapy, and significant advances (including clinical implementation) have been made, especially in Europe. For an overview of this topic, see van Elmpt et al. (2008). The central task here is to verify that the dose is delivered to the patient as planned using portal images taken during the treatment (i.e., with the EPID left on for image acquisition during the entire treatment fraction). There are two major approaches to this end. One is to model the detector response (i.e., pixel values) with the treatment geometry in the treatment planning system and generate a reference gray-scale EPID image at the detector plane during planning. This reference image is then compared with the measured gray-scale EPID image acquired during treatment to see if they match (i.e., if the greyascale pixel values of the acquired EPID image match those of the reference). If not, there is potentially an error in the treatment delivery, provided the modeling is correct. (It is also possible to reconstruct the actual dose distribution delivered inside the patient based on acquired EPID images by separating the scatter signal from the primary and compare the reconstructed dose distribution with the planned.) This approach requires accurate technical details of the EPID in order for an accurate modeling, which may not be available. The second approach is to convert the measured gray-scale EPID image into a dose-to-water (or dose-to-tissue) map (either at the detector plane or at a plane inside the patient) using a pixel value-to-dose calibration curve, and compare the measured dose with the planned from the treatment planning system. The calibration curve is usually obtained from measurements (with the help of other conventional dosimeters) under certain calibration conditions or calibration geometry. This approach requires that the pixel value-to-dose calibration curve is also valid for the treatment geometry, which may not be the same as the calibration geometry, and thus, the calibration curve may not be valid for the treatment geometry.

20.1.3 ISSUES WITH CURRENT PORTAL IMAGING TECHNOLOGY

Over the years, three major detector technologies have been developed for portal imaging application: film-screen cassettes, conventional EPIDs, and recently, flat panel-based EPIDs (Antonuk 2002). The flat panel-based EPIDs are the most advanced, and they typically consist of a metal buildup plate (usually made of copper), an energy conversion layer (usually made of phosphor that converts absorbed X-ray energy into optical signals), and then an active readout matrix for image readout (see Figure 20.2). The use of copper plate here is mainly (1) to prevent electrons generated in the patient from reaching the phosphor screen and thus blurring the image and (2) to improve X-ray absorption of the detector. In addition, some low-energy X-rays, including low-energy scattered X-rays, are also filtered out by the copper plate.

However, all these EPID systems developed so far have one common limitation, poor X-ray absorption, that is, low quantum efficiency (QE), typically on the order of 2%–4% for MV X-rays compared to the theoretical limit of 100%. This is because the total thickness of the energy conversion layer plus the metal buildup for all these detectors is only ~1 mm of lead equivalent. By contrast, the first half-value layer for 6 MV X-rays

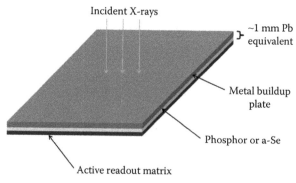

Figure 20.2 Schematic diagram showing the main components of a (low-QE) flat panel-based EPID.

is ~13 mm of lead. Thus, these detectors are too thin for MV X-ray imaging, resulting in a low QE, which is not desirable when used as an imaging tool (Pang and Rowlands 2004). In addition, the energy conversion layers in all these systems contain high atomic number (high-Z) materials, resulting in an over response of the EPID to low-energy X-rays (including low-energy scattered X-rays) compared to water; thus, they are not water equivalent when used as dosimeters (see Figure 20.3). Furthermore, in current flat panel-based EPIDs, there are significant ghosting and image lag (Siewerdsen and Jaffray 1999; Pang et al. 2001) due to charge trapping and the release of trapped charge in the photodiode layer of the active readout matrix. Image lag is a residual signal present in image frames subsequent to the frame in which the residual signal was generated. Ghosting, however, refers to the change of detector pixel sensitivity (or response) due to previous exposures to the detector. Neither image lag nor ghosting is desirable for imaging and dosimetry applications (McDermott et al. 2006).

Efforts have been made to develop high QE area detectors for portal imaging application, which include (1) the use of a scintillator to replace the phosphor screen in a mirror video-based EPID (Bissonnette and Munro 1996; Mosleh-Shirazi et al. 1998; Samant and Gopal 2006), (2) the use of metal converters and gas electron multipliers to detect MV X-rays (Ostling et al. 2000), (3) the use of a scintillating crystal matrix to replace the phosphor screen in a flat panel-based EPID (Sawant et al. 2005; Monajemi et al. 2006; Wang et al. 2009), and (4) the use of a large number of microstructured plates (with tungsten converters and microsize cavities filled with Xe gas) to achieve a high QE (Pang and Rowlands 2004). Recently, there have been also efforts to develop a high QE and water-equivalent area detectors for both imaging and dosimetry applications, which are based on plastic scintillating fibers to detect X-rays (Teymurazyan and Pang 2012b; Blake et al. 2013). Here, in this chapter, we will discuss a new approach that uses Čerenkov radiation for portal imaging and dosimetry applications in radiation therapy.

20.1.4 POTENTIAL APPLICATION OF ČERENKOV RADIATION IN PORTAL IMAGING DOSIMETRY

Čerenkov radiation (or Čerenkov light) was discovered in 1934 (Jelley 1958). It is an electromagnetic (EM) shock wave of light induced by a charged particle (e.g., an electron) passing through a dielectric medium with a speed greater than the speed of light in the medium. Based on the special theory of relativity, no material object travels faster than light *in vacuum*. However, in a medium such as water, electrons with

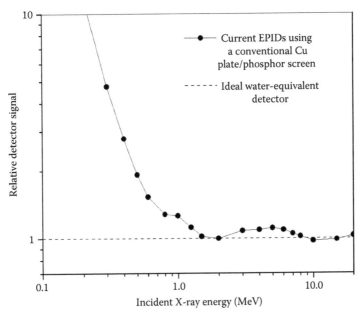

Figure 20.3 Relative detector response of current EPIDs using a conventional Cu plate/phosphor screen (1.5 mm-thick copper plate plus 134 mg/cm² Gd₂O₂S phosphor loading) for a given dose to water in comparison with that of an ideal water-equivalent X-ray detector. (Data from Teymurazyan, A. and Pang, G., *Med. Phys.*, **39**, 1518–1529, 2012b.)

energy higher than 260 keV travel faster than light *in water*, resulting in the generation of Čerenkov radiation. Several unique properties of Čerenkov radiation separate it from other sources of light, namely, (1) Čerenkov light is generated only when the following threshold condition is satisfied:

$$n\beta > 1 \tag{20.3}$$

where:

n is the refractive index of the dielectric medium in which the charged particle travels
β is the ratio of the charged particle's speed to the speed of light in vacuum

(2) The spectrum of Čerenkov light is broad and continuous. It spans the ultraviolet to infrared spectral regions. The relative number of light photons emitted in a given optical wavelength interval $\lambda \rightarrow \lambda + d\lambda$ varies according to λ^{-2}; (3) Čerenkov light emitted has a strong angle dependence: it is only emitted on the surface of a cone at an angle θ_c with respect to the charged particle's trajectory, where θ_c satisfies the equation:

$$\cos \theta_c = \frac{1}{n\beta} \tag{20.4}$$

This is in contrast to other sources of light, such as fluorescence, whose light output is expected to be uniform in all directions (Prasad 2004); (4) Čerenkov light is created instantaneously with a very short time delay of about 10^{-11} s between the onset of the interaction (between the charged particle and the dielectric medium) and the creation of Čerenkov light. This time delay is needed because any type of radiation is not emitted from a point but from a finite region whose size is on the order of the wavelength of the radiation (Ginzburg and Tsytovich 1990); and (5) when the charged particle's energy is above the threshold energy, the total number of Čerenkov light photons generated is proportional to the product of $\sin^2\theta_c$ and the charged particle's path length in the medium (Jelley 1958).

Čerenkov radiation has found many applications as particle counters for the studies of nuclear and cosmic ray physics (Jelley 1958). However, in radiation therapy, Čerenkov radiation has usually been detrimental to, for example, scintillator dosimeters coupled with optical fibers (Arnfield et al. 1996; Clift et al. 2002) (Chapter 5) because it was a major source of noise in these systems.

To demonstrate that Čerenkov radiation could actually be useful in radiation therapy applications, Mei et al. (2006) first modified a video-based EPID system by replacing the conventional Cu plate/phosphor screen with a 1 cm-thick clear acrylic plate (see Figure 20.4). Note that there is no phosphor in the acrylic plate. When MV X-rays incident on the acrylic plate, some of the incident X-rays will interact with the acrylic plate and generate fast electrons. These electrons with energy greater than the threshold energy for Čerenkov radiation will produce Čerenkov light along their tracks in the acrylic plate. The light image is then reflected by the mirror into the camera for readout. Figure 20.4(b) shows a Čerenkov image of a head phantom taken with this modified EPID using a 6 MV beam and a dose of ~10 Gy (Mei et al. 2006). This experiment demonstrates that it is possible to use Čerenkov radiation for portal imaging and portal dosimetry applications. However, in this experiment, the dose used to generate an image is too high. The reason is that the modified EPID system is not optimized for this application. In Section 20.3, we will discuss how to design an optimized Čerenkov detector for portal imaging and dosimetry applications.

20.2 ČERENKOV RADIATION AS THE DOMINANT LIGHT SOURCE IN IRRADIATED OPTICAL FIBERS

Regular (i.e., undoped) optical fibers are used in a new design for Čerenkov-based portal imaging devices (CPIDs). These regular optical fibers differ from scintillation fibers in that there is no or minimal scintillation light that would come out of these fibers under irradiation. We therefore want to confirm that (1) Čerenkov radiation is indeed the dominant light source in these regular optical fibers under irradiation and (2) the Čerenkov light output from these regular optical fibers is sufficient for portal imaging and dosimetry applications. If these are proved to be true, then we can indeed use these regular fibers in the

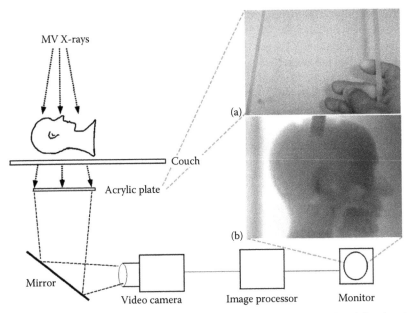

Figure 20.4 Schematic diagram of a modified video-based EPID where the conventional Cu plate/phosphor screen is replaced by (a) a clear acrylic plate of ~1 cm thick to demonstrate that images can indeed be obtained using Čerenkov radiation. The inset (b) shows a Čerenkov image of a RANDO phantom taken with the modified system using a 6 MV beam. (From Mei, X. et al., *Med. Phys.*, **33**, 4258–4270, 2006.)

design of CPIDs. In addition, measurements of characteristics (such as angular dependence and energy spectrum) of radiation-induced light in optical fibers will help us optimize the design.

20.2.1 MAGNITUDE OF RADIATION-INDUCED LIGHT IN OPTICAL FIBERS

Studies have shown that irradiated optical fibers can produce other sources of light aside from Čerenkov radiation, namely, fluorescence (de Boer et al. 1993). However, one of the strongest distinguishing factors of Čerenkov radiation is its energy dependence as described by Jelley (1958). Čerenkov radiation is only generated when the kinetic energy of the charge particle is larger than the threshold energy ε_c (i.e., 191 keV for electrons in silica). Thus, no Čerenkov radiation will be generated in optical fiber cores (made of silica) when irradiated with a 100 kVp X-ray photon beam because the energies of electrons generated will be below the threshold energy ε_c. By comparing the magnitude of light output created in optical fibers under high-energy irradiation (MV energy) with that under 100 kVp X-ray irradiation, we can determine the relative magnitudes of different light sources (i.e., fluorescence vs. Čerenkov) in optical fibers.

Table 20.1 shows the measured light output from an optical fiber under different sources of irradiation obtained by Silva and Pang (2012). A single fused silica optical fiber (JTFSH, Polymicro Technologies, Pheonix, AZ) with a core diameter of 600 μm and coated with a polymer cladding layer to an outer diameter of 630 μm was used in the measurement. The optical fiber was embedded in solid water and irradiated with a LINAC (Primus, Siemens, Concord, CA) for high-energy irradiation and an

Table 20.1 **Measured light output from an optical fiber under irradiation with difference sources**

IRRADIATION SOURCE	NORMALIZED PMT LIGHT OUTPUT (mV·min/cGy)
100 kVp X-ray beam	1.80
6 MeV electron beam	76.57
15 MeV electron beam	78.91

Source: Silva, I. and Pang, G., *Radiat. Phys. Chem.*, **81**, 599–608, 2012. Reprinted with permission from Elsevier.

orthovoltage X-ray machine (D3000, Gulmay Medical Ltd, Chertsey, England) for low-energy irradiation. The magnitude of the light output from the fiber was measured using a photomultiplier tube (PMT, type R6094, Hamamatsu, Bridgewater, NJ) operated at the gain of 6×10^5. The measured PMT values (in current) were normalized by the mean dose rate along the length of the fiber in the beam, and the results are shown in Table 20.1. It can be seen from Table 20.1 that the normalized light output of the fiber for the low-energy kilovoltage irradiation is only 2.35% and 2.28% of that for 6 and 15 MeV electron irradiation, respectively. Thus, combined with the results given in Sections 20.2.2 through 20.2.4, this indicates that the Čerenkov radiation is the predominant light source in the optical fiber irradiated with MV beams.

Of course, MV X-ray beams instead of MV electron beams are used for portal imaging and portal dosimetry because an electron beam is usually completely stopped within a patient before it can reach an EPID to generate any image signal. Thus, Mei et al. (2006) have measured the Čerenkov light output from an optical fiber under 6 MV X-ray irradiation and found that the Čerenkov light signal is detectable with a PMT at doses as low as one LINAC pulse (~0.026 cGy).

20.2.2 ANGULAR DEPENDENCE

Another feature of Čerenkov radiation that separates it from other sources of light is its strong angular dependence. Čerenkov radiation is emitted only within a cone at a specific angle with respect to the charge particle's trajectory. When MV X-rays interact with an optical fiber, Compton scattering and pair production processes will produce energetic electrons. Because the energetic electrons created by MV X-rays tend to be forward directed, it is anticipated that Čerenkov light output from an irradiated optical fiber would strongly depend on the incident angle of the radiation beam with respect to the center axis of the fiber. A number of research groups have measured the angular dependence of light output from optical fibres with radiotherapy beams (Beddar et al. 1992; de Boer et al. 1993; Mei et al. 2006; Law et al. 2007; Silva and Pang 2012). Figure 20.5 shows the result measured by Mei et al. (2006), that is, the Čerenkov light output from the optical fiber as a function of the beam incident angle φ for a 6 MV beam, in comparison with that measured with 100 kVp X-rays (Silva and Pang 2012). As φ increases from $0°$, the normalized light output in the MV case increases from unity, reaching the maximum intensity at $\varphi_{max} \sim 40°$, and subsequently decreases monotonically. By contrast, in the kilovoltage (kV) case the light-output curve shows no clear peak and is relatively constant over all angles suggesting no angular

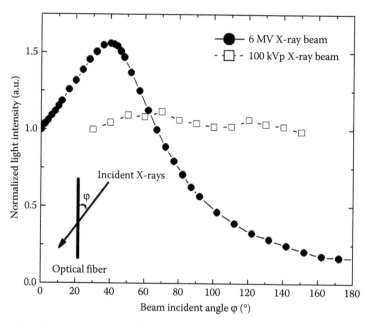

Figure 20.5 Measured angular dependence of Čerenkov light output from an optical fiber irradiated with a 6 MV beam (Data from Mei, X. et al., *Med. Phys.*, **33**, 4258–4270, 2006.) in comparison to that with a 100 kVp X-ray beam. (Data from Silva, I. and Pang, G., *Radiat. Phys. Chem.*, **81**, 599–608, 2012.)

dependence of light output in the kV case. This, combined with the result of the spectrum measurement given below, indicates that the radiation induced light for kV X-rays is not Čerenkov radiation.

The angular dependence measurement indicates that the light output in the fiber irradiated with MV X-rays reaches its maximum at about 40°. Thus, to maximize the light output in the optical fibers, the fibers should be placed with an angle of approximately 40° compared to the incident radiation beam. However, in CPIDs (see below) the fibers must be aligned with the incident X-rays to improve the spatial resolution. Fortunately, the light output in the irradiated fiber at 0° is only 35% less than the maximum light output for MV X-rays. Furthermore, Figure 20.5 shows that the light output decreases dramatically when the incident angle of the radiation beam is larger than 60°. Thus, scattered radiation through large angles will have less effect on the signal of a CPID using optical fibers. This, combined with the facts that (1) scattered X-rays usually have lower energies than the primary and (2) there is a threshold energy required for Čerenkov radiation, indicates that a CPID can be used as an antiscatter detector (Teymurazyan and Pang 2013).

20.2.3 SPECTRUM

The spectrum of Čerenkov radiation is also unique. A number of research groups have measured radiation-induced light output spectra using high-energy electron beams for some optical fibers (de Boer et al. 1993; Lambert et al. 2009). However, there is some confusion in the literature regarding whether the Čerenkov light intensity spectrum (i.e., the number of light photons as a function of wavelength) or the Čerenkov energy intensity spectrum (i.e., the total energy of emitted light photons as a function of wavelength) from an irradiated optical fiber follows the λ^{-3} law, where λ is the optical wavelength. We note here that it should be the energy intensity spectrum that is proportional to λ^{-3} (c.f. Chapter 5), where the energy intensity for a given wavelength is expressed by the product of the number of light photons for the wavelength and the energy of each photon (Silva and Pang 2012). Figure 20.6 shows the energy spectrum of light output from an irradiated optical fiber (JTFSH, Polymicro Technologies) measured by Silva and Pang (2012)

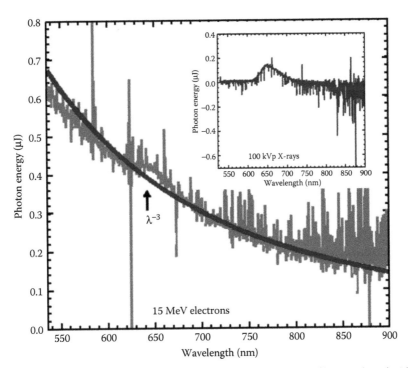

Figure 20.6 Measured energy spectrum of radiation-induced light in an optical fiber irradiated with 15 MeV electrons in comparison with the least-squares fitted, theoretical λ^{-3} curve (dark solid) for Čerenkov radiation. The inset shows the energy spectrum of radiation-induced light for the same optical fiber but irradiated with 100 kVp X-rays. (Reprinted from *Radiat. Phys. Chem.*, **81**, Silva, I. and Pang, G., Characteristics of radiation induced light in optical fibres for portal imaging application, 599–608, copyright 2012, with permission from Elsevier.)

using 15 MeV electrons, where the energy spectrum in photon energy (μJ) per wavelength is compared with the theoretical λ^{-3} curve. The theoretical curve was found by adjusting the coefficient κ in κ/λ^3 to fit the experimental data with the least-squares error. Although a MeV electron beam was used in the measurement, the result is not expected to be different if a MV X-ray beam was used. It can be seen from Figure 20.6 that the light output energy spectrum shows a definite wavelength dependence and an excellent fit with the theoretical λ^{-3} curve except for a small deviation for the peak at around 650 nm. The inset in Figure 20.6 shows the light output energy spectrum from the same fiber irradiated with 100 kVp X-rays, which has a completely different shape with a clear peak at approximately 650 nm due to light sources other than Čerenkov light. The 650 nm peak is fiber specific and a similar peak was observed by de Boer et al. (1993) for a different fiber.

The spectrum measurement shows that the emission of Čerenkov radiation is significantly stronger in the blue than in the red. This information will be useful in determining the type of active matrix flat panel imager (AMFPI) needed for CPIDs (see below). Because the detective QE of amorphous selenium (a-Se) is much higher for the blue light than for the red light, an a-Se-based AMFPI (Rowlands and Yorkston 2000) is a natural choice for CPIDs.

20.2.4 TEMPORAL PROPERTY

Because Čerenkov radiation is created instantaneously upon the interaction between the charged particle and the dielectric medium, it is expected that there should be virtually no delay between the onset of the interaction (between an optical fiber and a MV radiation beam) and the creation of Čerenkov light output signal from the optical fiber. To confirm this, Silva and Pang (2012) measured the temporal property of the radiation-induced light in an optical fiber using a LINAC. Radiation beams generated by a LINAC are not continuous but in pulses (each with typically a few microsecond duration) with a pulse frequency of typically ~160 Hz. The timing and shape of the X-ray pulses can be precisely monitored through the Beam-I connector in the virtual machine interface box of a LINAC. Figure 20.7 shows some sample Beam-I pulses measured by Mei and Pang (2005). The signal from Beam-I represents the current of electrons incident on the X-ray target to generate X-rays, and thus, the Beam-I pulse is proportional to the X-ray beam pulse. In the experiment conducted by Silva and Pang (2012), a small portion of an

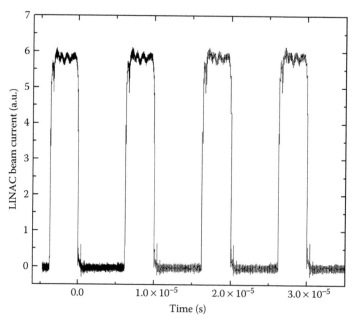

Figure 20.7 Sample Beam-I pulses from a 6 MV LINAC. (Reprinted from Mei, X. and Pang, G., *Med. Phys.*, **32**, 3379–3388, 2005, with permission from American Association of Physicists in Medicine.)

optical fiber was placed at the isocenter in line with an 18 MV X-ray beam from a LINAC, and the remaining portion of the optical fiber was extended outside the treatment room into a PMT for which the output as a function of time was measured on an oscilloscope. (The PMT was used to measure the light output of the irradiated optical fiber.) The Beam-I pulse was also measured on a different channel of the oscilloscope, so that the Beam-I pulse duration and the PMT output duration could be compared in time. The Beam-I signal was used as the reference time frame, that is, each measurement was triggered to the Beam-I output channel. (Any time difference between the PMT output and the Beam-I output would appear on the oscilloscope in the form of PMT output shifts relative to the Beam-I output.) Figure 20.8 shows the timing of the measured Beam-I and PMT outputs after corrections for the propagation delays in the fiber and PMT cables used (Silva and Pang 2012). It can be seen from Figure 20.8 that the Beam-I signal (or the X-ray pulse signal) and the PMT output signal (or the Čerenkov light output signal) are almost simultaneous in time. This information would be useful if we use a CPID to do time-of-flight measurements or to develop a new time-of-flight PET detector for other applications (Moses et al. 2010). A Čerenkov radiation-based PET detector would have a much better temporal resolution with an equal or better QE and spatial resolution, compared to current scintillators used for PET, for example, LSO ($Lu_2(SiO_4)O:Ce$). However, the light yield of the Čerenkov detector would be much less (about 2–3 orders of magnitude less) than that of LSO (Silva and Pang 2012).

To further demonstrate the feasibility of using optical fibers for portal imaging applications, Silva and Pang (2012) constructed a prototype array detector (Figure 20.9a). It consists of 100 optical fibers with a 30 cm end segment of the fibers enclosed as a single array in a solid buildup material, which should be placed in a radiation beam inside a LINAC room during image acquisition. The remaining ~10 m segment of optical fibers is mapped to a double-array steel ferrule connected to a high-field avalanche rushing photoconductor (HARPICON) camera and placed outside the LINAC room for image readout. The HARPICON camera uses a-Se as the camera target and operates in an avalanche mode to overcome camera noise (Tanioka et al. 1987). Figure 20.9b shows an open-field image taken with the prototype Čerenkov detector array at 6 MV using one LINAC pulse (about 0.026 cGy at the detector surface).

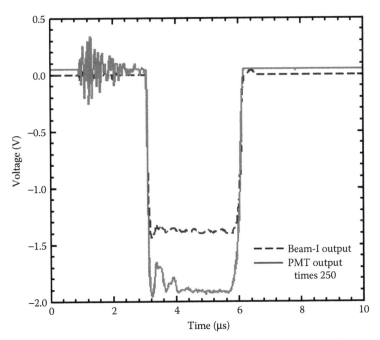

Figure 20.8 Measured Beam-I and PMT outputs as a function of time after corrections for the propagation delays in the optical fiber and PMT cables used. (Reprinted from *Radiat. Phys. Chem.*, **81**, Silva, I. and Pang, G., Characteristics of radiation induced light in optical fibres for portal imaging application, 599–608, copyright 2012, with permission from Elsevier.)

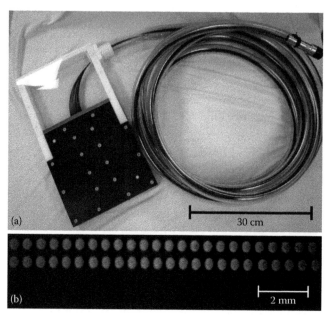

Figure 20.9 (a) Picture of a prototype Čerenkov detector array. (b) Open-field image taken with the Čerenkov detector array at 6 MV with a dose of one LINAC pulse (about 0.026 cGy at the detector surface). (Reprinted from *Radiat. Phys. Chem.*, **81**, Silva, I. and Pang, G., Characteristics of radiation induced light in optical fibres for portal imaging application, 599–608, copyright 2012, with permission from Elsevier.)

These white regions represent the light output from the optical fibers irradiated with MV X-rays. This image demonstrated the feasibility of using optical fibers for MV X-ray imaging.

20.3 DESIGN OF ČERENKOV DETECTORS FOR PORTAL IMAGING AND DOSIMETRY APPLICATIONS

In Section 20.2, we have shown that it is feasible to use optical fibers for MV X-ray imaging based on Čerenkov radiation. In this section, we will discuss the actual design of CPIDs for portal imaging and dosimetry applications.

20.3.1 DESIGN PRINCIPLES

Similar to current EPIDs, the new CPID should have a large effective imaging area (e.g., 40 cm × 40 cm) so that it can be used for most disease sites. The new detector should be digital with a fast readout rate (approximately seven frames per second or higher) so that it can be used for image-guided radiotherapy and real-time monitoring of target motion. It should also have a spatial resolution comparable to that of current EPIDs. Furthermore, it should overcome the main limitations of current EPIDs: (1) the new CPID should have a much improved QE compared to current EPIDs. It should be able to image with the lowest dose possible on a LINAC machine, that is, one LINAC pulse (~0.026 cGy). By contrast, current clinical EPIDs require a minimum of ~1 cGy to get a good quality portal image; (2) the new CPID should be made of low-Z materials so that the problem of overresponse to low-energy X-rays in current EPIDs will be eliminated; and (3) the new CPID should have much less ghosting and image lag than current EPIDs.

20.3.2 OPTICAL FIBER-BASED DETECTOR

The first design is based on regular optical fibers and its concept was introduced by Mei et al. (2006). Figure 20.10 shows a schematic illustration of the proposed system. It consists of a large area (~40 cm × 40 cm) and thick (thickness d ~10–30 cm) fiber optical taper (FOT) directly coupled to an optically sensitive 2D AMFPI. The active matrix is made optically sensitive with either an a-Si p-i-n photodiode at every pixel or a continuous layer of a-Se (Rowlands and Yorkston 2000). The FOT is a matrix of optical fibers, each of

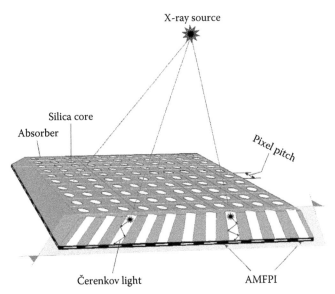

X-ray source

Silica core

Absorber

Pixel pitch

Čerenkov light

AMFPI

Figure 20.10 Schematic diagram showing the major components of a FOT-based CPID. (Adapted from Teymurazyan, A. and Pang, G., *Phys. Med. Biol.*, **58**, 1479–1493, 2013.)

which has a diameter of several hundred micrometers and is aligned with the incident X-rays (i.e., focused toward the X-ray source) to avoid blur due to oblique incidence of off-axis X-rays. This focusing toward the X-ray source can be achieved by an extra coating at the bottom of the otherwise uniform fibers. Gaps between optical fibers are filled with light absorber materials (generally referred to as extramural absorption [EMA]) to absorb light that is not guided by the fibers but instead escapes from the sides of the fibers. This new design will be referred to as a FOT-based CPID.

When MV X-rays interact with the FOT, energetic electrons will be produced primarily via the Compton effect and pair production. These electrons with energy greater than the threshold energy ε_C required for Čerenkov radiation in the fiber material will produce Čerenkov light along their tracks. The light photons produced in a fiber core and emitted within the acceptance angle of the fiber will be guided toward the optical detector, that is, the *optically sensitive* AMFPI by total internal reflection. In the case when a-Si p-i-n photodiodes are used in the optical detector, these optical photons exiting the fibers are converted to electron–hole pairs in the photodiodes integrated into the active matrix array. Once collected on the capacitance of the photodiodes, the resulting charge image can be temporarily stored and later read out using the active components on the active matrix, switching them using thin-film transistors (TFTs). By contrast, when a continuous layer of a-Se is used in the optical detector, these optical photons exiting the fibers are converted to electron–hole pairs in the energy sensitive a-Se layer. The holes are collected on pixel electrodes, stored on the pixel capacitors, and then read out by the TFT matrix (Pang et al. 2001). The a-Se layer (or a-Si p-i-n photodiodes) in the optical detector can in principle be operated in an avalanche mode (Tanioka et al. 1987) if necessary to amplify the signal in order to overcome the electronic noise in the active matrix readout.

We note here that in the new design (1) there is neither phosphor nor metal buildup layer. The conventional Cu plate/phosphor screen in a flat panel-based EPID is now replaced by the FOT that is made of low-Z materials and (2) the thickness of the detector d (or the length of optical fibers in FOT) determines the QE, which can be made as large as possible, provided that the spatial resolution is still acceptable. In general, a thicker detector will have a lower spatial resolution due to an increase of X-ray scatter in the detector (Pang and Rowlands 2004). However, for CPIDs, X-ray scatter is less of a concern because CPIDs are less sensitive to scattered radiation (see below for more details); and (3) due to a significant increase of QE, there will be many more incident X-rays attenuated by the FOT (compared to current EPIDs) before the incident X-ray can reach and interact with the AMFPI. As a result, less charge will be generated and trapped in the AMFPI, resulting in much less ghosting and image lag in the new design compared to current (low-QE) flat panel-based EPIDs.

20.3.3 GLASS ROD-BASED DETECTOR

Although the above FOT-based design has much better imaging properties than current EPIDs (see below), it has been found that the new design requires a technology (i.e., an AMFPI operating in an *avalanche* mode) that is not yet available. This is due to (1) the relatively weak light yield of Čerenkov radiation (~2 orders of magnitude less) compared to that of scintillation and (2) the relatively low numerical aperture (NA) or small acceptance angles of typical optical fibers. As a result, too few Čerenkov light photons reach the AMFPI for every incident X-ray, resulting in a weak signal compared to readout electronic noise in the AMFPI. Thus, an AMFPI with an avalanche gain is required for a FOT-based CPID in order to amplify the signal to overcome the readout noise for portal imaging and dosimetry applications (Mei et al. 2006).

To solve this problem, Teymurazyan et al. (2014) introduced another design based on air-spaced light guiding taper (LGT) to maximize the acceptance angle θ or the NA of optical fibers. The acceptance angle of an optical fiber is given by

$$\theta = \frac{\pi}{2} - \arcsin\left(\frac{n_{clad}}{n_{core}}\right) \tag{20.5}$$

where:

n_{clad} and n_{core} are the refractive indices of the cladding layer and fiber core, respectively, and their values are usually larger than 1

By replacing the optical fibers in the CPID with light guides without a cladding layer (i.e., setting $n_{clad} = 1$), we can minimize the second term in Equation 20.5 and thus maximize angle θ. As a result, the light guides have a much higher NA than typical optical fibers, and significantly more Čerenkov light photons are transmitted through to reach the AMFPI. Therefore, an AMFPI with an avalanche gain is no longer needed.

Figure 20.11 shows a schematic cross section of the proposed detector based on light guides (i.e., glass rods). Just as in the previous design, the light guides are focused toward the X-ray source to avoid blur due to oblique incidence of off-axis X-rays. Gaps between light guides are filled with air. This can be achieved by extra coatings at the top as well as at the bottom of the otherwise uniform light guides. This is also a large area (40 cm × 40 cm) high QE detector but would be much less expensive than other designs due to the use of glass rods that are cheap to make. This new detector will be referred to as an air-spaced LGT-based CPID (or simply glass rod-based CPID).

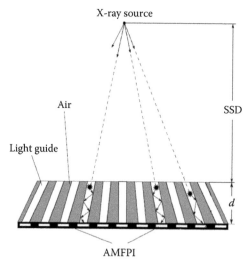

Figure 20.11 Schematic diagram showing the cross section of an LGT-based CPID. (Adapted from Teymurazyan, A. et al., *Med. Phys.*, **41**, 041907, 2014).

20.4 SIMULATION OF ČERENKOV DETECTORS FOR PORTAL IMAGING AND DOSIMETRY APPLICATIONS

20.4.1 METHOD OF SIMULATION

Teymurazyan and Pang (2012 and 2013) have developed a simulation program to study the imaging and dosimetric characteristics of the proposed CPIDs, based on the Geant4 toolkit (Agostinelli et al. 2003). The Geant4 toolkit contains physics models covering the interactions of most particle types over a wide energy range (down to 1 keV and below) for the simulation of the passage of particles through matter. Unlike most general-purpose Monte Carlo codes describing propagation of radiation in matter, it also includes models describing the transport and boundary processes for optical photons. In the Geant4 framework, optical photons can be produced by scintillation, Čerenkov effect, or wavelength shifting. For the detectors considered here, the optical photons are generated by Čerenkov effect only. The list of particles considered in simulation consists of photons, electrons, positrons, and optical photons. The list of physics processes controlling the particle interactions in the simulation, that is, photoelectric effect, Compton scattering, e^+e^- pair production, Rayleigh scattering, multiple scattering, ionization, bremsstrahlung, and annihilation, is largely based on the standard EM physics list supplied with Geant4 distribution. The EM list was extended to include the generation of optical photons via Čerenkov effect. The simulation accounts for optical photon interactions such as bulk absorption, optical Rayleigh scattering, and optical boundary processes, that is, reflection and refraction (Agostinelli et al. 2003).

During portal imaging and portal dosimetry, a patient is positioned between the X-ray source and the EPID. The presence of the patient both (1) alters the spectrum of primary X-rays exiting the patient (exit primary) and (2) results in generation of secondary particles (X-rays, electrons, and positrons) from the patient (exit scatter). To study the imaging and dosimetric characteristics of CPIDs, various X-ray energy spectra (see Figure 20.12) generated by a treatment planning system (Pinnacle3, Philips, Fitchburg, WI) for clinical machines were used in the simulation. The entrance spectra shown in Figure 20.12 are the X-ray energy spectra of a LINAC machine that would be incident on the EPID in the absence of a patient. The exit primary plus scatter spectra are the spectra that would be incident on the detector in the presence of a "patient" (i.e., 30 cm-thick water phantom in this case).

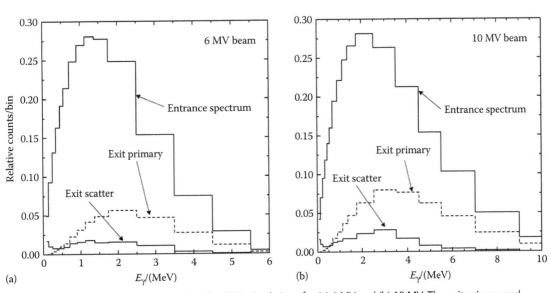

Figure 20.12 Various X-ray spectra used in the CPID simulations for (a) 6 MV and (b) 10 MV. The exit primary and scatter spectra were simulated by passing the incident beam (with the entrance spectrum) through a 30 cm-thick water phantom at the isocenter. (Data from Teymurazyan, A. and Pang, G., *Int. J. Opt.*, **2012**, Article ID 724024, 2012a.)

20.4.2 SIMULATED DETECTOR QUANTITIES

20.4.2.1 Quantum efficiency

When MV X-rays are incident on a CPID, not all of them interact with the detector and generate a signal in the detector. Based on Equation 20.1, only ΔI amount of X-rays will interact with the detector, and yet some of the interacting X-rays may not generate any image signal. The QE (sometimes also called detection efficiency) of a CPID is then defined as the probability of an incident X-ray to produce at least one optical (Čerenkov) photon that reaches the AMFPI, irrespective of the position of interaction of the primary X-ray in the detector, that is,

$$QE = \frac{\gamma \Delta I}{I_0} = \gamma(1 - e^{-\mu d}) \tag{20.6}$$

where:

γ is the percentage of interacting X-rays that generate image signals

In the case when the incident X-ray energy is lower than the threshold energy ε_c for Čerenkov radiation, γ is equal to zero. Therefore, both γ and μ (and thus QE) are dependent on the incident X-ray energy.

20.4.2.2 Spatial resolution

Spatial resolution is another important quantity of imaging detectors. It measures how closely two small objects (e.g., two lines) can be resolved in an image. The spatial resolution of the detector can be expressed by the modulation transfer function (MTF). The 1D MTF is the Fourier transform of the line spread function (LSF), that is,

$$MTF(f) = \int_{-\infty}^{\infty} e^{-2i\pi x f} LSF(x) \, dx \tag{20.7}$$

where LSF is defined as

$$LSF(x) = \int_{-\infty}^{\infty} S(x, y) \, dy \Big/ \int_{-\infty}^{\infty} \int_{-\infty}^{\infty} S(x, y) \, dy \, dx \tag{20.8}$$

Here, $S(x,y)$ is the image signal or the distribution of optical photons that reach the AMFPI under the irradiation of an infinitely small-size pencil beam (we assume that the AMFPI has 100% optical coupling efficiency and any optical photon incident on the AMFPI will generate an electronic signal in the AMFPI). To overcome the sampling limitation, as in any digital system (Dobbins III 1995), the location of the incident pencil beam within a detector pixel should be varied, and the LSF should be sampled over all uniformly distributed locations within a detector pixel (Teymurazyan and Pang 2012b, 2013).

20.4.2.3 Scatter fraction

The presence of a patient in the beam results in generation of secondary scattered particles that generally degrade image quality and make portal dosimetry calibration complicated. CPIDs have a unique feature of under-response to scattered X-rays (Teymurazyan and Pang 2013). Based on Equation 20.2, the image signal is proportional to the absorbed energy in the detector multiplied by the conversion efficiency η. This conversion efficiency differs between scattered radiation and primary radiation for CPIDs because the emission of Čerenkov radiation is angle dependent (see Section 20.2.2). Figure 20.5 has shown that scattered radiation through large angles will result in less detector signal than radiation with small incident angles (such as primary radiation).

A quantity to measure the response of a CPID to scattered radiation compared to the primary is the scatter fraction (Jaffray et al. 1994):

$$SF_{signal} = \frac{\overline{S}_s}{\overline{S}_s + \overline{S}_p} \tag{20.9}$$

where:

$\bar{S}_s$ and $\bar{S}_p$ are the mean signals per pixel due to the scattered and primary components of the exit beam from the patient, respectively

20.4.2.4 Energy dependence of dose response

Most patient dose calculation algorithms assume that the patient is similar to water, and the output of a LINAC machine is usually calibrated using water as the calibration medium. Thus, it is desirable to know the precise relationship between the detector signal (or gray-scale EPID image) and the dose to water for portal dosimetry application, that is,

$$\alpha = \frac{S}{D_w} \tag{20.10}$$

where:

S is the detector signal (for a given pixel)

D_w is the dose to water at the detector or, more precisely, the mean dose absorbed in a water slab that replaces the detector and has a thickness equivalent to that of the detector (Teymurazyan and Pang 2012a)

From Equation 20.2, Equation 20.10 can be rewritten as

$$\alpha = \frac{\eta D_d M}{D_w} \tag{20.11}$$

For conventional EPIDs where Cu plate/phosphor screens are used, the ratio of D_d to D_w differs from 1 and is highly dependent on the incident X-ray energy (D_d is significantly higher than D_w at low X-ray energies for conventional EPIDs). For detectors made of low-Z or water-equivalent materials (such as CPIDs), D_w is essentially equal to D_d. However, for CPIDs, the quantity η is not a constant and dependent on the incident X-ray energy because no image signal will be generated when the energy of electrons generated in CPIDs by X-rays is below the threshold energy ε_c for Čerenkov radiation.

Values of detector parameters used in the simulations are listed in Table 20.2. The X-ray source to detector surface distance was kept at 157 cm.

20.4.3 RESULTS

20.4.3.1 Quantum efficiency

Table 20.3 shows the QEs of CPIDs for different X-ray spectra (the entrance spectra in Figure 20.12) and detector thicknesses. The QEs of CPIDs are more than an order of magnitude higher than those of conventional EPIDs (typically 2%–4%). The main reason is that the CPID detectors are effectively much thicker than conventional EPIDs, and thus, many more incident X-rays are absorbed by the CPIDs.

Table 20.2 **Values of detector parameters used in the simulations**

DETECTOR TYPE	FIBER/GLASS ROD MATERIAL	PIXEL PITCH (mm²)	EFFECTIVE IMAGING AREA (cm²)
FOT based	FSHA optical fibers (Polymicro Technologies) with NA = 0.48	0.435 × 0.435	40.45 × 40.45
Glass rod based	Fused silica glass rod with refractive index = 1.46	0.505 × 0.505	40.45 × 40.45

Sources: Teymurazyan and Pang 2012b; Teymurazyan, A. et al., *Med. Phys.*, **41**, 041907, 2014.

Table 20.3 **QEs of CPIDs calculated for various clinically relevant X-ray spectra and detector thicknesses**

DETECTOR TYPE	DETECTOR THICKNESS d (cm)	6 MV BEAM (%)	10 MV BEAM (%)
FOT based	10	39.4	40.7
	20	58.3	62.0
	30	67.1	73.2
Glass rod based	20	55.3	56.0

Sources: Teymurazyan, A. and Pang, G., *International Journal of Optics*, **2012**, Article ID 724024, 2012a; Teymurazyan, A. et al., *Med. Phys.*, **41**, 041907, 2014.

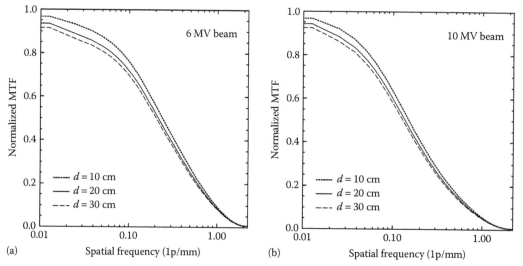

Figure 20.13 The MTF as a function of spatial frequency for various thicknesses of the FOT-based CPID at (a) 6 MV and (b) 10 MV. The entrance spectra of X-rays from Figure 20.12 were used in the simulation. (Data from Teymurazyan, A. and Pang, G., *Int. J. Opt.*, **2012**, Article ID 724024, 2012a.)

20.4.3.2 Spatial resolution

The spatial resolution of the FOT-based CPID expressed by MTF is shown in Figure 20.13. It can be seen that the MTF of the detector decreases slightly at all frequencies with the increase of the detector thickness, which is due to an increase of X-ray scatter in a thicker detector. However, in this case, the decrease of MTF with detector thickness is minimal compared to non-Čerenkov detector (Teymurazyan and Pang 2012b) because CPIDs are less sensitive to scattered radiation (see below).

Based on Figure 20.13, it can be seen that the values of f_{50} (i.e., the frequency at 50% of the MTF) of the CPID at 6 MV is ~0.21–0.24 lp/mm, which is comparable to that of conventional video-based EPIDs (f_{50} ~0.2 lp/mm) (Antonuk 2002), even though the energy conversion layer in the CPIDs is much thicker. This is because the resolution of an X-ray detector also decreases with the effective range of energetic electrons generated by incident X-rays in the detector. As an energetic electron travels in the CPID, it continuously loses its kinetic energy. Once its kinetic energy is below the threshold energy required for Čerenkov radiation, the electron can no longer generate Čerenkov light, and thus, any further spatial spread of low-energy electrons has no effect on the resolution of the CPID. As a result, the resolution of the FOT-based CPID is comparable to that of conventional EPIDs. The spatial resolution of a glass rod-based CPID is similar to that of a FOT-based CPID but can be significantly improved if we use glass rods with

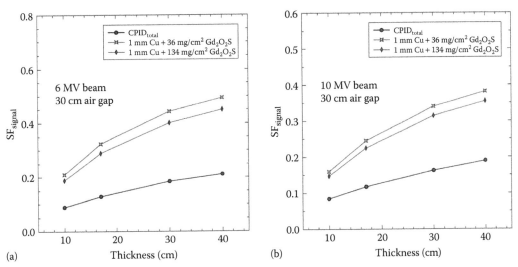

Figure 20.14 The scatter fractions as a function of phantom thickness for the FOT-based CPID (with 20 cm thick) in comparison with conventional EPIDs using Cu plate/phosphor screens at (a) 6 MV and (b) 10 MV. (Data from Teymurazyan, A. and Pang, G., *Phys. Med. Biol.*, **58**, 1479–1493, 2013.)

a higher density, resulting in even smaller effective range of energetic electrons (Teymurazyan and Pang 2014).

20.4.3.3 Scatter fraction

Figure 20.14 shows the calculated scatter fractions (or the percentage of contribution of scatter to the total signal) for the FOT-based CPID, compared to that of conventional EPIDs using Cu plate/phosphor screens, for various phantom thicknesses (with a 30 cm air gap between the phantom and the detector). The scatter fractions for both the CPID and conventional detectors exhibit similar qualitative behavior, that is, they increase with increasing phantom thickness and decrease with increasing air gap (not shown). However, the scatter fractions for the CPID are ~50% lower compared to those of conventional EPIDs for all phantom thicknesses. Results for a glass rod-based CPID should be similar.

20.4.3.4 Energy dependence of dose response

Figure 20.15 shows the energy dependence (for monoenergetic X-rays) of the dose response (for a given dose to water) of the FOT-based CPID (with detector thickness $d = 30$ cm) compared to that of a conventional EPID using a Cu plate/phosphor screen, which consists of a ~134 mg/cm^2 Gd$_2$O$_2$S phosphor loading (Kodak Lanex fast back) on a 1.5 mm-thick copper plate. Although the proposed CPID no longer overresponds to low-energy X-rays (as a conventional EPID does), it under-responds to low-energy X-rays, which could be a new challenge for dosimetry applications.

20.4.3.5 Sensitivity at extremely low doses

It has been shown (Teymurazyan and Pang 2014) that both types of CPIDs are sensitive to (i.e., can detect) dose levels as low as one single LINAC pulse (~0.026 cGy). This is due to the high QE that a CPID has. To overcome the electronic noises of AMFPI at low doses, the FOT-based CPID does need an AMFPI operating in an avalanche mode. However, the glass rod-based CPID does not require an avalanche gain in the AMFPI and is quantum noise limited at dose levels corresponding to a single LINAC pulse.

20.5 DISCUSSION AND CONCLUSION

In this chapter, we have discussed the feasibility of using Čerenkov radiation for portal imaging and portal dosimetry applications in radiation therapy. It has been shown that the CPIDs have several distinct features compared to current conventional EPIDs, including (1) CPIDs have an order of magnitude higher QE than current conventional EPIDs and yet with a comparable spatial resolution. As a result,

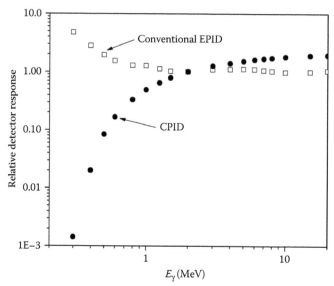

Figure 20.15 Energy dependence of the dose response (for a given dose to water) of a 30 cm-thick CPID (solid circles) and that of a conventional EPID using a Cu plate/phosphor screen as the energy conversion layer (open squares). (Data from Teymurazyan, A. and Pang, G., *Int. J. Opt.*, **2012**, Article ID 724024, 2012a.)

CPIDs are much more sensitive to low dose radiation than conventional EPIDs and can detect doses as low as one LINAC pulse; (2) CPIDs are antiscatter detectors, the first kind for MV X-ray imaging applications. This is the most important feature of CPIDs. As a result, the degradation of image quality due to patient scatter is expected to be significantly reduced. The calibration of CPIDs for dosimetry applications is also expected to be simpler because patient scatter has been a major issue for portal dosimetry; and (3) CPIDs no longer overrespond to low-energy X-rays as conventional EPIDs do. Instead, they under-respond to low-energy X-rays. As a result, CPIDs are not water equivalent, in contrast to plastic scintillating fiber-based EPIDs (Teymurazyan and Pang 2012b). In addition, it is anticipated that CPIDs would have much less ghosting or lag than current EPIDs. However, the proposed CPIDs could be bulky and heavier than current flat panel-based EPIDs, depending on the detector thickness chosen.

The development of Čerenkov detectors for portal imaging and portal dosimetry applications is still at its early stage. Large area prototype detectors have not yet been built at the time of writing. It is hard to predict if this development effort would lead to a commercial product. However, I am sure that the research and development on the use of Čerenkov radiation for radiation therapy applications will continue due to not only its potential applications but also the fascinating, rich physics involved. Some new development of using Čerenkov radiation for *in vivo* dosimetry is discussed in Chapter 21.

ACKNOWLEDGMENTS

This work was supported in part by the Individual Discovery Grant Program awarded by the Natural Sciences and Engineering Research Council of Canada.

REFERENCES

Agostinelli, S., Allison, J., Amako, K. et al. 2003. Geant4—A simulation toolkit. *Nucl. Instrum. Methods Phys. Res. A* **506**: 250–303.
Antonuk, L.E. 2002. Electronic portal imaging devices: A review and historical perspective of contemporary technologies and research. *Phys. Med. Biol.* **47**: R31–R65.
Arnfield, M.R., Gaballa, H.E., Zwicker, R.D., Islam, Q. and Schmidt-Ullrich, R. 1996. Radiation-induced light in optical fibers and plastic scintillators: Application to brachytherapy dosimetry. *IEEE Trans. Nucl. Sci.* **43**: 2077–2084.

Beddar, A.S., Mackie, T.R. and Attix, F.H. 1992. Cerenkov light generated in optical fibres and other light pipes irradiated by electron beams. *Phys. Med. Biol.* **37**: 925–935.

Bentel, G.C. 1996. *Radiation Therapy Planning*. New York: McGraw-Hill Professional.

Bissonnette, J.-P. and Munro, P. 1996. Evaluation of a high-density scintillating glass for portal imaging. *Med. Phys.* **23**: 401–406.

Blake, S.J., McNamara, A.L., Deshpande, S. et al. 2013. Characterization of a novel EPID designed for simultaneous imaging and dose verification in radiotherapy. *Med. Phys.* **40**: 091902.

Clift, M.A., Johnston, P.N. and Webb, D.V. 2002. A temporal method of avoiding the Cerenkov radiation generated in organic scintillator dosimeters by pulsed mega-voltage electron and photon beams. *Phys. Med. Biol.* **47**: 1421–1433.

de Boer, S.F., Beddar, A.S. and Rawlinson, J.A. 1993. Optical filtering in spectral measurements of radiation induced light in plastic scintillation dosimetry. *Phys. Med. Biol.* **38**: 945–958.

Dobbins III, J.T. 1995. Effects of undersampling on the proper interpretation of modulation transfer function, noise power spectra, and noise equivalent quanta of digital imaging systems. *Med. Phys.* **22**: 171–181.

Freund, L. 1904. *Elements of General Radiotherapy for Practitioners*. New York: Rebman.

Ginzburg, V.L. and Tsytovich, V.N. 1990. *Transition Radiation and Transition Scattering*. New York: Adam Hilger.

Hall, E.J. 1994. *Radiobiology for the Radiologist*. New York: Lippincott–Raven Publishers.

Herman, M.G., Balter, J.M., Jaffray, D.A. et al. 2001. Clinical use of electronic portal imaging: Report of AAPM Radiation Therapy Committee Task Group 58. *Med. Phys.* **28**: 712–737.

Holleb, A.I., Fink, D.J. and Murphy, G.P. 1991. *American Cancer Society Textbook of Clinical Oncology*. Atlanta, GA: The American Cancer Society.

Jaffray, D.A., Battista, J.J., Fenster, A. and Munro, P. 1994. X-ray scatter in megavoltage transmission radiography: Physical characteristics and influence on image quality. *Med. Phys.* **21**: 45–60.

Jelley, J.V. 1958. *Cerenkov Radiation and Its Applications*. London: Pergamon Press.

Lambert, J., Yin, Y., McKenzie, D.R., Law, S. and Suchowerska, N. 2009. Cerenkov light spectrum in an optical fibre exposed to a photon or electron radiation therapy beam. *Appl. Opt.* **48**: 3362–3367.

Law, S.H., Suchowerska, N., McKenzie, D.R., Fleming, S.C. and Lin, T. 2007. Transmission of Cerenkov radiation in optical fibres. *Opt. Lett.* **32**: 1205–1207.

McDermott, L.N., Nijsten, S.M., Sonke, J.J., Partridge, M., van Herk, M. and Mijnheer, B.J. 2006. Comparison of ghosting effects for three commercial a-Si EPIDs. *Med. Phys.* **33**: 2448–2451.

Mei, X. and Pang, G. 2005. Development of high quantum efficiency, flat panel, thick detectors for megavoltage X-ray imaging: An experimental study of a single-pixel prototype. *Med. Phys.* **32**: 3379–3388.

Mei, X., Rowlands, J.A. and Pang, G. 2006. Electronic portal imaging based on Cerenkov radiation: a new approach and its feasibility. *Med. Phys.* **33**: 4258–4270.

Monajemi, T.T., Fallone, B.G. and Rathee, S. 2006. Thick, segmented $CdWO_4$-photodiode detector for cone beam megavoltage CT: A Monte Carlo study of system design parameters. *Med. Phys.* **33**: 4567–4577.

Moses, W.W., Janecek, M., Spurrier, M.A. et al. 2010. Optimization of a LSO-based detector module for time of flight PET. *IEEE Trans. Nucl. Sci* **57**: 1570–1576.

Mosleh-Shirazi, M.A., Evans, P.M., Swindell, W., Symonds-Tayler, J.R.N., Webb, S. and Partridge, M. 1998. Rapid portal imaging with a high-efficiency, large field-of-view detector. *Med. Phys.* **25**: 2333–2346.

Nitske, R.W. 1971. *The Life of W. C. Röntgen: Discoverer of the X-Ray*. Tucson, AZ: University of Arizona Press.

Ostling, J., Wallmark, M., Brahme, A. et al. 2000. Novel detector for portal imaging in radiation therapy. *Proc. SPIE* **3977**: 84–95.

Pang, G., Lee, D. and Rowlands, J.A. 2001. Investigation of a direct conversion flat panel imager for portal imaging application. *Med. Phys.* **28**: 2121–2128.

Pang, G. and Rowlands, J.A. 2004. Development of high quantum efficiency, flat panel, thick detectors for megavoltage X-ray imaging: a novel direct-conversion design and its feasibility. *Med. Phys.* **31**: 3004–3016.

Prasad, P.N. 2004. *Nanophotonics*. Hoboken, NJ: Wiley.

Rowlands, J.A. and Yorkston, J. 2000. Flat panel detectors. In: Beutel, J., Kundel, H. L. and Van Metter, R. L. (eds.), *Handbook of Medical Imaging*. Bellingham, WA: SPIE, vol. 1, 223–328.

Samant, S.S. and Gopal, A. 2006. Study of a prototype high quantum efficiency thick scintillation crystal video-electronic portal imaging device. *Med. Phys.* **33**: 2783–2791.

Sawant, A., Antonuk, L.E., El-Mohri, Y. et al. 2005. Segmented crystalline scintillators: An initial investigation of high quantum efficiency detectors for megavoltage X-ray imaging. *Med. Phys.* **32**: 3067–3083.

Siewerdsen, J.H. and Jaffray, D. A. 1999. A ghost story: Spatio-temporal response characteristics of an indirect-detection flat-panel imager. *Med. Phys.* **26**: 1624–1641.

Silva, I. and Pang, G. 2012. Characteristics of radiation induced light in optical fibres for portal imaging application. *Radiat. Phys. Chem.* **81**: 599–608.

Tanioka, K., Yamazaki, J., Shidara, K. et al. 1987. An avalanche-mode amorphous selenium photoconductive layer for use as a camera tube target. *IEEE Electron Device Lett.* **8**: 392–394.

Other luminescence-based applications

Teymurazyan, A. and Pang, G. 2012a. Megavoltage X-ray imaging based on Cerenkov effect: A new application of optical fibres to radiation therapy. *Int. J. Opt.* **2012**: Article ID 724024. 1–13.

Teymurazyan, A. and Pang, G. 2012b. Monte Carlo simulation of a novel water-equivalent electronic portal imaging device using plastic scintillating fibers. *Med. Phys.* **39**: 1518–1529.

Teymurazyan, A. and Pang, G. 2013. An inherent anti-scatter detector for megavoltage X-ray imaging. *Phys. Med. Biol.* **58**: 1479–1493.

Teymurazyan, A., Rowlands, J.A. and Pang, G. 2014. A quantum noise limited Čerenkov detector based on air-spaced light guiding taper for megavoltage X-ray imaging. *Med. Phys.* **41**: 041907.

Van Dyk, Jacob (editor). 1999. *The Modern Technology of Radiation Oncology: A Compendium for Medical Physicists and Radiation Oncologists*. Madison, WI: Medical Physics.

van Elmpt, W., McDermott, L., Nijsten, S., Wendling, M., Lambin, P. and Mijnheer, B. 2008. A literature review of electronic portal imaging for radiotherapy dosimetry. *Radiother. Oncol.* **88**: 289–309.

Wang, Y., Antonuk, L.E., Zhao, Q., El-Mohri, Y. and Perna, L. 2009. High-DQE EPIDs based on thick, segmented BGO and CsI:Tl scintillators: Performance evaluation at extremely low dose. *Med. Phys.* **36**: 5707–5718.

Cherenkov imaging applications in radiation therapy dosimetry

Adam K. Glaser, Rongxiao Zhang, Brian W. Pogue,
and David J. Gladstone

Contents

21.1 OPTICAL DOSIMETRY AND THE CHERENKOV EFFECT

Cherenkov emission occurs in all media when photons, electrons, or charged particles are delivered for therapy (we refer the reader to Chapter 5 for a review of the basic theoretical concepts). Thus, there is potential to use this emission as a surrogate marker of radiation delivery and deposited dose in the medium. The theoretical and experimental underpinnings of how this signal could be used in radiation therapy dosimetry are outlined here along with the first water phantom and human imaging studies used to illustrate the concepts.

21.1.1 RELATIONSHIP BETWEEN CHERENKOV EMISSION INTENSITY AND RADIATION DOSE

In contrast to other forms of luminescence (e.g., fluorescence), which exhibit isotropic photon generation, optical photons generated via Cherenkov emission are released uniformly along a cone parallel to the axis of incident particle propagation with a half-angle known as the Cherenkov angle. This angle is a function of the particle energy and the index of refraction of the medium, as shown in Figure 21.1 [1]. The angle can be easily calculated by considering the kinematics of Figure 21.2(d).

As the particle moves from point A to B, it travels a distance $\overline{AB} = v\Delta t$, where v is the particle velocity and Δt is the travel time. Similarly, the electromagnetic pulse radiated from point A to C travels a distance $\overline{AC} = c/n\Delta t$, where c is the speed of light and n is the refractive index of the medium. Therefore, the Cherenkov angle can be calculated as

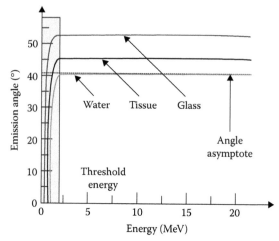

Figure 21.1 The energy dependence of the emission angle is plotted for water (n = 1.33), tissue (n = 1.41), and glass (n = 1.50).

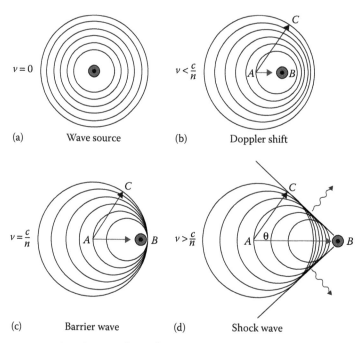

Figure 21.2 The electromagnetic pulses are shown for a particle moving with zero velocity (a), a velocity less than (b), equal to (c), and greater than (d) that of the local speed of light in the medium.

$$\cos\theta = \frac{\overline{AC}}{\overline{AB}} = \frac{c}{vn} = \frac{1}{\beta n} \qquad (21.1)$$

where:

$\beta = v/c$ is the relativistic phase velocity, defined in terms of kinetic energy as

$$\beta = \sqrt{1 - \left(\frac{mc^2}{E + mc^2}\right)} \qquad (21.2)$$

for a particle of energy E and rest mass mc^2 [2]. The energy dependence of Equation 21.1 is governed by Equation 21.2 and rises sharply at a threshold energy given by

$$E_{min} = mc^2 \left(\frac{1}{\sqrt{1 - \frac{1}{n^2}}} - 1 \right)$$ (21.3)

which approaches a constant angle asymptote in the limit that the phase velocity tends toward unity with increasing particle energy. The dependence of the threshold energy as a function of refractive index is plotted in Figure 21.3.

The number of Cherenkov photons emitted within the cone described by Equation 21.1 in a given wavelength range $[\lambda_1, \lambda_2]$ is given by the Frank–Tamm formula, which when combined with Equation 21.2 and assuming a constant refractive index can be used to express the number of emitted photons per unit path length, dN/dx, due to the Cherenkov effect as

$$\frac{dN}{dx} = 2\pi\alpha z^2 \left[1 - \frac{1}{\beta^2 n^2} \right] \frac{1}{\lambda^2} d\lambda$$ (21.4)

where the fine structure constant α is $\sim 1/137$ and z is the particle charge [3]. Per Equation 21.4, the radiation spectrum is inversely proportional to the square of the emission wavelength. As shown in Figure 21.4, the spectrum is broadband and most heavily weighted in the ultraviolet (UV) and blue wavebands, giving the effect its characteristic *blue* appearance, although it extends through the rest of the visible range and into the infrared.

There exists a lower limit below which Cherenkov radiation ceases to exist, as the spectrum depicted in Figure 21.4 would suggest that the emission increases infinitely at the shortest optical photon wavelengths. The limit is imposed by Equation 21.1, in which the Cherenkov angle is only defined for values of $\beta n \geq 1$ (i.e., cosine is undefined for values greater than 1). As $\beta < 1$, the above requirement cannot be satisfied in cases where $n \leq 1$. In general, for most dielectric materials, the refractive index fails to satisfy this condition in the UV or X-ray photon regions. Furthermore, in these regions, most materials are highly attenuating in that they immediately reabsorb photons emitted by the Cherenkov effect. A characteristic dispersion curve is shown in Figure 21.5.

In the case of the Cherenkov effect, Equation 21.4 can be rewritten in terms of optical photon energy, ϵ, rather than wavelength as

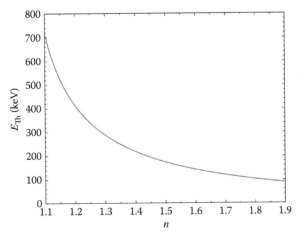

Figure 21.3 The dependence of the threshold energy on refractive index.

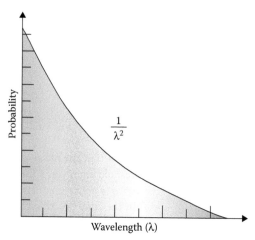

Figure 21.4 The emission spectrum of Cherenkov photons. The radiation emission probability exhibits an inverse square dependence with wavelength.

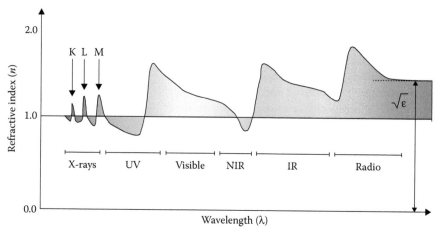

Figure 21.5 A dispersion curve for water. Cherenkov radiation ceases to exist at the lowest wavelengths where the refractive index is less than unity. At the longer wavelengths, the refractive index approaches an asymptote value of $\sqrt{\varepsilon}$, where ε is the dielectric constant of the medium. IR, infrared; NIR, near infrared. Characteristic absorption peaks of orbital, K, L, and M are shown in the X-ray region and are a function of atomic number.

$$\frac{\mathrm{d}N}{\mathrm{d}x} = \frac{2\pi\alpha z^2}{hc}\left(1 - \frac{1}{\beta^2 n^2}\right)\left(\epsilon_2 - \epsilon_1\right) \tag{21.5}$$

where the emission spectrum is constant with respect to photon energy, and therefore, the total energy loss per unit path length due to the Cherenkov effect, $-\mathrm{d}C/\mathrm{d}x$, can be expressed as

$$\frac{-\mathrm{d}C}{\mathrm{d}x} = \frac{\pi\alpha z^2}{hc}\left(1 - \frac{1}{\beta^2 n^2}\right)\left(\epsilon_2^2 - \epsilon_1^2\right) \tag{21.6}$$

where the mean emission photon energy is equated as $\left(\epsilon_1 + \epsilon_2\right)/2$.

In the context of dose in radiation therapy, the above is applicable to megavoltage X-ray photon beams, which irradiate a given volume liberating secondary electrons in the process. Assuming a volume of homogeneous atomic composition and density irradiated by a monoenergetic beam perpendicularly incident on a medium in the z-direction, the fluence of primary X-ray photons, Ψ, penetrating to a certain distance, z, will follow an exponentially decaying distribution:

$$\psi = \psi_0 e^{-\mu z} \qquad (21.7)$$

where:
ψ_0 is the fluence at the surface
μ is the linear attenuation coefficient of the medium

The corresponding kinetic energy released per unit mass (KERMA) at any point within the medium that leads to the production of secondary electrons, which deposit their energy locally through ionization events, K_c, can be expressed as

$$K_c = \psi \left(\frac{\mu_{en}}{\rho} \right) \qquad (21.8)$$

where:
μ_{en} is the mass energy absorption coefficient of the medium of density ρ
ψ is as defined in Equation 21.6

Although the collisional KERMA will decay exponentially per the X-ray photon fluence, for megavoltage beams the transfer of energy from the primary photons to electrons, and electrons to the medium will not occur at the same spatial location due to the nonzero propagation distance of the secondary electrons. Therefore, a buildup region will exist until a depth of $z = d_{max}$ (defined as the maximum penetration distance of secondary electrons produced at the surface of the medium), after which the medium will be in transient charged particle equilibrium (TCPE) and the energy absorbed per unit mass of the medium (i.e., imparted dose) will be proportional to K_c, and can be expressed as

$$D = \int \frac{\Phi}{\rho} \left(\frac{-dT}{dx} \right) dE \qquad (21.9)$$

where:
Φ is the polyenergetic electron fluence spectrum
$-dT/dx$ is the electron collisional stopping power

Similarly, the radiant energy, Q, and the total number of optical photons, N, released per unit mass of the medium due to the Cherenkov effect can be expressed as

$$Q = \int \frac{\Phi}{\rho} \left(\frac{-dC}{dx} \right) dE \qquad (21.10)$$

$$N = \int \frac{\Phi}{\rho} \left(\frac{dN}{dx} \right) dE \qquad (21.11)$$

where $-dC/dx$ and dN/dx are given by Equations 21.5 and 21.6, respectively. In regions of TCPE, the relative shape of Φ does not change appreciably, but rather scales in absolute magnitude by the X-ray photon fluence ψ. Therefore, the electron fluence spectrum at any point in the irradiated medium can be written as

$$\Phi = C(\psi)\Phi_0 \qquad (21.12)$$

Other luminescence-based applications

where:

$C(\psi)$ is a scalar parameter relating to the X-ray photon fluence

Φ_0 is the normalized electron fluence spectrum

Because $C(\psi)$ is independent of electron energy, Equations 21.9 and 21.11 can be rewritten as

$$D = C(\psi) \int \frac{\Phi_0}{\rho} \left(\frac{-\mathrm{d}T}{\mathrm{d}x} \right) \mathrm{d}E \tag{21.13}$$

$$N = C(\psi) \int \frac{\Phi_0}{\rho} \left(\frac{\mathrm{d}N}{\mathrm{d}x} \right) \mathrm{d}E \tag{21.14}$$

And so regardless of the form of $-\mathrm{d}T/\mathrm{d}x$ and $\mathrm{d}N/\mathrm{d}x$, the dose, D, and the number of emitted Cherenkov photons, N, are always proportional to one another, as given Equation 21.12. However, in regions of the irradiated medium where TCPE does not exist, Equation 21.12 fails (i.e., an air cavity and the penumbra of the beam) and disproportionality will exist.

When considering the characteristics of the surrogate Cherenkov photons, N, there are several important considerations. Per Equation 21.1 and given the distribution of scattered electron trajectories, the angular distribution of Cherenkov emission at any spatial location will be complex and anisotropic. Therefore, the Cherenkov photon irradiance generated by incident X-ray photons, and subsequent secondary electrons will exhibit a spatially dependent anisotropic phase function, $P = \mathrm{d}N/\mathrm{d}\Omega$, which describes the number of Cherenkov photons emitted per unit solid angle $\mathrm{d}\Omega$ at a given point in the irradiated volume. Similar to Equation 21.12, the phase function at any point within the irradiated medium is constant and scales in absolute magnitude by a scalar constant. Therefore, an accurate dosimeter can be constructed by either using a telecentric lens—which only accepts parallel light rays, that is, it samples the same portion of the phase function from all spatial locations being imaged—or using a conventional lens with Monte Carlo-derived calibration factors, or a conventional lens and a dilute dissolved fluorophore that may absorb and reemit the Cherenkov light isotropically, resulting in a uniform phase function [4,5].

21.1.2 PROJECTION IMAGING OF CHERENKOV EMISSION IN A WATER TANK

Given the theoretical discussion above, it is possible to utilize Cherenkov emission as a surrogate of the deposited dose in a medium, in the absence of other absorption or scattering effects in the medium, which would distort the Cherenkov signal. In particular, imaging dose in a water tank is a direct application of Cherenkov imaging to characterize linear accelerator (LINAC) beam dose deposition in water. Because the emitted Cherenkov light can be captured laterally from the direction of travel (see Figure 21.6), it is feasible to use this imaging as a projection of the dose laterally through the beam. This is illustrated conceptually and experimentally here.

The experimental setup for a Cherenkov optical dosimetry setup is shown in Figure 21.6 for a system utilizing either a telecentric or a conventional lens. There are two main components: (1) an irradiated medium and (2) an externally placed camera system. To comply with water equivalence for dosimetry, it is convenient that the irradiated volume can be a water tank analogous to those used in ionization chamber and diode detector dosimetry. In addition, although there are several possible geometric shapes for the water tank, to avoid refraction artifacts and corrections, it is convenient to choose a shape in which the camera will always image from a perpendicular and flat tank face. In a system composed of a single camera, a cubic volume suffices. However, if a system with multiple cameras were constructed, it would be feasible to construct a tank with a cross section resembling a higher order shape such as a hexagon or an octagon.

In addition, the definitions of the azimuthal (Φ) and polar (θ) angles are shown. As discussed in Section 21.1.1, the inherent Cherenkov radiation is extremely anisotropic. For a monodirectional megavoltage X-ray beam incident in the negative vertical direction, the phase function will be uniform azimuthally, but anisotropic with respect to the polar angle due to the Cherenkov angle. An example polar angle phase function, $P(\theta)$, of emission in water is shown in Figure 21.7. The distribution is peaked at 41°, the Cherenkov

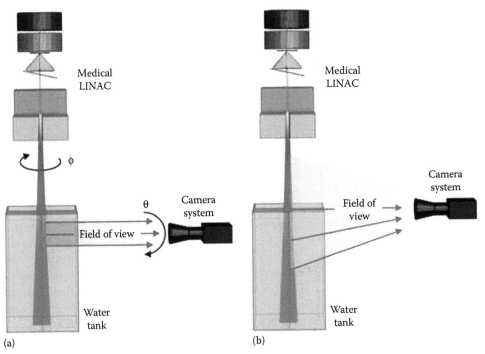

Figure 21.6 The generic experimental setup for an optical dosimetry system based on the Cherenkov effect is shown for a telecentric lens (a) and for a conventional lens (b).

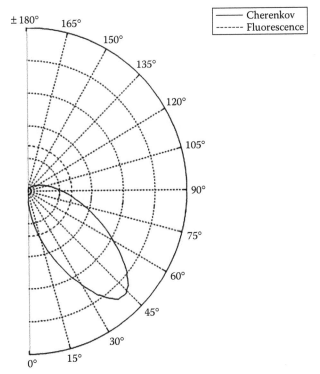

Figure 21.7 The polar angle phase function of emission for a monodirectional megavoltage X-ray photon beam in water, in comparison with an isotropic fluorescence phase function.

angle in water, but is broadly distributed due to electron scattering. This phase function is plotted in comparison with an isotropic fluorescence phase function.

The integrated emission for both functions plotted in Figure 21.7 is identical and would be related to the deposited dose by a scalar constant. However, in the case of the inherent Cherenkov phase function, the perceived optical intensity is heavily dependent on the viewing angle, that is, the perceived intensity will be high at 41° and low at 90° where a horizontally placed camera would typically sample the emission. As depicted in Figure 21.6, a telecentric lens will always sample the phase function at 90° from all spatial locations, and therefore, images of the inherent Cherenkov radiation will be indicative of the deposited dose. However, a regular lens will sample varying portions of the phase function from different spatial locations, resulting in a distorted image. These images can be calibrated by using scalar factors derived from Monte Carlo simulations or by dissolving a dilute fluorophore into the irradiated water tank, resulting in the isotropic fluorescence phase function, which is effectively independent of the sampling angle [4,5].

Example projection images of a 4 cm × 4 cm 6 MV radiation beam using a conventional lens and a color camera with and without dissolved fluorophore are shown in Figure 21.8. The image captured from the water with a dilute fluorophore appears brighter due to increased amount of Cherenkov radiation reemitted at 90° toward the camera. In addition, the relative amount of light captured from portions of the beam at larger depths reaching the camera from polar emission angles larger than 90° is increased. A cross plane profile and percentage depth dose (PDD) of both images in comparison with the treatment planning system (TPS) are plotted in Figure 21.9a and b, respectively. Due to the azimuthal emission independence of the inherent Cherenkov radiation, the cross plane profiles of both the Cherenkov and fluorescence projection images are indicative of the deposited dose predicted by the TPS. However, in the PDD, due to the polar angle dependence, the Cherenkov profile under predicts the calculated dose by the TPS. However, the PDD captured from the water volume with a dissolved fluorophore is in good agreement with that of the TPS due to the isotropic fluorescence polar phase function.

21.1.3 OPTICAL TOMOGRAPHY OF CHERENKOV FOR THREE-DIMENSIONAL BEAM PROFILES

Two-dimensional (2D) projection images are not of immediate use in dosimetry, as they represent a summation of the deposited dose with respect to the viewing direction of the camera, that is, they provide no information in this spatial direction. Of more use is the fully resolved three-dimensional (3D) dose distribution. Analogous to X-ray computed tomography (CT), this can be achieved with optical

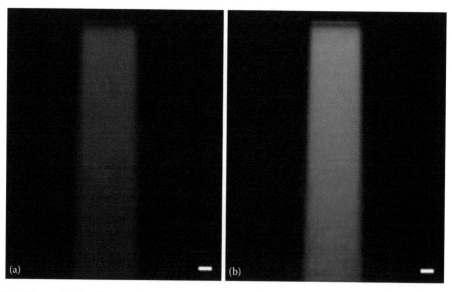

Figure 21.8 (a) The projection image of a pure water volume is shown, in comparison with (b) the same radiation beam in a water volume with a dilute dissolved fluorophore.

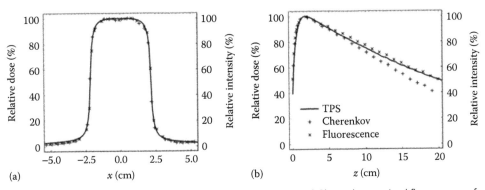

Figure 21.9 (a) The cross plane profiles for the inherent Cherenkov and Cherenkov-excited fluorescence for a 4 × 4 cm 6 MV beam are plotted in comparison with the TPS. (b) The PDD of each is plotted.

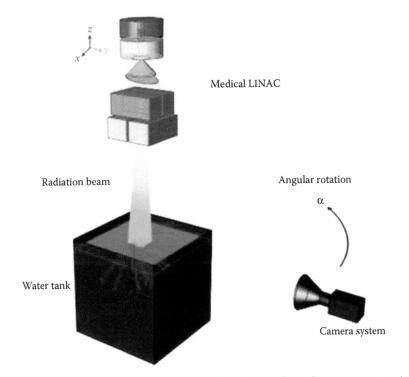

Figure 21.10 The experimental setup for optical tomography is shown, where the camera system is rotated about the beam central axis through an angle, α.

tomography by acquiring many projection images from varying angles about the beam. The experimental setup is depicted in Figure 21.10. The camera system is rotated about the beam through a rotational angle, α, to acquire many projection images for tomographic reconstruction. To avoid refraction artifacts at the tank face, the camera system and square tank are rotated together about the beam, such that the camera always faces a perpendicular tank face, although with proper calibration it may be possible to use a cylindrical tank.

In addition, the rotation may be provided by a stand-alone dosimetry system and rotational stage, by placing the tank and camera system on the patient bed, or by rotating the beam collimator. The number of angles necessary for a given reconstruction and resolution is commensurate with traditional X-ray CT.

Other luminescence-based applications

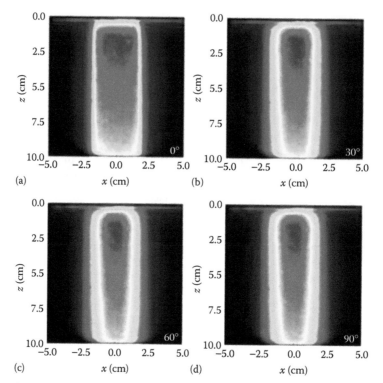

Figure 21.11 Optical projections of the induced Cherenkov light of a 4 × 4 cm 6 MV radiation beam at 0° (a), 30° (b), 60° (c), and 90° (d).

Similar to projection imaging, the details of tomographic reconstruction are specific to the type of imaging lens used in the experimental system. In the case of a telecentric lens, the optical light collection is analogous to parallel beam X-ray CT, and a simple parallel beam back projection algorithm may be used. Example projections of a 4 cm x 4 cm 6 MV radiation beam acquired using a telecentric lens-based camera system are shown in Figure 21.11 [6]. For each row of pixels, a sinogram can be constructed and used to reconstruct a slice of the 3D Cherenkov light volume. A sinogram corresponding to the projections in Figure 21.11 at a depth of 1.5 cm is shown in Figure 21.12 [6].

In the case of a camera system utilizing a regular lens, the conical field of view of the lens is analogous to a reverse cone-beam CT system and can be reconstructed using optical cone-beam tomography, a technique used previously in gel dosimetry [7,8]. Optical cone-beam tomography by the Cherenkov effect is currently an ongoing area of research.

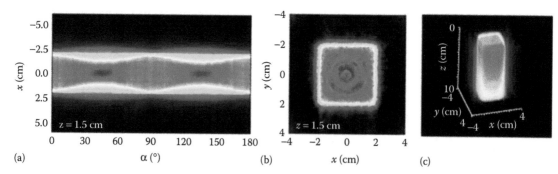

Figure 21.12 (a) A sinogram of the 4 cm × 4 cm 6 MV radiation beam at a depth of 1.5. (b) The resulting reconstructed cross section of the beam. (c) The full 3D reconstructed Cherenkov light volume.

21.1.4 FUTURE DIRECTIONS FOR CHERENKOV BEAM DOSIMETRY

The potential to use low-cost cameras or specially designed water tanks with internal cameras is good if the need for fast 3D dose characterization continues to grow beyond what is capable with ionization chambers. The validation of the dose accuracy is ongoing at this time, to ensure that 2D projection and 3D tomographic images of Cherenkov serve as truly accurate surrogates of the dose. The major benefits will likely be in the scalability of the imaging approach, which would allow characterization of complex beam shapes, unique delivery designs, and small beams where ionization chambers may be less accurate or time prohibitive.

21.2 CHERENKOV SURFACE IMAGING (CHERENKOSCOPY)

Potential applications of Cherenkov surface imaging have been explored in external beam radiotherapy for a number of reasons related to tracking the delivered dose. As illustrated in Figure 21.12a, the primary and secondary charged particles deposit energy through a range of ionization interactions with the tissue, and the Cherenkov photons are emitted along the path as part of the complex cascade. After Cherenkov generation, the signal is affected by the optical properties of the medium they are in, and thus, these optical photons will undergo elastic scattering and absorption prior to escaping the surface. Ultimately, only the Cherenkov photons generated in a thin layer (~5 mm at the entrance plane and 2–3 mm at the exit plane) of tissue at the surface can escape the surface to be detected [9,10], and the wavelengths of photons emitted are shifted to the red due to extremely high blood absorption in the blue–green spectrum.

21.2.1 RATIONALE FOR CHERENKOV SURFACE IMAGING IN PATIENTS

The rational for imaging Cherenkov surface emission is quality assurance of dose delivery and patient setup during daily treatment. Additionally, beam modifiers such as physical or dynamic wedges and multileaf collimators (MLCs) need to be verified and ideally might be monitored in real time during the treatment. This is especially poignant for modern treatment techniques employing fast dynamic field and intensity changes such as intensity-modulated radiation therapy. Cherenkov emission is not useful in proton therapy since energies higher than clinical relevance are necessary to produce this effect with heavy particles. Many factors, including patient movement, tissue heterogeneity, organ motion, deformation of the treatment region, weight loss, inaccurate image registration, mechanical malfunction, and false operation, can lead to inaccuracy or even errors in treatment delivery.

It is important to recognize that errors in the delivery of radiation therapy occur in less than 1% of these treatments [11] and are usually clinically insignificant events. However, serious errors do occur that can cause tissue injury, excessive skin burns, and potentially death [12], and with the increasing complexity of treatments, it will become even more important to monitor and verify that treatment is delivered appropriately for the entire course of therapy. Real-time Cherenkov emission imaging coupled with automated image analysis is a potential method of producing an on-patient interlock signal to reduce errors and improve reproducibility of treatments.

21.2.2 GATED DETECTION FOR CHERENKOV IMAGING

A significant technical challenge is that Cherenkov emission generated in patients during typical EBRT is of low intensity. Compared to normal incandescent light that has an irradiance of 10^{-1}–10^{-3} W/cm^2, the Cherenkov emission irradiated by radiation beams (interacting with tissue) from a medical LINAC is roughly 10^{-6}–10^{-9} W/cm^2, depending on the beam type and dose rate of irradiation [13]. These large differences in optical irradiance make continuous-wave detection of Cherenkov emission challenging in the presence of ambient light. For most LINACs, as shown in Figure 21.13b, radiation pulses are delivered in microsecond bursts (about 3 µs in length) at a fixed frequency of ~200 Hz. By taking advantage of a LINAC's inherent pulsed operation, time-gated detection of Cherenkov emission is possible to significantly improve (more than 1000×) the signal-to-ambient light ratio. As a result, under dimmed room lighting conditions, the ambient and Cherenkov optical signals are nearly equivalent in intensity during the pulse. Additionally, as shown in Figure 21.13a, by picking up light directionally from the tissue using a lens system coupled with sensitive detectors (such as intensified charge-coupled device [ICCD]), it is feasible to increase the Cherenkov signal over the ambient room light by 1–2 orders of magnitude [13]. It has been

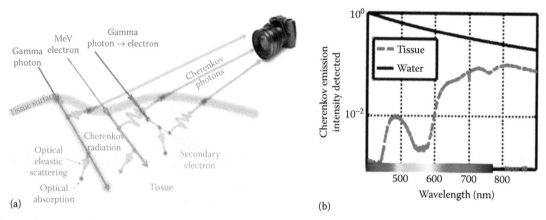

Figure 21.13 (a) High energy particles deposit energy to the environment during transportation. Cherenkov photons will be emitted along the path of primary and secondary charged particles, and the intensity of Cherenkov radiation emission is proportional to the deposited energy locally. Depending on the optical properties of the phantom or tissue, Cherenkov photons generated in a thin layer of surface will be scattered and finally escape the surface to be detected by the camera. (b) The results of Monte Carlo simulations show the emitted spectra that would be captured by the camera, which are broadband for water emission, although weighted to the blue, whereas tissue emission is weighted to the red/infrared wavelengths due to blood absorption.

shown that Cherenkov emission can be imaged from patient surfaces with the existence of reasonable ambient light by synchronizing an ICCD camera with the radiation pulses [10].

To detect enough Cherenkov photons, a frame of Cherenkov image is generally acquired by accumulating the Cherenkov emission from 50 to 100 radiation pulses, which is only a quarter to half a second of acquisition time [9,10]. Median filtering over several frames of repetitive measurements is an effective way to remove the saturated pixels caused by high energy photons hitting the detector directly [14]. Figure 21.14 shows a typical (100 accumulations, with median filter over five repetitive measurements and smoothed by a 10 × 10 median kernel) surface image of Cherenkov emission in the case of whole-breast radiotherapy at 10 MV. Imaging of Cherenkov emission from a single radiation pulse has been demonstrated by imaging with an ICCD whose intensifier is sensitive to the wavelength range (600–900 nm) of Cherenkov emission from the skin surface. In the case of single radiation pulse imaging, the rate of data acquisition could be as high as 30 frames per second with the highest sensitivity intensifiers currently available.

Imaging of Cherenkov photons escaping from the surface during megavoltage external beam radiotherapy at video rate led to this technique named Cherenkov surface imaging or Cherenkoscopy [10]. The potential applications of Cherenkoscopy are multifold, including validation of patient setup, quality assurance of treatment planning, monitoring of treatment delivery, superficial dose estimation, and skin reaction prediction.

21.2.3 REAL-TIME CHERENKOSCOPY AS A VALIDATION OF PATIENT SETUP

In Figure 21.14, a series of potential serious radiation errors due to patient setup (Figure 21.15b–d), mechanical malfunction of a dynamic wedge (Figure 21.15e and f), MLC (Figure 21.15g and h), or a missing physical wedge (Figure 21.15i and j) was simulated in the case of whole-breast radiotherapy with tissue-equivalent phantom. Phantom shifting and collimator rotation with small magnitude (3 mm and 3°) were detected in the Cherenkov images (Figure 21.15b–d). Statistical data from real patients demonstrated that by directly calculating the 2D correlations of Cherenkov images from different fractions of treatment, patient setup inaccuracy larger than ~2–3 mm in shifting and 3° in pitch angle rotation could be differentiated. Blood vessels, visible in the Cherenkov images (Figure 21.14), could also serve as an internal biological marker for patient setup and image registration in image-guided radiotherapy. Additionally, the obvious visual differences in the Cherenkov images would be observed during delivery of the incorrect treatment due to the malfunction of a dynamic wedge or the missing physical wedge (Figure 21.15) or inaccurate MLC control points. By viewing the Cherenkov surface images or with computational comparison to the treatment plan prediction, it could

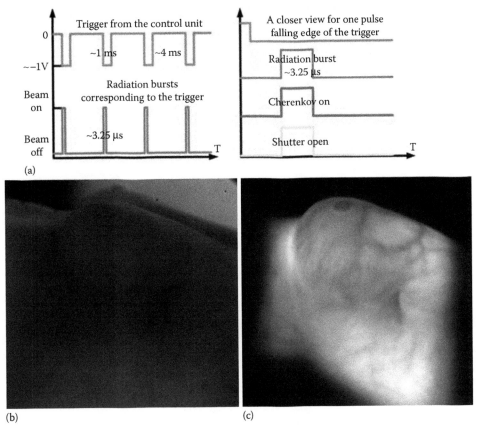

Figure 21.14 (a) The time line of radiation pulses and temporal gating. ICCD capture is synchronized to the LINAC pulses such that the images are reject much of the room light. (b) A photographic view of the treatment region from the imaging system. (c) A typical Cherenkov image is shown from a case of whole-breast radiotherapy (0.5 s acquisition, median filtered in time and spatially filtered).

be possible to alert the therapists if the treatment was delivered as planned or not in the first few pulses of radiation. If necessary, the treatment could be aborted in time without causing further radiation damage by incorrect delivery.

21.2.4 CHERENKOSCOPY AS A VALIDATION OF MLC MOVEMENT AND PLAN DELIVERY

The idea of treatment monitoring based on Cherenkoscopy is similar to portal imaging, except that, for Cherenkoscopy, the treatment field is directly viewed on the surface of the patient. Figure 21.16 shows that frames of Cherenkoscopy monitored the delivery of one treatment field for a typical whole-breast radiotherapy. Dynamic field-in-field blocks were designed for the treatment beams. MLC motions from a beam's eye view are shown in Figure 21.16a. These dynamic beam profile projections on the surface of the treatment region are listed in Figure 21.16b. By continuously monitoring the treatment, frames of Cherenkov images corresponding to the dynamic treatment field changes are captured. Comparison of Figure 21.16b and c validates that real-time Cherenkoscopy correctly monitored MLC motions as well as the corresponding beam field changes. With the aid of Cherenkoscopy, the radiation beam profiles on the surface of patient could be viewed straightforwardly, and comparison of the beam profiles with predictions from TPS could further validate dose delivery and thus increase the quality of treatments, especially for the dynamic changes due to fast MLC movements.

21.2.5 CHERENKOV DOSE IMAGING

Superficial dose is the radiation dose deposited within the first several millimeters underneath the skin surface. Knowledge of superficial dose would be beneficial in a range of treatments if it could be measured accurately and within the acceptable workflow of patient throughput for fractionated therapy. Depending

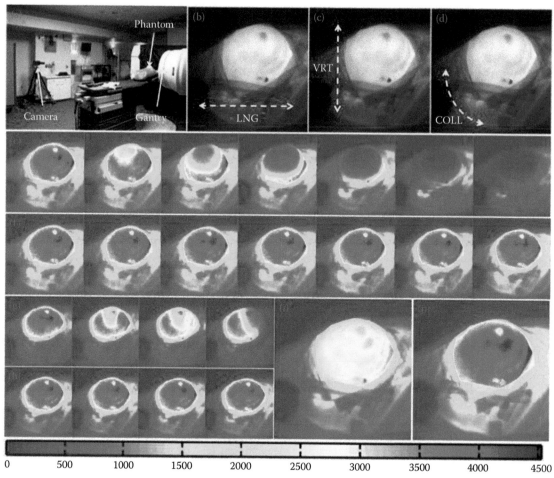

Figure 21.15 Detection of patient setup inaccuracy and radiation errors caused by hardware malfunction simulated in phantom studies. (a) Relative positions of the gantry, tissue-equivalent phantom, and imaging system. (b–d) Here LNG is Longitudinal and VRT is Vertical in patient coordinates Comparison of Cherenkov images of the phantom before and after shifting 3 mm in LNG and VRT directions, and rotating the collimator for 3°. (e) Cherenkov images of the entrance dose of the tangent field modulated by a dynamic wedge (standard MLC motions to account for tissue thickness differences to eliminate radiation *hot spots*. (f) Cherenkov images from the same radiation treatment fields as in (a) but simulated without the dynamic wedge motion. (g) Cherenkov images of the entrance region of the tangent field modulated by a dynamic field in field. (h) Cherenkov images from the same radiation treatment fields but simulated without the dynamic field in field. (i) Cherenkov image of the entrance field of modulated by a physical wedge (a device placed in the head of a LINAC to modulate the beam). (j) Corresponding Cherenkov image using the same tangent field in (e) but delivered without the physical wedge in the LINAC.

Other luminescence-based applications

on clinical goals, the skin may be included in the intended treatment volume or may be a dose-limiting organ at risk. Many factors such as source to surface distance (SSD), beam types (electron or photon), beam energy, field size, beam modifying devices, angle of incidence, complexities and deformation of the patients' surface profiles, and heterogeneities of the internal tissue lead to difficulty in achieving accurate superficial dosimetry estimates or measurements. In all of these factors, the incident angle with respect to the normal direction of the surface is one of the more complex issues, which affects skin dose. Irregular surface profiles of the treatment region decrease the accuracy of superficial dose prediction and may result in underdosing or overdosing compared to predictions of the treatment plan. Conventional superficial dosimetry methods such as radiochromic film, ionization chamber, diode, metal–oxide–semiconductor field-effect transistors (MOSFETs), and thermoluminescent dosimeters (TLDs) have been used to measure

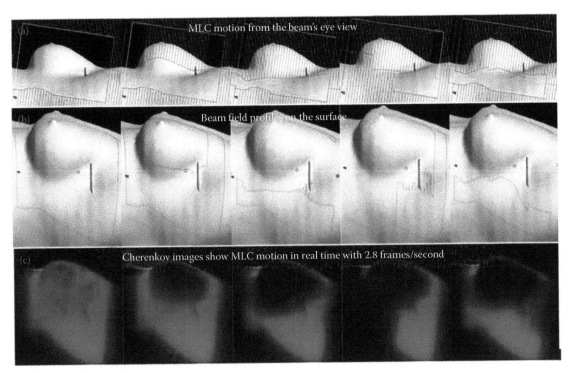

Figure 21.16 Cherenkoscopy detected dynamic changes in radiation treatment fields during a treatment delivery, with (a) a beam's eye view of the beam field designed with dynamic MLC motion used for partial blocking of radiation dose during treatment, to minimize radiation *hot spots* in the breast. Treatments start with an open-field, leftmost image, and then the MLCs (blue) move, blocking the radiation from areas behind the blue lines. Red line is the lumpectomy scar. (b) Brown lines show entrance profiles created by the MLC motion on the surface of the patient. (c) Cherenkov images taken during each MLC position capturing the changing doses in real time (overlaid with the photographic image of the treatment region in the blue channel).

superficial dose; however, these techniques require clinical intervention and additional personnel time for use. Each technique is limited by small fixed region measurements, and sensitivity is often a function of angular orientation of the detector with respect to the incident beam. Film and TLDs have longer off-line processing procedures that prevent superficial dose monitoring in real time.

Although TPSs can be used to calculate the superficial dose, factors such as patient motion, deformation, and unpredicted weight loss during fractionated radiotherapy contribute to inaccuracies between the calculated superficial dose and the superficial dose actually delivered. Due to lack of electronic equilibrium, tissue air interface, and surface obliquity, TPS algorithms provide their least accurate predictions at the patient surface. Commercial TPSs have recently included Monte Carlo algorithms with the potential of more accurate superficial dose. For practical reasons, these algorithms have been optimized for calculation speed and still loose accuracy near the surface. Finally, planning models are fit to measured dosimetry data, which may introduce uncertainties or inaccuracies at the surface due to measurement instrumentation. Thus, an accurate, real-time, superficial dose measurement method such as Cherenkoscopy is desirable.

It has been shown that, above the threshold energy for Cherenkov radiation (~220 KeV in biological tissue), under the approximation of charged particle equilibrium, the dose deposited by megavoltage radiotherapy, and the number of Cherenkov photons released locally are directly proportional; therefore, beam profiling and superficial dosimetry imaging based on Cherenkov radiation are feasible. Radiation dose is calculated by $D = \int_0^{E_{max}} S(E)P(E)dE$, where D represents the radiation dose, $S(E)$ represents the mass stopping power ($J \cdot m^2/kg$) f the medium and $P(E)$ represents the fluence spectra of charged particles ($1/m^2$). Similarly, the local intensity of Cherenkov radiation can be calculated by $I = \int_{E_c}^{E_{max}} C(E)P(E)dE$, where I represents the local intensity of Cherenkov radiation, E_c represents the threshold energy of Cherenkov radiation in the medium, $C(E)$ represents the number of Cherenkov photons emitted by a charged particle (such as electron) with kinetic energy of E per unit path length of the propagation of the charged particle, and $P(E)$ represents the spectra of charged particles. Typically, $S(E)$ and $C(E)$ have very different profiles; however, as long as $P(E)$ is spatially independent, that is, the spectra of charged particles are a constant distribution in the region of interest, radiation dose is directly proportional to the local intensity of Cherenkov radiation above the threshold energy. Reference data show that the continuous slowing down approximation (CSDA) range of electrons, with kinetic energy below the threshold energy in the medium (take water as example, $E_c \cong 0.263$ MeV) is around 0.1 mm. Due to scattering of electrons, the absolute distance of travel below the threshold energy of Cherenkov radiation is actually smaller than the CSDA range. This means that, to a resolution of 0.1 mm, with the assumption that $P(E)$ is spatially independent in the region of interest, the dose contributed by those charged particles below the threshold energy will be a constant offset of the dose contributed by charged particles above the threshold energy. As long as $P(E)$ is spatially invariant (charged particle equilibrium) in the region of interest, to the resolution of the CSDA range of charged particles below the threshold energy, local intensity of Cherenkov radiation will be directly proportional to radiation dose. This observation, demonstrated by Monte Carlo simulations [9], is the theoretical underpinning of why Cherenkov emission can be considered to be proportional to deposited dose. As illustrated in Figure 21.16a, depending on the optical properties, Cherenkoscopy could sample the deposited dose within several millimeters underneath the skin surface. Monte Carlo simulations in human skin models showed that the sampling depth is approximately 5 mm for the entrance plane and 2–3 mm for the exit plane. Spectral filtering of Cherenkov emission could be utilized to tune the sampling depth from submillimeter to about 6 mm [9,10].

Cherenkov radiation is intrinsically generated in tissue during irradiation. Different from conventional superficial dose measurement techniques, superficial dosimetry imaging based on Cherenkoscopy does not require any detector to be placed on patient or any clinical intervention within the process of EBRT. Instead of small region measurement, this technique is able to image a large field of view or focus on the region of interest, which provides global as well as a detailed information about superficial dose distribution. Figure 21.17a shows the false-colored Cherenkov image of a typical whole-breast radiotherapy. Resulted from the deposited superficial dose, radiation dermatitis was documented during the course of therapy (Figure 21.17b). Similar hot and cold regions could be observed in the Cherenkov and skin reaction images, which validated that Cherenkoscopy could estimate deposited superficial dose and predict the resulted skin reactions.

Although Cherenkoscopy has shown certain advantages for superficial dose assessment, several important issues exist which need to be clarified. First, local intensity of Cherenkov radiation is proportional to radiation dose under the approximation that energy spectra of charged particles are spatially independent. This approximation was validated with maximum discrepancy within 5% and average discrepancy within 2% for flat phantom and most of entrance and exit regions for curved phantoms. It is worth noting that the images could under- or overestimate radiation dose several percentages, especially near the edge of the beam field or in regions where the beam is tangential to the surface. Calibration of this issue requires detailed information about energy spectra of charged particles at different regions, which is potentially possible but computationally intense. In practice, the easiest solution is that regions such as beam edges and tangential surfaces with respect to the direction of the incident radiation beam could be eliminated from the image, or interpreted with caution for superficial dose estimation.

21.2.6 CHERENKOV SIGNAL ORIGIN

The sampling depth and detected Cherenkov intensity are correlated to optical properties of the tissue. One potential way to solve this issue is to include noninvasive optical property measurement techniques such as reflectance spectroscopy to measure optical properties of the skin accurately and then correct the detected Cherenkov signal in order to more closely correlate it to dose.

For complex surface profiles (breast or head and neck tumor treatment), the angular correction of Cherenkov images is important. For highly scattering media such as human tissue, optical photons could be scattered sufficiently and lose their initial angular distributions. Monte Carlo simulations in

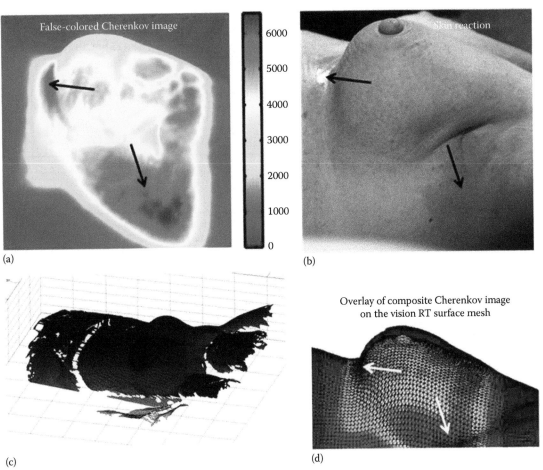

Figure 21.17 Cherenkoscopy was examined for imaging superficial doses, and thus, (a) a composite Cherenkov image for the four treatment beams shows high superficial doses in areas typical for higher grade radiation skin reactions, inframammary fold, and axilla. (b) The skin reaction of the treatment region of patient at the end of radiation. The arrow pointing to the inframammary fold indicates skin desquamation. (c) The 3D surface profile measured by the 3D surface mapping system (Vision RT). (d) An overlay of the composite Cherenkov image on surface mesh created by the surface scanning system shown for 3D visualization.

human skin models show that the angular distribution of emission is close to the theoretical Lambertian distribution [9,10]. Lambertian distributions essentially simplify the angular correction to a trivial monotonic function with a known analytic expression, which is especially important for complex curved surfaces, because the intensity changes due to the curvature of the surfaces will be exactly compensated by the corresponding changes of the solid angle. Preliminary study suggests that Lambertian correction would be a reasonably good normalization factor to correct the emission intensity [10]. Although complete angular correction is a challenge because angular distributions are affected slightly by the surfaces profiles combined with all the potential factors, many of these issues can be potentially addressed by 3D surface capture techniques and systems. As shown in Figure 21.16c, 3D surface profiles of the treatment region were obtained from the 3D surface profiling system (AlignRT®, Vision RT, London, England) or isolated from the patient CT scan by the TPS (Eclipse) Varian Medical Systems, Palo Alto, CA. Preliminary registration of the Cherenkov image to the 3D surface profile (Figure 21.16d) was implemented based on ray tracing algorithm, which allows Cherenkov images to be directly compared with 3D superficial dose distribution calculated by TPSs in CT scan. By coupling 3D surface profiles and measurements of optical properties with Cherenkov images, factors such as curvatures of the surfaces, incident angles, and optical properties could be decided for the treatment region and adopted to simulate the angular distribution, which could be further used for intensity corrections due to viewing angles and curvatures.

21.3 FUTURE DIRECTIONS OF *IN VIVO* CHERENKOV IMAGING

Initial trials with clinical imaging of Cherenkov are ongoing now, which will help define the role of Cherenkov imaging from patients during therapy. Choices around the application and value will affect the cameras used and the geometry of imaging chosen, whether it be onboard the gantry or fixed in the room. The signal is detectable, and the repeatability and accuracy need to be fully evaluated as part of these initial trials of systems.

ACKNOWLEDGMENTS

This work was funded by the National Cancer Institute research grants R01CA109558 and R21EB17559.

REFERENCES

1. Tearney, G. J., M. E. Brezinski, J. F. Southern, B. E. Bouma, M. R. Hee, and J. G. Fujimoto. 1995. Determination of the refractive index of highly scattering human tissue by optical coherence tomography. *Opt Lett* 20:2258.
2. Jackson, J. D. 1962. *Classical Electrodynamics.* Wiley, New York.
3. Tamm, I. E., and I. M. Frank. 1937. Coherent radiation from a fast electron in a medium. *Dokl Akad Nauk SSSR* 14:107–112.
4. Glaser, A. K., S. C. Davis, D. M. McClatchy, R. Zhang, B. W. Pogue, and D. J. Gladstone. 2013. Projection imaging of photon beams by the Cherenkov effect. *Med Phys* 40:012101.
5. Glaser, A. K., S. C. Davis, W. H. Voigt, R. Zhang, B. W. Pogue, and D. J. Gladstone. 2013. Projection imaging of photon beams using Cherenkov-excited fluorescence. *Phys Med Biol* 58:601–619.
6. Glaser, A. K., W. H. Voigt, S. C. Davis, R. Zhang, D. J. Gladstone, and B. W. Pogue. 2013. Three-dimensional Cherenkov tomography of energy deposition from ionizing radiation beams. *Opt Lett* 38:634–636.
7. Olding, T., and L. J. Schreiner. 2011. Cone-beam optical computed tomography for gel dosimetry II: Imaging protocols. *Phys Med Biol* 56:1259–1279.
8. Olding, T., O. Holmes, and L. J. Schreiner. 2010. Cone beam optical computed tomography for gel dosimetry I: Scanner characterization. *Phys Med Biol* 55:2819–2840.
9. Zhang, R., C. J. Fox, A. K. Glaser, D. J. Gladstone, and B. W. Pogue. 2013. Superficial dosimetry imaging of Cherenkov emission in electron beam radiotherapy of phantoms. *Phys Med Biol* 58:5477–5493.
10. Zhang, R., A. K. Glaser, D. J. Gladstone, C. J. Fox, and B. W. Pogue. 2013. Superficial dosimetry imaging based on Cherenkov emission for external beam radiotherapy with megavoltage x-ray beam. *Med Phys* 40:101914.
11. Margalit, D. N., Y. H. Chen, P. J. Catalano, K. Heckman, T. Vivenzio, K. Nissen, L. D. Wolfsberger, R. A. Cormack, P. Mauch, and A. K. Ng. 2011. Technological advancements and error rates in radiation therapy delivery. *Int J Radiat Oncol Biol Phys* 81:e673–e679.
12. Bogdanich, W. 2010. The radiation boom: Radiation offers new cures, and ways to do harm. *The New York Times,* January 23.
13. Glaser, A. K., R. Zhang, S. C. Davis, D. J. Gladstone, and B. W. Pogue. 2012. Time-gated Cherenkov emission spectroscopy from linear accelerator irradiation of tissue phantoms. *Opt Lett* 37:1193–1195.
14. Archambault, L., T. M. Briere, and S. Beddar. 2008. Transient noise characterization and filtration in CCD cameras exposed to stray radiation from a medical linear accelerator. *Med Phys* 35:4342–4351.

Index

Note: Locators followed by '*f*' and '*t*' denote figures and tables in the text

Milton Keynes UK
Ingram Content Group UK Ltd.
UKHW050455071024
449327UK00015B/394